BASIC METHODS
OF
STRUCTURAL GEOLOGY

BASIC METHODS
OF
STRUCTURAL GEOLOGY

Part I
Elementary Techniques

by

Stephen Marshak
University of Illinois

Gautam Mitra
University of Rochester

Part II
Special Topics

edited by

Stephen Marshak

Gautam Mitra

Prentice-Hall, Inc.
Upper Saddle River, New Jersey 07458

Library of Congress Cataloging-in-Publication Data

Marshak, Stephen (date)
    Basic methods of structural geology.

    Bibliography: p.
    Includes index.
    1. Geology, Structural. I. Mitra, Gautam. II. Title.
QE601.M365 1988   551.8'028   88-4044
ISBN 0-13-065178-8

Editorial/production supervision
 and interior design: *Kathryn Gollin Marshak*
Cover design: *Amy Scerbo*
Manufacturing buyer: *Paula Massenaro*

Cover drawing of the Himalaya Mountains, Nepal, by J. Knox
from a photograph by S. Marshak.

Printed in the United States of America

20 19 18 17 16 15 14

ISBN   0-13-065178-8

PRENTICE-HALL INTERNATIONAL (UK) LIMITED, *LONDON*
PRENTICE-HALL OF AUSTRALIA PTY. LIMITED, *SYDNEY*
PRENTICE-HALL CANADA INC., *TORONTO*
PRENTICE-HALL HISPANOAMERICANA, S.A., *MEXICO*
PRENTICE-HALL OF INDIA PRIVATE LIMITED, *NEW DELHI*
PRENTICE-HALL OF JAPAN, INC., *TOKYO*
PEARSON EDUCATION ASIA PTE. LTD., *SINGAPORE*
EDITORA PRENTICE-HALL DO BRASIL, LTDA., *RIO DE JANEIRO*

CONTRIBUTING AUTHORS

Terry Engelder
Department of Geosciences
Pennsylvania State University
University Park, Pennsylvania

Arthur Goldstein
Department of Geology
Colgate University
Hamilton, New York

Mark Helper
Department of Geological Sciences
University of Texas
Austin, Texas

Stephen Marshak
Department of Geology
University of Illinois
Urbana, Illinois

Gautam Mitra
Department of Geological Sciences
University of Rochester
Rochester, New York

Sharon Mosher
Department of Geological Sciences
University of Texas
Austin, Texas

Lucian B. Platt
Department of Geology
Bryn Mawr College
Bryn Mawr, Pennsylvania

Carol Simpson
Department of Earth & Planetary Sciences
Johns Hopkins University
Baltimore, Maryland

Steven Wojtal
Department of Geology
Oberlin College
Oberlin, Ohio

Nicholas Woodward
Department of Geological Sciences
University of Tennessee
Knoxville, Tennessee

CONTENTS

PREFACE

Basic Methods of Structural Geology is a textbook designed to serve two purposes. First, it is intended to serve as an accompaniment to techniques-based courses in structural geology or as an accompaniment to the laboratory portion of an undergraduate structural geology course. Second, the book is intended to serve as a reference source for information on structural geology methods. Thus, it should continue to be useful to undergraduates in other courses and to graduate students and professionals. The book provides detailed explanations of methods and worked-out examples of problems. Our intention is to focus on the "how-to" part of structural geology, and thus we do not exhaustively duplicate the definitions and theory covered in general texts like *Principles of Structural Geology* (Suppe, 1985).

Throughout *Basic Methods of Structural Geology*, the description of techniques is presented in a problem/method format. A specific problem is addressed and the step-by-step method for how to solve that problem is outlined; these examples are best understood when the student works through the steps and tries to duplicate the solution. Realistic exercises are included at the ends of the chapters to allow students to perfect their understanding and to see the application of specific methods. Chapters 10 and 16 challenge the student to complete the interpretation of the data presented in the body of the text. Chapters are arranged approximately in order of increasing of difficulty and/or complexity of subject matter; it is intended that the information available in the earlier portions of the book will provide a foundation of experience that the student can use to help in understanding the later chapters.

The text is divided into two parts. Part I begins at an elementary level so that the book is accessible to students early during their geological training. It discusses measurement and description of lines and planes, the use of a compass, analysis of contour maps, the use of trigonometry and orthographic projection for the solution of geometric problems in structural geology, and the use of stereonets and equal-area nets. We hope that Part I will hone the student's ability to visualize structures in three dimensions and to communicate descriptions of structures to others. Appendix 1 can serve as an introduction to Part I, as it provides a concise review of the concepts of maps and cross sections, in case the student is rusty on these subjects.

Part II includes eight contributed chapters, each dealing with the methods used in a subdiscipline of structural geology. This part covers map interpretation, analysis of rock-deformation experiments, analysis of fracture arrays, analysis of mesoscopic and microscopic structures, construction and balancing of cross sections, strain analysis, and interpretation of polydeformed terranes. The chapters of Part II include both introductory and advanced material and are self contained. Despite the diversity of subject matter in Part II, we have attempted to achieve a degree of uniformity in style of presentation to make the book easier to use; towards this goal, SM revised and reformatted much of the contributed material for Part II.

In order for the book to be comprehensive, it intentionally includes more material than can be covered in the laboratory portion of a standard one-semester structural geology course. Therefore, we suggest that instructors provide focused assignments from the book rather than swamp the student with reading. For example, the book covers several different approaches to the same problem, but a student can understand the concept by studying only one. It is not intended that the student work through each technique; the instructor should assign only one or two that exemplify the concept. The remaining techniques should be considered a resource for future reference.

A standard introductory structural-geology course should cover most of the material in Part I and a selection of material from four or five of the chapters in Part II. Material not covered in an introductory course could be assigned in a more advanced structural geology course. The book contains many exercises, and instructors can, if they wish, design a curriculum that uses only exercises from the book. However, the book can also be used accompany original exercises that are put together by individual instructors. For example, Chapter 9 on geologic map interpretation can serve as an introduction to a series of exercises involving published U.S.G.S. geologic quadrangle maps, and Chapter 10 could be used as an introduction to a laboratory demonstration.

As this is a first edition, we would appreciate comments and corrections provided by users of this book.

ACKNOWLEDGMENTS

The creation of this book would not have been possible without the efforts of the contributors. We thank them for their willingness to meet the many deadlines at stages during the development of this book and for their cooperation as their chapters were edited. We also wish to thank the individuals and organizations that allowed us to adapt illustrations from previously published material. A book such as this necessarily incorporates and adapts much material from existing texts, and we wish to acknowledge particularly the books by Turner and Weiss (1963), Ramsay (1967), Dennison (1968), Billings (1972), Hobbs et al. (1976), Ramsay and Huber (1983 & 1987), and Ragan (1985). We greatly appreciate the efforts of many current and former students, particularly Istvan Barany, Snehal Bhagat, Nancye Dawers, Karen Fryer, Allison Macfarlane, Mark McNaught, Nebil Orkan, Gretchen Protzman, and Scott Wilkerson, who worked through portions of the book and suggested improvements. The manuscript was greatly improved in response to critical reviews by D.N. Bearce, T. Byrne, G. H. Davis, J. G. Dennis, K. Hodges, C. E. Jacobson, W. D. Means, and S. Wojtal. We particularly thank the production editor, Kathryn Marshak, for her efforts in directing the manuscript through an accelerated production schedule. We are also indebted to Holly Hodder (aquisitions editor), Barbara Liguori and Linda Thompson (copy editors), and the pasteup artists of Precision Graphics, especially Jim Gallagher. Randy Cygan generously made his Apple laserwriter available for printing the camera-ready copy.

In order to save production costs and make this book affordable to students, the manuscript was typeset by SM and all art was author-provided. (As a consequence the style of illustration is not entirely uniform through the book, and the symbol ∂ was used in place of δ in early chapters for the proper font was not available.) Jesse Knox, Bill Nelson, and David Phillips ably assisted SM in the drafting of illustrations for Chapters 1-4, 10, and 14, and helped revise some figures from contributed chapters. GM illustrated Chapters 5-8 and 11, and thanks Margrit Gardner for typing early versions of these chapters. Finally, we wish to thank our wives, Kathryn Marshak and Judy Massare, for their patience and support during the time that this book was a member of our respective households. The long hours that Kathy generously spent entering text and changes into the Macintosh will not be forgotton.

Stephen Marshak
Urbana, Illinois

Gautam Mitra
Rochester, New York

ILLUSTRATION CREDITS

1-14, A1-9: Adapted from Compton, 1962, Manual of Field Geology: John Wiley & Sons, New York, Figs. 2-1, p. 21; 1-3, p. 19. Used by permission.

1-16: Adapted from Judson, Kauffman, and Leet, 1987, Physical Geology, 7th ed.: Prentice-Hall, Inc., Englewood Cliffs, NJ, Fig. 11.7, p. 210. Used by permission.

1-17: Adapted from Brunton Pocket Transits instruction flyer, Fig. 2. Used by permission.

2-2: Adapted from Radian, Inc., CPS/PC flyer, Isopach Contour figure. Used by permission.

2-7: Adapted from Hamblin and Howard, 1986, Exercises in Physical Geology, 6th ed.: Burgess, Minneapolis, Fig. 8.4, p. 81; original source U.S. Geological Survey.

2-15: Adapted from Bader, 1949, Geophysical history of the Anahuac oil field, Chambers County, Texas: in Nettleton, ed., Geophysical Case Histories, Vol. I: Soc. Expl. Geophysicists, Fig. 6, p. 71. Used by permission.

3-13, 4-16:   Adapted from Palmer, 1919, New graphic method for determining the depth and thickness of strata and the projection of dip: in Shorter Contributions to Geology, 1918, U.S. Geol. Survey Prof. Pap. 120, p. 122-128. Reprinted by permission.

3-14:   Adapted from Satin, 1960, Apparent-dip computer: Geol. Soc. Am. Bull., v. 71, p. 231-234. Used by permission.

5-2:   Adapted from Raisz, 1962, Principles of Cartography: McGraw-Hill, New York, Fig. 18-1, p. 179. Used by permission.

5-4b, c, 5-8b, 5-10a, 6-1, 8-5b: Adapted from Hobbs, Means, and Williams, 1976, An Outline of Structural Geology: John Wiley & Sons, New York, Figs. A6a, p. 491; A1a, p. 484; A46, p. 489; A3a, p. 487; A11, p. 498. Used by permission.

5-9: Adapted with permission from Berry and Mason, 1959, Mineralogy: Concepts, Descriptions, Determinations: Fig. 2-16, p. 38. Copyright © 1959, 1968, 1983, W.H. Freeman & Co., San Francisco.

5-10c, 5-11c, 8-4, 8-16, 11-16, 15-28:   Adapted from Ragan, 1985, Structural Geology: An Introduction to Geometrical Techniques, 3rd ed.: John Wiley & Sons, New York, Figs. 15-3, p. 273, 15-7, p. 278, 5-3, p. 60, 5-46, p. 61, final fig. Used by permission.

6-M2:   Adapted from Profett, 1977, Cenozoic geology of the Yerrington district, Nevada, and implications for the nature and origin of Basin and Range faulting: Geol. Soc. Am. Bull., v. 88, p. 247-266, Fig. 10. Used by permission.

6-M3:   Adapted from Harwood, 1983, Stratigraphy of upper Paleozoic volcanic rocks and regional unconformities in part of the northern Sierra terrane, California: Geol. Soc. Am. Bull., v. 94, p. 413-419, Fig. 2. Used by permission.

7-5:   Adapted from Badgley, 1959, Structural Methods for the Exploration Geologist: Harper & Brothers, New York, Figs. 273, p. 212; 274, p. 213. Used by permission.

7-8b: Adapted from Phillips, 1971, An Introduction to Crystallography, 4th ed.: John Wiley & Sons, New York, Fig. 37, p. 25. Used by permission.

8-6, 8-7, 8-21, 8-26, 8-32, 16-5, A4-7:   Adapted from Turner and Weiss, 1963, Structural Analysis of Metamorphic Tectonites: McGraw Hill, New York, Figs. 3-8, p. 60; 3-9, p. 61; 5-18, p. 171; 4-30, p. 127; 5-23, p. 176; 5-25, p. 180; 4-40, p. 140. Used by permission.

8-14, 8-27, 8-28, 8-29, 8-30, 11-15, 11-17, 16-2, 16-13, 16-15: Adapted from Ramsay, 1967, Folding and Fracturing of Rocks: McGraw Hill, New York, Figs. 7-17, 7-18, 7-24, 7-25, 7-26, 9-13, p. 496; 9-14, p. 496; 9-2, p. 492; 9-6, p. 494; 8-8, p. 470; 8-3, p. 464; 8-2, p. 463; 1-14, p. 12; 10-21, p. 539; 10-22, p. 539; 10-3, p. 522; 10-8, p. 527; 10-13, p. 531; 10-15, p. 533.   Used by permission.

8-20, 16-19:   Adapted from Ramsay, 1965, Structural investigations in the Barberton Mountain Land, Eastern Transvaal: Geol. Soc. S. Africa Trans., v. 66, p. 353-401, Figs. 27, 28, 29. Used by permission.

10-2b: Adapted from Handin, 1966, Strength and ductility: in Clark, Handbook of Physical Constants: Geol. Soc. Am. Mem. 97, p. 223-289, Fig. 11-2, p. 226. Used by permission.

10-3, 10-10d: Adapted from Heard, 1963, Effect of large changes in strain rate in the experimental deformation of Yule Marble: J. Geol., v. 71, p. 162-195, Figs. 1, p. 163; 11, p. 177. Used by permission of the University of Chicago Press.

10-7: Adapted from Handin and Hager, 1957, Experimental deformation of sedimentary rocks under confining pressure: Tests at room temperature on dry samples: Am. Assoc. Petrol. Geologists Bull., v. 41, p. 1-50, Fig. 22, p. 21. Used by permission of American Association of Petroleum Geologists.

10-9:   Adapted from Handin et al., 1963, Experimental deformation of sedimentary rocks under confining pressure: Tests at room temperature on dry samples: Am. Assoc. Petrol. Geologists Bull., v. 47, p. 717-755, Fig. 4, p. 730. Used by permission of American Association of Petroleum Geologists.

10-10a: Adapted from Edmond and Paterson, 1972, Volume changes during the deformation of rocks at high pressure: Int. J. of Rock Mech. and Mining Sci., v. 9, p. 161-182, Fig. 4, p. 168. Used with permission of Pergamon Journals, Ltd.

10-10b, c: Adapted from Heard, 1960, Transition from brittle fracture to ductile flow in Solenhofen limestone as a function of temperature, confining pressure and interstitial fluid pressure: Geol. Soc. Am. Mem. 79, p. 193-226, Figs. 3A, p. 200; 14, p. 224. Used by permission.

10-16, 10-17: Adapted from Shimamoto and Logan, 1981, Effects of simulated fault gouge on the sliding behavior of Tennessee Sandstone: Nonclay gouges: J. Geophys. Res., v. 96, p. 2902-2914, Figs. 3a, p. 2905; 6b, p. 2907. Copyright by the American Geophysical Union.

11-21a: Adapted from Huntoon, 1974, The post-Paleozoic structural geology of the eastern Grand Canyon, Arizona: in Breed and Rout, Geology of the Grand Canyon, Museum of Northern Arizona and Grand Canyon Natural History Assoc., p. 82-115, Fig. 8-c. Used by permission.

11-21c: Adapted from Malavieille, 1987, Extensional shearing deformation and kilometer-scale "a"-type folds in a Cordilleran metamorphic core complex (Raft River Mountains, northwestern Utah): Tectonics, v. 6, p. 423-448, Fig. 17. Copyright by the American Geophysical Union.

11-22a: Adapted from Anderson, 1964, Kink bands and related geological strkuctures: Nature, v. 202, p. 272-274. Used by permission.

11-22b: Adapted from Weiss, 1980, Nucleation and growth of kink bands: Tectonophys., v. 65, p. 1-38. Used by permission.

11-23, 11-25: Adapted from Sibson, 1977, Fault rocks and fault mechanisms: J. Geol. Soc. Lond., v. 133, p. 191-213, Fig. 8. Used by permission.

11-24f, 11-31a, 11-31d, 11-34a, 11-34b: From Lumino, 1987, Deformation within the Diana Complex along the Carthage-Colton mylonite zone: M.S. thesis, Univ. of Rochester, Rochester, NY, Figs. 2.4, 4.2, 5.5, 5.6, 5.9a. Used by permission.

11-27c, 11-49: Adapted from Durney and Ramsay, 1973, Incremental strains measured by syntectonic crystal growths: in DeJong and Scholten, eds, Gravity and Tectonics, John Wiley & Sons, New York, p. 67-96, Figs. 12, 18. Used by permission.

11-28: Marshak et al., 1982, Mesoscopic fault array of the northern Umbrian Apennine fold belt, Italy: Geometry of conjugate shear by pressure-solution slip: Geol. Soc. Am. Bull., v. 93, p. 1013-1022. Used by permission.

11-29a: Adapted from Hansen, 1971, Strain Facies: Springer-Verlag, New York, Figs. 15, 17. Reprinted by permission.

11-30d, 11-31b: From Gilotti, 1987, The Role of Ductile Deformation in the Emplacement of the Sarv Thrust Sheet, Swedish Caledonides: Ph.D. dissertation, Johns Hopkins Univ., Baltimore, MD, Figs. 2-9a, 2-9d. Used by permission.

11-31c: Adapted from Passchier and Simpson, 1986, Porphyroclast systems as kinematic indicators: J. Struc.

Geol., v. 8, p. 831-843, Fig. 4a. Used by permission of Pergamon Journals, Ltd.

11-32: From Vernon, 1976, Metamorphic Processes, Reactions and Microstructure Development: Allen & Unwin, Winchester, MA, Fig. 8-4. Used by permission.

11-33, 11-35b,c: Adapted from Simpson, 1986, Determination of movement sense in mylonites: J. Geol. Ed., v. 34, p. 246-261, Figs. 9, 13. Used by permission.

11-37a,b: Adapted from Engelder and Marshak, 1985, Disjunctive cleavage formed at shallow depths in sedimentary rocks: J. Struc. Geol., v. 7, p. 327-343, Figs. 1, 2. Used by permission of Pergamon Journals, Ltd.

11-45c: From S. Mitra, 1979, Deformation at various scales in the South Mountain anticlinorium of the central Appalachians: Summary: Geol. Soc. Am. Bull., pt. I, v. 90, p. 227-229, Fig. 1. Used by permission.

11-46a: Adapted from Smith, 1975, Unified theory of the onset of folding, boudinage, and mullion structure: Geol. Soc. Am. Bull., v. 86, p. 1601-1609, Fig. 11. Used by permission.

12-13, 14-4a: Adapted from Suppe, 1985, Basic Methods of Structural Geology: Prentice-Hall, Englewood Cliffs, NJ, Figs. 8-34, p. 293; 9-47, p. 351. Used by permission.

12-16a, c: Adapted from Aleksandrowski, 1985, Graphical determination of principal stress directions for slickenside lineation populations: An attempt to modify Arthaud's method: J. Struc. Geol., v. 7, p. 73-82, Figs. 6, 7, p. 77. Used with permission of Pergamon Journals, Ltd.

12-M1: SAR SYSTEM® imagery, courtesy of Aero Service Division, Western Geophysical Company of America and Goodyear Aerospace Corporation, Arizona Division. SAR SYSTEM is a registered service mark of Aero Service Division, Western Geophysical Company of America.

13-10, 13-11a, b: Adapted from Faill, 1969, Kink band folding, Valley and Ridge Province, Pennsylvania: Geol. Soc. Am. Bull., v. 84, p. 1289-1314, Figs. 5, 6, 20. Used by permission.

13-M1, 13-M2: Adapted from Conlin and Hoskins, 1962, Geology and mineral resources of the Mifflintown Quadrangle: Penn. Geol. Survey Atlas 126, Plate 1. Used by permission.

13-M5: Adapted from Dyson, 1967, Geology and mineral resources of the southern half of the New Bloomfield

Quadrangle: Penn. Geol. Survey Atlas 137cd, Plate 1. Used by permission.

13-M6: Adapted from Nickelsen, 1956, Geology of the Blue Ridge near Harpers Ferry, West Virginia: Geol. Soc. Am. Bull., v. 67, p. 239-269, Plate 1. Used by permission.

14-1: Adapted from Price, 1981, The Cordilleran foreland Thrust and fold belt in the southern Canadian Rock Mountains, in Mclay and Price, Thrust and Nappe Tectonics, Geol. Soc. Lond. Spec. Pub. 9, p. 428ff, Fig. 2. Used by permission.

14-4b: Adapted from Laubscher, 1962, Dis Zwiephasen-hypothese der Jurafaltung: Eclog. Geol. Helv., v. 55, p. 1-22, Fig. 1, p. 3. Used by permission.

14-6: Adapted from Perry, 1978, Sequential deformation in the central Appalachians: Am. J. Sci., v. 278, jp. 518-542, Fig. 2, p. 520. Used by permission.

14-7: Adapted from Boyer, 1978, Structure and origin of Grandfather Mountain window, North Carolina: Ph.D. dissertation, Johns Hopkins Univ., Baltimore, MD. Used by permission.

14-13: Adapted from Elliott, 1976, The energy balance and deformation mechanisms of thrust sheets, Phil. Trans. R. Soc., Lond. A., v. 283, p. 289-312, Fig. 2, p. 293. Used by permission.

14-26: Adapted from Geiser, 1988, The role of kinematics in the construction and analysis of geological cross sections in deformed terranes: Geol. Soc. Am. Mem., in press. Used by permission.

11-45, 15-1, 15-2, 15-9, 15-17: Adapted from Ramsay and Huber, 1983, The Techniques of Modern Structural Geology, Volume 1: Strain Analysis: Academic Press, London, Figs. 1.1, 1.8a, 4.1a, 4.1b, 4.6, 6.6. Used by permission.

15.23: Adapted from Wellman, 1962, A graphical method for analysing fossil distortion caused by tectonic deformation: Geological Mag., v. 99, p. 348-352, Fig. 1. Used with permission of Cambridge University Press.

15-30, 15-42, 15-43, 15-44, A4-9: Adapted from De Paor, 1988, R_f/ϕ_f strain analysis using an orientation net: J. Struc. Geol., in press, Figs. 1, 2, 3, 7. Used with permission of Pergamon Journals, Ltd.

15-36, 15-37, 15-38, 15-39, 15-40, 15-41, 15-42: Adapted from Lisle, 1985, Geological Strain Analysis: A Manual for the R_f/ϕ Technique: Pergamon Press, Oxford, Figs. 2.1, 2.2, 2.5, 6.1, fig. on p. 56. Used with permission of Pergamon Books, Ltd.

16-1: Adapted from Powell et al., 1985, Megakinking in the Lachlan Fold Belt, Australia: J. Struc. Geol., v. 7, no. 3, p. 281-300, Figs. 4, 6. Used with permission of Pergamon Journals, Ltd.

16-7: Adapted from Bell, 1981, Vergence: An evaluation: J. Struc. Geol., v. 3, no. 3, p. 197-202, Figs. 2, 3. Used with permission of Pergamon Journals, Ltd.

16-8b: Adapted from Wilson, 1982, Introduction to Small-Scale Geologic Structures: Allen & Unwin, London, Fig. 6.4a, p. 45. Reprinted by permission.

16-9: Adapted from Borrodaile, 1976, "Structural facing" (Shackelton's Rule) and the Palaeozoic rocks of the Malaguide complex near Velez Rubio, SE Spain: Proc. Koninklijke Nederlandse Akad. van Wetenschappen, Amsterdam, ser. B, v. 79, p. 330-336, Fig. 2. Used by permission.

16-16: Adapted from Anderson, 1971, Kink bands and major folds, Broken Hill, Australia: Geol. Soc. Am. Bull., v. 82, p. 1841-1962, Figs. 2, 7a-d, 9a-d. Used by permission.

16-23: Adapted from White and Jahns, 1950, The structure of central and east-central Vermont: J. Geol., v. 58, p. 179-220, Fig. 9. Used by permission of The University of Chicago Press.

A1-3, A1-5: Adapted from Greenhood, Mapping: University of Chicago Press, Chicago, pp. 9, 130, 139, 134. Used by permission.

A1-15: Roberts, Introduction to Geological Maps and Structures: Pergamon Press, Oxford, Fig. 4.8, p. 95. Used with permission of Pergamon Books, Ltd.

A4-5: Kalsbeek, 1963, A hexagonal net for the counting-out and testing of fabric diagrams: Neus Jahrbuch fur Mineralogie, Monatshefte, Fig. 1, p. 174. Used by permission.

I

ELEMENTARY TECHNIQUES

This part of the book introduces the fundamental tools of structural geology. The first four chapters are designed to accustom students to visualizing the attitude, location, and dimensions of geologic structures. (Appendix 1 outlines elementary aspects of maps and cross sections and thus provides an optional introduction to these chapters). We discuss how to measure and describe lines and planes, how to use a compass, how to create and interpret contour maps, how to calculate the attitude of planes from point data, and how to calculate the thickness and depth of layers. Through the study of these subjects, the student learns how to apply descriptive geometry and trigonometry to problems in structural geology. The second four chapters focus on the use of equal-angle (stereographic) and equal-area projections for the solution of geological problems and for the representation of geological data. These chapters discuss practical applications of these projections to the field study of fabrics and folds, and to the analysis of drill-hole data.

CHAPTER

1

MEASUREMENT OF ATTITUDE AND LOCATION

1-1 INTRODUCTION

Imagine that you are a field assistant on an expedition to map the remote highlands of Brazil. On the second day the chief geologist of the expedition sends you on a solo traverse to find the *contact*, or boundary surface, between a white sandstone unit and a grey limestone unit in the northeast corner of the map area. All morning you trudge through the brush of a broad plateau on which there are only isolated outcrops of rock. By studying the outcrops, you discover that the limestone is more weathered than the sandstone. Then at lunchtime you come to a deep north-south trending gorge and descend to the stream at its bottom to cool your feet and eat. The rock along the stream bed at your lunchspot is sandstone. Looking upstream, you see a weathered ledge and think, based on your experience, "It's probably limestone. . . LIMESTONE! Wait a minute! That contact must be between me and that ledge." You run upstream and find the contact perfectly exposed in the wall of the gorge (Fig. 1-1). The bedding on opposite sides of the contact is not parallel, and the contact appears to be covered with scratches (slip lineations) and is bordered by a thin zone of breccia; you conclude that the contact is a fault. Happy with your discovery, you sit down to write notes, and ask yourself, "What important features about this outcrop will the chief want to know?" Your list includes the following:

1. *Location* (Where is the exposure of the fault?)
2. *Attitude* (What is the orientation of the fault?)
3. *Appearance* (What does the fault look like?)

The discipline of structural geology frequently deals with such questions. Now the challenge (and the fun) begins; how do you answer them? You are a bit worried because this is only the second day on the job and your compass skills are minimal. Nevertheless, you decide to rely on a very useful asset - common sense, and quickly get to work.

This chapter focuses on the first two questions in the preceding list by introducing the methods and conventions used by geologists to describe the attitude and location of geologic structures. We begin with the concept of a *reference frame*, which is implicit in all such descriptions. Then we discuss the format that geologists use to specify attitude, and we illustrate how a compass is used to measure attitude. Finally, we show how a compass can be used to find locations. Our discussion assumes that you are familiar with the basic concepts of maps and cross-sections, and that you can read a map to find a location. If not, please study Appendix 1. Suggestions for describing the appearance of a structure are presented in Chapter 11.

Perhaps the most important skill of a structural geologist is to be able to visualize objects or features in three-dimensional space. We will emphasize again and again that when you describe the attitude of a geologic structure, you must create an image of the structure in your mind, and you must keep track of whether the structure is a volume, a plane, or a line.

1-2 REFERENCE FRAME

A *reference frame* in three-dimensional space is a set of three mutually orthogonal coordinate axes. The point at

North ◄━━━━

Figure 1-1. Geologic discovery! A fault exposed in a stream cut. Note that the marble layers to the left (north) of the zone are not parallel to the fault or to the sandstone layers to the right. The stream flows due south. The fault surface is covered with scratches (slip lineations) that are parallel to the intersection between the fault plane and the vertical gorge wall.

which the three axes join is the *origin*. A plane containing any two axes is called a *coordinate plane*. In this context we can define the *location* of a point by specifying its coordinates with respect to the three axes, and the *attitude* of a line or plane as the angle that a line or plane makes with respect to each coordinate axis. In a three-axis reference frame, a line can be resolved onto a coordinate plane (Fig. 1-2) by tracing the tip of the line along a path parallel to the axis that is perpendicular to the coordinate plane. The resulting line, which lies on the coordinate plane, is called the *projection* of the line.

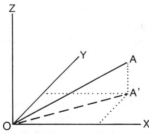

Figure 1-2. The orientation of a line (OA) in space can be described with reference to three mutually perpendicular axes (X, Y, and Z). The projection of line OA onto the horizontal (X-Y) plane is labeled OA'. Point A moves down along the dotted line to point A'. Line AA' is parallel to the vertical (Z) axis.

For a given point on or near the earth's surface, the three axes that are used to define the reference frame are (1) the line of *longitude* (which trends north-south; see Appendix 1), (2) the line of *latitude* (which trends east-west; see Appendix 1), and (3) a *vertical* line. The coordinate plane containing the lines of latitude and longitude at a point is the *horizontal plane* at the point. A "vertical line" is parallel to the radius of the earth at the

point (Fig. 1-3) and is, of course, perpendicular to the horizontal plane. Positions along a vertical line are specified by elevations. Because of the curvature of the earth's surface, the absolute orientation of the three axes changes from point to point around the globe. Remember the fault exposure mentioned in Section 1-1? To describe the location of this outcrop in your notes, you record its latitude, longitude, and elevation. This information can be read from a map (see Appendix 1).

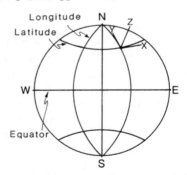

Figure 1-3. Three coordinate axes defining a reference frame at the surface of the earth. Line X is tangent to a line of latitude, line Y is tangent to a line of longitude, and line Z is perpendicular to the surface of the earth and is parallel to a radius vector.

1-3 ATTITUDES OF PLANES

Many geologic structures (e.g., faults, beds, joints, veins, cleavages, foliations, dikes, contacts, and unconformities) can be represented as planes. The attitude of a plane can be specified simply by a pair of numbers. Two alternative number pairs can be used; the first is *strike and dip* and the second is *dip and dip direction*. The use of dip and dip direction measurement is treated in Section 1-4.

Strike of a Plane

A horizontal line on a plane is called a *strike line*. A strike line on a structure can be visualized as the intersection between an imaginary horizontal plane and the structure. Remember that the intersection between two planes is a line; in geology, the line of intersection is called a *trace*. To help visualize a strike line, imagine a cliff rising from a calm sea; the intersection of the sea surface with the cliff is a strike line on the cliff face (Fig. 1-4). The trace of the breccia zone on the horizontal bed of the stream in Figure 1-1 is a strike line. The *strike* of a plane at a given location is the angle between the strike line and true north. In other words, strike is the angle between a horizontal line on a plane and true north. Memorize this definition! Strike is an angle that is measured in degrees with a compass. Any angle measured with a compass is called an *azimuth*.

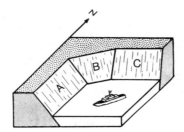

Figure 1-4. Intersection of the sea surface (horizontal plane) with a cliff face. The shoreline defines a strike line on the cliff face. Cliff A strikes north-south, cliff B strikes northeast-southwest, and cliff C strikes east-west.

The strike of a plane can be described in two ways. The first way to describe strike is known as the *quadrant convention*. In this convention, the range of possible directions is divided into four quadrants (NE, SE, NW, and SW) of 90° each (Fig. 1-5a), and the strike is specified by a given number of degrees east or west of north. If the strike line on a plane is parallel to the N-S compass direction, the plane has a strike, in the quadrant convention, of N00°E. If the strike line on the plane is parallel to the E-W compass direction, the plane has a strike of N90°E (or N90°W). A strike line that points NE is oriented N45°E.

A strike of N32°E is read, "north thirty-two degrees east." Note that a strike of N20°W is exactly the same as a strike of S20°E, because there is no need to differentiate between the ends of a horizontal line. It is common practice, however, to specify strikes in the quadrant system with respect to north. Look again at Figure 1-1. The trace of the fault on the stream bed is perpendicular to the north-south stream. Thus, even without using a compass, you were able to estimate that the fault strikes N90°E.

The second way to represent strike is known as the *azimuthal convention*. In this convention the range of possible directions on a horizontal plane is divided into 360°, with the direction of due north being assigned a value of 000° or 360° (Fig. 1-5b). Strike in the azimuthal convention can be specified entirely by a number. For example, if the strike line points exactly northeast, the strike is 045°. An azimuth of N00°W in the quadrant convention translates to 000° in the azimuthal convention. A strike of N32°E is identical to a strike of 032°, a strike of N32°W is identical to 328°, and a strike of S24°E is identical to 156°. Notice that in the azimuthal convention, a strike should always be specified by three digits, even if some of the digits are 0 (e.g., 056°). You can indicate the strike of the fault in Figure 1-1 as 090°.

Dip of a Plane

The *true dip* of a plane is the angle between the plane and a horizontal plane as measured in a unique vertical plane. This unique vertical plane is oriented such that it is exactly perpendicular to the strike line (Fig. 1-6a). In Figure 1-1 the dip of the fault in the vertical wall of the gorge is the true dip of the fault, because the strike of the fault is perpendicular to the wall. You could probably estimate the dip of the fault if you did not know how to measure it exactly; the fault looks like it dips about 70°. The true dip is always the steepest possible slope on the given plane, and the *true dip direction* is the azimuth that is exactly perpendicular to the strike. The true dip direction is always specified as the downslope direction; the fault in Figure 1-1 dips south (downstream). A dip angle measured in any vertical plane that is not exactly perpendicular to the strike line is called an *apparent dip* (Fig. 1-6b). The dips of the limestone and sandstone beds that you see in the gorge wall

Figure 1-5. Conventions for specifying strike. (a) Quadrant convention; (b) azimuthal convention. Items in parentheses are alternative expressions of the same direction.

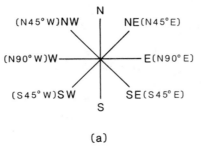

(a)

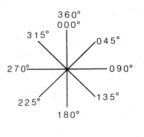

(b)

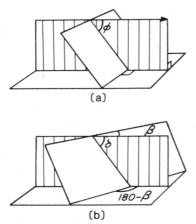

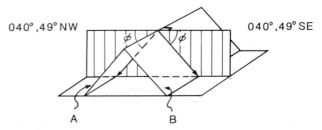

Figure 1-8. Convention for specification of dip direction. Note that the two inclined planes have opposite dips but the same strike.

Figure 1-6. Block diagram showing the meaning of dip. The vertical reference plane is ruled. (a) True dip (ϕ), with arrowhead indicating dip direction; (b) apparent dip (∂). The angle ß is the angle between true strike and the bearing of the plane in which the apparent dip was measured.

of Figure 1-1 are apparent dips, because the beds do not strike perpendicular to the gorge wall. The magnitude of an apparent dip must always be less than that of the true dip; the apparent dip measured in the vertical plane that contains the strike line is always equal to 00°.

Dip is specified as an angle between 00° and 90°. A plane with a 00° dip is a horizontal plane, whereas a plane with a 90° dip is a vertical plane. Generally, dips in the range of 00° to 20° are considered shallow dips, those in the range of 20° to 50° are moderate dips, and those in the range of 50° to 90° are steep dips (Fig. 1-7). These divisions are arbitrary and vary depending on author. In circumstances where the stratigraphic *younging direction* (the direction in which the beds get younger) of a sequence of rocks is known, and the beds have been tilted past vertical, the beds are said to be *overturned*. In such cases, the specified dip is still a number less than 90°, but a different map symbol is used.

Specification of the strike and the dip angle alone does not uniquely define the attitude of a plane. For example, an east-west striking plane can dip either north or south, and a plane that strikes N40°E can dip to the southeast or the northwest (Fig. 1-8). If the fault in the gorge of Figure 1-1 dipped to the north instead of to the south, its surface

would slope upstream instead of downstream. If planar orientations are specified by strike and dip, the *general direction of dip* must be specified. The exact direction is not needed, for the true dip direction is always exactly 90° from the strike. Thus, it is sufficient to say that a N30°E plane is dipping, say, 24°NW. The true dip direction of this plane is automatically known to be N60°W. Note that it is impossible for a plane to dip in the same direction that it strikes. The N30°E-striking plane cannot dip northeast or southwest; the dip direction must lie in one of the quadrants to either side of the strike quadrant. Visualize a plane and convince yourself that this rule holds! The fault in Figure 1-1 cannot dip east or west.

Representation of Planar Attitudes

The attitude of a plane is completely specified when the strike, dip, and general dip direction are indicated. For example, the attitude of the east-west striking plane that dips 30°N can be written as 090°,30°N or as N90°E,30°N. Some geologists prefer to substitute a semicolon or a slash for the comma (e.g., N90°E;30°N). Note that the strike number is written first and the dip number second. Generally, you should specify the quadrant toward which the plane is dipping (e.g., N42°W, 23°NE) unless the strike is within about 10° of north-south or east-west (e.g., N08°E,34°E). You should now be able to concisely specify the approximate orientation of the fault in Figure 1-1; it is N90°E,70°S.

Planar attitudes can be specified not only by pairs of numbers but also by symbols on a map. The use of such symbols makes the geometry of a structure on a map easier to visualize. Symbols for various planar features are displayed in Figure 1-9. The strike is indicated by a short

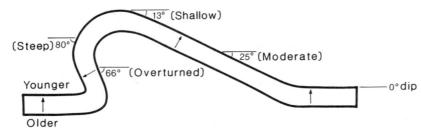

Figure 1-7. Adjectives used to describe dip of a layer. The example shows an overturned fold. The arrows indicate stratigraphic younging direction.

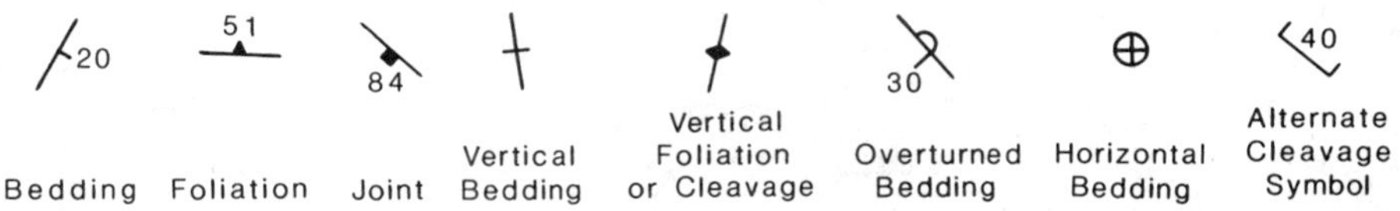

Figure 1-9. Basic symbols commonly used for specification of strike and dip of a planar structure on a map. Note that the numbers are always written in the same orientation.

Bedding	Foliation	Joint	Vertical Bedding	Vertical Foliation or Cleavage	Overturned Bedding	Horizontal Bedding	Alternate Cleavage Symbol

line segment drawn parallel to the strike line, and the dip is indicated by a tick pointing in the dip direction. The angle of dip is written next to the tick. Dip numbers on a map should all be written in the same orientation (usually parallel to the base of the map and to the words of the legend), regardless of strike, so that they are easy to read without having to constantly rotate the map. Symbols for joints and cleavage are used differently by different authors. If several sets of foliations are present, the author of a map may invent symbols. Because of the variety of symbols that are used, it is important that symbols be defined in the map explanation.

Notice that in the azimuthal system strikes are always specified by three-digit numbers (with no letters needed) and dips by two-digit numbers plus a dip-direction specification. Some geologists use a shorthand system of specifying strike and dip, called the *right-hand rule*. When following this rule, you must choose the strike azimuth such that the plane dips to your right when you are facing in the direction of the azimuth (Fig. 1-10a). On the dial of the compass, this rule is equivalent to saying that the dip direction is found by moving 90° clockwise around the dial

from the strike azimuth to the dip direction (Fig. 1-10b). The advantage of the right-hand rule is that attitudes can be expressed entirely by numbers, which is especially convenient when attitude data are to be entered in a computer file.

1-4   ATTITUDES OF LINES

Many geologic features (e.g., scratches on a fault surface, the intersection of two planes, elongate minerals and pebbles, flute casts, fold hinges) can be pictured as lines. Linear structures related to deformation of rock are called *lineations*. The attitude of a linear structure cannot be represented by strike and dip. Instead, linear attitudes are represented by a pair of numbers called *plunge and bearing*. If the line occurs on a plane of known attitude, its orientation may be give by a single number called the *rake* or *pitch*.

Plunge and Bearing of a Line

The *plunge* of a line is the angle that the line makes with respect to a horizontal plane as measured in a vertical plane (Fig. 1-11). Values for plunge range between 00° and 90°; a plunge of 00° refers to a horizontal line, and a plunge of 90° refers to a vertical line. If the bearing of the lineation is exactly parallel to the dip direction of the plane, the plunge must equal the dip (visualize the scratches on the fault in Figure 1-1). Generally, plunges of between 00° to 20° are considered shallow, those between 20° to 50° are considered moderate, and those between 50° to 90° are considered steep.

The *bearing* (also called *trend*) of a line is the azimuth of the projection of the line onto a horizontal coordinate plane. The line and its projection must both lie in the same vertical plane (Fig. 1-11). A bearing can be specified using either the quadrant or azimuthal conventions, depending on preference. A line that is exactly parallel to a strike line on a plane has a bearing that is equal to the strike.

When specifying a bearing, it is very important that the azimuth indicated gives the direction in which the line plunges. A line plunging due east is not the same as a line plunging due west; these two lines plunge in opposite directions. The scratches on the fault surface in Figure 1-1 are perpendicular to the strike of the fault and are parallel to

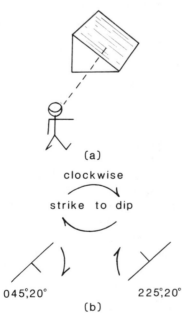

(a)

clockwise

strike to dip

045°,20°   225°,20°

(b)

Figure 1-10. Illustration of the right-hand rule convention for specification of strike and dip. (a) Plane dips to the right of the line of sight. (b) Dip number lies to the right of the strike number on the compass.

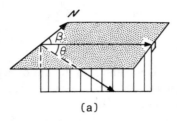 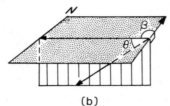

(a)   (b)

Figure 1-11. Definition of the plunge and bearing of a line. The horizontal plane is shaded and the vertical plane is ruled. ß is the angle of bearing. (a) Line plunging to the east; (b) line plunging to the west. Note that the bearing of the two lines is different even though the magnitude of the plunge (θ) is the same and both lines lie in the same plane.

the dip direction of the fault. The scratches therefore plunge due south; they could not possibly plunge due north.

Dip and Dip Direction

As noted earlier, strike and dip are not the only means by which the attitude of a plane can be specified. The attitude of a plane can be specified by giving the plunge and bearing of the line on the surface of the plane that is exactly perpendicular to the strike. The values of the plunge and bearing for this line are the *dip and dip direction*. We could specify the orientation of the fault in Figure 1-1 by saying that its dip is 70° and its dip direction is 180° (i.e., its dip and dip direction are: 70°,180°).

Rake of a Line

The *rake* of a line (sometimes referred to as the *pitch* of a line) is the angle between the line and the horizontal as measured in the plane on which the line occurs (Fig. 1-12). The rake is an angle between 00° and 90°. If the bearing of the lineation is parallel to the strike of the plane, the rake must equal 00°. If the bearing of a lineation is perpendicular to the strike, the rake is 90°. The scratches on the fault surface of Figure 1-1, for example, have a rake of 90°. Any lineation whose bearing is between the strike and dip direction of the plane on which it occurs must have

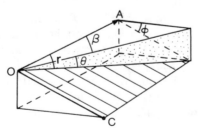

Figure 1-12. Block diagram illustrating the meaning of rake and the relation of rake to plunge and bearing. Ruled plane is inclined and the stippled plane is vertical. r = rake (measured in the inclined plane); ß = bearing (measured in the horizontal plane); ø = true dip of the plane, θ = plunge of the line.

a rake that is intermediate in value between 00° and 90°. Try to visualize why this rule is true. The direction of rake must be indicated. Imagine that the plane shown in Figure 1-12 strikes northeast-southwest (i.e., the arrow that points from O to A points northeast). The line in the dipping plane that runs from O to D pitches to the northeast and a line from A to C (not shown) pitches southwest.

A rake angle alone does not completely describe the attitude of a line in space. To completely specify the attitude of the line both the rake of the line and the strike and dip of the plane on which it lies must be indicated. We will see in Chapters 3 and 6 how to calculate the plunge and bearing of a line if its rake and the strike and dip of the plane on which it occurs are known.

Representation of Linear Features

The attitude of a line is completely specified by the plunge and bearing. The plunge (a two-digit number) is written first, followed by the bearing (a three-digit number). For example, a linear attitude would be written 48°,021° or 48°,N21°E (meaning a plunge of 48° in the direction north 21° east). The scratches on the fault surface in Figure 1-1 are oriented 90°,180°. Many geologists substitute an arrow or a semicolon for the comma. Remember, in contrast, that a planar attitude by the right-hand rule would be written with the three-digit number first (e.g., 021°,48°).

The map symbol for a linear attitude is an arrow drawn parallel to the bearing. A number is written at the tip of the arrow to indicate the angle of plunge. Often, the arrow is drawn to originate from a planar attitude symbol that indicates the strike and dip of the plane on which the lineation was observed (Fig. 1-13). Rakes are rarely shown on maps. If rakes are measured in the field, they are usually converted to plunge and bearing before being transferred to a map (see Chapters 3 and 5).

1-5   USE OF A COMPASS

In the scenario presented in Section 1-1 we suggested that your compass skills were minimal. Thus, you relied on common sense to determine a way to describe the attitude

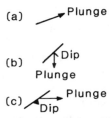

Figure 1-13. Common symbols used for repre- senting the plunge and bearing of a line on a map. Orientation of the arrow gives the bearing; the number at the end of the arrow gives the plunge. (a) Lineation alone; (b) lineation on a bedding plane; (c) lineation on a foliation plane.

of the fault to the chief. Realizing that the north-south stream trace and the vertical gorge wall provided an ideal three-axis reference frame, you estimated the attitude of the fault. The chief is pleased with your effort, but requires more exact measurements in the future, and thus spends the next few hours training you in the use of a compass.

The traditional instrument used by geologists for measurement of the attitudes of structural features is the *Brunton compass,* though in recent years other types of compasses (e.g., the Silva compass) have come into favor, and in areas of magnetic rocks a sun compass must be used. The discussion that follows is keyed to use of the Brunton-style compass, but the principles can be applied to any compass. Practice with a compass will help you develop the ability to visualize lines and planes in three-dimensional space.

Elements of a Compass

A compass (Fig. 1-14) is composed of a magnetized needle that is balanced on a pin so that the needle can rotate easily and becomes aligned with the magnetic field lines at the

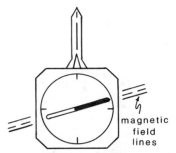

Figure 1-15. Sketch showing the orientation of a compass with respect to a magnetic field line.

location of measurement (Fig. 1-15). The white painted end of the needle points to the north *magnetic pole.* A magnetic pole (there are two, north and south) is a point on the surface of the earth where the lines of magnetic force are vertical (Fig. 1-16). On the outer circumference of the compass face is a scale graduated in degrees. This scale is called a *compass card.* On old-style "mariner's" compasses, the compass card was divided into 16 increments (N, NNE, NE, ENE, E, ESE, etc.). More modern "surveyor's" compasses are divided by degrees in one of two ways. The compass card of *quadrant compasses* is divided into four quadrants of 90° each; north and south are each assigned a value of 00°, and east and west are each assigned a value of 90°. On an *azimuthal compass*, the card is divided into 360°, with 000° (360°) coinciding with north, 090° corresponding to east, 180° corresponding to south, and 270° corresponding to west.

A fold-out metal *pointer* projects from the Brunton compass. When the white end of the compass needle lies on 000°, this pointer, when folded out, is pointing due north. Likewise, when the white end of the needle lies on 045°, the pointer is pointing northeast, and so forth.

Though values for azimuth increase clockwise from north on the surface of the earth (e.g., if you are facing

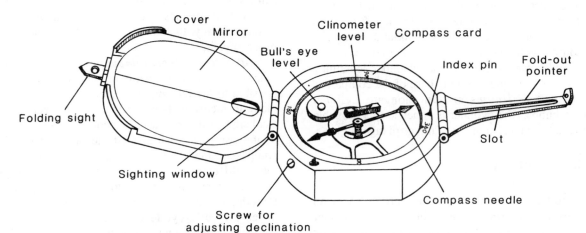

Figure 1-14. Sketch of a Brunton compass, with the key components labeled. Adapted from Compton, 1962.

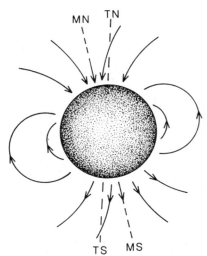

Figure 1-16. Earth's magnetic field lines. MN = magnetic north; TN = true north; MS = magnetic south; TS = true south. (Adapted from Judson, Kauffman, & Leet, 1987.)

north and want to face east, you turn clockwise), the numbers representing azimuth on the compass card increase in the counterclockwise direction. Likewise, on the compass card of a quadrant compass, east and west are reversed. While this convention may seem confusing at first, it actually makes use of the compass much easier. This is because as you rotate the compass (and therefore the compass card and pointer) clockwise from north to east, the compass needle actually remains fixed in space; the needle continues to be aligned with the magnetic field line (Fig. 1-15). Therefore, in the reference frame defined by the compass body, the needle appears to rotate counterclockwise. In order for the white end of the needle to lie on 090° when the compass pointer is pointing due east, the azimuthal numbers on the compass card *must* increase in a counterclockwise direction. On a quadrant compass, imagine that the compass pointer is directed exactly NE. On the compass card, you simply read off "north 45° east." The word "east" is written on the card to the left of north so that you can read off the word east without thinking.

In addition to the compass needle, the compass also contains a *"bull's-eye" level* (a circular chamber containing a bubble), which tells you when the base of the compass is horizontal, and a *clinometer* (an elongate cylinder containing a bubble; the cylinder is attached to a movable arm), which allows measurement of dip or plunge angles.

Magnetic Declination

The magnetic field of the earth can be represented by an array of lines that run from one magnetic pole to the other (Fig. 1-16). At a given locality on the earth, the moving element of the compass, the magnetized needle, aligns

itself with the magnetic field line at that locality. The needle is usually balanced so that it lies parallel to the horizontal plane at the point of measurement and therefore gives the horizontal component of the magnetic field. Averaged over long periods of time, the magnetic dipole of the earth corresponds to the spin axis of the earth, so that the magnetic poles are the same as the *geographic poles* (the geographic poles are the points at which the spin axis pierces the earth). At any given time, however, the magnetic poles may be located at a distance from the true poles. Today, for example, the north magnetic pole is located in northern Canada. The acute angle between the direction of true north (a line of longitude) and the direction that the compass needle points in the present-day magnetic field is called *magnetic declination*. A declination of 12° east means that the angle between true north and magnetic north is 12°, and that true north lies 12° counterclockwise from magnetic north. Values for magnetic declination at a given time in the United States can be plotted on a map (Fig. 1-17). The magnetic pole drifts slightly every year, so such maps must be constantly updated.

As we noted earlier, the reference frame used to specify locations and orientations on the earth's surface is keyed to the geographic poles. Therefore, a correction must be made in order to account for magnetic declination. By making this correction, the compass pointer is pointing to true north when the white end of the needle is lying on 0°, even if the needle is not parallel to the pointer. A Brunton compass may be set for the magnetic declination of a map area by turning the screw on the side of the compass; this screw rotates the compass card with respect to the pointer. Figure 1-18 shows compasses set for two different magnetic declinations.

Measurement of Planes with a Compass

In this section we describe the practical methods that you can use to measure the attitude of a plane with a compass. You will learn these methods more easily if you work through them with someone who is experienced in the use of a compass.

(a) Direct Measurement of Strike: If the plane you are measuring is well exposed and fairly smooth, it is possible to lay the compass directly on the surface of the plane to measure its strike. Make sure your hammer or steel clipboard is not near the compass. With the side edge of the compass flush against the bed surface, move the compass so that the level bubble is in the bull's-eye. Note that a different edge is used depending on whether the surface is upward-facing or downward-facing (i.e., use the top edge of the compass to measure under an overhang). When the bull's-eye level indicates that the plane of the compass is horizontal, the edge of the compass in contact with the surface defines the horizontal intersection line

Figure 1-17. Map of the declination lines for the United States for 1980. At a location along one of these lines the declination is equal to the number of degrees indicated. (Adapted from Brunton compass instruction book.)

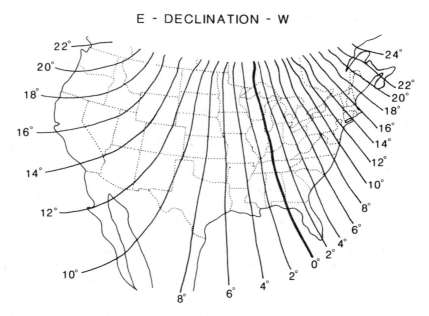

E - DECLINATION - W

LINES OF EQUAL MAGNETIC DECLINATION 1980

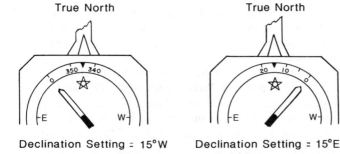

True North   True North

Declination Setting = 15°W   Declination Setting = 15°E

(a)   (b)

Figure 1-18. Sketches showing the dial of a compass set to correct for magnetic declination. (a) Declination of 15°W; (b) declination of 15°E. Each compass is shown pointing due north. Note that the needle is not parallel to the fold-out pointer. Angles are exaggerated.

Figure 1-19. Sketch illustrating the position of a compass during measurement of strike. Note that the bottom side edge is flush with the dipping surface. (a) Block diagram. Stippled plane is vertical and is perpendicular to strike. (b) View of compass looking along strike for an upward-facing surface. True dip is ø. (c) View of compass looking along strike for a downward-facing surface. (d) Top view of compass showing bubble centered in bull's-eye.

between the compass and the surface and is, therefore, a strike line on the surface (Fig. 1-19). Either end of the compass needle gives the value of strike, though usually the end closer to north is specified. Remember, if the compass needle reads 315° (= N45°W), the pointer is pointing 315°.

Because of the design of the Brunton compass (a circular metal ridge projects from the bottom to protect the clinometer adjustment lever), it is difficult to measure strike directly for planes with dips of less than about 12°. For such shallowly dipping planes, it is easier and more accurate to first determine dip direction and then calculate strike. Very slight undulations of a shallowly dipping plane can drastically change the strike, so extra care must be taken in measuring such planes. Remember that a unique strike cannot be specified for a horizontal plane.

(b) Direct Measurement of Dip: Direct determination of the dip of a surface can be done in two ways.

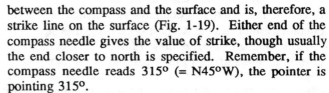

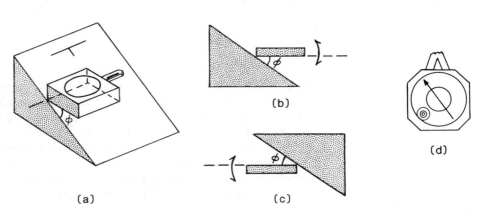

(a)   (b)

(c)   (d)

The first way is to indicate a perpendicular to the strike line on the plane using string or a stick. Be sure not to make a permanent mark that would disfigure the outcrop! Place the side of the compass on the surface, making sure that the compass is not upside-down. Note that a different side of the compass is placed against the surface depending on whether the surface is upward-facing or downward-facing, because the clinometer dial is only on one half of the compass face (Fig. 1-20). Using the lever on the back of the compass body, move the clinometer so that the bubble is centered. The angle indicated by the clinometer is the dip. The second way, which does not require prior knowledge of the strike direction, is to lay the side of the compass on the surface parallel to your best estimate of the dip direction, center the clinometer bubble, and swing the compass back and forth slightly (all the while keeping the side in contact with the surface) so that it swings through a narrow range of apparent dip directions (indicated by the arrows in Fig. 1-20a). If, during this operation, the bubble moves out of center such that you must adjust the clinometer to a steeper dip to recenter the bubble, your original estimate was an apparent, not a true dip. The direction in which the compass is oriented when the clinometer indicates the steepest slope is the true dip.

(c) Use of a Compass Plate: If the exposed portion of the surface to be measured is too small or is slightly irregular, such that it is not possible to lay the edge of the compass directly on the surface, a direct strike measurement may still be possible with the aid of a *compass plate*. A compass plate is a smooth sheet of wood or aluminum that provides an adequate base for the compass to contact. When making a compass plate, it is best to cut a large notch out of one corner (Fig. 1-21) in order to facilitate measurement of planes that intersect the corner of an outcrop. Standard clipboards, which have steel

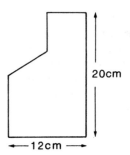

Figure 1-21. The surface of a compass plate. The notch is to facilitate measurement of planes that intersect corners. If the plate is made of aluminum, it should be about 0.3 cm thick.

clips, or soft notebooks do not make appropriate compass plates. If only a small ledge is available for measurement, the plane of the ledge can be extended by holding the plate firmly against the ledge (Fig. 1-22a). The compass can then be placed on the plate. If no ledge is available, but the intersection between the plane and the outcrop face is visible on two nonparallel planes that join at a corner, a measurement can be made by aligning the two edges of the notch in your compass plate with the two intersection lines on the corner (Fig. 1-22b). The two lines define the plane to be measured. Make sure the two lines lie in the plane of the compass plate, and then make a measurement.

(d) Shooting a Strike and Dip: The attitude of a plane can also be determined from a distance, using the following steps (the procedure is commonly called "shooting a strike and dip;" Fig. 1-23): (1) Position yourself so that your are able to sight along a strike line on the plane. This means that your line of sight should be a strike line on the plane and you should not be looking down on the surface of the plane or up to the backside of

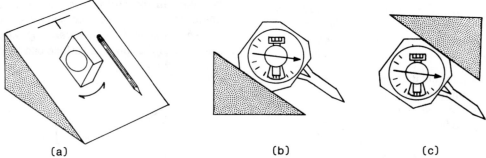

(a)   (b)   (c)

Figure 1-20. Sketch illustrating the position of a compass during measurement of true dip. (a) Block diagram. Stippled plane is perpendicular to strike. The arrows indicate movement of the compass during the operation to confirm that the dip measured is the steepest possible dip on the surface. The pencil points in the direction of true dip; (b) view looking down strike showing the proper position of the clinometer for an upward-facing surface; (c) view looking down strike showing the proper position of the clinometer for a downward-facing surface.

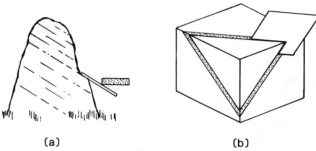

Figure 1-22. Use of a compass plate. (a) Extension of the surface of a bed at a ledge. The stippled box is the edge of a compass; (b) measurement of a plane defined by two lineations on a corner.

the surface. (2) Hold the compass away from your eyes (about half arm's length). (3) Fold up the mirrored cover of the compass so that you can see through the small window at the base and can see the reflection of the compass dial. Level the base of the compass with the bull's-eye level; (4) Line up the tip of the metal pointer, the tick mark in the window, and the strike line on the plane with your line of sight. (5) By looking in the mirror, check the bull's-eye bubble and relevel if necessary. Realign your line of sight with the pointer, the tick mark, and the strike line and read off the strike; (6) To determine the dip, maintain your position with your line of sight parallel to a strike line. Hold the compass at arm's length perpendicular to the strike direction. Make sure it is at the same elevation as your line of sight. Tilt the compass so that the edge of the compass parallels the plane being measured, center the clinometer bubble, and read off the dip. Shooting a strike and dip is inherently less accurate than making a direct measurement on a surface but may be necessary because an outcrop is inaccessible or because the layering to be measured is wavy. If the the layering is wavy, a single measurement directly on the surface may not indicate the average attitude of the layer.

Measurement of Lines with a Compass

There are three approaches to measurement of the bearing of a lineation with a Brunton compass. The first two methods work best for lineations that are on shallowly dipping planes, and the third method works best for lineations that are on steeply dipping planes.

(a) Bearing Method A: Fold out the metal pointer (Fig. 1-24a). Notice that there is a slot in the pointer. Hold the compass at chest height and align the compass with the line to be measured such that the line is visible in the slot and the pointer is pointing in the plunge direction. If the line is hard to see, you may lay a pencil along it. Do not draw on outcrops. Level the compass with the bull's-eye level. The bearing is the azimuth indicated by the white end of the needle .

(b) Bearing Method B: Align the edge of your compass plate along the line and place the side of the compass on the surface of the plate such that the metal pointer is pointing down-plunge. Adjust the orientation of the plate so that it is vertical and the bull's-eye level in the compass indicates horizontal (Fig. 1-24b). With the compass in this position, the needle indicates the bearing of the line.

(c) Bearing Method C: Place two points of the edge of the compass on the lineation (Fig. 1-24c); one point should be a corner of the compass body and the other a corner of the compass cover. The contact point on the body should be down the plunge of the line from the cover contact point. Center the bull's-eye level and read the bearing. The edge of the compass defines a vertical plane. Therefore, the azimuth indicated on the compass dial is the bearing of the line. This method works only for lineations that are on overhangs.

(d) Shooting a Bearing: If it is necessary to determine the bearing of large linear feature (such as a highway, a river, or the path between two points), you may shoot a bearing. One way to do this is to configure your compass as shown in Figure 1-23b. Level the compass and point it toward a point in the distance along the line that you are measuring. The point should be at eye level (e.g., it could be your field partner standing in the distance. Look through the window of the compass cover so that you see the distant point. Read the black end of the needle (because the compass is pointing toward you) to determine the bearing of a line pointing away from you. An easier, but less accurate, way of shooting a bearing is to hold the horizontal compass at waist level or chest level and simply point it toward the distant point. The white end of the needle gives an approximate bearing to the point.

(e) Plunge Measurement: To determine the plunge of the line, lay the side of the compass along the

Figure 1-23. Shooting a strike and dip. (a) Position of observer with respect to plane; (b) configuration of compass.

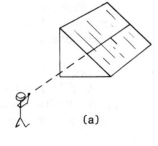

(a)

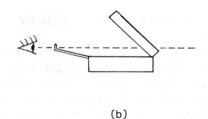

(b)

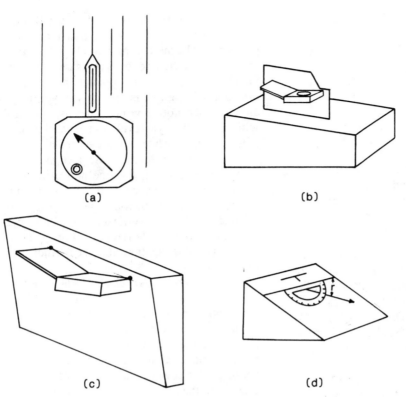

(a)

(b)

(c)

(d)

Figure 1-24. Measurement of a lineation. (a) Looking down on a compass with the lineation in the pointer slot; (b) use of a compass plate; (c) two-point contact method for overhangs; (d) determination of rake (r is the angle of rake).

lineation (or along the edge of the compass plate that is aligned with the lineation). Make sure that the plane of the compass is vertical, then use the clinometer to measure the plunge. Be sure that the scale of the clinometer is right-side up. Notice that bearing is usually measured before plunge, even though plunge is written in front of the bearing.

(f) Rake Measurement: Measurement of rake is done with a protractor. Use your compass to determine the strike line. Position the protractor so that it is lying against the surface and so that its base is parallel to the strike line (Fig. 1-24d). Lay your pencil on top of the protractor so that it passes through the center point of the protractor and is parallel to the lineations. Measure the rake off the protractor scale. Use your compass to determine the direction of rake. On a steeply dipping surface, it is easier to measure the rake of a lineation than it is to measure its plunge and bearing. Remember that plunge and bearing can be calculated from rake only if the strike and dip of the plane on which the line occurs is known.

1-6 LOCATING POINTS WITH A COMPASS

After you have learned how to make measurements with a compass, the chief asks you to produce a detailed map showing the positions of limestone and sandstone outcrops in the region near the gorge described in Section 1-1. Such a map will help you to trace the fault across the countryside. Unfortunately, a detailed topographic map of the area does not exist, so you have no *base map* on which to plot the outcrop locations. A base map is any map at an appropriate scale on which geologic measurements can be plotted. The chief suggests that you use your compass and do a simple survey. So, armed with this book, you set out through the brush once again.

Below, we introduce a few simple surveying methods that can be done with a Brunton compass. Simple surveying with a compass helps students to practice compass skills.

Tape and Compass Mapping

A map showing the approximate positions of points on the ground surface can be constructed using only a tape and a compass. Using a tape and compass, you can determine the distance and direction between a starting point and a second point.

Problem 1-1 (Tape-and-compass mapping)
Construct a map showing the relative positions of four outcrops (A, B, C, and D). The ground surface in the map area is horizontal.

Method 1-1
Step 1: Plot the position of outcrop A on a sheet of paper. Position point A so that all other points can be represented on the paper. In this example, we place point

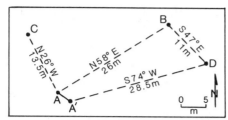

Figure 1-25. Construction of a tape and compass map. Point A is the starting point, and the positions of points B, C, and D must be located. The distance A - A' is the closure error.

A in the corner of the proposed map area (Fig. 1-25). Estimate the size of the area that you are to map, and choose an appropriate scale so that you can fit the map on the sheet of paper. Draw a north arrow and the scale.

Step 2: Have your partner stand at point A (or place a visible marker on point A). Then walk to outcrop B. Stretch a tape between A and B to determine the length of line AB. The length of line AB is 26 m. If a tape is not available, you can estimate the length by counting the number of paces that it took for you to walk from A to B. If you know the length of your pace, you can determine the length of the traverse line by simple multiplication. This modification is called the *pace-and-compass method.*

Step 3: Next, shoot the bearing of line between outcrops A and B. Be sure to read the correct end of the needle! The bearing of a line running from A to B is N58°E. The bearing of a line running from B to A is S58°W.

Step 4: Once you know the bearing and length of line AB, plot the position of outcrop B on your map sheet. Use a protractor to determine the orientation of line AB with respect to the north arrow (the angle should be 58°), and use the map scale to determine the length of line AB. If you make an attitude measurement at outcrop B, plot the structural symbol at the position of point B on your map sheet.

Step 5: You may locate other outcrops from point A (e.g., the position of outcrop C), or you may stay at outcrop B and locate additional points from outcrop B (e.g., outcrop D).

Step 6: It is best that your traverse ultimately loops back, so that the final outcrop that you locate is your starting outcrop (A). Such a loop allows you to assess the accuracy of your map. The author of the map in Figure 1-25 shot a bearing from outcrop D to outcrop A and obtained a measurement of S74°W, then he paced from D to A and found the distance to be 28.5 m. However, when he drew a 28.5- m-long S74°W bearing line from point D on the map, he did not return to point A, but instead located point A'. The discrepancy between the location of the original point A on the map and the position of point A measured from your final surveyed location (i.e., point

A') is called the *closure error* of the map. On the map, the closure error is line AA', which is about 3 m.

The accuracy of a tape-and-compass map depends on the care with which the measurements are made. The best tape-and-compass maps are possible in areas where ground surface is level, and there are no obstacles between points. If the ground surface is sloped, the line between two points on the ground surface does not represent the horizontal distance between two points, and calculation of the map distance between two points is more complicated (see Chapter 2).

Two-Point or Three-Point Sighting

Sometimes it is not feasible to directly measure the length of an oriented traverse line. It is still possible, fortunately, to determine the position of a point on the ground, if you have a few landmarks on your map. This is done by sighting from the unknown point to two or, better, three landmarks (Fig. 1-26). The procedure is as follows:

Problem 1-2 (Three-point sighting)
You have a map on which the localities of three landmarks (a house, a telephone pole, and a sign) are located (Fig. 1-26). You are standing on an outcrop in the map area but do not know exactly where the outcrop is. Call the position of this outcrop "point X." Determine the location of the outcrop with respect to the three landmarks, so that you can plot point X on the map

Method 1-2
Step 1: Stand on the outcrop. Point the compass toward you, level the compass and sight through the cover window at the house (landmark A). Read the white end of the needle. The bearing that you read (S43°E) is the bearing of a line that points from the house to your position. Draw this line on your map, starting at landmark A (Fig. 1-26). You must estimate the length of the line. The position of the outcrop on the map must lie on this

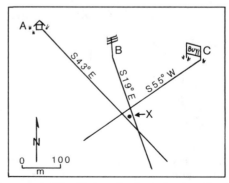

Figure 1-26. Determination of a location by three-point sighting. Point X is the unknown point, and points A, B, and C are landmarks.

line, but you do not know the exact distance between the house and the outcrop.

Step 2: Shoot a bearing to the telephone pole (landmark B). Draw a line from landmark B parallel to this bearing (S19°E) back toward the outcrop (Fig. 1-26). The position of point X on the map should be at the location where the line from B crosses the line from A. The position of point X is the position of the outcrop.

Step 3: Usually, your measurements are not exact, so a better constrained position is obtained if you shoot a bearing to a third landmark. A line from the sign (landmark C) to the outcrop has a bearing of S55°W. If all your measurements are perfect, the three bearing lines intersect at point X. Usually, however, the intersections of the three lines define a small triangle (Fig. 1-26). Assume that the position of the outcrop is at the center of this triangle.

Triangulation

Most surveyed maps are constructed using the technique of *triangulation* (see Appendix 1). Accurate triangulation requires accurate surveying instruments, but rough triangulation can be accomplished with a Brunton compass.

Problem 1-3 (Triangulation)
Locate the map position of an outcrop at Z, given the positions of two landmarks. One landmark is at A and the other is at B.

Method 1-3
Step 1: Define the line between landmarks A and B as a *base line* (Fig. 1-27). A base line is a line of known position, orientation, and length. Draw line AB on your map; define a scale and north arrow. The outcrop at point Z is too far away to be located by one step of triangulation, so you must first find two intermediate points, called X and Y.

Step 2: Place a flag at point X so that it is easily visible. Shoot a bearing from A to X, and shoot a bearing from B to X.

Step 3: As shown in Figure 1-27, your bearing

measurements define the angles ø and ß. From these measurements you can calculate the angle μ:

$$\mu = 180° - (\phi + \beta) \qquad \text{(Eq. 1-1)}$$

and from μ you can calculate the length of AX (or BX), using a simple trigonometric identity (see Appendix 2):

$$AB \sin \beta = AX \sin \mu$$

$$AX = (AB \sin \beta)/\sin \mu \qquad \text{(Eq. 1-2)}.$$

Step 4: Once you have determined the position of X, you can calculate the position of a new point (Y) by establishing XB as a new base line. You can then establish line XY as a base line from which you can establish the location of the outcrop at point Z (Fig. 1-27). Note that when you have finished the procedure, there are several points on the map. Such a network of points is called a *triangulation net*.

We return one last time to your experience as a field assistant in the brush of the Brazilian highlands. You have decided that the most appropriate technique for locating outcrops in this region is the pace-and-compass method. You return to the gorge in which the fault was exposed and trace the fault up the east side of the gorge to the plateau. You place this point at the west edge of the map, define a scale, and draw a north arrow. Then you shoot a bearing at an outcrop in the distance, mark the bearing in your notes, and pace toward the outcrop. You measure the strike and dip of the outcrop and mark the appropriate symbol on your map. By the end of the day you have located 20 outcrops and have the *beginnings* of a map that shows the trace of the fault across the countryside as well as a fold whose presence had been unknown before drawing the map (Fig. 1-28).

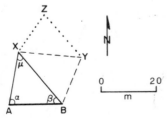

Figure 1-27. Determination of positions by triangulation. AB is the original base line. Point X is the first unknown location. XY is the second base line. Z is the outcrop position.

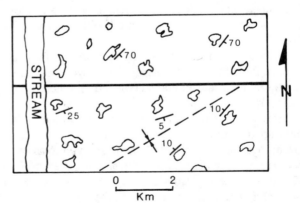

Figure 1-28. Simple outcrop map of the area in the vicinity of the outcrop shown in Figure 1-1. Heavy line is the trace of a fault, dashed line is the trace of a fold hinge, and the irregular shapes represent outcrops.

EXERCISES

Questions 1-8 are intended to give you a little practice in thinking about and visualizing attitude measurements. In these questions it may help to use your hand to simulate the plane or line in question. Lay a pencil on the table in front of you to simulate north.

1. Translate from the azimuthal convention to the quadrant convention, or vice-versa, as is necessary.

    (a) N43°E   (b) N43°W   (c) N90°W
    (d) 087°   (e) S20°E   (f) 355°
    (g) S62°W   (h) N62°E   (i) 127°
    (j) 241°   (k) 270°   (l) due S
    (m) 617°   (n) 264°   (o) 092°
    (p) 180°   (q) S58°E   (r) 000°
    (s) 112°   (t) S47°W

2. Circle those attitudes in the list below that are impossible (i.e., a bed with the indicated strike cannot possibly dip in the direction that is indicated).

    (a) N23°W,57°SE   (b) N46°W,56°NE   (c) N45°W,78°NW
    (d) 089°,43°W   (e) N34°W,14°N   (f) 089°,43°E
    (g) 089°,43°N   (h) 341°,84°NE   (i) 324°,67°NW

3. Translate the following attitudes into number pairs according to the right-hand rule.

    (a) N30°W,34°NE   (b) N48°E,56°SE   (c) 078°,76°SE
    (d) 067°,74°NW   (e) 234°,43°NW   (f) 117°,21°NE

4. Draw an isometric block diagram of a cube (see Appendix 1). Within the volume of this cube, draw a plane whose attitude is 045°,30°NW. Next to the drawing indicate the orientation of the three coordinate axes that define your reference frame. Use a ruler to keep your lines straight.

5. Translate the following strike and dip measurements into equivalent dip and dip direction measurements.

    (a) N34°W,38°NE   (b) 087°,21°N   (c) N48°E,57°SE
    (d) 245°,41°NW   (e) 117°,33°NE   (f) S64°E,21°SW

6. Draw an isometric block diagram of a line whose attitude is 60°,045°. Your drawing should include three Cartesian axes to define three-dimensional space (vertical, north-south, and east-west). Use a ruler to keep your lines straight. Indicate the bearing angle (ß) in the horizontal plane and the plunge angle (∂) in the vertical plane.

7. Imagine a fault surface on which there are four different overprinted sets of slip lineations. The surface is oriented N39°W,47°NE. A geologist recorded the following measurements to describe the four sets of lineations.

    47°,N51°E   (lineation 1)
    68°,due N   (lineation 2)
    47°,N51°W   (lineation 3)
    34°,due N   (lineation 4)

(a) Assuming that the planar attitude was measured correctly, determine which lineation measurements are impossible. In other words, which lineation(s) cannot possibly lie in the specified plane. (Indicate why for each).

(b) Assuming that the measurement of lineation 1 is a correct, what is its rake?

(c) Assuming that the measurement of lineation 1 is correct, does it indicate movement parallel to the strike of the fault (strike-slip movement) or movement parallel to the dip of the fault (dip-slip movement)?

8 . Construction of a simple field geologic map. Your instructor will arrange a set of five or six rock slabs in an open space of about 400 m^2. Each rock slab is a measurement station. (If real rock outcrops are accessible, they will be used instead). Construct a map showing the distribution of the rock slabs (outcrops). Use a scale that is appropriate to draw your map on a single sheet of paper. Be sure to draw a scale and a north arrow on your map to provide the reference frame. Determine the orientation of each slab, and using appropriate symbols, plot the strike and dip of each slab on your map in the correct position. If a simple structure is indicated by the map pattern, your instructor will help you to interpret it. You may use whichever method of locating points suits you.

CHAPTER

2

INTERPRETATION AND CONSTRUCTION OF CONTOUR MAPS

2-1 INTRODUCTION

In many applications it is important to be able to describe the spatial variation of a physical or a statistical parameter. Such variations can be illustrated on maps through the use of *contour lines*. Recall from earlier courses in geology that a contour line is a line representing the locus of points in the map area of equal value for a specific parameter. For example, on the familiar topographic map each contour line represents the locus of points with a given elevation. A contour line on a topographic map can be envisioned as the intersection of a horizontal plane with the ground surface (Fig. 2-1). Any map that employs contour lines to represent spatial variations in the value of a parameter is called a *contour map*. In some books, the term *isoline map* is used as a general term for a map that employs contours, and the term *contour map* is restricted to maps that show variation in elevation. We prefer to use contour map as the general term.

In this chapter we review the principles of contour mapping, explore a range of applications of contour maps to structural geology, and learn how to create contour maps from point data. Work with contour maps provides an excellent opportunity to develop three-dimensional perception. With practice you may be able to visualize the shape portrayed by the spacing and form of the contour lines on a map. These days, computers can assist in creating a visual image of a contour map by producing block diagrams from topographic data; on such diagrams the form of the contoured surface is simulated by a grid of lines (Fig. 2-2). (*Note:* You may wish to defer studying portions of this chapter until you have learned the terminology used to describe folds and faults. It is beyond the scope of this book to provide a detailed discussion of these terms).

2-2 ELEMENTS OF CONTOUR MAPS

Gradients and Contour Intervals

The difference in the value of a parameter represented by adjacent contour lines is called the *contour interval*. For example, on a topographic map the contour interval represents the difference in elevation between two contour lines. The usefulness of a contour map is greatly increased if the contour interval is constant because then the *gradient* (rate of change) of a parameter in a given direction is directly proportional to the spacing of the contour lines. Closely spaced contours represent steep gradients, and widely spaced contours represent gentle gradients (Fig. 2-3).

The contour interval must be selected so that variations in the morphology of the parameter can be represented in adequate detail. The choice of a contour interval for a specific map depends on three factors: the detail that you wish to portray, the quality of the data that you have to work with, and the scale of the map.

The required detail on a contour map is a value judgement. If the contour interval is too large and there are

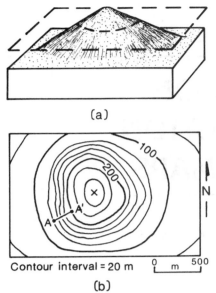

(a)

(b)

Contour interval = 20 m

Figure 2-1. The shape of a hill represented by a topographic contour map. (a) Block diagram of a hill. The horizontal plane indicated by dashed lines intersects the hill at an elevation of about 200 m. (b) Contour map of a hill. Note that the index contours are darker.

too few contours in the map area, small variations in the parameter will not be resolved. If the contour interval is too small, there will be so many contours on the map that they will merge with one another, and it will become impossible to distinguish adjacent contours from one another. The quality of the data available for constructing the map controls the contour interval in that if *control points* - the points at which a direct measurement of the parameter has been made - are widely spaced and are at greatly different elevations, you will not be able to resolve local topographic details even if you use a small contour interval. Finally, scale affects your choice of a contour interval. For example, on a standard U.S.G.S. 1:24,000-scale topographic quadrangle map, the width of the ink line of a contour on the map represents about a 4-m-wide belt on the ground; clearly, there is no point in trying to resolve features that are less than about 10 m in diameter.

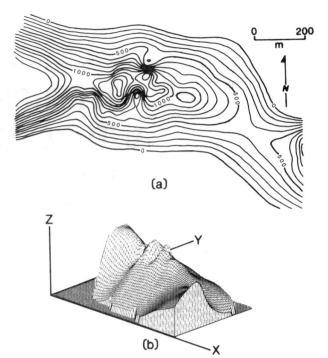

(a)

(b)

Figure 2-2. Computer simulation of topography. (a) Computer-generated contour map; (b) perspective diagram of the same area employing a grid of lines to create the illusion of relief. Y-direction is north, X-direction is east, Z is the vertical axis. (Courtesy of Radian, Inc.)

Commonly, to make a contour map easier to read, every fifth contour line is defined as an *index contour*. An index contour is indicated by a thicker ink line and is labeled with the value of the parameter. The contour interval is usually indicated at the bottom of a map, but if not, the contour interval may be determined by dividing the difference between two adjacent labeled index contours by 5.

To calculate the gradient along a specific *traverse line* (*Note:* the term *traverse line* is used in this book to refer to any line drawn between two points on a map) of a known length, draw the traverse line, measure its length, and count the number of contour lines that cross it. Multiply the number of contour lines by the contour interval to obtain

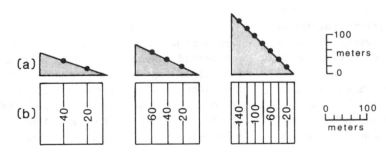

Figure 2-3. Illustration of the relationship between gradient and the spacing of contour lines. (a) Cross section of slope; (b) contour map of slope.

the change in value of the parameter. If the ends of the traverse line do not fall precisely on a contour line, the change in value of the parameter will be a little greater than the simple multiplication yields, so you must increase the value accordingly. The gradient can be represented either by the *angle of slope*, by a *slope fraction*, or by the *grade* (Fig. 2-4). An angle of slope is the angle between the horizontal and the sloping surface; it can be determined either by making a scaled drawing or by a simple trigonometric formula

$$\arctan \frac{(\text{parameter change})}{(\text{horizontal traverse length})} = \text{slope angle} \qquad (\text{Eq. 2-1}).$$

A slope fraction is merely the ratio of rise (vertical change) over run (horizontal change). A grade is a percentage that specifies the number of units of rise for every 100 units of horizontal distance.

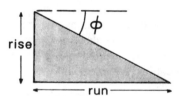

Figure 2-4. Definition of slope angle and grade. ø = slope angle; rise/run = slope fraction; (rise X 100)/run = grade.

Problem 2-1

A traverse line (AA') is indicated on the topographic map of Figure 2-1. Determine the gradient along this line and express this gradient as angle of slope, grade, and slope fraction.

Method 2-1

The length of the traverse line, by the map scale, is 210 m. The traverse line starts on the 120-m contour line and terminates on the 200-m contour line, so it crosses a vertical elevation change of 80 m.

slope angle = arctan(80 m/210 m) = 20.9°

slope fraction = 80 m/210 m = 1/2.63

grade = (80 m X 100)/210 = 38.1%.

Problem 2-2

The scale of a map is 1:10,000 (i.e., 10 cm = 1 km) and the contour interval is 200 m. A N45°E-trending traverse crosses five contour lines over a distance of 20 cm on the map. The lowest elevation is at the northeast end of the traverse. What is the slope of the traverse line?

Method 2-2

The elevation change (rise) is 1 km over a horizontal distance (run) of 2 km, so the average slope in this interval is arctan(0.50) = 26.6°.

General Constraints on Contour Maps

There are several general rules that constrain the construction of contour maps. We list the major rules and describe acceptable exceptions. For additional constraints see Bishop (1960) and Badgley (1959).

(a) The contour interval on a map is constant: The difference in the value of a parameter represented by any two adjacent contour lines should be the same everywhere on a map. If the contour interval is not constant, it is difficult to calculate gradients. The only exception to this rule occurs for maps that encompass domains of both very steep and very shallow gradients. On such maps two contour intervals can be used. A large contour interval is selected to accommodate domains of the map in which there are steep gradients. In the domains where gradients are shallow, however, intermediate contours can be added to provide greater resolution of features. For example, imagine that a topographic map covers a region in which steep hills border a flat flood plain. A 40-ft contour interval might be used in the hilly domain of the map, but such a large contour interval could not be used to define features in the flood plain. Intermediate contours could be added in the flood plain domain to make the contour interval in the flood plain only 20 ft. Intermediate contours should be dashed.

(b) Contour lines generally should not merge or cross: If contour lines cross one another or join and become one, the map may be wrong. There are only two acceptable situations in which contour lines can merge or cross: (1) Contour lines appear to cross on topographic maps or structure-contour maps where there is an overhang (Fig. 2-5); the contours on the underside of the

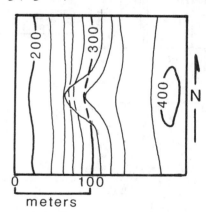

Figure 2-5. Pattern of contours for an overhang. Note that the contours below the overhang are dashed.

overhang should be dashed. (2) Contour lines merge into a single line if a gradient is infinite (i.e., the contoured surface on the map is vertical). Contour lines *appear* to merge on many maps where there are steep gradients, because of the finite width of inked contour lines on the map sheet.

(c) Contour lines either close within the area of the map or are truncated by the edge of the map or by a structure: It is impossible for a contour line to simply stop if the surface being represented is a continuous feature. For example, on a topographic map of a hill, you will always be able to trace a line of constant elevation around the hill so that it connects with itself or *closes* on the other side. If this line extends off the end of the map, it will be truncated by the edge of the map. Exceptions to this rule occur on certain types of contour maps. For example, on a structure-contour map we will see that the contour lines can be truncated within the map area by a fault, and contour lines on a fault surface need not close.

(d) Contour lines are repeated to indicate reversals in gradient direction: If, along a traverse line, the direction of a gradient reverses, the lowest (or highest) contour crossed before the reversal must be repeated after the reversal. This statement is best illustrated by an example on a topographic map. If you walk down a slope, cross the 200-m contour, cross a saddle, and then walk up another slope, you must again cross the 200-m contour line.

(e) A reference frame for a contour map is defined by specifying a datum plane: A *datum plane*, also called a reference plane, for a contour map is an imaginary surface on which the parameter described by the

map has a value of 0. For example, on a topographic map, the datum plane is mean sea level. Elevations are specified by a distance above or below mean sea level.

2-3 INTERPRETATION OF TOPOGRAPHIC MAPS

We have described *topographic maps* already at several points in this chapter. In this section we discuss how geologic structures can be studied with the aid of topographic maps.

Structures on Topographic Maps

The shape of the ground surface commonly indicates the distribution of lithologies, which, in turn, are controlled by the geometry of geologic structures. Different units may have different topographic expressions, and thus contacts between units can be mapped by identifying the boundary between two topographic domains. A particularly resistant unit (e.g., a quartzite) may stand out in relief and trace out a structure. Geologic mapping using topographic maps is often done in conjunction with study of stereo pairs of air photographs.

Characteristic topographic patterns are associated with certain structural geometries (Fig. 2-6). Horizontal strata may be indicated by flat-topped plateaus or mesas bounded by *steplike escarpments*. On such escarpments, steep cliffs are backed by resistant strata, and gentle slopes are underlain by nonresistant strata. Dipping beds lead to the formation of asymmetric ridges. If the strata are steeply dipping, the asymmetry is not pronounced and the ridge is

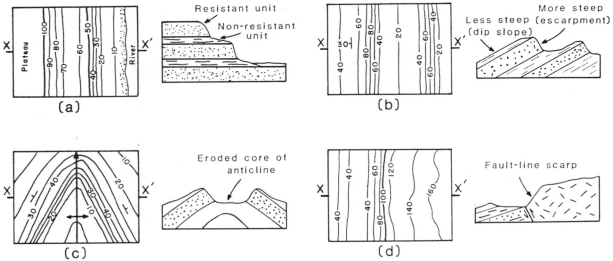

Figure 2-6. Topographic patterns of simple structures. (a) Horizontal strata; (b) dipping strata; (c) broken-crested anticline; (d) fault-line scarp between granite and tilted strata.

called a *hogback*. If the strata are shallowly dipping, the ridge is highly asymmetric, with one steep slope and one shallow slope, and is called a *cuesta*.

On many asymmetric ridges the topographic surface of the more gently dipping face of the ridge corresponds to the plane of bedding or foliation. Such a surface is called a *dip slope*. If the surfaces of synclines and anticlines are dip slopes, synclines will cup valleys, and anticlines will arch over ridges. Commonly, however, the crests of anticlines erode away, so an anticline will appear as two oppositely facing ridges that are separated from one another by a valley. If the fold is plunging, the ridges will join and define a single U-shaped or V-shaped ridge, depending on the shape of the fold hinge zone. The Valley and Ridge Province of Pennsylvania contains spectacular examples of topographically defined plunging folds (Fig. 2-7).

Igneous structures may also be reflected by topographic patterns. For example, the composition of a dike rock is usually very different from that of the country rock that it intruded. If the dike rock is less resistant to weathering, it will preferentially erode and underlie a trough. If the dike rock is more resistant, it will stand up as a ridge. Granitic intrusions have a very distinctive knobby topographic expression because of their tendency to weather by exfoliation.

Topographic expression of faults occurs for several reasons. Faulting breaks up the rock, thereby creating a zone of weak rock, and this zone of weakness preferentially erodes. Alternatively, if the fault was a zone of fluid circulation and mineralization, the fault breccia may become better indurated than surrounding country rock and therefore will stand out in relief. If a fault displaces the ground surface, it results in a *fault scarp* that will have a topographic expression. Even if the fault scarp itself is no longer visible, topographic features, such as faceted mountain fronts, uplifted terraces, or a rejuvenated stream, may attest to fault movements. The traces of strike-slip faults can be recognized by the offset of other topographic features, such as stream beds, and may be marked by local ridges and depressions. Significant cumulative displacements on faults can result in creation of mountain ranges that stand high relative to adjacent areas and

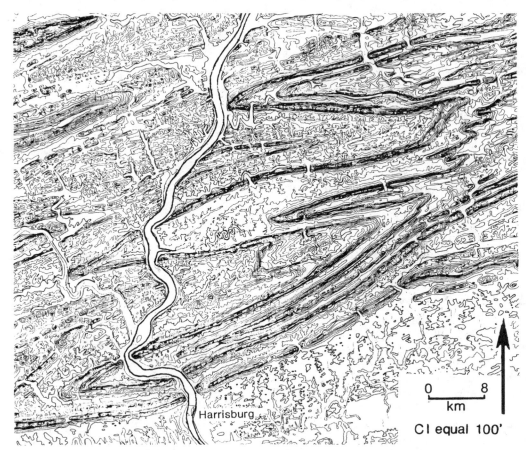

Figure 2-7. Topography of portion of the Valley and Ridge province of Pennsylvania, at the Susquehana River near Harrisburg. The resistant ridges define the form of plunging upright folds. (Adapted from Hamblin and Howard, 1986.)

represent horsts, tilted fault blocks, or thrust sheets. Grabens may be represented by topographic troughs. Likewise, windows and klippen in thrust-faulted regions are topographically delineated if the hanging-wall sheet is composed of a different rock type than the footwall sheet. If a fault juxtaposes lithologies of different erodability, the fault trace can become an erosional scarp, called a *fault-line scarp,* in which the land underlain by the more resistant rock steps down to the land underlain by the less resistant rock.

Joints represent planes of weakness along which blocks of rock preferentially break off. As a consequence, the faces of ridges are often parallel to joints, and streams often take sudden angular bends as they follow joints. Joints also zones of enhanced weathering, and thus may evolve into narrow linear troughs.

The *structural grain* of a region refers to the orientation of the dominant deformation elements in an area. This grain may represent the trends of folds, fractures, or metamorphic foliation. Because topography reflects structure, the structural grain of a region may stand out on a topographic map. In Figure 2-7 the structural grain of the Appalachian foreland is defined by the topographic pattern.

Finally, note that the geometry of sea-floor structures (e.g., spreading ridges, transform faults, and trenches) is indicated by topographic patterns. Maps that show the topography of sea floor or of a lake floor are called *bathymetric maps.* Contour lines on bathymetric maps are called *isobaths* and are specified not as elevations above mean sea level, but rather as depths beneath the surface of the overlying body of water.

Intersection of Planes with Topography (Rule of V's)

The trace of the intersection of one plane with another is a straight line. For example, where a planar contact intersects a perfectly horizontal ground surface, the trace of the contact is a straight line. If the contact is not planar (e.g., it is folded) the trace is a curved line. Likewise, if the ground surface is not planar but wraps around hills and valleys, the trace of a non-vertical contact on the ground surface is a curved line even if the contact itself is planar. In other words, the trace of a contact shown on a map is determined both by the shape of the contact and the topography of the map area.

For some students visualization of the pattern resulting from the intersection of a contact with the ground surface comes naturally, but for others it does not. The easiest way to develop the ability to visualize such intersections is to use your hands. Start by trying to visualize the outcrop pattern that results where an imaginary bed of sandstone crosses a valley. Let your right hand represent the bed of sandstone (call this your "bed hand") and let your left hand represent a V-shaped valley (call this your "valley hand"). Imagine that the floor of the valley is a gently plunging line and that a stream runs down it (Fig. 2-8). As you read the following paragraphs, physically use your hands to duplicate the described situation.

Figures 2-8 and 2-9 present several intersection patterns between the bed and the the valley. The strike of the bed is perpendicular to the bearing of the valley axis. Note that in several examples the intersection is V-shaped. As a consequence, the relationship between bed dip and valley-floor plunge is commonly referred to as the *Rule of V's.* Please do not memorize the patterns of these figures

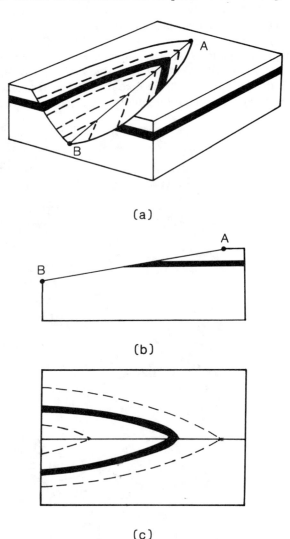

(a)

(b)

(c)

Figure 2-8. Intersection of a horizontal plane with topography. (a) Block diagram; (b) cross section along the axis of the valley; (c) map view showing that the outcrop trace is parallel to contour lines.

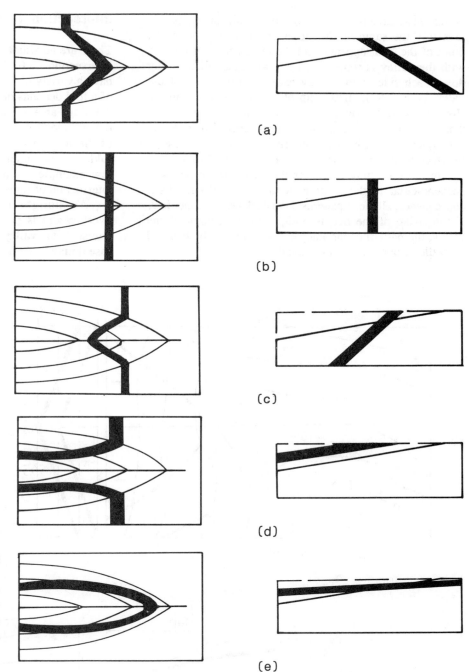

Figure 2-9. Intersection of non-horizontal planes with topography. Left column is map view, right column is cross-sectional view. (a) Plane dipping upstream; (b) vertical plane; (c) plane dipping downstream at an angle greater than the stream gradient; (d) plane dipping downstream at an angle equal to the stream gradient; (e) plane dipping downstream at an angle less than the stream gradient.

as if they are a rule; rather, practice visualizing the geometry of bed-valley intersections.

Start by holding your hand horizontally and allow it to intersect the valley hand. Your hand traces a V that is identical to the trace of a topographic contour line (Fig. 2-8). Remember that a topographic contour line, by definition, represents the intersection of a horizontal plane with the ground surface. Rotate your bed hand around the strike so that it dips into your valley hand (i.e., dips in the direction opposite to the flow of the stream). Notice that

the intersection of your hands now forms an upstream-pointing V (Fig. 2-9a). Keep rotating your bed hand until it is vertical, then look straight down on the intersection. The trace of the intersection, were it to be projected on a map plane, would be a straight line (Fig. 2-9b). Vertical planes "ignore" topography and will always appear as a straight line on the map. Continue rotating your bed hand until it is dips downstream and dips more steeply than the plunge of the stream. The intersection of your bed hand with your valley hand is a V, but the V points downstream (Fig. 2-9c). If the dip of the bed is precisely the same as the slope of the valley floor, the point of the V disappears, and the intersection of the bed with the valley is represented by two lines, one running down each side of the valley parallel to the floor of the valley (Fig. 2-9d). If the dip of the bed is less than the slope of the valley floor, the intersection of the two is a V that points upstream (Fig. 2-9e). V patterns also arise as a consequence of the intersection between a layer and a ridge. In effect, a ridge can be visualized as an inverted valley.

In the preceding examples the strike of the bed was perpendicular to the bearing of the valley floor. As a consequence, all the V patterns described were symmetrical. If the strike of the bed is oblique to the trough of the valley, the pattern of the intersection between the bed and the valley floor is not symmetrical.

Intersection of Folds with Topography

There are no simple "rules" to follow when describing the intersection of folds with topography because folds can have a variety of forms and orientations. The map pattern of a fold depends on the attitude of the hinge, the shape of the hinge area, the amplitude and wavelength of the fold, the attitude of the axial plane, the angle between the limbs of the folds, variation in thickness of a unit around the fold, and the pattern of topography. We can give only a few examples to help you see how to think about the intersection of folds with topography (Fig. 2-10).

An upright nonplunging anticline whose hinge overlies and is parallel to the trace of the valley floor intersects the valley as an upstream-pointing V. In such a case it may be difficult to recognize the map pattern as that

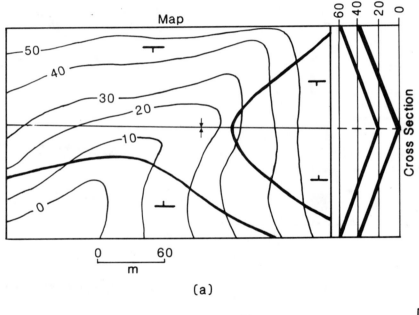

(a)

Figure 2-10. Examples of maps showing the intersection of a fold with topography. Cross-sectional view of each map is also shown. (a) Syncline intersecting a valley. Hinge of the fold is oblique to the valley floor. Both top and bottom contacts of the unit are shown. The trace of the vertical axial plane is also shown. (b) Nonplunging anticline intersecting a valley. The trace of the axial plane is coaxial with the trace of the valley floor. (c) Nonplunging anticline intersecting a valley. Trace of the axial plane is perpendicular to the trace of the valley floor.

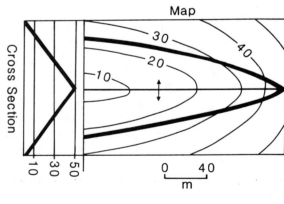

(b)

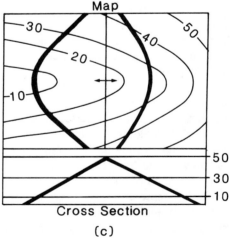

(c)

of a fold instead of just a dipping planar bed. A plunging fold whose hinge coincides with the trace of the valley floor also defines a V, but the direction in which the V points depends on the magnitude and direction of the plunge. For example, a plunging anticline whose axis plunges in the direction opposite to the plunge of the valley floor will form a V upstream. If the axis of the anticline plunges in the same direction as the valley floor, then the V points downstream. The pattern for a syncline is the reverse of that for an anticline. An asymmetric fold defines an asymmetric V in a valley. A unit defining a fold whose axis is perpendicular to the axis of a valley appears as two outcrop belts. Depending on the limb dips, one limb may form a V upstream and the other may form a V downstream. If the axis of a fold is oblique to a stream, the outcrop pattern may be quite irregular.

In regions of complex topography, folds may be difficult to recognize, especially if a single unit cannot be traced around the hinge. In such cases, it is perhaps easiest to study attitude data plotted on the map to determine if any folds are present. If attitude data are not available, the occurrence of a repeated unit (i.e., the same unit crosses a valley or a ridge twice) is a clue that a unit has been folded. Of course, reverse faulting can also repeat a unit, so it is important that you check for evidence of faulting in the map area before concluding that a fold is present. Remember that another clue to the presence of a fold can be obtained by application of the rule of V's. Limbs with different dips and/or dip directions will display different V patterns.

The interplay of topography and structure can sometimes lead to very unusual outcrop patterns that do not resemble the pattern that the structure would have on a featureless plain. This is particularly true in regions where there are dip slopes. In such regions the erosion pattern of the bed becomes the dominant factor in controlling the outcrop pattern.

2-4 STRUCTURE-CONTOUR AND FORM-LINE CONTOUR MAPS

Representation of Key Horizons on Structure-Contour Maps

A *structure-contour map* is a map on which contour lines represent lines of equal elevation on a structurally significant surface. The surface that is being contoured is called the *key horizon* and may be a marker bed, contact, unconformity, or fault. A *marker bed* is simply a distinctive bed that can be easily recognized. The contour lines of a structure-contour map are similar in meaning to topographic contours, in that they represent distances above or below a datum plane, but the geometry that is portrayed on a structure-contour map is that of a geologic feature, not that of the ground surface. Generally, structure-contour maps are used to describe subsurface features. Thus, the data used to construct a structure-contour map can be presented either in terms of elevations above (or below) sea level or in terms of depths below the ground surface. Because of the variety of data sources used to define structure contours (e.g., measurements of depth below ground surface, depth below sea level, or elevation above sea level), it is critical that you define the datum plane on every structure-contour map. If the ground surface is not horizontal, and data are provided as depth below ground surface, a correction must be applied so that all points represent depth below the same datum plane. Mean sea level is the most common datum plane for structure-contour maps.

On a structure-contour map a horizontal key horizon does not have any contour lines crossing it. A dipping, but planar, key horizon looks like a slope and is represented by parallel contour lines spaced in proportion to dip. Structural domes look like hills (Fig. 2-11), synclines look like valleys, and anticlines look like ridges. Basins

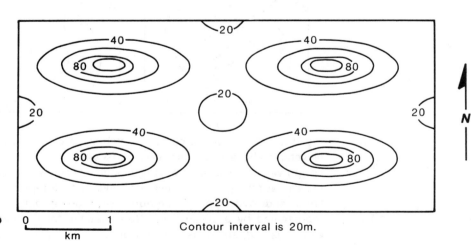

Figure 2-11. Structure-contour map of ideal dome-and-basin structure.

0   1
km

Contour interval is 20m.

appear as bowl-shaped depressions. Often the key horizon used to define a major sedimentary basin is the nonconformity between crystalline basement and sedimentary cover.

The form and extent of *closure* of a structure can be calculated from the contour map of the structure. The term "closure" in this context refers to the distance between the highest point on the structure and the lowest contour line on the structure that is closed.

Problem 2-3

Calculate the closure of a dome shown in Figure 2-11.

Method 2-3

The highest point on a dome is at an elevation of about 110 m, and the lowest closed contour line around a dome is at an elevation of 40 m. Thus, the closure of a dome is about 70 m.

Representation of Faults on a Structure-Contour Map

Faults are the most challenging structure to portray on a structure-contour map. Vertical faults appear as single lines at which contours are missing and/or are truncated (Fig. 2-12a, b). If the fault trace is parallel to the

contours, offset on the fault results in a step in the elevation of the key horizon; if the vertical component of movement, called the *throw* of the fault, is greater than the contour interval, one or more contour lines will be absent. The number of missing contour lines reflects the throw. If the fault trace is oblique to the contours, contours are truncated (Fig 2-12a, b).

If the fault is not vertical, the trace of its intersection with a nonplanar key bed is not a straight line (Figs. 2-12c, d, e). The pattern of a dipping fault intersecting a nonplanar key bed is comparable to the map pattern of a dipping plane intersecting hilly ground. Nonvertical normal faults are indicated on structure-contour maps by two parallel lines, one defining the intersection of the fault plane with the downthrown block and one indicating the intersection of the fault plane with upthrown block (Fig. 2-12c). No contour lines on the key horizon can be present in the area between the two traces of the fault. This gap in contours is called the *fault gap*. Nonvertical reverse faults can be indicated by either two parallel lines or by one line. If two lines are used to represent the fault, the contours of the hanging wall overlap the contours of the footwall in the interval between the two fault traces (Figs. 2-12d). If a single line is used to represent a reverse fault, it is drawn where the fault intersects the key horizon of the hanging wall (Fig. 2-12e).

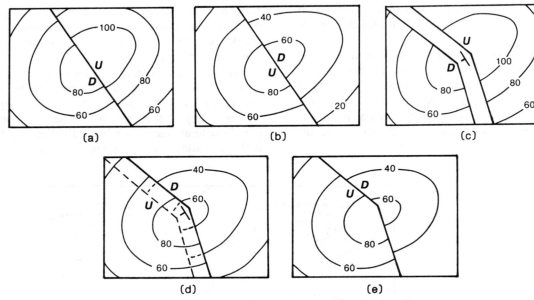

Figure 2-12. Structure-contour sketch maps of faults. Contours are elevations above mean sea level. (a) Vertical fault, northeast side is up; (b) vertical fault, southwest side is up; (c) nonvertical normal fault, showing a fault gap; (d) nonvertical reverse fault represented by two lines and overlapping contours; (e) nonvertical reverse fault represented by a single line. Note that the fault traces shown in examples c, d, and e are bent, because the fault plane is dipping and the key horizon is not planar.

The placement of a fault on a structure-contour map may be required by drill-hole and outcrop observations or by seismic profiling. On many structure-contour maps, however, the drawing of a fault trace merely represents an interpretation of data on depth to the key horizon. A map of the same data could be constructed by substituting a zone of closely spaced contours for a fault. Commonly, a structure-contour map is first constructed with no faults and then is reexamined to determine if any faults may be present. Geologic intuition or knowledge of regional structural style may suggest that sudden changes in contour spacing or in contour strike on the map indicate the presence of a fault. If a linear zone of very closely spaced contours must be drawn in order to accommodate data, it is possible that the linear zone is a fault parallel to the contours along which the key horizon was displaced (Fig. 2-13a). If contours suddenly change strike along a linear zone, it is possible the zone represents a fault oblique to the contours on the key horizon (Fig. 2-13b).

The point of intersection between a drill hole and a fault is called a *cut point*. Recognition of a cut point is definitive in describing a fault. The fault plane itself can be contoured from a number of cut points. Contour lines on the fault plane are dashed to distinguish them from the contours on the key horizon. Fault-plane contours may be open-ended; where they are, they intersect the trace of the intersection between the fault plane and the key horizon (Fig. 2-14). The trace of the fault plane may close or change trend if the fault plane is curved or if the fault cuts topography (Fig. 2-14). The fault trace may also close if the displacement of the fault decreases to zero along the strike of the fault in the map area.

Certain fault-related structures yield distinctive patterns on structure-contour maps. For example, grabens may be indicated by elongate depressions, and horsts by elongate highs. Complex faulting affecting the strata above salt domes shows up very clearly on structure-contour maps (Fig. 2-15).

Form-Line Contour Maps

For some places, data on the attitude of a unit are available but the actual depth of a key horizon is not known; it may be that the stratigraphic sequence contains no marker beds that can be recognized with confidence in drill holes or outcrops. The form or shape of the structure involving the unit cannot be indicated by a structure-contour map because the data necessary for construction of such a map are not available, but the structure can be indicated by a *form-line contour map*. On a form-line contour map the contour lines represent approximate lines of equal elevation but cannot be assigned specific values. Therefore, visually, a form-line contour map indicates the form of a structure, just as a structure-contour map does, but a form-line

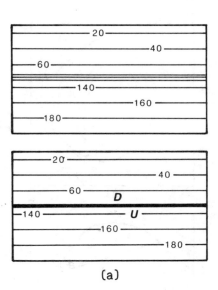

(a)

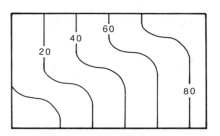

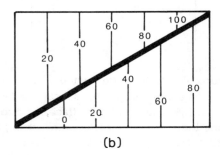

(b)

Figure 2-13. Alternative interpretation of faults. Contours are elevations above mean sea level. (a) Closely spaced contours versus fault; (b) bent contours versus oblique fault. Note that the contour map alone does not provide sufficient data to determine direction of slip on the fault.

contour map is qualitative, whereas a structure-contour map is semiquantitative.

Form-line contour maps are constructed from attitude data. The contour lines at a given locality are parallel to strike, and the spacing of the contour lines is roughly proportional to dip (Fig. 2-16). If contour lines are more closely spaced, the dip of the interval being contoured is steeper.

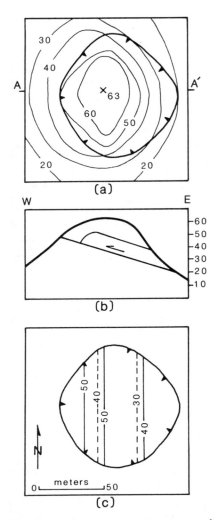

Figure 2-14. Structure contours on a fault. Contours are elevations above mean sea level. (a) Geologic map showing the trace of a dipping planar thrust fault intersecting a hill slope; (b) cross section along line AA'. A folded bed is shown by the thin solid line above the fault; (c) contours truncated at the trace of the fault. Dashed contours are on the fault plane; solid contours are on the bed.

2-5 ISOPACH AND ISOCHORE MAPS

Isopach and isochore maps are used to indicate variations in the thickness of a unit, and thus the contours on these maps are quite different in meaning than those on either topographic or structure-contour maps. An *isopach map* represents variations in true thickness of a specified unit as measured perpendicular to the bedding in the unit (Fig. 2-17a, b). The contour lines on an isopach map represent lines along which the true thickness of a unit is constant, and the contour interval represents a change in true thickness. An *isochore map* represents apparent thickness

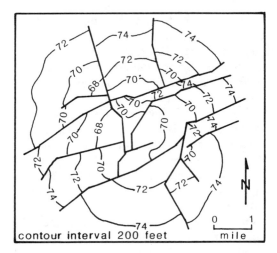

Figure 2-15. Structure-contour map of the top of a sandstone unit overlying a salt dome. Contours are depths below mean sea level (adapted from Bader, 1949). The faults, shown by heavier lines, developed to accommodate extension as the sandstone was arched over a rising salt diapir.

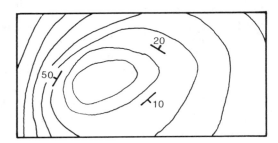

Figure 2-16. A form-line contour map. Note that the spacing of the contour lines is roughly proportional to the dip of the contoured layer, and that no exact elevation values are indicated.

of a unit as measured in the vertical direction (Fig. 2-17a, c). The contour lines on an isochore map represent lines along which the vertical thickness of a unit is constant, and the contour interval represents a change in the vertical thickness. On some isopach or isochore maps the mapped unit disappears within the area of the map, so that there are domains in the map area where the thickness of the unit is zero. The boundary between the area where the unit is present and the area where it is absent is called the *zero isopach* or *zero isochore*, depending on which type of map is being used.

Original variations in thickness of a unit represent variations in the pattern of deposition during creation of a unit. The thickness of a unit may be modified during deformation by faulting (Fig. 2-18) or by ductile flow. The true thickness of a unit indicated on an isopach map

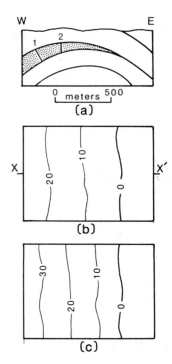

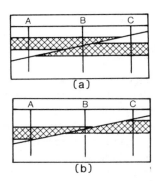

Figure 2-18. Modification of unit thickness. (a) Cross section showing thickening of a unit by thrust faulting; (b) cross section showing thinning by normal faulting. In each case the thickness in well B of the unit will not be the original depositional thickness. Thicknesses measured in wells A anc C are correct.

Figure 2-17. Thickness representation with a contour map. (a) Cross section showing a sand layer that pinches out to the east. Line 1 is perpendicular thickness. Line 2 is vertical thickness; (b) isopach map of of the sand layer; (c) isochore map of the sand layer. Note that at a given locality the isochore shows a greater thickness than does the isopach.

reflects both variations in original depositional thickness and variations in thickness due to deformation.

If the beds of the unit under consideration are horizontal, then the thicknesses indicated by an isochore map will be identical to those on an isopach map at a given locality. However, if the beds of a unit are dipping, the thickness of the unit as indicated on an isochore map will be greater than the thickness of the same unit indicated on an isopach map. In fact, if the unit is folded, an isochore map may indicate variations in thickness that are not a consequence of original variations in unit thickness, but rather reflect variations only in dip (Fig. 2-19).

Isopach maps are usually preferable to isochore maps because variations in original thickness are of more interest in subsurface mapping studies. But isopach maps are more difficult to construct than are isochore maps because they cannot be constructed until variations in thickness resulting from variations in dip have been corrected for. Dip variations are indicated by a structure-contour map of the area or by direct measurement of subsurface attitude using a dipmeter (see Chapter 7). Where dips are low, the difference between an isochore and isopach is not large. For example, a dip of 5° results in only a 4% error in

thickness, and most information on the stratigraphic thickness of a unit is not even known to within 4%.

Isopach maps are of great value in regional geologic studies. They are commonly used to provide data on variations in unit thickness resulting from the original pattern of deposition. Such information defines the shape of sedimentary basins. Detailed isopach maps may display the patterns of ancient river systems and of *paleo-topography*, which is the shape of the ground surface at the time of deposition. Isopach maps may also lead to the discovery of important unconformities and faults and of *stratigraphic traps*. Stratigraphic traps occur where a reservoir bed such as a porous sandstone thins and finally pinches out against an impermeable bed and thus are represented by the position of the zero isopach.

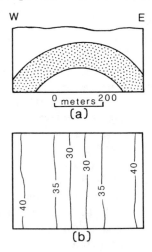

Figure 2-19. The effect of folding on the isochore pattern. (a) Cross section of a sandstone layer of constant original thickness; (b) isochore map of the top of the sand layer.

2-6 CONSTRUCTION OF CONTOUR MAPS

Data Sources

The method used to construct a contour map depends on the data that are available. If the map is to represent ground-surface topography, two data sources are available; air-photo images and surveyed *point data* (measurements at one point). If the map is to represent subsurface features, it must be constructed primarily from point data.

Most modern topographic maps are constructed using techniques of *photogrammetry* (the study of aerial photographs). The cartographer obtains stereoscopic pairs of vertical aerial photographs of the map area. Such photographs are usually made during precisely navigated flights that follow an array of parallel flight lines. These flight lines are designed so that the photos overlap by about 60%-80% along the line of flight and by 20%-30% perpendicular to the line of flight (Dickinson, 1979).

Once stereopairs of vertical air photos are available, the stereopairs are viewed with an instrument called a *stereoscopic plotter*. When using a steroscopic plotter, the observer sees a three-dimensional image of the ground surface as well as a light spot called a *floating point*. The apparent elevation of the floating point above a surveyed reference point, called a *datum*, can be adjusted with a micrometer scale. The observer adjusts the floating point so it is at the elevation of a desired contour line, then manually moves the floating point laterally to make the point appear to lie on the ground surface in the field of view. A contour line can be traced by moving the floating point so that it remains in contact with the image of the ground surface. Either the floating-point controls are mechanically connected to a pen, which traces out a contour line on a map sheet, or the movements of the floating point are digitized and stored in a computer file, which can subsequently be used to reproduce the map electronically.

For localities where air photos are not available or where more detail is needed than can be provided by air photos, data for construction of surface topography maps are obtained by on-ground surveying. The surveyor determines the location and elevation of points on the ground surface and plots the surveyed *control points* as dots on the map sheet with an elevation written next to each dot. The contour map is produced by drawing lines that best accommodate the point data, using methods discussed later. The surveyor measures more elevation points in topographically complex areas, in order to achieve better control, and measures fewer points in simpler areas, to save time. The surveyor places extra points along topographically significant features, such as ridge crests or valley floors.

Structure-contour and isopach/isochore maps are always constructed from point data. If the unit described by the map is exposed in outcrops, the point data can be obtained by field examination of the outcrops. In many situations, however, it is necessary to construct maps of units that are present only in the subsurface. In such circumstances point data are obtained primarily from well logs or seismic-reflection records.

A key fact to remember when constructing a contour map from point data is that it is impossible to produce an absolutely true map. In order to have perfect control and no uncertainty on the map, it would be necessary to have an infinite number of data points. Real contour maps must always be constructed from limited data. The lack of adequate control is particularly troublesome when constructing maps of subsurface data, because the spacing of subsurface control points may not reflect the complexity of the surface. As a consequence, more than one map can be drawn to accommodate a given set of point data, and the choice of which map best reflects reality may depend only on intuition.

Contouring Point Data

Once point data have been plotted and a contour interval selected, the next task in construction of the the contour map is the production of the contour lines themselves. It is possible to digitize the point data and have a computer draw the contour lines according to a specified set of rules, but in this book we consider methods for contouring point data by hand so that you will understand the basis of contouring. Different contour maps can be produced from the same set of point data, depending on what technique is used.   Three common techniques of contouring are described below. For additional description and alternative methods see Rettger (1929), Bishop (1960), and Dennison (1968).

(a) Objective ("mechanical") contouring: The basis of this technique is the assumption that the slope between two adjacent control points is constant. Therefore, between any two control points it is possible to *interpolate* to determine the position of specific elevations. To determine interpolated elevation points, draw a traverse line between each pair of points. Assume that the gradient between the two points along the traverse line is constant, then use Method 2-4a or 2-4b to locate specific elevations along the line. Once you have located the points that fall on designated contour lines throughout the map area, you can draw smoothed contours.

(b) Parallel contouring: Contours are drawn so that adjacent contour lines are as parallel to one another as possible. The spacing of contours between two control points need not be constant.

(c) Interpretive contouring: In this technique the author of the map recognizes that point data are merely

a random sampling of information and that there is no need for gradients to be constant between adjacent points or for contours to be parallel to one another. The author therefore draws the map freehand, taking care that contour lines accommodate the control points. On an interpretive map the contour pattern is drawn to emphasize geologic or topographic features that are thought to occur in the map area. For example, on a structure-contour map, a fault line that truncates contour lines can be drawn in place of a zone of very closely spaced contours.

Problem 2-4 (Interpolation between two points)

Figure 2-20a shows the elevation of two control points, X and Y. A contour map, with a contour interval of 20 m, is to be drawn of the area that includes these control points. Interpolate to determine the position of the contour lines that lie between X and Y.

Method 2-4a (Use of an arbitrary scale)

The difference in elevation between points X and Y in Figure 2-20a is 100 m. The contour interval is 20 m, so the 20-m, 40-m, 60-m, 80-m, and 100-m contours must pass between points X and Y.

Step 1: Draw a traverse line between points X and

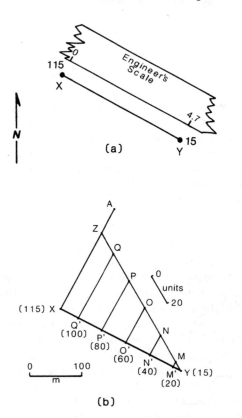

(a)

(b)

Figure 2-20. Interpolation between points. (a) Arbitrary-scale method of interpolation; (b) graphic method of interpolation.

Y. Assume that the gradient of the traverse line is constant.

Step 2: Measure the distance between the endpoints of the traverse line on the map with a convenient engineer's scale. The number of units between X and Y measured on the scale depends on the scale of your drawing and on the spacing of the units of the engineer's scale that you happen to use. In Figure 2-20a the distance between points X and Y happens to be 4.7 units. Thus, 4.7 units represent 100 m of elevation difference, 0.047 unit represents 1 m of elevation difference, and 0.94 unit represents 20 m of elevation difference.

Step 3: From the scale created in step 2 the spacing between contour lines in the interval must be 0.94 unit. The 20-m contour line crosses the traverse line at a distance of 0.047 X 5 = 0.24 unit from the 15-m endpoint. The 40-m contour line intersection is at a distance of 0.24 + 0.94 = 1.18 units from the 15-m endpoint, and so forth. Note that the map scale is not used in the calculation. You can now indicate the positions at which the contour lines will cross the traverse line between X and Y.

Method 2-4b (Graphic interpolation)

Step 1: Draw line XA from point X; line XA is perpendicular to XY. Select a convenient scale to represent 100 units; the length of a line that is 100 units long at your chosen scale should be slightly longer than the length of line XY in your drawing.

Step 2: Draw a line from point Y that is 100 units long at the chosen scale. This line crosses line XA at point Z (Fig. 2-20b). The angle between YZ and XY is a function of the scale used and need not be measured.

Step 3: Line YZ forms the hypotenuse of the right triangle XYZ. Each unit along line XZ is automatically equal to 1 m. Mark the position of point M at a distance of 5 units along YZ from Y, N at a distance of 25 units, O at a distance of 45 units, P at a distance of 65 units, and Q at a distance of 85 units from Y. Draw MM', NN', OO', PP', and QQ' parallel to ZX (i.e., perpendicular to XY). The points M', N', O', P', and Q' correspond to the interpolated positions of 20-, 40-, 60-, 80-, and 100-m contour lines respectively. Think about why the method works!

Problem 2-5 (Construction of a simple contour map)

A surveyor determined the elevation and relative positions of seven control points on the ground in the vicinity of Rymer Pass along Katcubb Ridge in the Appalachian Valley and Ridge province (Fig. 2-21a). He gives the data to you and asks you to produce two contour maps, to display different interpretations of the data. You produce one map by objective contouring, and one map by parallel contouring.

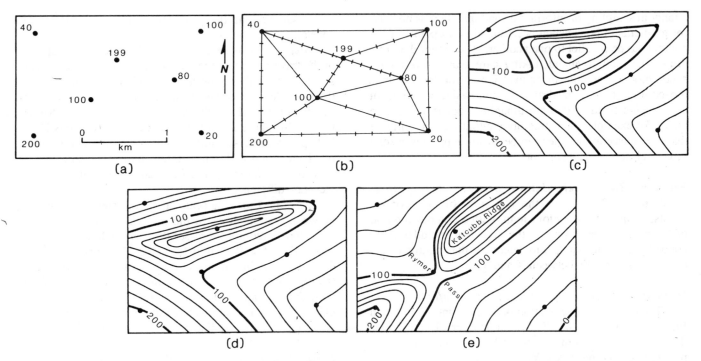

Figure 2-21. Variability of contour maps from the same data set. (a) Positions of control points on the ground. Numbers next to the control points are elevations in meters. (b) Interpolated contour positions; (c) objective contour map; (d) parallel contour map; (e) interpretive contour map.

Method 2-5a (Objective contouring)

Step 1: Draw lines between each pair of control points. Select an appropriate contour interval. In this case, you choose a contour interval of 20 m.

Step 2: Using either Method 2-4a or 2-4b, determine the interpolated positions of the contours between each pair of the control points. These positions are indicated by tick marks on the traverse lines in Figure 2-21b. Clearly, if there are many control points, interpolation is tedious; that is why it is usually done by computer.

Step 3: Draw contour lines that pass through the appropriate interpolated positions. Inititally, draw the contours in pencil, because you will probably find it necessary to modify contour lines as you work.

Step 4: Examine your initial map. Make sure that no "contouring rules" are violated unnecessarily. Also, if you are making a structure-contour map, at this stage think about whether any faults are indicated. If you are satisfied with your interpretation, ink in the map; use a heavier line for every fifth contour, and label every fifth contour (Fig. 2-21c).

Method 2-5b (Parallel contouring)

If you have an objectively contoured map to start with, it may be easiest to produce the parallel-contoured map by modifying the objectively contoured map until the contour lines are as parallel to one another as possible. If an objectively contoured map is not available, start by trying to draw one complete contour through the map area, such that it accommodates the control points. Sketch in other contour lines so that an appropriate number of contours fit between control points. By trial and error, smooth out the contours and modify them so that adjacent contours are parallel. Figure 2-21d provides a parallel-contoured map of the control points from Figure 2-21a. Note how different this map looks from the one shown in Figure 2-21c.

Method 2-5c (Interpretive contouring)

The surveyor returns to check your maps and is dismayed because the maps give a completely erroneous impression of the topography at Rymer Pass. The map of Figure 2-21d does not even show a pass, and neither map shows Katcubb Ridge! The surveyor shows you where the ridge and the pass are, relative to the control points, and asks you to produce an interpretive contour map that emphasizes the ridge and the pass. You produce the map of Figure 2-21e by simply drawing contour lines freehand so that they accommodate the point data and show Rymer Pass and Katcubb Ridge. During the process of drawing the map, you find it necessary to make many modifications by trial and error. You finish this exercise with the solid understanding that there is no unique contour map for a given finite set of control points.

· EXERCISES

1. Figure 2-M1 shows the topography along Catskill Creek in New York State.

(a) Draw a section line that runs from point D to the eastern edge of the map (D') and follows a trend of S60ºE. You must locate D'.
(b) What is the contour interval of this map?
(c) What is the maximum relief crossed by section DD'?
(d) On a sheet of tracing paper construct a topographic profile with no vertical exaggeration along the line of section. What is the steepest slope along your line of section? Express your answer as grade, slope fraction, amd slope angle. Use a protractor to check your slope angle. Indicate the position of the steepest slope on your section line.
(e) Construct a second profile along the same line of section. This profile should have **4X** vertical exaggeration. Go to the same point along your profile that you used for your measure of slope in problem 1d. Use a protractor and measure the angle in this exaggerated profile. By what factor must you divide this angle to obtain the true slope at this locality?

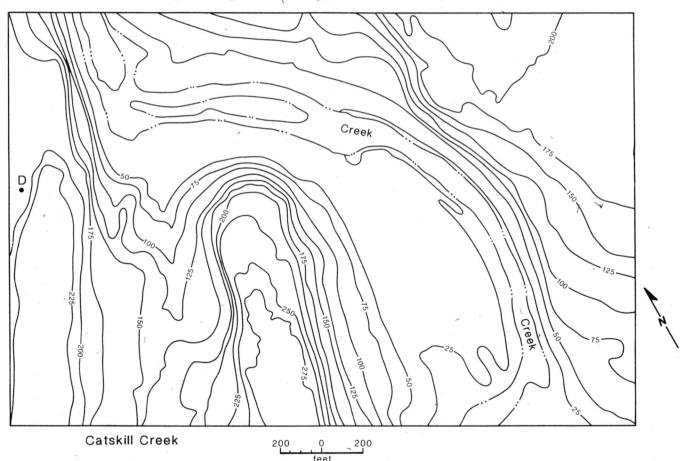

Catskill Creek

200   0   200
feet

Figure 2-M1. Portion of a contour map of an area along Catskill Creek, New York.

2. Figure 2-M2 shows the Kiskatom Escarpment. J. Bass & Co., an engineering geology company, must determine how to run the rail line from Flatville to Bonview so that the grade of the railbed does not exceed 4%. Draw the route by connecting 500 m-long line segments. The obtuse angle between two connecting

line segments should not be less than 120º. The route does not need to be the shortest possible route. What is the gradient of the steepest part of the path represented as slope fraction?

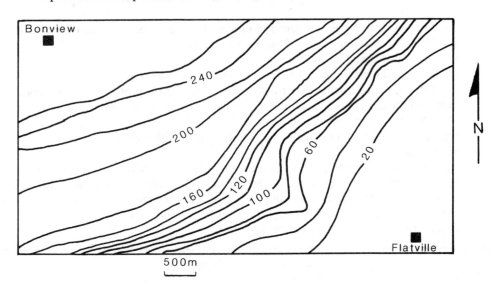

Figure 2-M2. Topographic map of the region between Bonview and Flatville.

3 . Figure 2-M3 shows the topography of an area near Kingston, New York, where topography is controlled to some extent by the structure of the underlying strata.

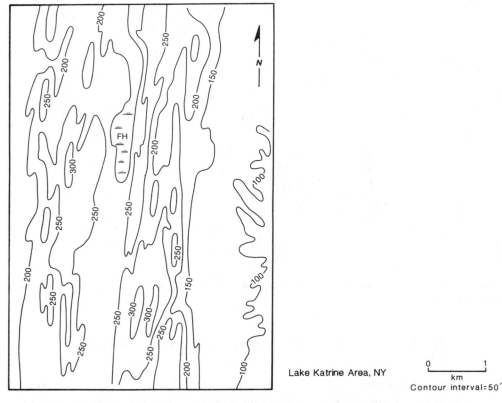

Lake Katrine Area, NY

Contour interval=50´

Figure 2-M3. Topography near Lake Katrine, north of Kingston, New York. FH = Fox Hollow.

(a)   What is the trend of the "structural grain" in the map area?

(b)   Field studies indicate that the ground surface on either side of Fox Hollow (FH) is a dip slope. Based on this observation, what is the major structural feature that surrounds and underlies Fox Hollow?

4.   The structure-contour maps shown in Figure 2-M4 represent basic types of structures. The numbers given are depths below mean sea level. Interpret each map and identify the structure depicted by the map.   Estimate the attitude of the contoured surface at point X on each map.

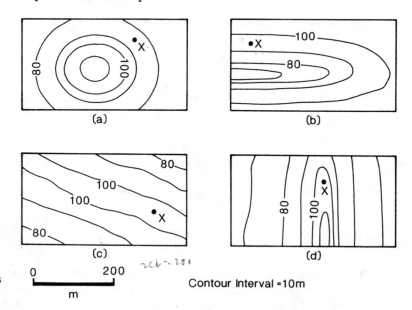

Figure 2-M4. Structure-contour maps of simple structures. Contours are depths below mean sea level.

0   200
m

Contour Interval = 10m

5.   Points in Figure 2-M5 are plumb-line measurements of water depth in Dawersport Harbor. Numbers given are depth below mean sea level.

(a)   Interpolate between points and produce an objective bathymetric map.

(b)   Construct a bathymetric map of the same points using parallel contouring.

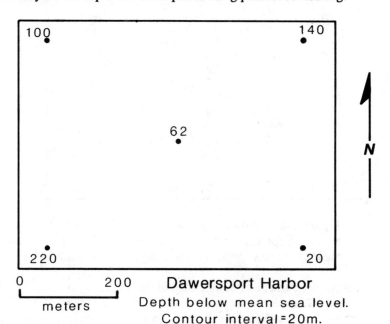

Figure 2-M5. Plumb-line measurements of depths in Dawersport Harbor. Numbers are depths below mean sea level.

0   200
meters

Dawersport Harbor
Depth below mean sea level.
Contour interval = 20m.

6. The numbers on Figure 2-M6 represent the depths below sea level of an unconformity separating crystalline metamorphic basement from Mesozoic cover strata in Bloomer County. The cover strata are horizontal.

(a) Construct an interpretive (freehand) structure-contour map of the basement/cover unconformity.
(b) What is the approximate attitude of the unconformity below well C1?
(c) At what depth does well C1 penetrate basement?
(d) Explain the geologic meaning of this structure-contour map.

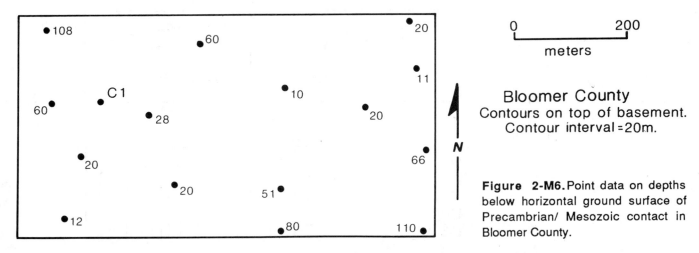

Bloomer County
Contours on top of basement.
Contour interval = 20m.

Figure 2-M6. Point data on depths below horizontal ground surface of Precambrian/ Mesozoic contact in Bloomer County.

7. The numbers on Figure 2-M7 are the depths below sea level of the base of the Bayou Sandstone intersected by drilling above a salt dome in the Sidi Bashrig field. Make an interpretive structure-contour map of this area using a 20-m contour interval. Assume any faults in the area are vertical, and indicate on the map which is the upthrown and which is the downthrown side.

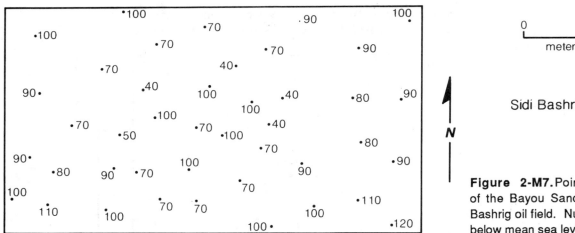

Sidi Bashrig Field

Figure 2-M7. Point data to the top of the Bayou Sandstone in the Sidi Bashrig oil field. Numbers are depths below mean sea level.

8. Figure 2-M8 shows a portion of the Dry Gully Quad. The solid lines are formation contacts, and the dashed lines are the traces of kink-fold hinges. Below the map is a cross section of the quad drawn perpendicular to strike.

(a) Construct an isochore map of the Snehal Shale.
(b) Construct an isopach map of the Snehal Shale.
(c) Explain the difference between your isopach and isochore maps.

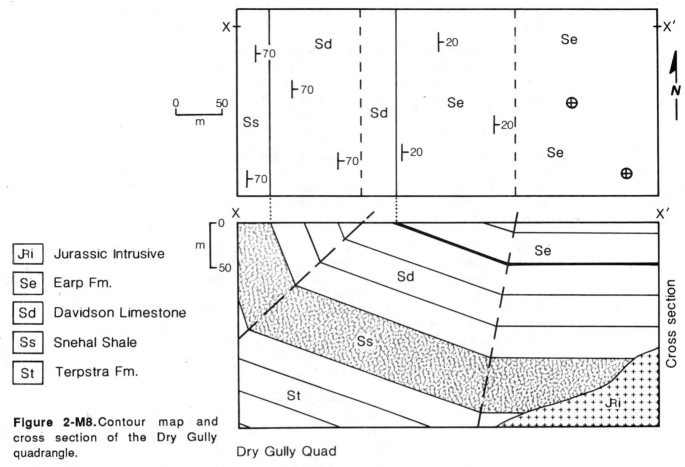

JRi   Jurassic Intrusive

Se   Earp Fm.

Sd   Davidson Limestone

Ss   Snehal Shale

St   Terpstra Fm.

Figure 2-M8. Contour map and cross section of the Dry Gully quadrangle.

Dry Gully Quad

9. Figure 2-M9 shows strike and dip data collected from outcrops of the James Formation (composed dominantly of mudstone, which forms a good trap) in the vicinity of Stanton. The James Formation overlies the Michael Sandstone, which is a potential reservoir rock and has a true thickness of 80 m. Construct a form-line contour map of the Michael/James contact, based on the attitude data.

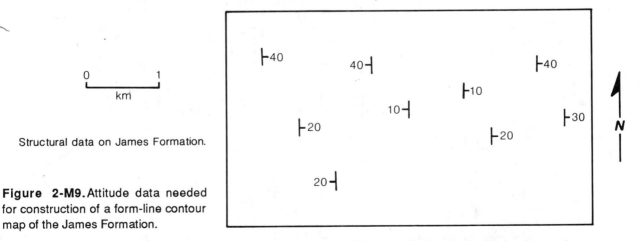

Structural data on James Formation.

Figure 2-M9. Attitude data needed for construction of a form-line contour map of the James Formation.

10. Geologists of Flyhi Oil & Gas are checking an oil play in the Wainesborough Field. None of the rocks are exposed, and the ground surface in the field is essentially horizontal and is at sea level. The geologists have assembled the drilling data and have plotted it on Figure 2-M10. The dots are the localities of the wells, and the numbers next to the wells provide the following information. The

numbers are depths measured from the ground surface. (*Hint:* All units strike east-west).

> * Depth to the base of the Cowlick Volcanics (Jrc)
> * Depth to the base of the Franklin Shale   (Df)
> * Depth to the base of the Boneyard Sandstone (Db)
> * Depth to the base of the Churchville Shale (Oc)
> * Depth to the base of the Figaro Sandstone (Of)
> * Depth to the bottom of the drill hole (basal unit of the hole)

The dash (-) means that either the unit was not found in the hole *or* that the base of the unit was not intersected, even though the unit was present. The unit at the bottom of the hole is indicated in parentheses next to the last number in the list. The symbol Ot stands for the Treyne Formation, which is stratigraphically below the Figaro Sandstone.

(a)   On separate sheets of tracing paper, create the following maps.
  * Interpretive structure-contour map of the base of Cowlick Volcanics.
  * Interpretive structure-contour map of the base of the Figaro Sandstone.
  * An isochore map of the Figaro Sandstone.

(b)   What is the attitude of unit Jrc beneath point X? Assume that the bedding of this unit is parallel to its basal contact.

(c)   Draw a stratigraphic column to scale for the stratigraphic section that appears beneath point X. Use the apparent thicknesses that appear in the well.

(d)   What type of structure defines the base of the Cowlick Volcanics?

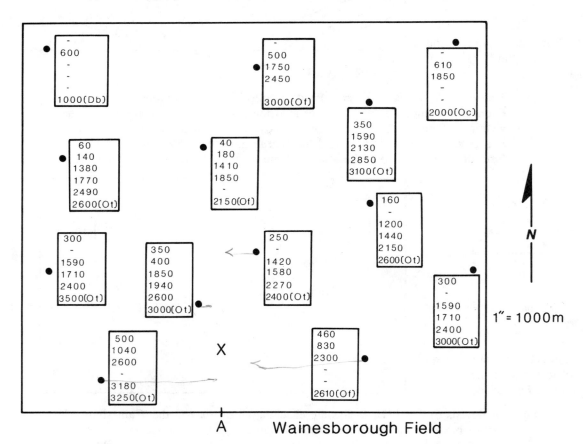

Figure 2-M10. Drilling data from the Wainesborough Field.
Numbers are depths below horizontal ground surface.

(e) Draw a cross section of the map area that starts at point A and runs perpendicular to the structural grain of rocks in the map area. Remember that all units strike E-W. Your cross section should have no vertical exaggeration. Show your line of section on the map.

(f) Describe the geologic history that led to the development of the relationships that occur in the subsurface of the map area.

(g) Which well or wells have the best prospect of yielding oil, and at what depth would the oil be? (Ask your instructor for information on oil reservoirs and traps if this information is unfamiliar to you).

11. Figure 2-M11 is a structure-contour map of the top of the Katgol Quartzite. Notice the prominent fault within the map area. The dashed contour lines are contours on the fault surface, and the solid contour lines are at the top of the quartzite. The contours are depths below sea level.

(a) Draw a cross section perpendicular to strike to help you visualize the structure.

(b) Does this fault display normal or reverse offset?

(c) What is the dip of the fault plane?

(d) What is the throw of the fault?

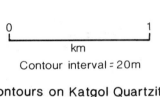

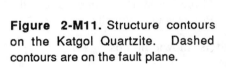

Contour interval = 20m

Contours on Katgol Quartzite
Solid contours on Katgol Quartzite.
Dashed contours on fault plane.

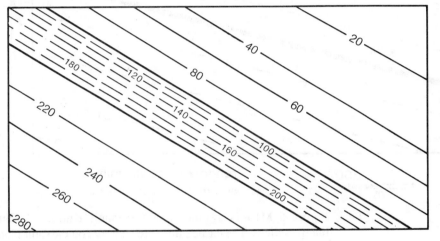

Figure 2-M11. Structure contours on the Katgol Quartzite. Dashed contours are on the fault plane.

12. The numbers in Figure 2-M12 are depths below sea level of the Ransome Bentonite in the vicinity of Edmundale. Remember that there are many ways to draw a contour map that accommodates the given data. Construct two interpretive contour maps of this data.

km

Contours on top of Ransome Bentonite
Contour interval = 50m
Numbers given are depth below sea level.

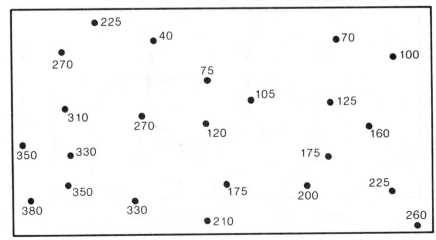

Figure 2-M12. Point data on the top of the Ransome Bentonite.

(a) Construct a map that contains a north-northwest-trending fault.
(b) Construct a map that does not contain any faults.
(c) Which map do you think is a more reasonable interpretation? Why?

13. Figure 2-M13 shows the trace of a fault on a structure-contour map. The contour lines are on the Prendow unconformity. The cut point at which Well 46 intersects the fault is at a depth of 50 m below the level at which the well crosses the unconformity. The cut point at which Well 50 intersects the fault is at a depth of 300 m below the unconformity.

(a) What type of fault is shown?
(b) What is the attitude of the fault?
(c) Assuming that the slip direction on the fault is parallel to the dip of the fault, what is the magnitude of displacement on the fault?

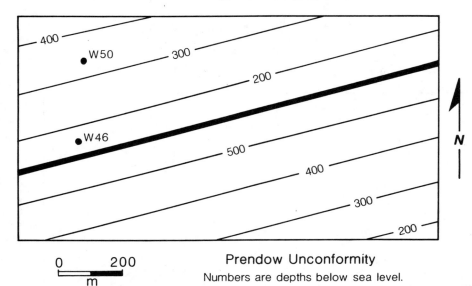

0   200
m

Prendow Unconformity
Numbers are depths below sea level.

Figure 2-M-13. Contours on the Prendow unconformity.

14. Figure 2-M14 is a topographic Bonnell Knob. A fault is shown in this interpretation. The dashed lines are contours on the fault plane.

(a) What is the attitude of the fault plane?
(b) Draw a cross section perpendicular to strike of the fault plane showing topography and the fault plane.

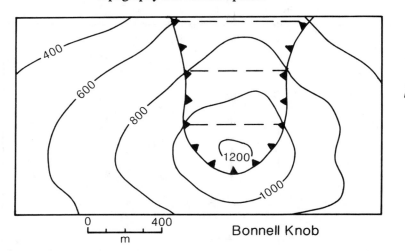

0   400
m

Bonnell Knob

Figure 2-M14. Topographic map of Bonnell Knob.

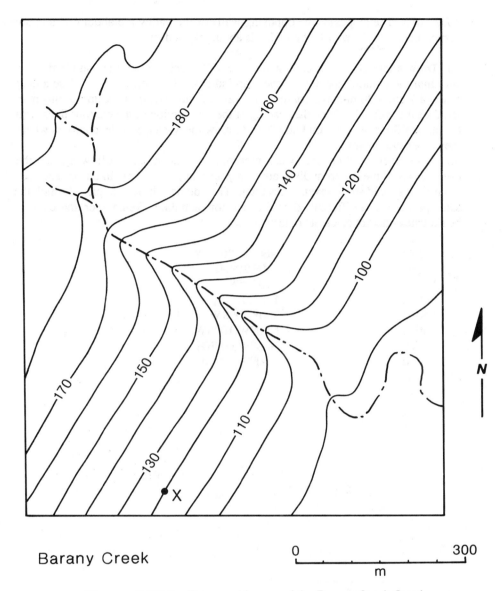

Barany Creek

0 ⸻ 300
m

Figure 2-M15. Topographic map of the Barany Creek Quad.

15. On the accompanying topographic map of the Barany Creek area (Figure 2-M15) *sketch* the outcrop traces of veins with the orientations listed below. Assume that each vein passes through point X, and assume that each vein is so thin that it can be represented by a single line. Label each vein on the map. Your answers should be quick sketches; do not calculate outcrop traces. (*Hint:* Determine the stream gradient first).

Vein A:	N30°E,20°NW
Vein B:	N30°E,10°SE
Vein C:	N50°E,90°
Vein D:	N30°E,50°SE

16. The course of the Pohz River in Prajikistan is quite straight for a distance of 1 km between the villages of Nimla and Gradu. The two villages lie on the river bank. The river flows down a V-shaped valley. The plunge and bearing of the river bed is

20°,210°. The slopes of the opposing walls of the valley are about 45° and are perpendicular to the river. The valley is about 100 m deep.

(a) Draw a topographic sketch map of the valley between Nimla and Gradu. Be sure that the contour spacing on the map indicates the proper slopes. Use a map scale of 1 cm = 20 m and a contour interval of 20 m. (*Hint:* Creation of this map requires that you determine the spacing of the intersection of the contour lines with the stream floor, so that you know how to draw the V-shape of the contours where they cross the stream.)

(b) A number of dike sets were measured in the area. Following are the orientations of the different dike sets. On your map, *sketch* one dike from each set. Your map should show seven dikes total. They do not all have to pass through the same point. Use a different colored pencil for each dike. Remember, these are to be sketches - do not calculate outcrop traces.

Set A:	N60°W,20°SW
Set B:	300°,20°NE
Set C:	240°,90°
Set D:	110°,01°
Set E:	300°,40°SW
Set F:	N60°W,10°SW
Set G:	N60°W,60°NE
Set H:	N60°W,10°NE

CHAPTER

3

GEOMETRIC METHODS I: ATTITUDE CALCULATIONS

3-1 INTRODUCTION

In the previous chapter we developed the concept of a reference frame and showed how the position and attitude of lines and planes can be specified with respect to a reference frame. So far, however, we have primarily described situations in which linear or planar geologic elements could be directly observed and measured. In many circumstances, direct measurement of a geologic feature is not possible, and the attitude of the feature must be calculated by other means. Say, for example, that you are mapping a limestone formation in a humid region where outcrops are weathered. In such a location bedding planes are not distinct, and the strike and dip of the formation cannot be measured directly with a compass. In this chapter we introduce basic geometric methods that can be used to calculate the attitudes of lines and planes when direct measurement is impossible. You will find that the stereographic techniques introduced in Chapter 5 permit more efficient solution of many of the problems posed in this chapter. We encourage you to study the geometric methods introduced here, however, because they help you to further develop the skill of visualizing shapes and attitudes, and they will ultimately make it easier to understand stereographic techniques and to appreciate the relative ease of using them.

3-2 PROJECTIONS AND DESCRIPTIVE GEOMETRY

A *projection*, like a shadow, is a representation of a three-dimensional object on a two-dimensional plane. It is constructed by drawing *projection lines* from points on the object to a *projection plane* (the surface on which the projection is being created). The shape of the projection is affected by the orientation of the projection lines with respect to the projection plane. These lines may emanate from a point source (e.g., light rays from a nearby small bright bulb; Fig. 3-1a), or they may be parallel to one another (e.g., light rays from a distant star; Fig. 3-1b). Parallel projection lines can be perpendicular to the projection plane (Fig. 3-1b) or oblique to the projection plane (Fig. 3-1c). If the projection lines are parallel to one another and are perpendicular to the projection plane, the resulting projected *image* is called an *orthographic projection* (Fig. 3-1b). The use of orthographic projections for solving problems involving the lengths of lines, the areas of planes, and the angles between lines and planes is the subject of *descriptive geometry*. The solution of problems in descriptive geometry involves measurement of angles and lengths in a *scaled drawing*, which depicts the geometry of a structure to scale.

It is most common for projection planes to be either horizontal or vertical; the former is called a *map projection* or a *plan-view projection*, and the latter is called a *cross-sectional projection*. In regions of plunging structures, structures may be projected onto a nonvertical projection plane that is perpendicular to the plunge of the structure so that the geometry of the structure is not distorted (see Appendix 1 and Chapter 13). Vertical cross-sectional projections are typically oriented either parallel to or perpendicular to the strike or bearing of a given geologic structure.

Two nonparallel projection planes join along a *folding line*, which can be pictured as a hinge connecting the two

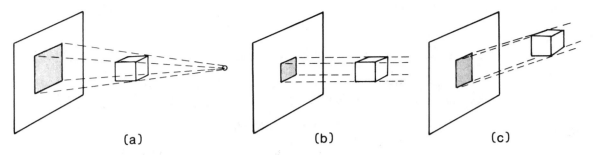

Figure 3-1. Projections of a cube onto a plane. (a) Point source of projection lines; (b) parallel projection lines that are perpendicular to the screen; (c) parallel projection lines that are inclined to the screen.

projection planes (Fig. 3-2a). A step in the solution to many problems of descriptive geometry involves the rotation of a vertical cross-sectional projection plane around a folding line by 90° so that it lies in the same plane as the horizontal map projection plane (Fig. 3-2b). In some cases it is useful to specify the altitude of the map projection plane and therefore the altitude of the folding line. When a rotation around a folding line has been completed, the representation of the once-vertical cross-sectional plane on the horizontal plane is called a *rotated projection*; lines that connect a point on the original map projection with the the equivalent point on the rotated projection are called *connecting lines* (Fig. 3-3). Connecting lines must be perpendicular to the folding lines that they cross.

Imagine an object suspended at the center of a bottomless cardboard box. An image of the object can be projected onto the top and onto the four sides of the box (Fig. 3-4a). Each intersection of the top of the box with a side of the box is a horizontal folding line, and the edges of the box are vertical folding lines. Notice that when the

sides have been rotated around horizontal folding lines so that all sides lie in the same horizontal plane as the top of the box, once-adjacent sides are no longer connected by a folding line. In other words, two vertical projection planes that join along a vertical folding line cannot be connected by a horizontal folding line. Thus, the rotated projections on the two planes cannot be joined by connecting lines. The images on two rotated projection planes can be connected, however, by segments of circular arcs (Fig. 3-4b), here called *connecting arcs*. The center of the connecting arcs is the intersection of the two orthogonal horizontal folding lines.

Graphic solutions to some problems requires use of two *reference planes*. A reference plane (RP) is merely an imaginary horizontal plane parallel to the map projection plane. For example, we can let RP1 be the ground surface and RP2, which is parallel to RP1, lie at a depth d below the ground surface. It is possible to locate the position of the intersection of a structure with each reference plane as we will see in subsequent problems.

The solution of problems involving descriptive

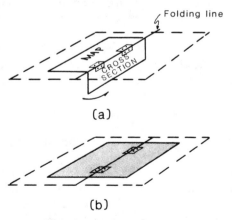

Figure 3-2. Concept of a folding line. (a) Map projection and cross-sectional projection connected along a horizontal "hingelike" folding line; (b) rotation of cross section into the map projection plane.

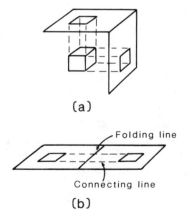

Figure 3-3. Concept of connecting lines. (a) Cube projected onto two orthogonal planes; (b) connecting lines between the map projection and the rotated cross-sectional projection.

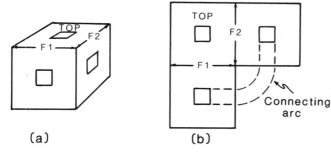

Figure 3-4. Concept of connecting arcs. (a) Projections of a cube onto three mutually orthogonal planes; (b) connecting arcs between two rotated projections.

geometry requires selection of a projection plane in which an angle or a line length is not distorted. For example, measurement of a dip angle must be done in a vertical projection plane that is perpendicular to strike, and measurement of a true line length must be done in a vertical projection plane that is parallel to the bearing of the line (Fig. 3-5). It is also useful to keep in mind that solutions obtained using descriptive geometry are limited in accuracy by the care used in constructing scaled drawings. To improve the accuracy of your calculations, use a sharp, hard pencil and well-made protractors and scales for making scaled drawings, and make your drawing big enough to work with. As a rule of thumb, drawings used to answer problems in this book should fit on about a half of a sheet of paper.

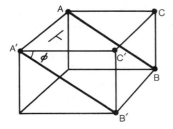

Figure 3-5. Significance of the orientation of a projection plane with respect to a structure. The true dip of the plane (ϕ) can be represented only on a plane that is perpendicular to the strike of the plane. Line AA' is parallel to strike. The length of line AB can be measured only on a projection plane that is parallel to a plane that contains the line. Therefore, the length of AB can only be measured in a vertical plane that is perpendicular to strike; it can not be measured in the map plane. [AB] = [A'B'], but [AB]≠ [AC]; the square brackets mean "length."

3-3   THREE-POINT PROBLEMS

Use of Point Data

In this section we learn how to calculate the attitude of planes from knowledge of the map coordinates and elevation of three points on the plane. Such problems, appropriately, are called *three-point problems*. The data necessary to set up a three-point problem may be obtained from a geologic map on a topographic base or from survey measurements. Calculation of planar attitudes from three points is based on the fundamental theorem of geometry that three points define a plane. We will treat two cases: first, the case where two points on the plane are at the same elevation, and second, the case where all three points are at different elevations. The method is presented in somewhat of a "cookbook fashion" so that it is easier to follow the steps, but please do not treat it as a cookbook. Think through each step and be sure you understand *why* it is done.

Problem 3-1 (Two points at same elevation)
    Imagine a distinctive white tuff bed that is interlayered between dark massively bedded volcanic agglomerates. The sequence is *homoclinally dipping* (i.e., there are no changes in layer attitude within the area of concern), but because the tuff is friable, meaning that it breaks up easily into little pieces, it is impossible to find a well-defined bedding plane in the unit on which to make a direct compass measurement. The locations and elevations of three points on the basal contact of the tuff bed have been surveyed (Fig. 3-6a). Points X and Y are at an elevation of 100 m, and point Z is at an elevation of 60 m. Determine the attitude of the basal contact of the tuff layer. Plane XYZ is the plane that defines the base of the tuff layer.

Method 3-1
    Step 1: Make a scaled drawing that depicts the three points projected onto a map plane that lies at an elevation of 100 m (Fig. 3-6a). Be sure to indicate your scale and north arrow. Label the three points X, Y, and Z'. We use Z' instead of Z because the real point Z does not lie in the projection plane, but points X and Y do.
    Step 2: Connect X and Y with a straight line (line XY). Because these points are at the same elevation, this line is a strike line on plane XYZ, and its orientation is the strike of the plane (Fig. 3-6b).
    Step 3: Using a protractor or a triangle, draw the perpendicular to line XY so that it passes through point Z'. Let the point at which this perpendicular line intersects line XY be called Q. Line Z'Q, which runs from a higher point toward the projection of the lower point, is, by definition, parallel to the dip direction. Create folding line F1 parallel to Z'Q and rotate the cross-sectional plane up to horizontal

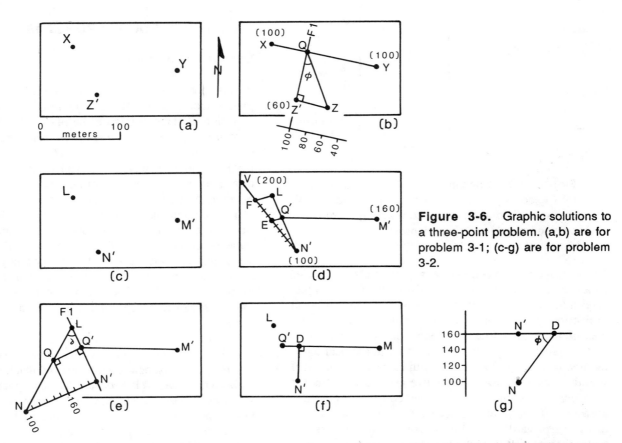

Figure 3-6. Graphic solutions to a three-point problem. (a,b) are for problem 3-1; (c-g) are for problem 3-2.

(Fig. 3-6b). The vertical scale in the cross section must be the same as the horizontal scale in the map. Locate point Z on the rotated projection plane; Z is at a depth of 40 m below Z'Q. Draw line ZQ, which is the trace of the contact in the rotated cross-sectional projection plane. The angle ø, the true dip, can be measured with a protractor (= 34°) or can be calculated with the equation

$$\text{ø} = \arctan([ZZ']/[Z'Q]) = \arctan(40/60) = 34° \qquad \text{(Eq. 3-1)}$$

where the numbers in the square brackets are line lengths in your scaled drawing.

Problem 3-2 (Three points at different elevations)

Determine the attitude of a homoclinal bed given the location and elevation of three points on the bed. The three points are L, M, and N. L is at an elevation of 200 m, M is at an elevation of 160 m, and N is at an elevation of 100 m.

Method 3-2

In this problem the three points are at different elevations. Therefore, in order to determine the orientation of a strike line on the bed, it is necessary to first define two points at the same elevation.

Step 1: Make a scaled drawing that depicts the map projection of the three points. Let the projection plane be at an elevation of 200 m (Fig. 3-6c). Therefore, the true position of point L is depicted, but only the *projections* of points M and N can be depicted; the projections of these points are labeled M' and N', respectively.

Step 2: Draw line LN' connecting the highest point and the projection of the lowest point. Somewhere along line LN' there must be a point Q' that is the projection of a point Q onto the 200-m-elevation projection plane. Point Q, whose location is not yet known, is defined to be a point at the same elevation as point M. Since point Q is at the same elevation as point M, line QM is a strike line on the bed at an elevation of 160 m. Line Q'M' is the projection of this strike line onto the map plane. Steps 3a and 3b provide alternative methods of locating point Q'.

Step 3a: Draw a line N'V at any orientation starting from N', which is the projection of the point of lowest elevation (Fig. 3-6d). It is best if line N'V is oriented at an angle of about 20°-40° from LN' and is a little longer, say 20%, than line LN'. Using your engineer's scale, carefully tick off a distance on line N'V that represents the difference in elevation between point L (highest) and point N (lowest); the scale you use can be arbitrary. Note that in Figure 3-6d the difference in elevation between L and N is 100 m, so we started at N' and located 10 ticks along N'V, each representing a change of 10 m. Connect the point on N'V that represents the elevation of L, call it point F, to point L. Line FL is not necessarily perpendicular to LN'. Now find the point along line N'V that, according to your arbitrary scale, represents the elevation of point M. Call this point E. Draw a line from point E to line LN'; this line must be parallel to line FL. The intersection between the line drawn from point E and line LN' is point Q'. You can now draw line Q'M', which is the strike line. Remember, be sure to understand why this method works!

Step 3b: Create a folding line (F1) that passes through point L and runs along LN' (Fig. 3-6e). Rotate the cross-sectional projection into the horizontal projection plane. F1 is a horizontal line at an elevation equal to point L (the highest point). On the rotated cross-sectional plane draw a line perpendicular to F1 through point N. Using the same scale as your map view, lay off increments until the true depth of point N can be shown. Mark point N, and draw line LN. The angle ∂ between LN' and LN is an apparent dip angle. Now, find the point along line LN that is at the same depth as point M, and call this point Q. Draw a connecting line from point Q parallel to NN' to where it intersects line LN'. This intersection defines point Q', and you can now draw line Q'M', which is the projection of the strike line.

Step 4: Once the projection of the strike line (Q'M') has been determined, it is necessary to determine the dip direction. The dip direction is perpendicular to the strike line and points in the direction of the point with the lowest elevation. Draw a line from N' (the projection of the point at lowest elevation) that is perpendicular to Q'M' (Fig. 3-6f). This line intersects line Q'M' at D. Line DN', which is perpendicular to the projection of the strike line on the map plane, gives the direction of true dip.

Step 5: The final step is to determine the dip angle. To avoid confusion, do this with a separate cross section. Draw a cross-sectional projection along line DN' at the same scale as the map (Fig. 3-6g). Put the horizontal reference line at 160 m. On this reference line draw line DN', so that it is the same length as DN' in Figure 3-6f. Point N' is the projection of N onto a horizontal plane at an elevation of 160 m. Locate point N at an elevation of 100 m directly below point N'. Draw line DN, which is the trace of the plane in cross section. The true dip angle ϕ can be measured directly from this figure.

Steps 3a and 3b reflect the fact that the position of Q' on line LN' is determined by the equation (square brackets indicate length):

$$[N'Q'] = [LN'] \frac{\text{altitude of M - altitude of N}}{\text{altitude of L - altitude of N}} \qquad \text{(Eq. 3-2)}.$$

A proof of the preceding equations is not immediately obvious (see Dennison, 1968, pp. 62-64, for the derivation).

Use of Outcrop Patterns for Three-Point Problems

So far we have used point data for calculating layer attitude. A map pattern, if carefully drawn, also provides sufficient data for calculation of layer attitude. Three points on the line of intersection between a contact and topography can be used to set up a three-point problem, since the position and elevation of each point is known. Once the three points on the plane have been selected, the same procedures as described above can be used to calculate the attitude.

Problem 3-3

The trace of the contact between two formations is shown on a map (Fig. 3-7). From the map pattern of the contact, determine the orientation of the contact.

Method 3-3

Choose three points along the contact (points A, B, and C). To make the solution easier, it is best to choose points at locations where the contact crosses contour lines. If possible, choose two points so that they lie on the same contour line. Once you have located the points, follow the same procedure used in the three-point problem. In the example of Figure 3-7 two points (A and B) were chosen to lie on the 40-m contour, and the third point (C) lies on the 60-m contour. A line connecting A and B strikes east-west, so the contact strikes east-west. Completion of the three-point calculation yields a dip of 38°N.

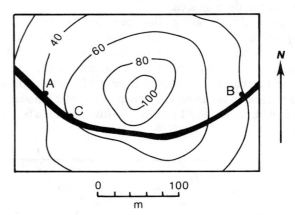

Figure 3-7. Map showing three points on a contact to illustrate problems involving calculation of attitude from a map pattern.

3-4 CALCULATION OF OUTCROP TRACE FROM ATTITUDE DATA

In regions where exposure is poor and outcrops are sparse, it may not be possible to walk out a contact and thereby determine its map trace. However, if you are able to measure the attitude of the contact (or of adjacent units) at a single outcrop and are confident that the attitude does not change in the area of interest, it is possible to *calculate* the trace of the outcrop belt.

Problem 3-4

A topographic base map of an area is available. A distinctive sandstone bed crops out at point A in the northwest corner of the map (Fig. 3-8). The attitude of the bed is 090°,20°S. Assuming that the attitude of the bed is uniform throughout the map area, plot the outcrop belt of the bed on the map.

Method 3-4

Step 1: Draw a folding line (F1) perpendicular to the strike direction of the bed and rotate the cross-sectional view into the plane of the map projection (Fig. 3-8). Put the folding line outside the map area so that the rotated cross-sectional view does not overlap the map.

Step 2: On the rotated view of the cross section, draw a scale perpendicular to the folding line at the same scale as the map. The cross-sectional scale represents elevations. Draw lines parallel to the folding line at intervals along the scale separated by a distance equivalent to the contour interval on the map. Each of these lines represents a reference plane equivalent in elevation to a contour line in the map area. The highest elevation on this section should be above the highest elevation in the map area, and the lowest elevation should be below the lowest elevation in the map area.

Step 3: Draw a line parallel to strike (i.e., perpendicular to F1) from point A so that it intersects the one reference plane in the rotated cross section that is equivalent in elevation to the elevation of the ground surface at point A. Label the point of intersection A'. Draw a line in the cross-sectional plane so that it makes an angle with respect to F1 equal to the true dip of the bed and

so that it passes through point A'. This line represents the trace of the bed in cross section. Be sure the bed dips in the correct direction.

Step 4: Mark a dot at each point where the bed crosses a reference plane in the cross section. From one of these dots, extend a line that is parallel to strike back across the map area. Mark a dot on the map at each point where this line crosses a contour line equivalent in elevation to the reference plane at which the dot occurs in cross section. Repeat the procedure for all other dots.

Step 5: You can now construct the outcrop trace by connecting the dots. Do not connect the dots blindly. Take care to account for local variations in topography between the dots by remembering the rule of V's.

If your task is to draw the outcrop pattern of a bed of known thickness, make sure that your original point A lies on either the top or bottom contact of the bed. Draw the true thickness of the layer on your cross-sectional plane. Mark the points along both top and bottom contacts where the contacts cross the reference planes. Finally, draw the chords from these points across the map area to locate the map traces of the top and bottom contacts.

3-5 TRUE AND APPARENT DIPS

In Chapter 1 we defined the *true dip* of a bed as the dip angle measured in a vertical plane that is oriented perpendicular to strike, and the *apparent dip* as the dip angle measured in a vertical plane that is not perpendicular to strike. In many circumstances true dip cannot be measured directly, but an apparent dip can be measured. As an example, consider a quarry in which dipping dikes are exposed on vertical walls (Fig. 3-9). The angle that the dikes make with horizontal in a quarry wall that is not perpendicular to strike is an apparent dip. If the quarry wall

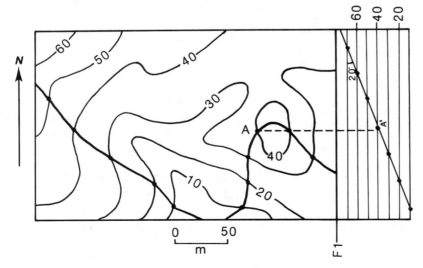

Figure 3-8. Map used to illustrate the calculation of outcrop patterns. Heavy dark line is the outcrop trace.

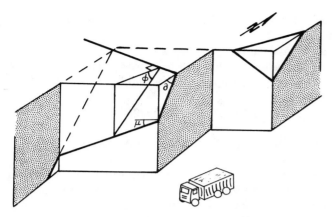

Figure 3-9. Intersection of dikes (heavy lines) with the walls of a quarry. Note that the strikes of the dikes are not perpendicular to the quarry walls; therefore, the angle between the trace of the dikes and horizontal line in the quarry walls is an apparent dip. ø is the true dip; μ is an apparent dip in an east-west trending wall; ∂ is an apparent dip in a north-south trending wall.

is parallel to strike, the apparent dip is 0°. We will introduce graphical, trigonometric, and nomographic methods for calculation of true dip from apparent-dip data. We provide several different examples to illustrate different situations and different methods, but really all the examples are simply variations on the same theme, namely, that if you know the orientation of two lines on a plane, you can calculate the attitude of the plane. The two lines can be either a strike line and the trace of the plane in a vertical plane (e.g., quarry wall) or two nonparallel traces of the plane in vertical planes.

True Dip From Strike and Apparent Dip

Imagine that a quarry with vertical walls has been cut in a region where the original ground surface was horizontal. The trace of the intersection between a dipping bed and the ground surface is, by definition, the strike of the bed. The true dip can be calculated if this strike is known, if the apparent dip in one vertical wall is known, and if the angle in map view between the wall and the strike line is known. The direction of apparent dip is the trend of the quarry wall.

Problem 3-5
Given the strike of a bed (330°) and the apparent dip (25°) in the direction (260°), determine the true dip. Three different methods are presented.

Method 3-5a (Descriptive geometry)
Step 1: Visualize the problem (Fig. 3-10a). Define two reference planes, RP1 and RP2, at a distance d apart (i.e., BB' = CC' =d). RP2 is below RP1. Let ø be the true dip, ∂ the apparent dip, and ß the angle in RP1 between the true dip direction and the apparent dip direction.
Step 2: Make a graphic construction. Start by drawing north-south and east-west coordinate axes in a map-view plane at the level of RP1 (Fig. 3-10b). Let

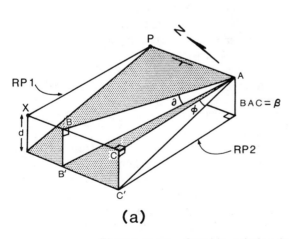

(a)

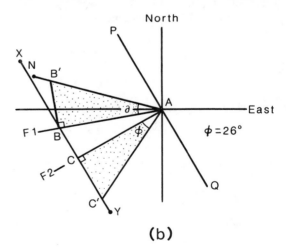

(b)

Figure 3-10. Graphic solution for calculation of true dip from strike and apparent dip. (a) Block diagram of the problem. Shaded surface is the dipping bed; RP1 and RP2 are upper and lower reference planes, respectively. ∂ is the apparent dip; ø is the true dip; ß is the angle between the true dip direction and the apparent dip direction; d is the distance between reference planes; (b) solution on a map projection plane. Stippled triangles are the rotated cross-sectional views.

point A be located at the intersection of these axes. Draw a line PQ representing a strike line on the bed at the same elevation as point A. Draw line AB parallel to the apparent dip direction.

Step 3: Let AB become folding line F1, and rotate the cross-sectional projection (quarry wall) into the map projection plane. Draw line AN so that it makes an angle ∂ with respect to AB, and draw the perpendicular to AB so that it intersects AN and defines the position of point B'. Point B lies in RP1, and B' lies in RP2. The length of BB' defines the distance d.

Step 4: Draw line XY so that it is parallel to the strike line and passes through point B. Draw a line from A so that it is perpendicular to strike and intersects XY. The intersection defines point C (line AC is parallel to the true dip direction).

Step 5: Find point C', which lies a distance d below C. Let AC be folding line F2, and rotate the cross-sectional plane around this folding line into the map projection plane. In this rotated projection, C lies in RP1. Point C', which is in RP2, must lie along XY because AC (true dip direction) is perpendicular to PQ (strike line). The length of CC' must equal the length of BB' (= d). Use your scale and measure a distance d along XY to find C'. Draw a line from A to C'. The angle CAC', which you now measure with a protractor, is the true dip angle ϕ (= 26°).

Method 3-5b (Trigonometry)

From the block diagram (Fig. 3-10a) it is possible to derive a trigonometric formula for calculation of true dip. One solution is provided here; this solution employs the fact that BB' = CC'.

$$BB'/AB = \tan \partial$$

$$CC'/AC = \tan \phi$$

$$AB \tan \partial = BB' = CC' = AC \tan \phi$$

$$(AC \tan \phi)/AB = \tan \partial$$

$$\tan \phi = (\tan \partial)AB/AC$$

$$AC/AB = \cos ß$$

$$AB/AC = 1/\cos ß$$

$$\tan \phi = \tan \partial(1/\cos ß)$$

$$\tan \phi = \tan \partial/\cos ß$$

$$\phi = \arctan(\tan \partial/\cos ß) \qquad \text{(Eq. 3-3)}$$

where, again, ϕ is the true dip, ∂ the apparent dip, and ß the angle in RP1 between the true dip direction and the apparent dip direction (i.e., 20°).

To determine the true dip for Problem 3-5, you can substitute the appropriate values into Equation 3-3:

$$\phi = \arctan(\tan 25°/\cos 20°) = 26°.$$

True Dip from Two Apparent Dips

If the strike of a plane is not known, it is possible to determine the true dip and strike of the plane if the apparent dips in two nonparallel vertical sections are known. The graphical procedure is similar to the preceding one and works both when the true dip direction lies between the two apparent dip directions and when it does not.

Problem 3-6

Given the apparent dip (25°) in the direction 240° and the apparent dip (20°) in the direction 170°, determine the true strike and dip. Three different methods are presented for the solution of this problem.

Method 3-6a (Descriptive geometry)

Step 1: Visualize the problem (Fig. 3-11a). Line AC is perpendicular to strike, so its orientation is the true dip direction. Lines AB and AD are apparent dip directions; λ is the angle between AB and true strike, and Δ is the angle between AB and AD. Both λ and Δ are measured in RP1. ∂ is the apparent dip in the direction of AB, μ is the apparent dip in the direction of AD, and ϕ is the true dip.

Step 2: Draw north-south and east-west coordinate axes so that they intersect at point A (Fig. 3-11b). Draw line AB parallel to the first apparent dip direction and line AL parallel to the second apparent dip direction (these lines are of arbitrary length).

Step 3: Using AB as folding line F1, rotate the cross-sectional plane that contains the first apparent dip up into the map projection plane. In the rotated projection draw line AN so that it makes an angle of ∂ (= 25°) with respect to AB.

Step 4: Draw a line from B so that it is perpendicular to AB and intersects AN at B'. The distance BB' defines d (the distance between RP1 and RP2).

Step 5: Using AL as a folding line F2, rotate the cross-sectional plane parallel to the second apparent dip direction into the map projection plane. Draw line AM in the rotated projection so that it makes an angle μ (= 20°) with respect to AL.

Step 6: You must now find the position of D. To do this, draw a line between lines AL and AM that is perpendicular to AL and is the same length as BB'. This line is DD'. Note that the positions of D and D' are not arbitrary.

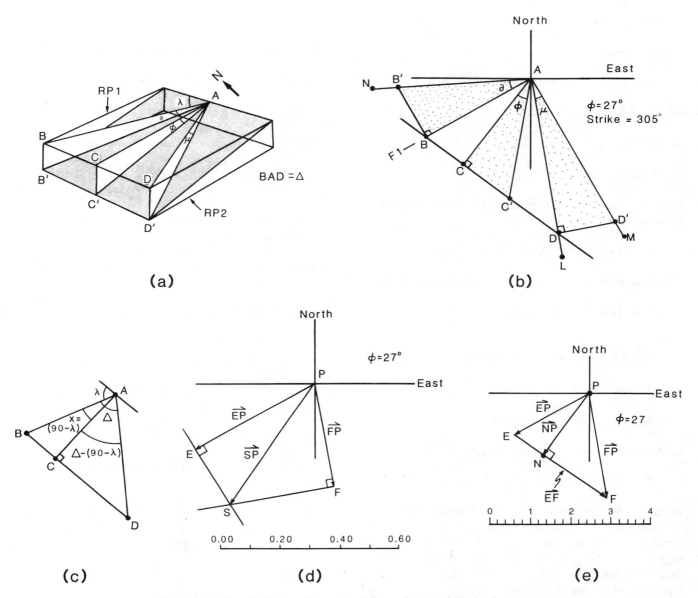

Figure 3-11. Solutions for calculation of true attitude from two apparent dips. (a) Block diagram of the problem. Shaded surface is the dipping plane. ∂ and μ are apparent dips, ø is the true dip, λ is the angle between strike and one apparent dip direction; (b) graphic solution, on map projection plane; (c) angles used in trigonometric solution; (d) tangent vector method; (e) cotangent vector method.

Step 7: You can now draw strike line BD on the upper reference plane; the orientation of BD with respect to your north coordinate axis is the strike.   Draw AC, the direction of true dip, perpendicular to the strike line. Let AC be a folding line, and rotate the cross-sectional plane that contains the true dip around AC into the map projection plane. Lay off line CC' so that it is the same length as BB' and DD'. Angle CAC' (ø = 27º) is the true dip.

Method 3-6b (Trigonometry)
   In this method we determine the strike by calculating the angle between one of the apparent dip directions (in this case, AB) and the strike line (angle λ in Figure 3-11a). Below is an equation for determining the angle λ, given two apparent dips (∂ in the direction AB and μ in the direction AD). To simplify the calculation, first calculate the angle x (Fig. 3-11c), where x = (90 - λ). The distance between RP1 and RP2 is d.

$AB = d/\tan \partial$

$AD = d/\tan \mu$

Now, remember the identity $\cos(90 - \lambda) = \sin \lambda$.

$AC/AB = \cos x = \sin \lambda$

$AC/AD = \cos(\Delta - x)$

$AC = AB \cos x = AD \cos(\Delta - x)$

Now, use the formula for the cosine of the difference between two angles.

$AB \cos x = AD(\cos \Delta \cos x + \sin \Delta \sin x)$

$(d/\tan \partial)\cos x = (d/\tan \mu)(\cos \Delta \cos x + \sin \Delta \sin x)$

$(1/\tan \partial)\cos x = (1/\tan \mu)(\cos \Delta \cos x)$
$\qquad\qquad\qquad\qquad + (1/\tan \mu)(\sin \Delta \sin x)$

$1/\tan \partial = (1/\tan \mu)[\cos \Delta + (\sin \Delta \tan x)]$

$\tan \mu/\tan \partial = \cos \Delta + \sin \Delta \tan x$

$\sin \Delta \tan x = (\tan \mu/\tan \partial) - \cos \Delta$

$\tan x = \tan(90 - \lambda) = [\tan \mu/(\sin \Delta \tan \partial)] - \cot \Delta$

$\lambda = 90^o - \arctan\{[\tan \mu/(\sin \Delta \tan \partial)] - \cot \Delta\}$ (Eq. 3-4)

where λ is the angle between strike and the apparent dip in the direction of AB, μ is the apparent dip angle (20°) in the direction of AD (bearing of 170°), ∂ is the apparent dip angle (25°) in the direction AB (bearing of 240°), and Δ is the angle between AB and AD (= 70°). Now that we have Equation 3-4, we can illustrate the solution.

Step 1: To determine the strike, first calculate the angle λ using Equation 3-4.

$\lambda = 90^o - \arctan\{[\tan 20^o/(\sin 70^o\tan 25^o)] - \cot 70^o\}$

$\lambda = 65^o$

strike = $240^o + 65^o = 305^o$.

Step 2: Once strike has been determined, the true dip can be determined by using Equation 3-3. First, calculate ß, which is the angle between the true dip direction and the appparent dip in the direction of AB. The true dip direction is

$305^o - 90^o = 215^o$

$ß = 240^o - 215^o = 25^o$.

Now, substitute the appropriate values into Equation 3-3:

true dip = $\emptyset = \arctan(\tan 25^o/\cos 25^o) = 27^o$.

Method 3-6c (Tangent vector method):

Hubbert (1931) and Ragan (1985) showed that the tangent of the apparent dip can be treated like a vector and thus that the true dip can be calculated as follows.

Step 1: Draw north-south and east-west coordinate axes to intersect at point P (Note: We use P instead of A to indicate the origin in this figure to emphasize that it displays vectors, not lines and planes, as did Figure 3-11b) and define an arbitrary scale (Fig. 3-11d). Draw vector **EP** in the direction of AB. The length of **EP** is equal to the tangent of the apparent dip angle in the direction of AB. In other words,

$[EP] = \tan \partial = \tan 25^o = 0.47$ units

where the square brackets indicate length. The line representing **EP** in Figure 3-11d is 0.47 unit long at the scale of the figure.

Step 2: Draw vector **FP** in the direction of AD. The length of **FP** is equal to $\tan \mu$.

$[FP] = \tan \mu = \tan 20^o = 0.34$ unit.

The line representing FP in Figure 3-11d is 0.34 unit long at the scale of the figure.

Step 2: Draw perpendiculars to both vectors. These perpendicular lines cross at point S. Vector **SP** indicates the direction of the true dip. The length of SP is 0.51 unit. The value of true dip is given by

true dip = $\emptyset = \arctan[SP] = \arctan(0.51) = 27^o$ (Eq. 3-5).

Using this tangent vector method will yield more accurate results than the graphical method, especially for small dip angles.

Method 3-6d (Cotangent vector method)

This method is similar to that of Method 3-5c, but now we use the cotangents of the apparent dip angles rather than the tangents.

Step 1: Draw coordinate axes that intersect at P (Fig. 3-11e). Draw vector **EP** so that it is parallel to AB and is equal in length, using the scale of the drawing, to the cotangent of the apparent dip ∂ in that direction (cot $25^o = 2.15$). Draw vector **FP** so that it is parallel to the direction of AD and is equal in length to the cotangent of the apparent dip μ in that direction (cot $20^o = 2.75$).

Step 2: Draw vector **EF**, which connects the ends

of **EP** and **FP**. Draw a vector from P so that it is perpendicular to **EF**; call this vector **NP**. The length of vector **NP** (= 1.96), measured by the scale, is the cotangent of the true dip. Therefore,

$$\phi = \text{arccot}[\mathbf{NP}] = \text{arccot}(1.96) = 27^\circ \qquad \text{(Eq. 3-6)}.$$

Apparent Dip Determined From True Dip

Perhaps the most common application of calculations involving true and apparent dips occurs in the construction of geologic cross sections. As noted earlier, if the line of section is drawn to be perpendicular to strike of the beds that cross the line, then the dip of the beds shown in the section will be true dips. If the line of section is oblique to strike, then the dip of the beds shown in the line of section must be an apparent dip. Next we offer a few geometric methods for determination of apparent dip if the trend in which the apparent dip is desired is given, and the true strike and dip are known.

Problem 3-7

Given the strike and dip of the bed (N45°W,30°SW), determine the apparent dip (∂) in the direction N80°W. The trend of the requested apparent dip direction may represent a line of section. Three alternative methods to solve this problem are given below. To visualize the problem, refer to the block diagram of Figure 3-12a. We wish to determine ∂, given $\phi = 30^\circ$ and $\beta = 55^\circ$, where ϕ is the true dip and β is the angle between the apparent dip direction and the true dip direction.

Method 3-7a (Descriptive geometry)

Step 1: Draw north-south and east-west coordinate axes that intersect at point A (Fig. 3-12b). Draw line AC of arbitrary length parallel to the true dip direction (i.e., perpendicular to the strike). Draw line SR so that it passes through point C and is parallel to the direction of strike. SR is the projection of a strike line onto RP1.

Step 2: Let AC be a folding line F1. Rotate the cross-sectional view into the map projection plane. Draw

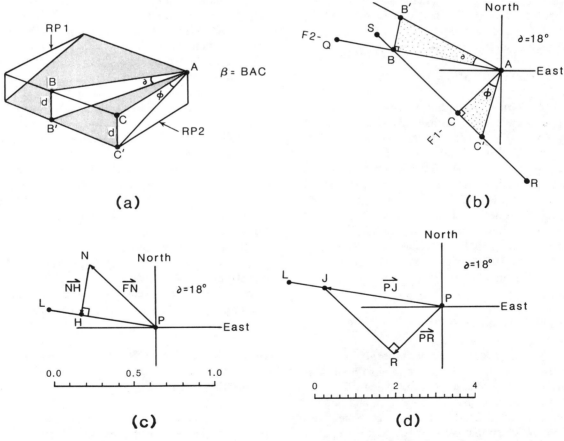

(a)

(b)

(c)

(d)

Figure 3-12. Solutions for calculating apparent dip from true attitude. (a) Block diagram. ∂ is the apparent dip; ϕ is the true dip; β is the angle between true dip and apparent dip directions; (b) graphic solution; (c) tangent vector method; (d) cotangent vector method.

line AC' so that it makes an angle of ø (=30º) with AC. Point C' in the rotated projection must lie along line SR. The distance CC' defines the distance d between the upper and lower reference horizons (RP1 and RP2).

Step 3: Draw line AQ so that it is parallel to the specified direction of the apparent dip (N80ºW) and crosses line SR at B. Consider AB to be folding line F2. Rotate the cross section around F2 into the map projection plane. On the rotated projection draw line BB' so that it is perpendicular to AB and has a length of d. Now that B' is located, you can draw AB'. The angle between AB and AB' is the apparent dip (∂ = 18º) in the direction AB.

Method 3-7b (Trigonometry)

We can derive a trigonometric formula by referring to Figure 3-12a.

$$\tan \partial = \tan ø \cos ß$$

$$\partial = \arctan[\tan ø \cos ß] \qquad \text{(Eq. 3-7)}$$

where, again, ∂ is the apparent dip in the direction AB, ø is the true dip, and ß is the angle between the true dip direction and the apparent dip direction. Substituting appropriate angles into Equation 3-7 yields

$$\partial = \arctan(\tan 30º \cos 55º) = 18º.$$

Method 3-7c (Tangent vector method)

Step 1: Draw coordinate axes that intersect at P, and specify a scale (Fig. 3-12c; note that the scale provided in this figure is arbitrary and units do not matter.).

Step 2: Draw vector **PN** in the direction of true strike, equal in length to the tangent of the true dip (tan 30º = 0.577) at the scale of your drawing. Draw line PL in the direction of the desired apparent dip direction.

Step 3: Draw a line from N that joins PL and is perpendicular to PL. This line is vector **NH**. The length of **NH**, at the scale of the figure, is the tangent of the apparent dip in the direction of PL. The length of **NH** is 0.32, so,

$$\partial = \arctan[\mathbf{NH}] = \arctan(0.32) = 18º \qquad \text{(Eq. 3-8).}$$

Method 3-7d (Cotangent vector method)

A construction using cotangents may also be used in the solution of this problem (Ragan, 1985). To help visualize this solution, refer to Figure 3-12a, and imagine that d (=BB' and CC') is set to be 1 unit long. Note that if d = 1, tan ø = 1/AC, so that cot ø = AC. Likewise, cot ∂ = AB. This fact permits a quick construction to determine the apparent dip (Fig. 3-12d).

Step 1: Draw coordinate axes so that they intersect at P. Draw vector **PR** perpendicular to the strike and in

the direction of true dip; make **PR** equal in length to the cotangent of the true dip (cot ø = 1.73) according to the scale of the figure.

Step 2: Draw line PL parallel to the desired apparent dip direction. Draw a vector from the tip of **PR** that is perpendicular to **PR** and joins PL at J. The length of vector **PJ** is the cotangent of the apparent dip in the direction of PL. PJ is 2.97 units long in Figure 3-12d at the scale of the figure, therefore,

$$\partial = \text{arccot}[\mathbf{PJ}] = \text{arccot}[3.0] = 18º \qquad \text{(Eq. 3-9).}$$

Nomograms for Apparent Dip Calculations

A *nomogram* is a graphical tool that permits quick solution to equations. It is basically a graphical solution to a single formula (see Palmer, 1919; Billings, 1972). The type of nomogram that we introduce here is an alignment diagram used for true and apparent dip calculations (Fig. 3-13). The diagram consists of three columns; column one represents true dip, column two represents apparent dip, and column three represents the angle between the apparent dip direction and the strike. The plane in which the apparent dip is measured is called the projection plane. These columns represent quantities in an equation that relates true dip to apparent dip. To use the apparent dip nomogram of Figure 3-13, simply mark off values on two of the columns and draw a straight line through these two points so that it crosses the third column. The intersection of the line with the third column gives the value of the third variable. Any two values can be used to calculate the third.

A circular nomogram, the apparent dip computer (Fig. 3-14), is used in a similar way. Mount the outer scale on a piece of cardboard and cut out the inner scale so that it can rotate. Knowledge of any two values allows the third to be determined.

These nomograms are most useful when many calculations are to be done quickly, such as when constructing a geologic cross section that is not perpendicular to strike. They cannot be used for determining true dip from two apparent dips.

3-6   CALCULATION OF LINEAR ATTITUDES

Determination of Rake

It is often possible to measure the rake of a lineation directly in the field. However, if direct measurement is not possible, the rake of a lineation can be calculated either trigonometrically or by descriptive geometry if the attitude of the plane on which the lineation occurs is known and the bearing of the lineation is known.

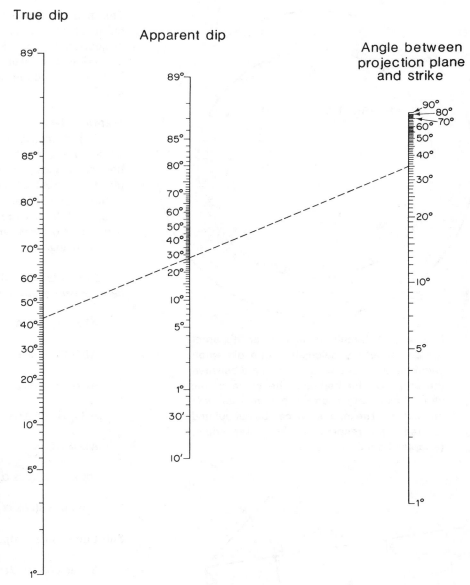

Figure 3-13. Linear nomogram for calculation of true dip given the apparent dip on a vertical plane (projection plane) that is inclined to the strike direction. (Adapted from Palmer, 1919).

Problem 3-8

The attitude of a slip lineation on a fault surface is 20°,S38°W. The fault-plane attitude is N10°E,40°NW (i.e., ∅ = 40°). The angle between the fault strike and the bearing of the lineation is, therefore, α = 28°. What is the rake (Δ) of the lineation?

Method 3-8a (Descriptive geometry)

Step 1: Draw a map representing the fault trace and the projection of the lineation on a horizontal plane (Fig. 3-15a). Specify points A, B, C, and D, which all lie in the map projection plane. Consider the map projection to define an upper reference plane (RP1). Line AD represents the fault trace in RP1, line AC is the projection of the lineation onto RP1, lines AB and DC are drawn parallel to the dip direction of the fault. Line BC is the projection of

the trace of the fault in a lower reference plane (RP2) up onto RP1. The plane ABCD is the projection of a dipping plane on the horizontal map plane. In order to determine the rake (an angle measured in the fault plane), first rotate the fault plane up to horizontal.

Step 2: Let line AB be a folding line (F1). Draw line AJ (Fig. 3-15b), which represents the trace of the fault on the rotated plane (∅ is the true dip of the fault). Locate point K along AJ; K lies in RP2 and therefore lies vertically below point B.

Step 3: Now rotate the fault plane into the map projection plane. To do this, use a compass and draw a connecting arc from K to L; point L lies along folding line F1 and point A is the center of the circle. The length of line AL equals the length of line KL.

Step 4: Construct rectangle ALMD (Fig. 3-15b).

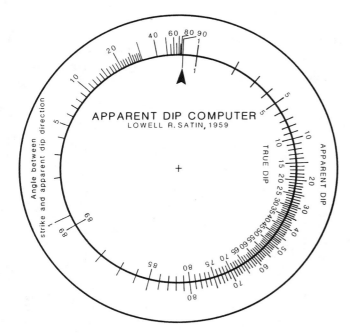

Figure 3-14. Circular nomogram or "apparent dip computer" for calculating true dip when given the apparent dip and the angle between the strike of the bed and the strike of the vertical plane on which the apparent dip was mentioned. The inner scale can be cut out and rotated with respect to the outer scale. (Adapted from Satin, 1960.)

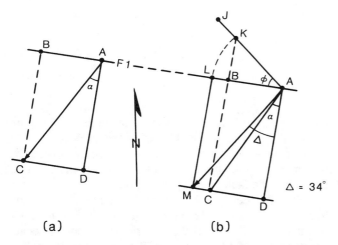

(a)   (b)

Figure 3-15. Graphic solution for calculation of rake. (a) Map projection of four points on the fault plane. α is the bearing of lineation; (b) use of connecting arcs to permit rotation of fault plane into map projection. ø is the true dip; Δ is the rake; α is the angle between the bearing of the line and the strike of the plane.

This plane is the portion of the fault between the two reference planes after it has been rotated into the map projection plane. Remember, ABCD was only the projection of the plane. Line AM is therefore the lineation in the plane, and the angle Δ is the rake of the lineation; Δ = 34°.

Method 3-8b

Trigonometric formulas for specifying rake can be written in two ways. The formulas written below refer to lines and angles in Figure 3-16. Plane ABC is a horizontal plane, plane ABDE is the plane on which the lineation occurs, AD is the lineation, BA is a strike line, AC is the bearing of the lineation, ø is the true dip, Δ is the rake, α is the angle between strike and bearing of the lineation, and ∂ is the plunge of the lineation.

Formula 1 (Rake in terms of plunge and angle between bearing and strike):

$$AB/AD = \cos \Delta$$

$$AB/AC = \cos \alpha$$

$$AC/AD = \cos \partial$$

$$AB = AC(\cos \alpha)$$

$$AD = AC/\cos \partial$$

$$\cos \Delta = AC(\cos \alpha)/(AC/\cos \partial) = \cos \alpha \cos \partial$$

$$\Delta = \arccos(\cos \alpha \cos \partial) \qquad \text{(Eq. 3-10)}.$$

Substitution of appropriate values for the angles yields

$$\Delta = \arccos(\cos 28° \cos 20°) = 34°$$

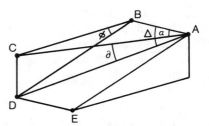

Figure 3-16. Angles used for trigonometric calculation of rake from plunge and bearing. AD is the line in question; Δ = rake; ø = true dip; α = angle between strike of plane ABDE and the bearing of line AD; ∂ = plunge of line AD.

Formula 2 *(Rake in terms of dip and angle between bearing and strike):*

BD/BA = tan Δ

CB/BA = tan α

CB/BD = cos ø

BA = CB/tan α

BD = CB/cos ø

tan Δ = (CB/cos ø)/CB/tan α) = tan α/cos ø

Δ = arctan(tan α/cos ø)   (Eq. 3-11).

Substitution of appropriate values for the angles yields

Δ = arctan(tan 28°/cos 40°) = 34°

where, again, ø is the true dip, Δ is the rake, α is the angle between strike and bearing of the lineation, and ∂ is the plunge of the lineation.

Attitude of the Intersection of Two Planes

If two planes of known attitude cross, the line of intersection is called an *intersection lineation*. The attitude of an intersection lineation can be determined both from descriptive geometry and from trigonometry. We provide only the descriptive geometry solution here, because it helps students to visualize the problem.

Problem 3-9

Two nonparallel dikes intersect each other at point A. Dike 1 is oriented N40°E,30°SE and dike 2 is oriented N70°W,60°NE. What is the attitude of the line defined by the intersection of the two dikes?

Method 3-9 (Descriptive geometry)
Step 1: Draw the map traces of the two dikes at a convenient scale so that they intersect at point A (Fig. 3-17a). Indicate the dip directions with tick marks. The map trace of the intersection lineation must lie between the traces of the two dikes, and the dip tick marks on each dike point toward the trace of the lineation.

Step 2: Draw folding line F1 perpendicular to dike 1 and folding line F2 perpendicular to dike 2. Rotate the cross-sectional planes into the map projection plane. Define a lower reference plane at a distance d below the map plane and draw the cross-sectional representations of dikes 1 and 2. The dip angles shown in the rotated

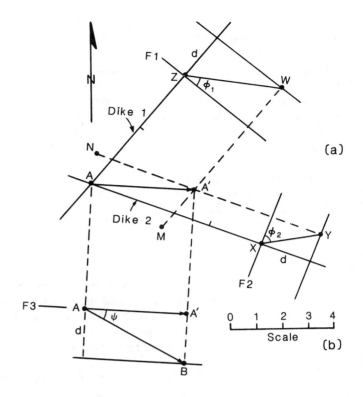

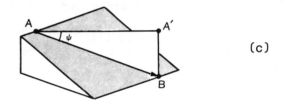

Figure 3-17. Graphic solution for determining the orientation of an intersection lineation formed where two dikes cross.   (a) The dikes and their rotated cross-sectional projections; (b) rotated projection of the plane containing the intersection lineation; (c) angles used in trigonometric calculation of plunge.

projections are the true dips of the dikes. In Figure 3-17a the trace of dike 1 in the rotated projection is line ZW, and the trace of dike 2 in the rotated projection is line XY.

Step 3: Draw lines YN and WM. These lines represent the projection on the map plane of strike lines along the dikes in the lower reference plane. YN and WM intersect at point A'. Point A' is the projection on the map plane of the point at which the dikes intersect in the lower reference plane. Line AA' is therefore the projection in the map plane of the intersection lineation. The bearing of this line (S85°E) is the bearing of the lineation.

Step 4a: You now know the bearing of the intersection lineation, but you still need to determine its plunge. To determine the plunge of AA', let AA' be folding line F3. F3 is offset and redrawn in Figure 3-17b to simplify the figure. Rotate the cross-sectional plane into the map projection plane, and locate the lower reference plane at a distance d below F3. Draw AB, which is the profile of the lineation. Angle ψ, which can be measured from the figure by using a protractor, is the plunge of the intersection lineation; ψ = 25°.

Step 4b: It is faster to use a simple trigonometric formula to determine the plunge. In Figure 3-17c if you set d = 1 (i.e., A'B = 1), then

$$\tan \psi = A'B/AA'$$

$$\psi = \arctan(1/AA') \tag{Eq. 3-12}$$

where ψ is the plunge and AA' is measured from Figure 3-17b. Return to Figure 3-17b. At the scale of this figure, A'B is 2.1 units, and AA' is 4.25 units. You must divide by 2.1 to get the appropriate value of AA' for Equation 3-17; 4.5/2.1 = 2.1.

$$\psi = \arctan(1/2.1) = 25°.$$

Thus, the attitude of the intersection lineation is 25°,S85°E.

EXERCISES

1. A distinctive sandstone bed crops out at three localities in a corner of the Edmundsville Quadrangle. Outcrops A and B are on the 340-m contour line, and point C is on the 280-m contour line. Outcrop B is 400 m to the N40°E of outcrop A, and outcrop C is 240 m to the N20°W of outcrop A. Assuming that the sandstone bed is homoclinal, what is its strike and dip?

2. A basalt sill is exposed at three localities within an area being surveyed by a geologist. The geologist collected the following data concerning the three outcrops of the sill. (Locations are specified with respect to a reference point, X. The first number is the distance from X, and the second number is the azimuth from X):

Locality	Location	Altitude
A	200 m; 070°	700 m
B	100 m; 330°	900 m
C	100 m; 210°	1200 m

    (a) What is the strike and dip of the sill?
    (b) Repeat the problem using a different method.

3. Three wells have been drilled in Chatalkqua County by the Beanbody Coal Company in order to find the thick Queen Mother coal seam. In order to track the seam into the next county, the company geologists must know the strike and dip of the seam. The following data were obtained by the well-site crew. The wells were positioned at the corners of a square that is 500 m on a side. Two edges of this square trend north-south, and two edges trend east-west.

Well number	Location	Ground elevation	Depth to top of coal
459	NE	730 m	220 m
460	NW	850 m	410 m
461	SE	760 m	340 m

    What is the strike and dip of the coal seam?

4. Fred Spear is attempting to determine the regional tilt of a peneplain surface (a *peneplain* is a region that has been beveled flat by erosion). Such information will tell him about postunconformity epeirogenic movements. The term *epeirogeny* refers to gentle regional vertical movements of continental crust. He accurately

measured the altitude of the peneplain at three corners of a square mile section, and his data are listed below.

Locality	Altitude
NE corner	1400 ft
NW corner	1700 ft
SW corner	1900 ft

(a) By how much has the peneplain tilted since it formed, assuming that it was initially horizontal? Assume no curvature to the earth.
(b) Around what axis did it rotate?

5. Some rock units display *massive bedding*, meaning that, as a consequence of bioturbation or of other characteristics of the depositional environment, the unit does not contain distinct bedding planes. The upper 10 m of the Becram Limestone is massively bedded, and its attitude cannot be measured with a compass. Three outcrops of the upper Becram Limestone are located on Figure 3-M1.

(a) What is the strike and dip of the unit in the map area, assuming it is homoclinal in the map area?
(b) Detailed mapping in the map area indicates that between points A and B there is a syncline hinge. The amplitude of the syncline is 80 m. Keeping this observation in mind, reconsider your preceding calculation. Do you think your answer to (a) is worth plotting on the map? (Explain.)

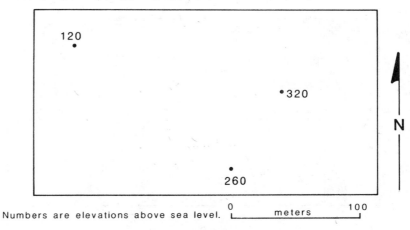

Figure 3-M1. Outcrop positions of the Becram Limestone. Elevations of the outcrops are indicated.

Numbers are elevations above sea level.

Outcrop locations of Becram Limestone

6. The true attitude of a basalt dike is N40°E,30°NW. What is the apparent dip of this dike as exposed in a vertical cliff face that trends N70°E?

7. The strike of bedding on the horizontal floor of a limestone quarry is N43°E. The apparent dip of the bedding in a north-south trending quarry wall is 32° toward the south. What is the true dip of the bedding?

8. A prospector has dug two small trenches, which are not parallel to one another, at the base of Tabor Ridge in order to expose a thick vein of gold-bearing bull quartz. The walls of the pits are vertical. In the first pit the apparent dip of the vein is 24°,N17°E, and in the second pit the apparent dip of the vein is 56°,N39°W. What is the orientation of a shaft that lies in the plane of the vein and parallel to the true dip of the vein?

9. Cross beds can be used to determine paleocurrent directions. Such information is of great value in regional stratigraphic study. Often, however, cross beds are not fully exposed, and their true attitude must be calculated from data on apparent dip. Consider a cross bed in sandstone that is exposed on two vertical nonparallel joint faces. On one face the apparent dip of the cross bed is $10°,016°$, whereas on the other it is $28°,082°$.

(a) What is the attitude of the cross bed?
(b) Assume that the current direction is parallel to the true dip direction of the cross bed. What was the orientation of the current direction during deposition of the sandstone?
(c) What is the orientation of a joint face on which the apparent dip of the cross bed is the maximum possible value?

10. An east-west line of section in the Rusty Ridge Quadrangle cuts across a folded sequence of Carboniferous strata. A portion of the line of section is shown in Figure 3-M2. In order to draw a cross section along this line, it is necessary to determine the apparent dip in the direction of the line of section and project these measurements down-plunge onto the fold.

(a) Calculate the apparent dip in the line of section for each measurement given on the map. Identify your results by specifying station number. (A *station* is merely a location where a measurement was made).
(b) Draw a cross-sectional sketch along the line of section, using the above data and projecting onto the line. Assume the ground surface is horizontal.

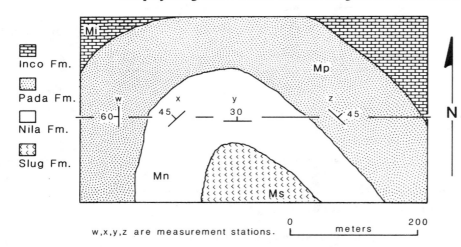

Inco Fm.

Pada Fm.

Nila Fm.

Slug Fm.

w,x,y,z are measurement stations.

0   200

meters

Rusty Ridge Quadrangle

Figure 3-M2. Geologic map of a portion of the Rusty Ridge Quadrangle. Attitude measurements at points x, y, z, and w are given.

11. In an exposure of Precambrian granite in the Hudson Highlands north of New York City, there are many mesoscopic faults on which there are well-developed slip lineations. The slip lineations are composed of fibrous chlorite. A geologist measured the rake of lineations on the faults. Below are some measurements from her field notebook.

(a) Calculate, using trigonometry, the plunge and bearing of the lineations on the faults and fill in the table. Do you think that the movement on the different faults might have occurred during the same tectonic event? Why or why not?
(b) Divisions between the fault classes can be made according to the following criteria: dip slip ($90° \geq$ rake $\geq 70°$), oblique slip ($70° \geq$ rake $\geq 20°$), or strike slip ($20° \geq$ rake $\geq 0°$). Indicate how the geologist would classify the faults.

Fault attitude	Rake	Plunge and bearing	Fault class
359°,72°E	80° S	_____	_____
315°,84°NE	68° SE	_____	_____
272°,80°S	72° E	_____	_____
313°,08°SW	90° NW	_____	_____
032°,40°NW	12° SW	_____	_____
076°,12°SE	35° SW	_____	_____

(b) About 3 km southeast of the outcrop that contains the above mesoscopic faults, there is an exposure of the Rakesh fault. The Rakesh fault, which is a major structure in the region, is oriented N45°E,70°SE. There is evidence for two periods of movement on the fault; the first period is strike slip and is evidenced by slip lineations that are oriented 08°,S42°W; the second period is dip slip and is evidenced by slip lineations oriented 69°,S62°E. Do you think there is a relationship between movement on the Rakesh fault and the movement on the mesoscopic faults described above? (Why?)

12. Foliation at a locality near the town of Ouro Preto in the highlands of Brazil is oriented N10°E,70°E. There are two lineations visible on foliation planes. One is a mineral lineation that trends N20°E, and the other is a crenulation that trends N75°E.

(a) What is the rake of each lineation?   (Use descriptive geometry for your calculation).
(b) Does the rake increase or decrease as the angle between strike of the plane and bearing of the lineation increases?
(c) What is the bearing of the lineation that has the maximum possible plunge on this foliation plane?

13. A chevron fold is one in which the hinge is very angular, so that in profile the fold has the shape of a V. A beautiful chevron fold crops out in Four Day Canyon in British Columbia. The attitude of one limb of this fold is N20°W,30°NE, and of the other limb is N50°E,60°NW. What is the plunge and bearing of the hinge?

14. A high concentration of uranium occurs at the intersection of a 040°,60°NW fault and a 350°,40°NE sandstone bed. The intersection of the bed and the fault crops out in a wash north of the True Blue Mine in western Arizona. The owners of the mine have decided to explore the uranium play by drilling it. If they start the hole at the outcrop, what should be the bearing and plunge of the drill hole such that the hole follows the intersection lineation (and stays in the play)?

15. A thick quartz vein can be observed in Carlisle Canyon.   The vein occurs in granitic gneiss and is parallel to the foliation in the gneiss. In the map area there is no surface on which a compass can be placed to make a direct measurement of foliation. The only way to determine the attitude of foliation in the gneiss is to use the quartz vein as a marker horizon and to calculate the attitude of the vein from its outcrop pattern. A map showing the trace of the quartz vein is provided in Figure 3-M3. From this map, determine the attitude of the vein (and therefore of the foliation in the gneiss). The dot-dash line is the trace of a stream.

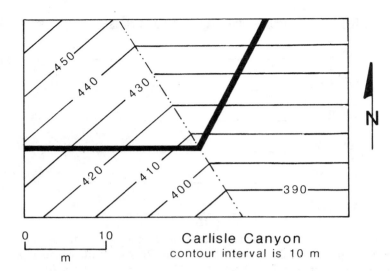

Figure 3-M3. Map of a portion of the Carlisle Canyon region. The thick line is the outcrop trace of the quartz vein.

Carlisle Canyon
contour interval is 10 m

0   10
m

16. Figure 3-M4a shows the map of the Sheep Hollow anticline as it may have appeared in the vicinity of Cresty Ridge before the area was dissected by rivers. The black lines represents the traces of a marker bed on opposite limbs of the fold. Figure 3-M4b shows the current topography of the Cresty Ridge area. Draw the map pattern of the fold as it would appear on Figure 3-M4b. Assume that the ground above the 160 m contour is still horizontal and is at an elevation of 161 m.

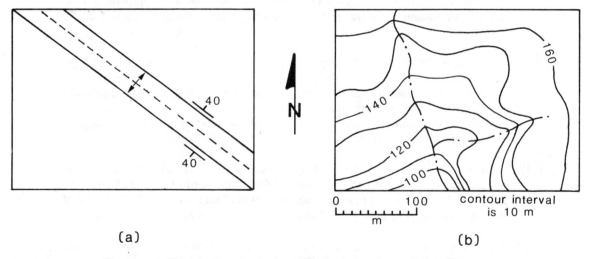

(a)   (b)

Figure 3-M4. (a) Hypothetical geologic map of the Sheep Hollow anticline in the Cresty Ridge area before river erosion (assume that the ground surface is horizontal). The two lines are the traces of a marker horizon on opposite limbs, and the dashed line is the trace of the fold hinge; (b) present-day topography of the Cresty Ridge area.

17. The base of the Plower Formation is exposed at the 1000-m contour interval on the slope of Jacob's Peak (Figure 3-M5). At this locality, and throughout the map area, the formation is 100 m thick and has a dip of 0°. The Duke's Ranch fault is oriented N40°W,90°NE and passes through BM 800 on the map. The net slip on the fault is 240 m along a vector oriented 60°,S40°E. The southwest side of the fault is down. Complete the map by showing the trace of the Duke's Ranch Fault and the outcrop belt of the Plower Formation.

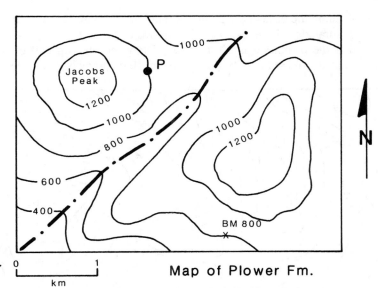

Figure 3-M5. Map of a portion of the Jacob's Peak area. The Plower Formation crops out at point P.

Map of Plower Fm.

18. A map of the portion of the Burnish Corners quadrangle is presented as Figure 3-M6. The trace of the Green Hollow fault is shown as well as mapped outcrops of the contact between the Freiburg Gneiss and the Baxter Schist. The Green Hollow fault offsets a basaltic sill.

(a) Assume that the Freiburg/Baxter contact is homoclinal throughout the map area. From the outcrop data given, calculate the attitude of the the contact.
(b) Complete the map of the Freiburg/Baxter contact. (Calculate the contact position.)
(c) What is the attitude of the Green Hollow fault? Measured slip lineations indicate that movement on the fault is parallel to the dip on the fault. Which side of the fault moved down? What is the approximate magnitude of net slip on the fault?

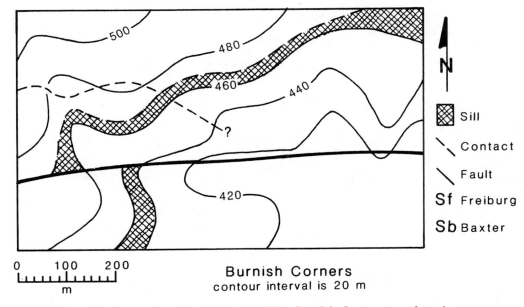

Burnish Corners
contour interval is 20 m

Figure 3-M6. Map of a portion of the Burnish Corners quadrangle. Part of the Freiburg/Baxter contact is shown.

19. Assume the base of the Judith Formation is exposed at point P in the Twin Peaks quadrangle (Figure 3-M7). The attitude of the formation is N90°E,35N. Assume that the formation is 30 m thick. Draw the outcrop belt of the formation. A rotated cross section plane graduated in 10 m intervals is provided.

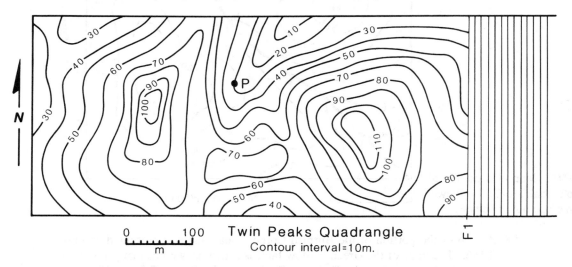

0 ___ 100   **Twin Peaks Quadrangle**
  m   Contour interval=10m.

Figure 3-M7. Map of a portion of the Twin Peaks Quadrangle. The base of the Judith Formation crops out at point P.

4

GEOMETRIC METHODS II: DIMENSION CALCULATIONS

4-1 INTRODUCTION

We noted earlier that it is often easier to use stereographic projections to solve attitude problems than it is to use geometric methods. Stereographic projections, however, cannot be used to represent dimensions (e.g., lengths and areas) of geologic features. Thus, when calculating thickness of a bed, depth to a horizon below the ground surface, length of a line, or area of a plane, you must use descriptive geometry or trigonometry. Problems requiring such calculations arise quite commonly during the course of both resource exploration and academic studies. For example, construction of stratigraphic columns depends on knowledge of unit thickness; the thickness of a unit can be measured directly at some localities, but commonly it is necessary to calculate stratigraphic thickness from indirect measurements. In this chapter we outline methods for solving problems in structural geology that require specification of a dimension. As before, remember to visualize the problems that are described before trying to solve them. For simplicity, we will refer to sedimentary beds in the problems described in this chapter, although the same methods can be used with reference to any type of rock layer.

4-2 DEPTH TO A PLANE

Depth in a Vertical Hole

If the attitude of a planar structure (e.g., a bedding surface or a fault) is known, the depth at which the structure will be reached below the ground surface at a given locality can be determined from two pieces of information. First, we must know the location of one point where the plane intersects the ground surface (e.g., the location of an outcrop) and second, we must know the attitude of the plane. The formulas and constructions used for solving *depth problems* depend on whether the ground surface is horizontal or not and on whether the *traverse line* (a line on the ground surface between two points) connecting the outcrop and the point at which the depth is to be determined is perpendicular to or oblique to the strike of the plane. Different situations will be handled individually below. Note that in our descriptions we use the ground surface and the position of an outcrop to provide a reference frame. For some of the situations we provide only a trigonometric solution. Problems 4-1 through 4-4 all describe variations of the same general theme.

Problem 4-1 (Ground surface horizontal; traverse line perpendicular to strike)

A sandstone bed crops out at point O (Fig. 4-1a). The attitude of the bed is N-S,30°W. A vertical hole is drilled at point J. At what depth will this hole intersect the bed? Assume that the ground surface is horizontal and that point J is 100 m due west of point O. The traverse line is OJ.

Method 4-1a (Descriptive geometry)
Step 1: Visualize the problem (Fig. 4-1a).
Step 2: Draw a map-view projection showing the relative positions of O and J at a convenient scale (top half of Fig. 4-1b). For convenience, locate the points at the south edge of the map area. Line OJ is the traverse line.
Step 3: Draw folding line F1 parallel to the traverse line and rotate the cross-sectional view into the plane of the

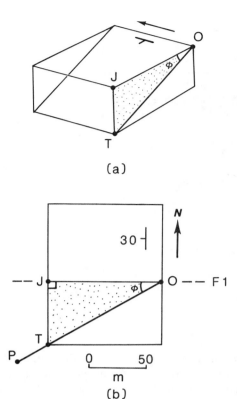

(a)

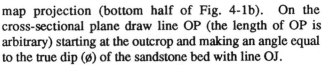

(b)

Figure 4-1. Depth calculation with traverse line perpendicular to strike and ground surface horizontal. (a) Block diagram; (b) map view and rotated cross-sectional view. ø = true dip, JT = true depth, OJ = traverse line.

map projection (bottom half of Fig. 4-1b). On the cross-sectional plane draw line OP (the length of OP is arbitrary) starting at the outcrop and making an angle equal to the true dip (ø) of the sandstone bed with line OJ.

Step 4: Draw line JT so that it is perpendicular to line OJ and intersects line OP at point T. The length of line JT, measured with your map scale, is the depth to the horizon. JT = 58 m.

Method 4-1b (Trigonometry)

In the cross-sectional plane (Fig. 4-1b) note that triangle OJT is a right triangle. Therefore, the length of line JT can be determined trigonometrically from the dip angle ø and the length of line OJ using Equation 4-1.

$$JT/OJ = \tan ø$$

$$\text{depth} = JT = 100 \tan(30º) = 58 \text{ m} \qquad \text{(Eq. 4-1).}$$

Problem 4-2 (Ground horizontal; traverse not perpendicular to strike)

A sandstone bed crops out at point O (Fig. 4-2a). The attitude of the bed is N-S,30ºW. A vertical hole is drilled at point K. At what depth will this hole intersect the bed? The ground surface is horizontal, and the traverse line KO is oriented N50ºW and is 139 m long.

Method 4-2a (Descriptive geometry)

Step 1: Visualize the problem (Fig. 4-2a).

Step 2: Draw a map-view representation of the problem (top half of Fig. 4-2b). Using one of the methods

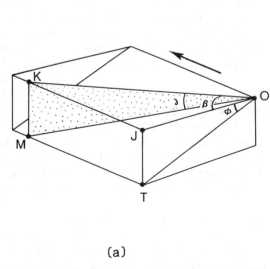

(a)

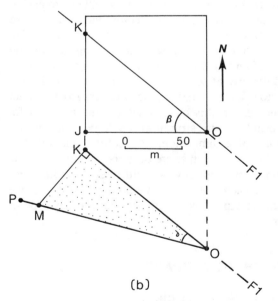

(b)

Figure 4-2. Depth calculation with traverse line oblique to strike and ground surface horizontal. (a) Block diagram; (b) map view and rotated cross-sectional view. ∂ = apparent dip, ø = true dip, KM = true depth, KO = traverse line.

described in Chapter 3, determine the apparent dip (∂) of the horizon in the traverse direction. $\partial = 24°$.

Step 3: Draw folding line F1 parallel to KO, and rotate the cross-sectional plane into the map projection plane (bottom half of Fig. 4-2b). Draw line OP so that it makes an angle of ∂ (= apparent dip) with respect to KO.

Step 4: Draw line KM perpendicular to KO so that it intersects OP at point M. The length of line KM is the depth to the horizon. KM = 61 m.

Method 4-2b (Trigonometry)

Two formulas can be applied to this problem (refer to Fig. 4-2a). First, the depth can be calculated in terms of the apparent dip by the equation

$$\text{depth} = \text{KM} = \text{KO}(\tan 24°) = 61 \text{ m} \qquad \text{(Eq. 4-2)}$$

where KO is the traverse length (= 138 m by the map scale), and ∂ is the apparent dip. Second, the depth can be calculated in terms of the true dip. Note that in this derivation, KM = JT. Note that the length of OJ in this problem is not the same as the length of OJ in Problem 4-1.

$$\text{JT/OJ} = \tan \phi$$

$$\text{JT} = \text{OJ} \tan \phi$$

$$\text{OJ/KO} = \cos \beta$$

$$\text{OJ} = \text{KO} \cos \beta$$

$$\text{depth} = \text{KM} = \text{JT} = (\text{KO} \cos \beta)(\tan \phi) = 61 \text{ m} \qquad \text{(Eq. 4-4)}$$

where KO is the traverse length, β is the angle between dip direction and the traverse direction (40°), and ϕ is the true dip (30°).

Problem 4-3a (Ground surface inclined in the direction opposite to dip; traverse perpendicular to strike)

A sandstone bed crops out at point O (Fig. 4-3a). The attitude of the bed is N-S,30°W. A vertical hole is drilled at point J, 104 m upslope in a direction due west of point O. At what depth will this hole intersect the bed? The ground surface slopes at an angle of $\Delta = 15°$ to the east. In this situation the total depth must incorporate both the distance from the ground surface to a horizontal plane as well as the distance from the horizontal plane to the bed.

Method 4-3a

If the ground slopes in the direction opposite to the dip of the bed, reference to Figure 4-3b yields the following formulas. (A scale drawing like Figure 4-3b could also provide the answer; if line ON = 100 m, then line OJ = 104 m, and line JT = 85 m):

$$\text{JN/OJ} = \sin \Delta$$

$$\text{JN} = \text{OJ}(\sin \Delta)$$

$$\text{NT/ON} = \tan \phi$$

$$\text{NT} = \text{ON}(\tan \phi)$$

$$\text{ON/OJ} = \cos \Delta$$

$$\text{ON} = \text{OJ}(\cos \Delta)$$

$$\text{depth} = \text{JT} = \text{JN} + \text{NT} = \text{OJ}(\sin \Delta) + \text{OJ}(\cos \Delta)(\tan \phi)$$

$$\text{depth} = \text{JT} = \text{OJ}[(\sin \Delta) + (\cos \Delta)(\tan \phi)] = 85 \text{ m} \quad \text{(Eq. 4-6)}$$

where Δ is the slope of the traverse line, OJ is the traverse line on the ground surface, ON is the projection of the

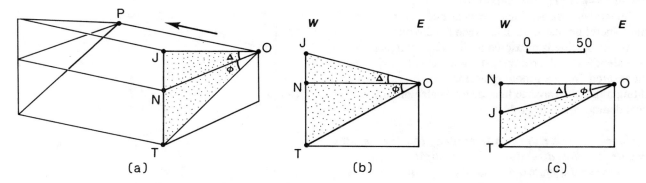

Figure 4-3. Depth calculation with traverse line perpendicular to slope. (a) Block diagram; (b) cross-sectional view of case in which slope is opposite to dip; (c) cross-sectional view of case in which slope is in the same direction as the dip but is shallower than the dip. Δ = slope of ground, ϕ = true dip, JT = true depth, JO = traverse line.

traverse line on a horizontal plane, and ø is the true dip of the sandstone bed.

Problem 4-3b (Ground surface inclined in the same direction as dip; traverse perpendicular to strike)

This problem is identical to Problem 4-3a, except that the ground surface (OJ) slopes in the same direction as the bed (i.e., OJ slopes 15° W, and the bed dips 30°W). Again, ON = 100 m, and OJ = 104 m, Δ = 15°, and ø = 30°.

Method 4-3b

Reference to Figure 4-3c yields the following formulas:

NJ = OJ sin Δ

NO = OJ cos Δ

NT = NO tan ø = OJ(cos Δ)(tan ø)

depth = JT = NT - NJ = [OJ(cos Δ)(tan ø)] - OJ sin Δ

depth = JT = OJ[(cos Δ)(tan ø) - (sin Δ)] = 31 m (Eq. 4-7).

Note that Equation 4-7 differs from Equation 4-6 only in the sign of one term.

Problem 4-4 (Sloping ground surface; traverse not perpendicular to strike)

A sandstone bed crops out at point O (Fig. 4-4). The attitude of the bed is N-S,30°W. A vertical hole is drilled at point K. At what depth will this hole intersect the bed? The bearing of the traverse line is N50°W, the length of the traverse line (KO) is 100 m, and the traverse line (KO) plunges at Δ = 15° toward the southeast.

Method 4-4 (Trigonometry)

Remember that bearings are measured in a horizontal plane. Therefore, the angle ß between the dip direction and the traverse bearing is measured in the horizontal plane, not on the slope. The slope angle, along traverse line KO, can be measured from a topographic map, or by a compass sighting along the traverse line, or by a calculation similar to that demonstrated in Problem 3-7.

Case A (The ground slopes to the east, opposite to the direction of bed dip):

Reference to Figure 4-4 yields the following formulas.

KP/KO = sin Δ

KP = KO sin Δ

OP/KO = cos Δ

OP = KO cos Δ

ON/OP = cos ß

ON = OP cos ß = (KO cos Δ)cos ß

NT/ON = tan ø;

NT = ON tan ø = [(KO cos Δ)cos ß]tan ø

depth = KM = KP + PM;

however, PM = NT, so:

depth = NT +KP = [(KO cos Δ)cos ß]tan ø + (KO sin Δ)

depth = KO[(cos Δ cos ß tan ø)+ sin Δ] = 62 m   (Eq. 4-8).

Case B (The ground surface slopes to the west, in the same direction as bed dip):

The sign of (sin Δ) used in Equation 4-8 is opposite, so the formula to be used is

depth = KO[(cos Δ cos ß tan ø) - sin Δ] = 10 m   (Eq. 4-9).

In Equations 4-8 and 4-9, KO is the traverse length, Δ is the slope of the traverse line, ß is the angle between dip direction of the bed and the bearing of the traverse line, and ø is the true dip of the bed.

Depth in Inclined Drill Holes

The geometry of depth problems becomes more complicated if the distance from the ground surface to the structural plane (e.g., bedding surface) of interest is measured in an *inclined drill hole* (i.e., a hole that is not vertical). We consider two cases: first, the case in which

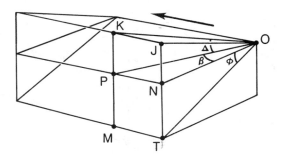

Figure 4-4. Block diagram for depth calculation with traverse line oblique to strike, and ground surface sloping in the direction opposite to the dip direction of the layer. Δ = plunge of the traverse line, ø = true dip, ß = angle between true dip direction and traverse line, KM = true depth.

the plunge of the hole is in the same direction as the apparent dip of the bed, and second, the case in which the plunge of the hole is opposite to the apparent dip of the bed. In the two problems described below, the ground surface is assumed to be horizontal. The "apparent dip" refers to the apparent dip of the bed in the direction parallel to the bearing of the drill hole.

Problem 4-5 (Inclined hole; bearing of the hole is in the same direction as the apparent dip of the plane)

A sandstone bed intersects the ground surface at point N and has an attitude of N-S,30°W (Fig. 4-5a). A hole is drilled at point O (it will intersect the bed at point J). Point O is 105 m in the direction N42°W from point N (i.e., line ON, which is not shown, is 105 m long). The hole attitude is 50°,N60°W. How far must the hole be drilled before it intersects the sandstone bed?

Method 4-5a (Descriptive geometry)

One method to solve this problem combines descriptive geometry with trigonometry (following Ragan, 1985).

Step 1: Calculate the apparent dip of the bed (∂) in the direction parallel to the bearing of the hole (using the methods of Chapter 3); $\partial = 27°$.

Step 2: Construct a map view showing the position of points O and N (middle of Fig. 4-5b). Let N be at the corner of the map area. Line NV defines the intersection of the bed with the ground surface (the trace of the bed). Extend a line parallel to the bearing of the drill hole in both directions from point O. This line intersects the outcrop belt at point Q. Line OQ makes an angle of $\alpha = 60°$ with respect to line NV.

Step 3: Draw folding line F1 parallel to OQ, and rotate the cross-sectional plane into the map projection (bottom of Fig. 4-5b). In the rotated cross-sectional plane draw line QS to represent the trace of the bed. QS makes an angle of ∂ (= apparent dip) with respect to the ground surface. Draw line OT so that it makes an angle of Ω (= plunge of drill hole) with respect to the ground surface.

Step 4: Lines OT and QS intersect at point J in the subsurface. Point J is, therefore, the point at which the drill hole intersects the bed. The projection of point J onto the ground surface is point J'. Therefore, line JJ' (at the map scale) is the depth below the ground surface, measured along a vertical line, at which the drill hole intersects the bed. JJ' = 66 m. Line OJ is the required length of the drill hole (90 m).

Method 4-5b (Trigonometry)

Reference to Figure 4-5b yields the following formulas. Note that in order to solve the problem, it is necessary to draw folding line F2 (parallel to the dip direction) through point O (top of Fig. 4-5b). Rotate the

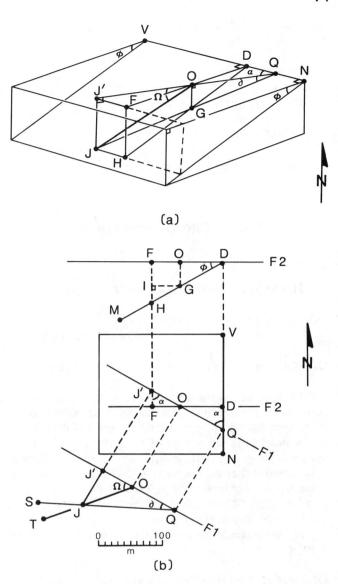

(a)

(b)

Figure 4-5. Depth calculation from data in an inclined hole. The bearing of the hole is in the same direction as the apparent dip direction. (a) Block diagram; (b) map view and rotated cross-sectional views. Ω = plunge of the hole, ∂ = apparent dip of the bed in the direction of the hole, α = angle between the bearing of the hole and the strike of the bed, $\varnothing$ = true dip, JJ' = true depth, OJ = the length of the drill hole.

cross-sectional plane around F2 into the map projection. In this rotated plane the trace of the bed intersects the ground surface at D. Draw line DM so that it makes an angle of $\varnothing$ (true dip) with respect to folding line F2. In the map view, point J' projects along strike onto F2 at point F. Plot point F on the rotated cross section. Point H lies on the trace of bedding vertically below F, and point G lies on the trace of bedding vertically below O. Point I lies vertically below F at an elevation equal to that of G. Note that JJ' = FH.

$$OJ'/OJ = \cos \Omega$$

$$OJ' = OJ(\cos \Omega)$$

$$OF/OJ' = \sin \alpha$$

$$OF = OJ'(\sin \alpha) = OJ(\cos \Omega)(\sin \alpha)$$

$$IH/OF = \tan \phi;$$

$$IH = OF(\tan \phi) = OJ(\cos \Omega)(\sin \alpha)(\tan \phi)$$

$$JJ'/OJ' = \tan \Omega$$

$$JJ' = OJ'(\tan \Omega) = OJ(\cos \Omega)(\tan \Omega)$$

$$OG = JJ' - IH = [OJ(\cos \Omega)(\tan \Omega)]$$
$$- [OJ(\cos \Omega)(\sin \alpha)(\tan \phi)]$$

$$OJ = OG/(\cos \Omega)[(\tan \Omega) - (\sin \alpha)(\tan \phi)] \qquad \text{(Eq. 4-10)}.$$

In these equations ϕ is the true dip, α is the angle between the bearing of the hole and the strike of the bed, ∂ is the apparent dip of the bed in the direction of the bearing of the hole, Ω is the plunge of the hole, and OG is the vertical distance between the origin of the hole and the bed. It is assumed that the length of OG is easily determined using some other method (e.g., Method 1a or 1b); in this problem OG = 40 m. Note that the apparent dip ∂ is not needed in the trigonometric formula.

If we substitute the appropriate values into Equation 4-10, we find

$$OJ = 40/(\cos 50°)[(\tan 50°) - (\sin 60°)(\tan 30°)] = 90 \text{ m}.$$

Problem 4-6 (Inclined hole; bearing of the hole opposite to the apparent dip direction of the bed)

A sandstone bed intersects the ground surface at point N and has an attitude of N-S,30°W (Fig. 4-6a). A hole is drilled at point C (it will intersect the bed at point E). Point C is 215 m in the direction N53°W from point N. The hole attitude is 30°,S60°E. How far must the hole be drilled before it intersects the sandstone bed?

Method 4-6a (Descriptive geometry)

This method is almost identical to Method 5a, so we will outline the steps in abbreviated form.

Step 1: Construct a map view with the hole starting at point C (middle of Fig. 4-6b). The hole plunges toward point O. Folding line F1 is parallel to the true dip

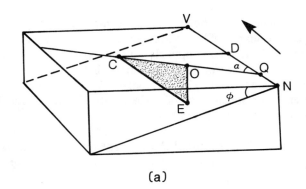

(a)

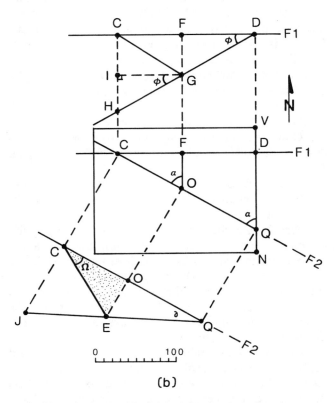

(b)

Figure 4-6. Depth calculation from data in an inclined hole. (a) Block diagram; (b) descriptive-geometry solution. The bearing of the hole opposite to the apparent dip direction. ϕ = true dip, α = angle between strike and bearing of the hole, CE = length of hole, OE = true depth, stippled plane is vertical.

direction of the bed, and folding line F2 is parallel to the bearing of the hole. The extension of line CO intersects line NV (the trace of the bed) at point Q. CQ makes an angle of α with respect to NV.

Step 2: Rotate a cross-sectional plane around F1 into the plane of the map projection (top of Fig. 4-6b; F1 has been moved so that the figure is not cluttered). In this cross-sectional plane draw the trace of the bed (DH) and the trace of the hole (CG). Note that CG does not represent the true length of the hole. Angle ø is the true dip of the bed in this cross-sectional plane.

Step 3: Now rotate a cross-sectional plane around F2 into the plane of the map projection (bottom of Fig. 4-6b). Draw line CJ down from point C; CJ is perpendicular to CQ and is the same length as CH. Now you can draw the trace of the bed (QJ) in this cross-sectional plane. QJ makes an angle ∂ with respect to QC; ∂ is the apparent dip of the bed in the direction parallel to the plunge direction of the hole.

Step 4: The hole intersects the trace of the bed (QJ) at point E. The projection of point E on the ground surface is at point O. The length of OE is the vertical distance from the ground to the intersection at the map scale (52m). The length of CE gives the distance between the ground and the bed in the hole at the map scale (104 m)

Method 4-6b (Trigonometry)

Reference to Figure 4-6b yields the following formulas. Note that CF = IG.

$$OC/CE = \cos \Omega$$

$$OC = CE(\cos \Omega)$$

$$CF/OC = \sin \alpha$$

$$CF = OC(\sin \alpha) = CE(\cos \Omega)(\sin \alpha)$$

$$IH/CF = \tan \phi;$$

$$IH = CF \tan \phi = CE(\cos \Omega)(\sin \alpha)(\tan \phi)$$

$$OE/OC = \tan \Omega$$

$$OE = OC(\tan \Omega) = CE(\cos \Omega)(\tan \Omega)$$

$$CJ = CH = CI + IH$$

$$OE = FG = CI$$

$$CJ = OE + IH = CE(\cos \Omega)(\tan \Omega)$$
$$+ CE(\cos \Omega)(\sin \alpha)(\tan \phi)$$

$$CE = CJ/(\cos \Omega)[(\tan \Omega) + (\sin \alpha)(\tan \phi)] \quad \text{(Eq. 4-11).}$$

In these equations CE is the length of the drill hole, and CJ is the vertical distance between the ground surface and the bed below point C. (The value of CJ must be determined by Method 4-1, once the length of traverse line CD has been measured; CJ = 97 m.) Ω is the plunge of the hole, α is the angle between the bearing of the hole and the strike of the bed, ø is the true dip of the bed. The angle

∂, the apparent dip of the bed in the direction parallel to the bearing of the hole, is not needed.

If we substitute the appropriate values into Equation 4-11, we find that

$$CE = 97/(\cos 30^\circ)[\tan 30^\circ + (\sin 60^\circ)(\tan 30^\circ)] = 104 \text{ m}.$$

4-3   CALCULATION OF LAYER THICKNESS

The most straightforward way of calculating layer thickness is by direct measurement. Direct measurement is, of course, only possible at locations where there is either a complete exposure of the unit on a plane that is perpendicular to the bedding of the unit, or a drill hole that is oriented perpendicular to the bedding of the unit (Fig. 4-7). If the bedding is inclined, the outcrop face or drill hole must also be inclined. Localities at which direct measurement is possible are relatively rare. Usually, the angle between the outcrop face or the drill hole and the bedding of the unit is not 90°. In such circumstances, the *true thickness*, which is measured perpendicular to the bedding, must be calculated. Thickness measurements can be obtained at the outrcrop either with a *Jacob's staff*, or with a tape and compass. Thickness can also be calculated from the outcrop pattern on a map or from drill data.

Thickness Determination Using a Jacob's Staff

Problem 4-7

A unit whose orientation is north-south is exposed on a slope that dips to the east. A profile of the unit, drawn perpendicular to strike, is illustrated in Figure 4-7c. Measure the unit thickness with a Jacob's staff.

Method 4-7

The base of the staff is placed on the outcrop at the base of the unit (point A) and the staff is inclined from perpendicular by an amount equal to the true dip of the strata (ø). The geologist visually sights in a direction perpendicular to the strike direction across the top of the staff to a point on the outcrop (point B). Point B is stratigraphically above point A by an amount equal to the height of the staff. The base of the staff is then moved up to point B, and the procedure is repeated until the top of the unit of interest is reached. The total thickness of the unit is the sum of the increments.

If it proves impossible to follow a traverse that is itself perpendicular to strike, the geologist can sidestep along the ground at a specific stratigraphic level until an appropriate place to proceed upsection can be found (Fig. 4-7d).

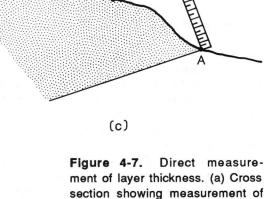

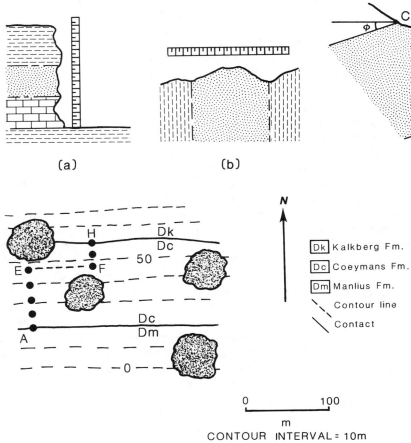

(a)   (b)   (c)

(d)

Figure 4-7. Direct measurement of layer thickness. (a) Cross section showing measurement of horizontal beds on a vertical scarp; (b) cross section showing measurement of vertical beds on a horizontal pediment; (c) cross section showing measurement with a Jacob's staff; (d) map showing offset of a traverse line. Dots represent measurement stations. Geologist sidesteps the hill from point E to point F.

Thickness Determination from Tape-and-Compass or Map Data

The thickness of a unit can be determined if the positions of the top and bottom of the unit are known, and the attitudes of the unit and of the traverse line are known. The distance between the top and bottom, measured with a tape (in the field) or with a scale (on a map) along the traverse line, is an *apparent thickness*. The attitude of the unit and of the traverse line is determined with a compass. The graphic and trigonometric solutions used for converting tape-and-compass or map-and-compass data into true layer thickness depend on the orientation of the traverse line with respect to strike and on whether the ground surface is horizontal or planar. Several situations are described separately below.

Problem 4-8 (Horizontal ground surface; traverse line is perpendicular to strike)

The base of a distinctive sandstone bed is exposed at point O and its top at point T in an area of no relief (Fig. 4-8). The bed attitude is N-S,30°W. A geologist uses a tape to measure the distance on the ground surface on a traverse line between O and T. The attitude of the traverse line between O and T is 00°,270°. What is the true thickness of the sandstone bed?

Method 4-8

Reference to Figure 4-8 yields the formulas

$$TB/OT = \sin \phi$$

$$TB = OT \sin \phi \qquad \qquad \text{(Eq. 4-12)}$$

where TB is the true thickness, OT is the apparent thickness measured along the traverse, and ϕ is the true dip.

Problem 4-9 (Horizontal ground surface; traverse line is oblique to strike)

A sandstone bed, whose attitude is N-S,30°W, is exposed in a region of no relief. A geologist measures its apparent thickness along traverse line KO, which trends N45°W (Fig. 4-9). What is the true thickness of the bed?

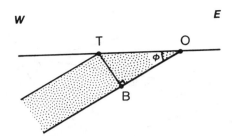

Figure 4-8. Thickness measurement with traverse line perpendicular to strike and ground surface horizontal. Cross-sectional view is shown. ø = true dip, TB = true thickness, OT = traverse length.

Method 4-9

Reference to Figure 4-9 yields the following formulas:

$$OT/KO = \sin \alpha$$

$$OT = KO \sin \alpha$$

$$TB = OT \sin \phi$$

$$TB = KO \sin \alpha \ \sin \phi \qquad \text{(Eq. 4-13)}$$

where TB is the true thickness, KO is the length of the traverse line, OT is the projection of the traverse line onto a line drawn perpendicular to the strike, ø is the true dip of the bed, and α is the angle between the traverse line and the strike of the bed.

Problem 4-10 (Traverse line slopes; traverse line is perpendicular to strike)

A sandstone bed is exposed on a slope (Fig. 4-10). A geologist measures the apparent thickness (OT) of the bed on a traverse that trends perpendicular to the strike of the bed. What is the true thickness (TB) of the bed?

Method 4-10

There are three variations of this problem, depending on whether the bed dip is greater or less than traverse-line slope, and on whether the dip direction is the same as or opposite to traverse-line slope. We will provide only trigonometric solutions. The equations refer to angles and lines shown in Figure 4-10. In these figures ø is the true dip, and Δ is the slope of the traverse line. Note that it does not matter whether the strike of the ground slope is parallel to the strike of the bed, as long as the bearing and slope of the traverse line are known.

Case A (Traverse line slopes in dip direction; plunge < dip):
The layer attitude is N-S,60°E. The base of the bed is exposed at point O and

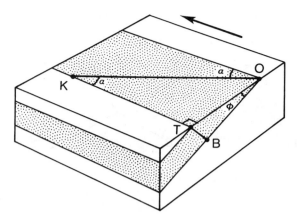

Figure 4-9. Thickness measurement with traverse line oblique to strike and ground surface horizontal. Block diagram is shown. KO = traverse length, α = angle between traverse bearing and the strike of the bed, ø = true dip, TB = true thickness.

the top of the bed at point T. The traverse line is oriented 30°,090°. Reference to Figure 4-10a yields the following formulas:

$$TB/OT = \sin(\phi - \Delta)$$

$$TB = OT[\sin(\phi - \Delta)] \qquad \text{(Eq. 4-14)}.$$

Case B (Traverse line slopes in the dip direction; plunge > dip):
The layer attitude is N-S,10°E. The base of the bed is exposed at point O and the top of the bed at point T. The traverse line is oriented 30°,090°. Reference to Figure 4-10b yields the following formulas:

$$TB/OT = \sin(\Delta - \phi)$$

$$TB = OT[\sin(\Delta - \phi)] \qquad \text{(Eq. 4-15)}.$$

Case C (Traverse-line slope is opposite to dip direction):
The layer attitude is N-S,30°W. A traverse line (OT) running from the base to the top of the bed is oriented 50°,090°. Note that O is at the base of the bed in this example. Reference to Figure 4-10c yields the following formulas:

$$TB/OT = \sin(\phi + \Delta)$$

$$TB = OT[\sin(\phi + \Delta)] \qquad \text{(Eq. 4-16)}.$$

Problem 4-11 (Ground surface is sloping; traverse is not perpendicular to strike)

A sandstone bed, whose attitude is N-S,30°E, is exposed on the face of a hill that slopes toward the west

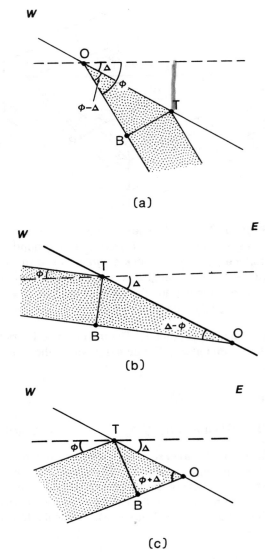

(a)

(b)

(c)

Figure 4-10. Thickness measurement on a slope with traverse line perpendicular to strike. Cross-sectional views are shown. (a) Dip is in the same direction as traverse-line slope, and dip is greater than slope; (b) dip is in the same direction as traverse-line slope, and dip is less than slope; (c) dip is in the direction opposite to traverse-line slope. Δ = slope, ∅ = true dip, OT = traverse length, TB = true thickness.

(Fig. 4-11). A traverse line (OT) running from the base to the top of the bed is oriented 50°,315° and is therefore inclined to the strike of the bed. What is the true thickness of the bed?

Method 4-11

Note that angle α (between the strike and the bearing of the traverse line) must be measured in a horizontal

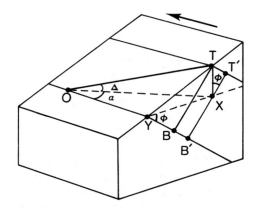

Figure 4-11. Thickness measurement on a slope with traverse line that is oblique to strike and dips in a direction opposite to the traverse-line bearing. Block diagram is shown. Δ = plunge of traverse line, α = angle between bearing of traverse line and strike, ∅ = true dip, TB = true thickness, OT = traverse length.

plane. In order to solve the problem, we must create line T'B', which is equal to the true thickness (TB) but does not intersect the ground surface. Reference to Figure 4-11 yields the following formulas (after Mertie, 1922 and Ragan, 1985):

$$TX/OT = \sin \Delta$$

$$TX = OT \sin \Delta$$

$$OX/OT = \cos \Delta$$

$$OX = OT \cos \Delta$$

$$YX/OX = \sin \alpha$$

$$YX = OX \sin \alpha = OT(\cos \Delta)(\sin \alpha)$$

$$XB'/YX = \sin \varnothing$$

$$XB' = YX \sin \varnothing = OT(\cos \Delta)(\sin \alpha)(\sin \varnothing)$$

$$T'X/TX = \cos \varnothing$$

$$T'X = TX \cos \varnothing = OT(\sin \Delta)(\cos \varnothing)$$

$$TB = T'B' = T'X + XB' = OT[(\cos \Delta)(\sin \alpha)(\sin \varnothing) + (\sin \Delta)(\cos \varnothing)]$$
(Eq. 4-17)

where TB is the true thickness, OT is the traverse length, Δ is the plunge of traverse, α is the angle between traverse bearing and strike, and ∅ = true dip of the bed. Note that if the bed dips in the same direction as the slope, the sign in Equation 4-17 becomes negative.

Thickness Determination from Drill Data

Modern down-hole logs (e.g., gamma-ray and electric logs) make it possible to recognize strata in a drill hole without requiring expensive core recovery. If strata are horizontal and the drill hole is vertical, the distance measured in the hole between the top and bottom of a unit is the true thickness of the unit. Below we discuss two additional situations: first, the case where a vertical hole intersects inclined bedding (which is identical to the case where an inclined hole intersects horizontal bedding) and second, the case where an inclined hole intersects inclined bedding.

Problem 4-12 (Thickness in a vertical hole cutting inclined bedding)

From field evidence it is known that the bedding beneath well C-6 is oriented N-S,30°E (Fig. 4-12). Well C-6 is a vertical hole that intersects the top of a distinctive sandstone bed at a depth of 100 m and the base of the bed at a depth of 220 m below ground surface. What is the true thickness of the bed?

Method 4-12

From a cross-sectional view drawn perpendicular to the strike of the layer (Fig. 4-12), we obtain the following formulas:

$$TB/OT = \cos \phi$$

$$TB = OT(\cos \phi) \qquad \text{(Eq. 4-18)}$$

where TB is the true thickness, OT is the thickness as measured in the drill hole, and ϕ is the true dip of the layer.

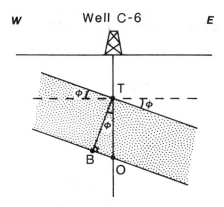

Figure 4-12. Thickness measurement of an inclined bed in a vertical hole. Cross-sectional view is shown. OT = thickness as measured in hole, TB = true thickness, ϕ = true dip of the bed.

Note that knowledge of the strike of the bed is not actually needed for the calculation, as long as the true dip angle is known.

Problem 4-13 (Thickness in an inclined hole cutting inclined bedding)

A bed of sandstone is oriented N-S,40°W. A hole is drilled on horizontal ground. The hole is oriented 60°,S30°W. The hole penetrates the bed at point M in the subsurface and passes through the bed entirely in the subsurface (Fig. 4-13). The thickness of the bed as measured in the hole (line ML) is 100 m. What is the true thickness of the bed?

Method 4-13

Figure 4-13 illustrates this problem. The top surface of the block shown in Figure 4-13 is a horizontal plane in the subsurface that intersects the top of the bed along line MP. The dashed line (ML) represents the segment of the drill hole that passes through the stippled bed. Line ML lies entirely within the stippled bed, though this could not be easily represented on the figure. As indicated in Figure 4-13,

$$TB/LT = \sin(90° - \phi)$$

$$TB = LT \sin(90° - \phi) \qquad \text{(Eq. 4-19)}$$

where TB is the true thickness, LT is the thickness in a vertical hole, and ϕ is the true dip of the bed. In the

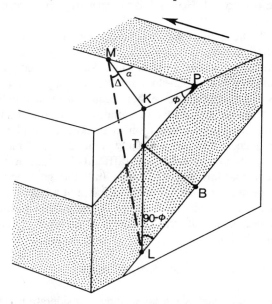

Figure 4-13. Thickness calculation of an inclined bed in an inclined hole. Block diagram is shown. ML = thickness measured in hole, TB = true thickness, ϕ = ture dip, Δ = plunge of the hole, α = angle between bearing of the hole and the strike of the bed.

problem, however, the value of LT is not known. It is calculated as follows:

$$MK = ML \cos \Delta$$

$$MP = MK \cos \alpha = ML \cos \Delta \cos \alpha$$

$$KP = MP \tan \alpha = ML \cos \Delta \cos \alpha \tan \alpha$$

$$KT = KP \tan \phi = ML \cos \Delta \cos \alpha \tan \alpha \tan \phi$$

$$KL = ML \sin \Delta$$

$$LT = KL - KT = (ML \sin \Delta)$$
$$- (ML \cos \Delta \cos \alpha \tan \alpha \tan \phi)$$
$$(Eq. 4\text{-}20).$$

Substitution of Equation 4-20 into Equation 4-19 yields

$$TB = [ML(\sin \Delta - \cos \Delta \cos \alpha \tan \alpha \tan \phi)][\sin (90° - \phi)]$$
$$(Eq. 4\text{-}21).$$

Note that Equation 4-20 involves only the true thickness (TB), the thickness measured in the drill hole (ML), the plunge of the drill hole (Δ), the angle between the bearing of the drill hole and the strike of the bed (α), and the true dip of the bed (ϕ), all of which are numbers that were provided in the problem.

Thickness of Folded or Nonuniform Layers

In all the situations described above, we assumed parallelism between the top and bottom of the layer whose thickness we wished to determine. Such an assumption is not always reasonable. In many localities a particular unit thins in a given direction either because of truncation by an unconformity or because of variation in sediment supply during deposition. Variation in thickness can also be a consequence of ductile stretching or shortening. Thickness measurements on folds are difficult, because it is not always clear how to specify dip. In the hinge zone of the fold the measured dip on the top surface of a layer may not be the same as the measured dip on the bottom surface of the layer, even if the true thickness of the layer is constant around the fold.

Nonparallel Layer Boundaries

If the boundaries of the layer under consideration are not parallel to one another (Fig. 4-14), then it is not possible to measure a true thickness that is perpendicular to both the upper and lower boundaries. An estimate of the thickness is possible by averaging the strike and dip of the upper and lower boundaries and using this as the layer attitude in the calculations described above:

$$\text{dip}_{\text{average}} = [\text{dip}_1 + \text{dip}_2]/2 \qquad (Eq. 4\text{-}22)$$

$$\text{strike}_{\text{average}} = [\text{strike}_1 + \text{strike}_2]/2 \qquad (Eq. 4\text{-}23).$$

The area over which this averaging is done depends on the degree to which the layer boundaries deviate from parallelism and thereby *converge*; the greater the deviation, the smaller the area for which an average value can be assumed.

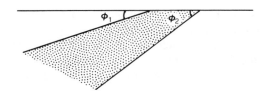

Figure 4-14. Cross section showing convergence of two layer boundaries. The dips of the two contacts are not the same. ϕ_1 = dip of top surface, ϕ_2 = dip of the bottom surface.

Thickness of a Folded Layer

Folded layers present a special problem because the layer boundaries are curved. First, it is necessary to decide what is meant by "thickness." Ideally, we would like to know the thickness of the layer before folding. For parallel folds, the layer thickness does not change, but for other types of folds (see Chapter 13) layers do change. Three measures of thickness are commonly used in reference to folded layers (Fig. 4-15). The first is *vertical thickness*, which is the thickness of a layer as it would occur in a vertical drill hole (such a measure will usually deviate greatly from the original thickness). The second is *orthogonal thickness*, which is the thickness along a line drawn perpendicular to both the upper and lower boundaries. The third is *isogonal thickness*, which is the thickness measured along a line connecting points on the boundaries that have equal dips.

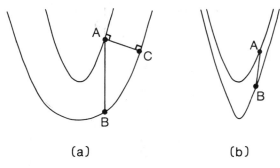

(a)   (b)

Figure 4-15. Thickness measurement in a fold profile. (a) Vertical thickness is line AB, and orthogonal thickness is line AC; (b) isogonal thickness is line AB. Dip at point A is the same as the dip at point B.

The second and third methods refer to thickness measured in a profile plane drawn perpendicular to the axis of the fold.

Because of the complications apparent from the preceding discussion, the thickness of a folded layer must generally be considered to be an estimate. Geometric and trigonometric solutions are possible for determining the thickness of folded layers (see Hewett, 1920; Mertie, 1940; Ragan, 1985), but generally an adequate approach is to draw an accurate down-plunge projection with no vertical exaggeration (see Chapter 13). Any desired measure of thickness can be derived by directly measuring such a drawing.

Nomograms for Calculating Layer Thickness

As was the case for solution of true and apparent dip problems, nomograms have been developed for solving certain types of thickness problems. The nomogram presented in Figure 4-16 permits calculation of thickness if the traverse is perpendicular to strike. *Note:* (1) If the ground surface is sloped, and the slope and dip directions are opposite to one another, then the dip angle to be used in column 1 must be the sum of the dip angle and the slope. (2) If the ground surface is sloped and the slope and dip directions are not opposite to one another, the dip angle to be used in column 1 must be the difference between the slope and dip.

4-4 DETERMINATION OF LINE LENGTH

It is often necessary to determine the distance in the subsurface between two features as measured along a line of known orientation. For example, imagine that the top of a mineralized horizon occurs at a known depth below the ground surface, but that for technical reasons it must be reached in an inclined mine shaft. The length of the shaft must be known in order to calculate its cost. Such problems can readily be solved with the geometric methods that you have learned already.

Problem 4-14

A fold hinge is exposed at point A on the ground surface and is intersected at point B in a mine tunnel at a depth of 400 m below the ground surface (Fig. 4-17). From existing maps of the mine tunnels it is known that

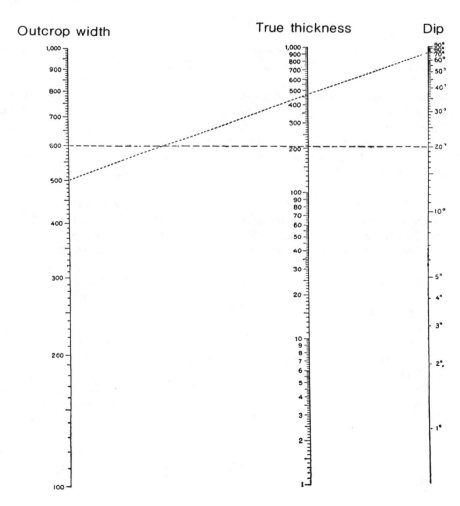

Outcrop width   True thickness   Dip

Figure 4-16. Nomograms for thickness calculations. (Adapted from Palmer, 1919.)

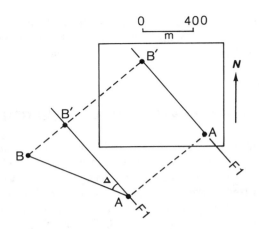

0   400
m

Figure 4-17. Map view and rotated cross-sectional view showing how to measure the length of a line. Δ = plunge of the line, AB = true length of the line.

point B lies directly below point B' (point B' is on the ground surface at a distance of 800 m to the N40°W of point A). How long is the linear geologic structure between points A and B?

Method 4-14a (Descriptive geometry)

Step 1: Draw a map to scale showing the positions of points A and B' (Fig. 4-17). Draw folding line F1 along AB'. Note that AB' is parallel to the bearing of the linear feature.

Step 2: Rotate the cross-sectional view around F1 into the plane of the map projection, and locate point B in the cross-sectional plane. Line BB' is 400 m long at the map scale. Line AB is a representation of the linear feature, and its length can be measured directly from the map scale. AB is 898 m long.

Method 4-14b (Trigonometry)

Step 1: In order to solve this problem, we must first determine the plunge (Δ) of the linear feature. Reference to Figure 4-17 yields the following formulas:

$$BB'/AB' = \tan \Delta$$

$$\arctan(B'B/B'A) = \Delta \qquad \text{(Eq. 4-24).}$$

Step 2: Now that the plunge is known, it is possible to determine the length of AB, using the following formulas:

$$\cos \Delta = AB'/AB$$

$$AB = AB'/(\cos \Delta) \qquad \text{(Eq. 4-25).}$$

Step 3: Applying Equations 4-24 and 4-25 to the data in this problem yields:

$$\arctan(400/800) = \Delta = 27°$$

$$AB = 800/\cos 27° = 898 \text{ m}$$

4-5   AREA OF A DIPPING PLANE

Problems involving calculation of the area of a plane arise in circumstances where, ultimately, the volume of a layer (e.g., a seam of coal) must be determined.

Problem 4-15

A bed is oriented N20°E,40° NW. The bed crops out in a region of no relief at point A (Fig. 4-18). At a distance of 200 m to the N45° E of point A there is an exposure of a vertical fault trending N60°W. At a distance of 200 m to the S45°W of point A there is an exposure of a vertical fault trending N90°W. What is the area of the bed between the two faults above a depth of 200 m below the ground surface?

Method 4-15

In order to solve this problem, you must first obtain a projection of the plane of the bed.

Step 1: Draw a map to scale showing the position of the points identified above and the traces of the two faults and of the bed on the ground surface (right half of Fig. 4-18). Let line XY represent the trace of the bed and lines YN and XM represent the traces of the two faults. The angle ß (= 10°) is the angle between YN and the perpendicular to strike, and angle ∂ (= 20°) is the angle between XM and the perpendicular to strike.

Step 2: Locate the map projection of the bed in the lower reference plane (at a depth of 200 m). To do this, draw folding line F1 perpendicular to the bed strike through point P. (In Figure 4-18 we have extended the bedding trace XY out to point P so that the figure will not be cluttered.) Rotate the cross-sectional view into the plane of the map projection and draw the cross-sectional trace of the bed to a depth of 200 m; the cross-sectional trace of the bed is line PK. Angle ø is the true dip of the bed. The projection of K onto the ground surface is K'; line KK' is 200 m long.

Step 3: To determine the area of the plane you must now rotate the bed itself into the map plane. To do this, let PK be folding line F2. Rotate the bed around F2 into the plane of the map projection (left half of Fig. 4-18). In this representation PR is the intersection of the ground surface with the bed, and SK is the intersection of the bed with the lower reference plane. Mark off a segment of line PR that is equal in length to XY; this segment of PR is labeled X'Y'.

Step 4: Draw lines Y'N' and X'M' so that they make the same angles with respect to PR as YN and XM do to PY, respectively. Y'N' intersects SK at Z', and X'M'

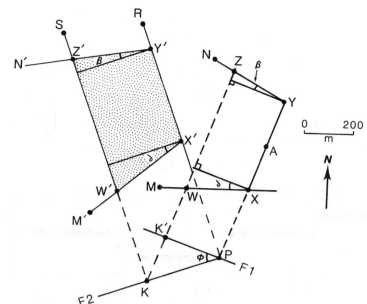

Figure 4-18. Map view, rotated cross-sectional view, and layer-normal view showing how to measure the area of a plane. XA is the trace of the bed, YN and XM are the traces of faults, the stippled area is the the area of the bed.

intersects SK at W'. The area Z'Y'X'W' (stippled) is the area of the plane in question. To measure this area quickly, divide it into two triangles and one square, as shown in the figure. Note that the projection of this plane onto the horizontal ground surface is ZYXW. At the scale of the figure, the area of plane Z'Y'X'W' is about 225,000 m^2.

4-6 DESCRIPTIVE-GEOMETRY ANALYSIS OF FAULT OFFSET

The *net slip* on a fault is the postmovement distance, measured in the plane of the fault, between two points that were adjacent prior to faulting but are now on opposite sides of the fault (Fig. 4-19a). The *dip-slip component* of movment is measured in the direction parallel to the dip of the fault, and the *strike-slip* component is measured in the direction parallel to the strike of the fault. A fault on which displacement is primarily dip-slip is a *dip-slip fault,* a fault on which the displacement is primarily strike-slip is a *strike-slip fault,* and a fault on which the displacement has both dip-slip and strike-slip components is called an *oblique-slip fault.*

If the fault plane is not vertical, the term *hanging-wall block* can be used to refer to the rock above the fault plane and the term *footwall block* can be used to refer to the rock below the fault plane. If the hanging wall moves up dip with respect to the footwall, then the fault is a reverse fault or thrust fault. If the hanging wall moves down dip with respect to the footwall, then the fault is a normal fault. Note that in these definitions, the ground surface is used as a reference frame. Recently, the term *contractional fault* has been used to refer to faults whose movement has resulted in shortening of the crust, and *extensional fault* has

been used to refer to faults that result in stretching or lengthening of the crust (Fig. 4-19b).

The term *separation*, when used in the context of describing movement on a fault, refers to the distance between displaced parts of a marker as measured in a specified direction (Dennis, 1987). *Strike separation*, for example, is the distance between the two displaced ends of a marker (e.g., an offset dike), as measured along the strike of the fault at a specified elevation (Fig. 4-20a).

The apparent displacement across a fault that is indicated by a map pattern is usually the strike separation. If the offset marker is vertical, the strike separation is the strike-slip component of displacement. It is not possible to determine strike-slip and dip-slip components of displacement from the separation of only one marker, if the dip of the offset marker is less than 90°. For example, imagine a normal fault that offsets a dipping bed. After erosion has removed the fault scarp, a map of the fault displays strike separation (Fig. 4-20b).

Measurement of the strike separation of two nonparallel markers that have been offset along a fault does, however, permit the net slip on a fault to be calculated. In this section we briefly outline a descriptive geometry procedure that can be used to calculate the true offset on a vertical fault plane. A more efficient method of solving such fault problems is presented in Chapter 6. In Chapter 6 we also introduce a technique for determining net slip on an inclined fault.

Problem 4-16

Imagine that a vertical fault occurs in a region of no relief. The fault strikes N70°W. The fault cuts a dike oriented N20°W,40°NE and a contact oriented N30°E,70°NW. The intersections of these structures with

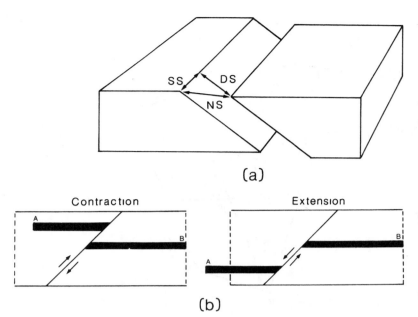

(a)

Contraction   Extension

(b)

Figure 4-19. (a) Terminology used to describe displacement on a fault. NS = net slip; DS = dip slip component; and SS = strike slip strike slip component; (b) cross sections illustrating the contrast between contraction faults and extension faults. Note that the distance between the end points A and B on the black marker layer as measured in a projection plane parallel to the marker layer decreases as a consequence of contraction faulting and increases as a consequence of extension faulting.

the fault are shown in Figure 4-21a. Prior to movement on the fault, points A and B were adjacent, and points C and D were adjacent. Determine the bearing, plunge, and magnitude of the net slip, and determine the dip-slip and strike-slip components of displacement.

Method 4-16

Step 1: Draw a scaled map-view of the fault and of the offset structures, and label points A, B, C, and D (Fig.

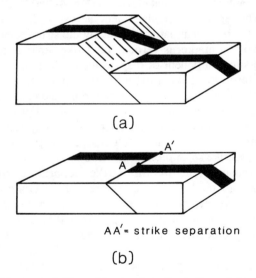

(a)

(b)

AA' = strike separation

Figure 4-20. Illustration that the strike separation across a fault indicated by a single offset marker does not uniquely define the net slip on the fault. (a) Displaced dike before erosion. Movement on the fault is pure dip-slip; (b) strike separation displayed after erosion. Points A and A' were originally adjacent.

4-21b). Extend the lines representing the dike and the contact on the south side of the fault out to a convenient distance.

Step 2: Draw folding line F1 perpendicular to the dike trace. Let the elevation of F1 (and of all folding lines in this problem) be the same as the ground surface. Rotate the cross-sectional view around F1 into the plane of the map projection. Draw a line representing the lower reference plane at a distance d below F1 in the cross-sectional view. Draw line NN' so that N lies at the ground surface and N' lies in the lower reference plane. Line NN' must make an angle of 40° (equal to the dip of the dike) with respect to F1. Remember, it must dip to the northeast. Line NN' represents the cross section of the dike.

Step 3: Draw folding line F2 perpendicular to the contact trace, rotate the cross-sectional view around F2 into the map projection plane, and draw a line representing the lower reference plane at the same distance d below F2. Construct line MM' to represent the cross section of the contact. Note that MM' must make an angle of 70° with respect to F2 and must dip toward the northwest.

Step 4: Let the fault trace be folding line F3. Rotate the cross-sectional view around F3 into the plane of the map projection and draw the trace of the lower reference plane at a distance d below F3. This cross section represents the plane of the fault.

Step 5: Draw a dashed line from point N' so that it intersects F3 at R. Draw a line perpendicular to F3 down to the lower reference plane in the fault plane and locate point R'. Repeat the procedure and locate P'. Point R' represents the position at which the cross-sectional trace of the dike crosses the lower reference plane in the plane of the fault on the southwest side of the fault. Point P'

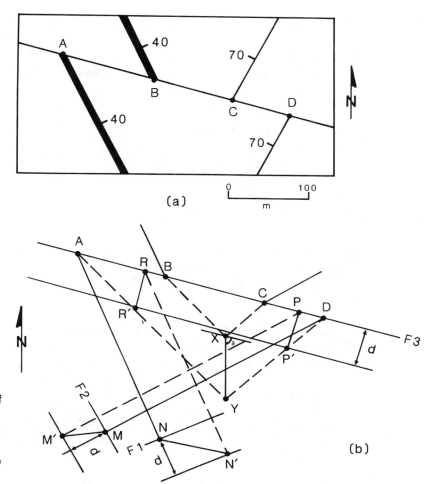

(a)

Figure 4-21. Determination of fault offset from the map pattern. (a) Map showing the fault truncating a dike and a contact (the fault plane is vertical); (b) descriptive geometry construction.

represents the point at which the contact crosses the lower reference plane in the plane of the fault on the southwest side of the fault.

Step 6: Extend a line from A through R', and extend a line from D through P'. AR' and its extension represent the trace of the intersection between the dike and the fault in the southwest wall of the fault. DP' and its extension represent the trace of the intersection between the contact and the fault in the southwest wall of the fault. These two lines intersect at Y, which represents the intersection of the dike and the contact in the southwest wall of the fault.

Step 7: Draw a line from B that is parallel to AY, and a line from C that is parallel to DY. These lines

represent the traces of the dike and the contact in the northeast side of the fault. The lines intersect at X. Prior to movement on the fault, points X and Y were adjacent. Thus, a line connecting X to Y represents the net slip on the fault. As the fault plane is vertical, the bearing of the net-slip line is S70ºE, and its plunge (∂) is measured from vertical. The dip slip and strike-slip components of displacement can be determined by resolving the net-slip line into components parallel to and perpendicular to the fault trace. The dip-slip component is $XY \sin \partial$, and the strike-slip component is $XY \cos \partial$. Note that in this example, the fault was an oblique-slip fault, and that the larger component of movement was dip-slip.

EXERCISES

1. A fault is exposed in outcrop at point A. For each of the situations defined below, determine the distance (depth) between the ground surface at point B and the fault plane. In each case, the traverse length betwen points A and B is exactly 200 m, and the traverse bearing between the two points and the elevation difference between the two points is specified.

Fault attitude	Traverse bearing	Elevation difference between A and B
(a) N-S,30°E	090°	0
(b) N-S,30°E	060°	0
(c) N-S,30°E	090°	50 (A is higher than B)
(d) N-S,30°E	060°	50 (A is higher than B)

2 . Karen Freer is mapping a heavily forested slope. The stratigraphic sequence on the slope is (from top to bottom) Gomez Creek Shale, Race Point Limestone, Hunter Formation. As a consequence of the vegetation, outcrops are hard to come by. She finds, however, an exposure of the contact between the Gomez Creek Shale and the Race Point Limestone at point A on the hillslope. The attitude of the contact is 300°,40°NE. From drilling data at a nearby well, she knows that the Race Point Limestone is 50 m thick. A straight path runs from the Race Point/Gomez Creek outcrop at point A down the slope of the hill. Karen sights down the path with her compass and finds that the path is oriented 15°,S40°E. How far must she walk down the path before she will find the contact between the Race Point Limestone and the Hunter Formation? (Assume that the bedding is homoclinal.)

3 . It is common during offshore drilling for many holes to be drilled from a single platform. The holes are inclined so that they fan out from the platform and thus cover a broader area of the reservoir. Consider a platform located at point A in the Straits of Vermouth. A new hole drilled from this platform is oriented 60°,340°. The drilling target is a reservoir sandstone formation that lies beneath a horizontal unconformity. A salt layer above the unconformity provides an effective seal. The unconformity lies at a depth of 1200 m below the sea floor. Water depth below the platform is 100 m. How long will the drill stem be when it penetrates the unconformity?

4 . Bill Nelson is mapping a small region in western Nevada (Figure 4-M1). He found the contact between the Figaro Sandstone and the underlying Jarbidge Volcanics at BM 98 (el. 98 m). The attitude of the contact is 360°,30°W. He walked 150 m due west across nearly horizontal ground and found the contact of the Figaro Sandstone with the overlying Franklin Shale. In order to cross the Franklin Shale, heavy brush forced him to run his traverse in the direction N45°W (also along horizontal ground). He found the top of the Franklin Shale at the 100-m contour line.   Bill then began to climb Pointop Ridge. The lowest unit on the ridge is the Rufus Springs Formation. Bill was able to traverse this formation in the direction N90°W and found its top at an elevation of 330 m. He then found a

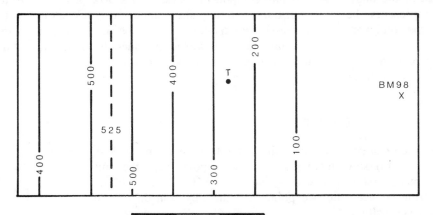

Figure 4-M1. Sketch map of the Pointop Ridge region in Nevada. Contours are shown by solid lines, and the crest of the ridge is indicated by the dashed line.

convenient sheep trail that crossed the overlying Milo Formation. The bearing of the trail is S50°W. Bill crossed 150 m of Milo Formation before finding its top contact with the Pointop Dolomite. The Pointop Dolomite is exposed all the way to the crest of Pointop Ridge. The crest of the ridge is indicated by the dashed line in Figure 4-M1, and is at an elevation of 525 m.

(a) Complete the geologic map (Fig. 4-M1) by adding the contacts and by labeling the different units. Assume that all contacts are oriented 360°,30°W.

(b) Calculate the thicknesses of the Figaro Sandstone, Franklin Shale, Rufus Springs Formation, and Milo Formation (show your work). Assume that bedding attitude is constant throughout the map area.

(c) Construct a scaled stratigraphic column of units exposed in the map area. Choose your own scale.

(d) Old man Thompson the hermit, whose cabin is located at point T on the map, wants to drill a water well. The Figaro Sandstone is known to be a good aquifer. How deep will Thompson have to drill a vertical well in order to reach the top of the Figaro Sandstone?

5. In western Nevada a recent fault offset the ground surface and displaced a fence. A somewhat forgetful geologist drove for five hours across rough dirt roads to get to the site of the offset fence in order to measure the *net slip* on the fault. The net slip, as defined by Ried et al. (1913) is "the distance, measured on the fault surface, between two originally adjacent points situated, respectively, on opposite sides of the fault." When the geologist arrived at the fault, he found that he had not brought a tape measure. Though forgetful, he was not stupid, because he cleverly determined the net slip by surveying the difference in elevation (6 m) of the ground surface on opposite sides of the fault and by measuring the plunge and bearing of a line connecting the two ends of the fence (70°,N30°W). What is the net slip in the plane of the fault?

6. A 3-m-thick conglomerate layer oriented N30°W,40°SW contains placer gold. Initial assays suggest that the conglomerate is worth $50.00 per cubic meter. Backyard Mining Company has the ability to excavate down to a depth of 50 m. The property that they own and the outcrop trace of the layer are shown in Figure 4-M2. What is the total value of the gold that they will be able to obtain, assuming they excavate the entire layer within their property?

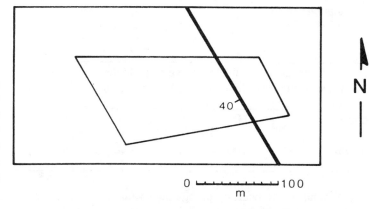

Figure 4-M2. Sketch map of the prospect being exploited by Backyard Mining Company. The thin lines indicate property boundaries, and the thick line represents the outcrop trace of the conglomerate layer (its thickness is exaggerated).

7. A geologist is completing a map of regional-scale folds involving the Itabirita Formation. This formation contains major iron deposits and has been drilled extensively during exploration. During folding, the ductile rocks of the formation did not maintain a uniform thickness. The geologist is trying to determine

whether regional variations in thickness are associated with the folding or predated the folding. To do this, she is calculating the thickness of the unit at various localities and plotting the results on a map showing the attitude of lithologic layering (it is not clear if original bedding has been preserved) and the position of the folds (Figure 4-M3).

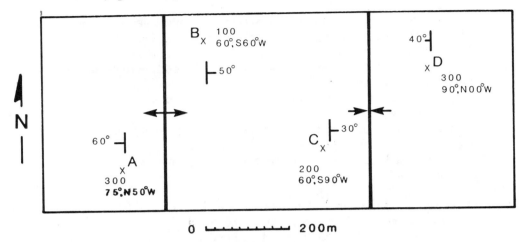

Figure 4-M3. Sketch map of the Itabirita Formation. Points A, B, C, and D are measurement stations where the attitude and thickness of the formation could be determined.

(a) Next to each attitude measurement on Figure 4-M3 we indicate the thickness (in meters) of the Itabirita Formation measured in the hole and the plunge and bearing of the hole. Assuming that the top and bottom contacts of the formation are parallel, calculate the thickness of the bed at each locality.

(b) Based on the above results, do you think that the thickness variations are associated with the development of the folds or that they developed during an earlier independent deformation event? Explain your answer.

8. A vertical fault strikes N90°E across a horizontal plain. A telephone pole happens to be planted exactly on the fault trace, and two nonparallel veins are offset by the fault. The strike separation of the veins as measured on the ground surface is described in the following table. All distances are measured to the west of the the telephone pole.

Dike	Attitude	Distance between telephone pole and intersection of the dike with the S side of fault	Distance between telephone pole and intersection of the dike with the N side of fault
A	340°,30°NE	600 m	400 m
B	040°,50°NW	100 m	300 m

(a) What is the bearing, plunge, and value of the net-slip line that characterizes movement on the fault?

(b) Which side of the fault moved relatively up, prior to erosion and creation of the present ground surface?

(c) Is this a dip-slip, oblique-slip, or strike-slip fault?

CHAPTER

5

INTRODUCTION TO STEREOGRAPHIC PROJECTIONS

5-1   INTRODUCTION

Representation and manipulation of structural data by the geometric methods introduced in the previous chapters becomes cumbersome and difficult if we have to analyze a large number of measurements. In this chapter we introduce the concept of the stereographic projection, which has become widely used by structural geologists during the last 50 years (Bucher, 1944) and provides a simple and quick alternative way to represent three-dimensional data in two dimensions.   Although data plotting using a stereographic projection may seem abstract at first, once you are used to it you will find that the methods are powerful and allow you to solve many types of structural problems easily. Computers are increasingly being used to plot structural data on stereographic projections, but you will not be able to interpret computer output if you are not adept at plotting data by hand. In fact, you will find that the cardboard stereonet itself is versatile and quick and can easily be carried with you to the field, even if you are backpacking into a remote area.

5-2 CONCEPT OF A STEREOGRAPHIC PROJECTION

We can understand stereographic projections more easily if we first think about spherical projections. Imagine an observer standing at the center of a large hollow glass sphere. Any *direction* can be specified by marking a dot on

the surface of the sphere. For example, the direction "due west" can be indicated by a dot on the equator of the sphere that is due west of the observer, and a point that is "straight up" will be a dot on the surface of the sphere that is directly over the head of the observer. Early astronomers displayed the relative positions of stars by plotting the stars as white dots on the surface of a blackened sphere whose center was the earth. The resulting representation was called the *celestial sphere*. Note that the relative *distances* of the stars from the earth can*not* be represented on the celestial sphere. A spherical surface on which positions are indicated is called a *spherical projection*. Remember, only orientations, not distances, can be represented on spherical projections!

Spherical projections can be used to represent the orientations of a line or a plane if the line or plane is positioned so that it passes through the center of the sphere. A line that is so positioned intersects the surface of the sphere at two points, and a plane that is so positioned intersects the surface of the sphere along a circle (Fig. 5-1). The intersection of the line or plane with the surface of the sphere is its spherical projection.

A spherical projection is three dimensional and is, therefore, not particularly portable. Fortunately, a sphere can be projected onto a two-dimensional plane. The most common planar projections of a sphere are called *azimuthal projections*.   An azimuthal projection is constructed by passing projection lines (see Chapter 3 and Appendix 1) from a point source through the sphere to where they intersect a projection plane (imagine that the point source

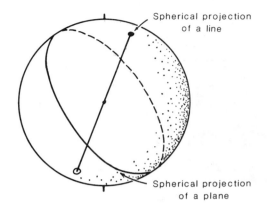

Figure 5-1. Spherical projections of lines and planes. Any line through the center of the sphere pierces the sphere at two points. Any plane through the center of the sphere intersects the sphere along a circle.

is a small light bulb; Fig. 5-2). The plane must be oriented such that the unique projection line that passes through the center of the sphere (the *central projection line*) is perpendicular to the projection plane. The projection plane can be tangent to the surface of the sphere, or it can be at a distance from the surface of the sphere, or it can pass through the center of the sphere. A change in the position of the projection plane merely changes the scale of the projection (Fig. 5-2). The projection plane can also

have any orientation, and this determines whether the projection is an equatorial, polar, or oblique one (Fig. 5-2). A *stereographic projection* is a special kind of azimuthal projection that was developed and refined by crystallographers. The special characteristic of a stereographic projection is that the point source used in its construction lies on the surface of the sphere. In geology the projection plane used to construct a stereographic projection is positioned to pass through the center of the sphere.

Now, let's try to visualize the construction of a stereographic projection. Imagine that a dot has been marked on the lower half of our glass sphere; the dot is a spherical projection of a point in space. A stereographic projection of the dot is constructed by drawing a projection line from a *zenith point,* placed at the top of the sphere, to the dot. This projection line must pass through the equatorial plane of the sphere. The intersection between the projection line and the equatorial projection plane is the stereographic projection of the dot (Fig. 5-3). In structural geology we always project points from the lower half of the sphere. Thus, in figure captions of structural geology papers you will usually see the words "lower hemisphere projection."

The intersection of the equatorial projection plane with the sphere is called the *primitive circle* (often abbreviated simply as *"the primitive"*). The primitive has the same radius as the original projection sphere, and all points on the surface of the lower hemisphere project to points inside the primitive on the projection plane (Fig. 5-3).

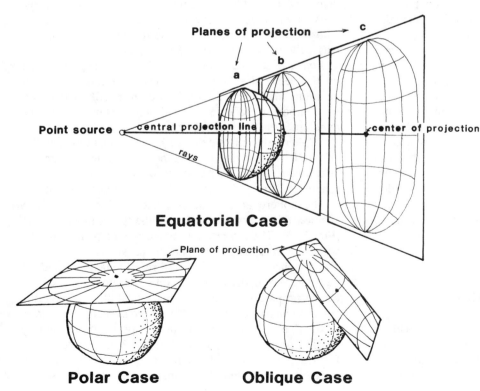

Figure 5-2. Three different azimuthal projections. The projections vary according to the position of the point source with respect to the sphere. The size of the projection depends on the relative distances among the point source, the sphere, and the projection plane. The orientation of the projection plane determines the type of projection. (Adapted from Raisz, 1962.)

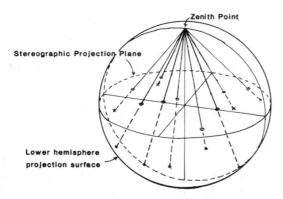

Figure 5-3. The basis of a lower-hemisphere stereographic projection on an equatorial plane. The projection of a point on the lower hemisphere lies along a line drawn from the point to the zenith.

Stereographic Projection of a Plane

Imagine a dipping plane that intersects the ground surface; the plane can be a bed, a fault, a joint, or any other planar structure. The trace of the plane on the ground surface is a straight line (Fig. 5-4a). Create a projection sphere of radius r that is centered at a point O on the outcrop trace of the inclined plane. The dipping plane (and its extension into the sky) intersects the sphere as a circle, whose radius is the same as that of the sphere (Fig. 5-4b). This circle is the spherical projection of the dipping plane and is called a *great circle* (see Appendix 1). A plane with a different attitude will intersect the sphere along a different great circle. In other words, there is a unique great circle for each different orientation of a plane.

To construct the stereographic projection of the plane join each point on the portion of the great circle projection

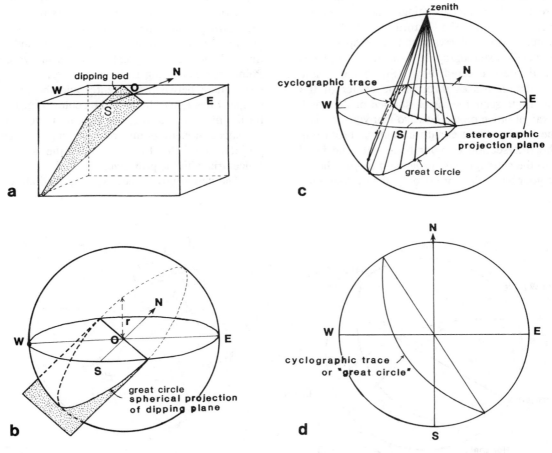

Figure 5-4. Stereographic representation of a plane. (a) Block diagram of a plane oriented 300°,50°SW; (b) projection sphere (radius r) set up on an outcrop trace that intersects the plane in a great circle (adapted from Hobbs et al., 1976); (c) lines from the zenith to the lower half of the great circle intersect the equatorial projection plane in a circular arc (the cyclographic trace) (adapted from Hobbs et al., 1976); (d) the completed stereogram of the dipping plane.

that lies on the lower half of the projection sphere to the zenith (Z) of the projection sphere (Fig. 5-4c). These straight lines generate part of a circular cone that intersects the equatorial projection plane as part of a circular arc. The arc is the *stereographic projection* of the dipping plane and is called a *cyclographic trace* of the plane (Fig. 5-4d). We will refer to the arc as a *great circle* because it corresponds to a great circle on the original spherical projection.

Note that a horizontal plane will project along the primitive circle (i.e., the radius of curvature equals the radius of the projection sphere), and a vertical plane will project as a straight line passing through the center of the primitive circle (i.e., the radius of curvature is infinity). A plane of intermediate dip projects as a cyclographic trace with a radius of curvature intermediate between that of the primitive and infinity; the radius of curvature of a plane's cyclographic trace increases as the dip angle increases (Fig. 5-5a). Also note that as the strike of the plane changes, the two intersections between the cyclographic trace and the primitive rotate around the primitive. The two intersections are, however, always exactly 180° apart.

To help visualize stereographic projections of planes imagine a hemispherical cereal bowl (Fig. 5-5b). Hold your hand so that it passes through the center of the hemisphere and intersects the inside surface of the bowl. Then, look straight down on the bowl so that you simulate projecting the bowl onto an equatorial plane. From this viewpoint the intersection of your hand with the bowl corresponds to the stereographic projection of your hand. As you make your hand dip more steeply, the cyclographic trace of your hand approaches the center of the bowl and starts to look like a straight line. As you make your hand dip more shallowly, the cyclographic trace of your hand approaches the rim of the bowl and starts to look like a circle.

Stereographic Projection of a Line

The orientation of any linear geologic structure (such as a mineral lineation, fold hinge, or the axis of a drill hole) can be represented on a stereographic projection in a manner similar to that described for planes. First, you must visualize a spherical projection of the structure. A linear structure can be represented as a line that passes through the center of the projection sphere and intersects the surface of the lower hemisphere at a point P (Fig. 5-6a), which is the spherical projection of the line. A straight line joining P to the zenith of the projection sphere intersects the projection plane at point P', which is the *stereographic projection* of the line. A vertical line will project as a point at the center of the primitive circle, and a horizontal line will project as a point on the primitive circle itself. Points corresponding to lines of intermediate plunge lie somewhere between the center and the primitive, with steeper lines lying closer to the center (Fig. 5-6b). If you fix the plunge of a line, but vary its trend through 360°, the succession of points representing the stereographic projection of the line will describe a circle that is concentric with the primitive.

The cereal bowl analogy, described above, will help

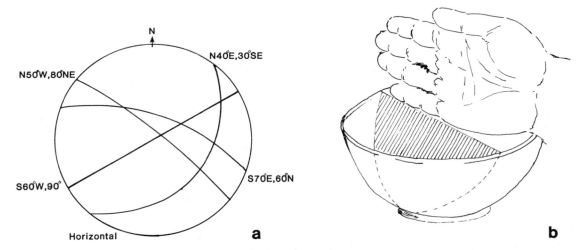

Figure 5-5. Cyclographic traces of planes. (a) Planes of different orientations represented as unique great-circle arcs; (b) a plane continuous with the inclined palm of your hand would intersect the inside of a cereal bowl in a semicircle. Imagine looking vertically down into the bowl to visualize how the semicircle projects as an arc on a horizontal equatorial plane.

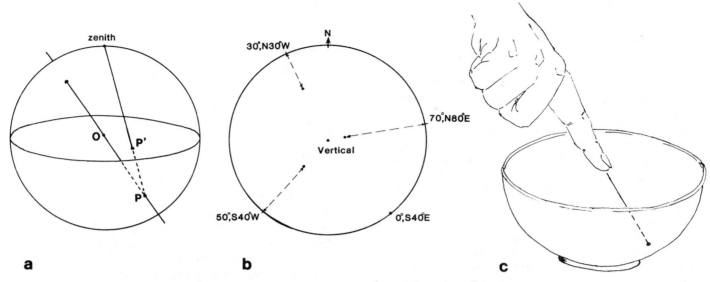

a   **b**   **c**

Figure 5-6. Stereographic representation of a line. (a) A line OP through the center of the sphere pierces the lower hemisphere at a point P that is the spherical projection of the line. A line from P to the zenith intersects the equatorial projection plane at P', which is the stereographic projection of the line; (b) lines of different orientations project to unique points on the stereogram; (c) the cereal bowl analogy for lines. Use your index finger to simulate lines of different orientations.

you to visualize the stereographic projection of a line. Use your index finger to simulate a line, and make sure that your finger passes through the center of the imaginary sphere of which the bowl is the lower half (Fig. 5-6c). Clearly, the intersection of your finger with the surface of the bowl is a point. As you move your finger to different orientations, note how the position of the point moves. If your finger is vertical, it touches the center of the base of the bowl, and if it is horizontal, it touches the rim of the bowl.

Stereographic Projection of a Small Circle

The intersection of a sphere with a plane that does not pass through the center of the sphere is a *small circle* (see Appendix 1). The radius of the small circle depends on the distance of the plane from the center of the sphere. The radius decreases as the distance between the plane and the center of the sphere increases (Fig. 5-7) and becomes zero when the plane is tangent to the sphere. Such small circles also result from the intersection of a circular cone with the sphere if the apex of the cone is positioned at the center of the sphere. Small circles appear on stereographic projections as circles (Fig. 5-7); the projection of a small circle is not coaxial with the primitive unless the central projection line passes through the center of the small circle. If the center of the stereographic projection of the

small circle is closer to the primitive than the radius of small circle, then the small circle will be truncated by the primitive; the other half of the small-circle projection will appear at the opposite edge of the primitive.

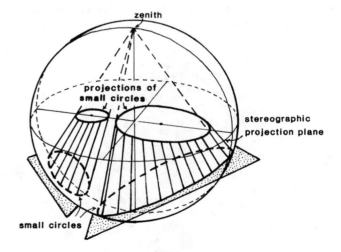

Figure 5-7. Stereographic representation of small circles. Planes that do not pass through the center of a sphere intersect the sphere in small circles. These circles project as circles (or circular arcs) on the stereographic projection plane.

5-3   THE STEREONET

Visualization of the Stereonet Grid

Clearly, stereographic projections would be of limited use if you first had to construct a spherical projection every time you wanted to create a stereographic projection of a line or plane. Fortunately, if a coordinate grid is projected onto the plane of the stereographic projection, the projections of lines and planes can be plotted directly. A coordinate grid (Fig. 5-8a) that is constructed using a stereographic projection is called a *stereonet*. The stereonet is also called an *equal-angle net* (for reasons that are described in Chapter 8) or a *Wulff net*, after the crystallographer who popularized its use. The representation of a point, line, or plane on a stereonet is called a *stereogram*.

A stereonet is constructed by drawing the great-circle cyclographic traces of a set of planes that all intersect along a line that runs between the poles of the primitive and passes through the center of the projection sphere (Fig. 5-8b). The angle between each plane and its neighbor is 2°. Thus, the cyclographic traces define a longitudinal grid with a 2° interval between adjacent grid lines. Every fifth grid line is darker, so it stands out and makes the grid easier to read.

Superimposed on the longitudinal grid is a latitudinal grid composed of the stereographic projections of small circles. These small circles are constructed from the spherical projections of a series of coaxial right circular cones with progressively decreasing ratios between the basal radius and the altitude. The axis of these cones is the line that runs between the north and south poles of the primitive and passes through the center of the projection sphere (Fig. 5-8b). The angle of intersection between a

longitudinal grid line and a latitudinal grid line on the stereonet is always a right angle.

Construction of a Stereonet Grid

One of the most useful properties of a stereographic projection is that any circle (great circle or small circle) on the projection sphere appears as a circle or a circular arc on a stereogram (Fig. 5-9). This simple geometric property allows us to construct the stereographic projection of any plane quite easily using either a graphical approach or an analytical expression.

First, we must calculate the shortest distance between the cyclographic trace of a plane and the center of the primitive circle. To do this, we draw a vertical cross section (Fig. 5-10a) through the center of the projection sphere and parallel to the dip direction of the inclined plane. The trace (OP) of the inclined plane on the section makes an angle ϕ (true dip of plane) with the horizontal. A straight line from P to the zenith (Z) intersects the trace of the projection plane on the cross section at Q. Point Q represents the position of the cyclographic trace of line OP in the profile. The length of OQ represents the shortest distance of the cyclographic projection of the dipping plane from the center of the primitive circle and can be measured directly from the construction. It can also be calculated, using the analytical expression

$$OQ = r \tan[(\pi/4) - (\phi/2)] \qquad \text{(Eq. 5-1).}$$

Thus, we have been able to determine the position of point Q. On the stereographic projection (Fig. 5-10b) the cyclographic trace of the inclined plane is a segment of a circular arc that must pass through point Q and two points (S, T) on the primitive; points S and T represent the

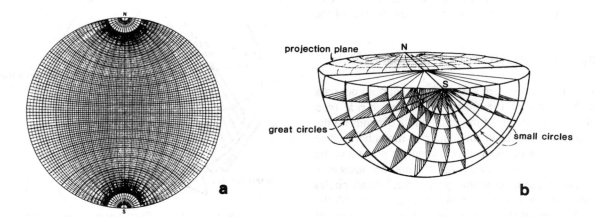

Figure 5-8.   The nature of the stereonet grid.   (a) The stereonet, Wulff net, or equal-angle net; (b) the great circles on the stereonet represent dipping planes that all intersect along the north-south axis. The small circles represent a family of coaxial right-circular cones (adapted from Hobbs et al., 1976).

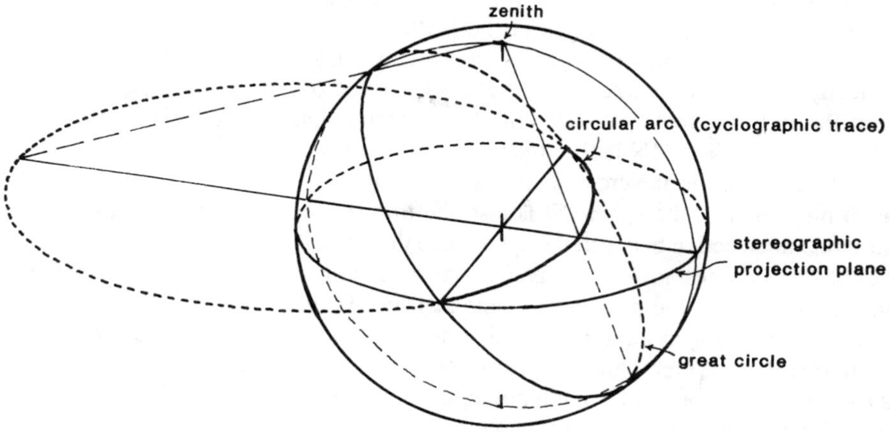

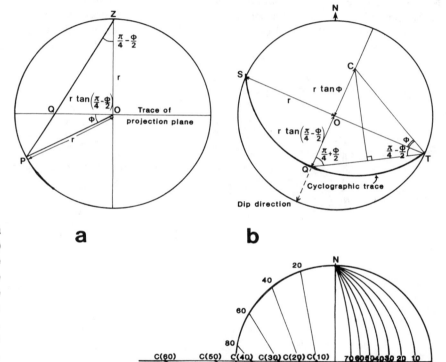

Figure 5-9. A great circle on a sphere projects as a circular arc on the stereographic projection. (Adapted from Berry & Mason, 1959.)

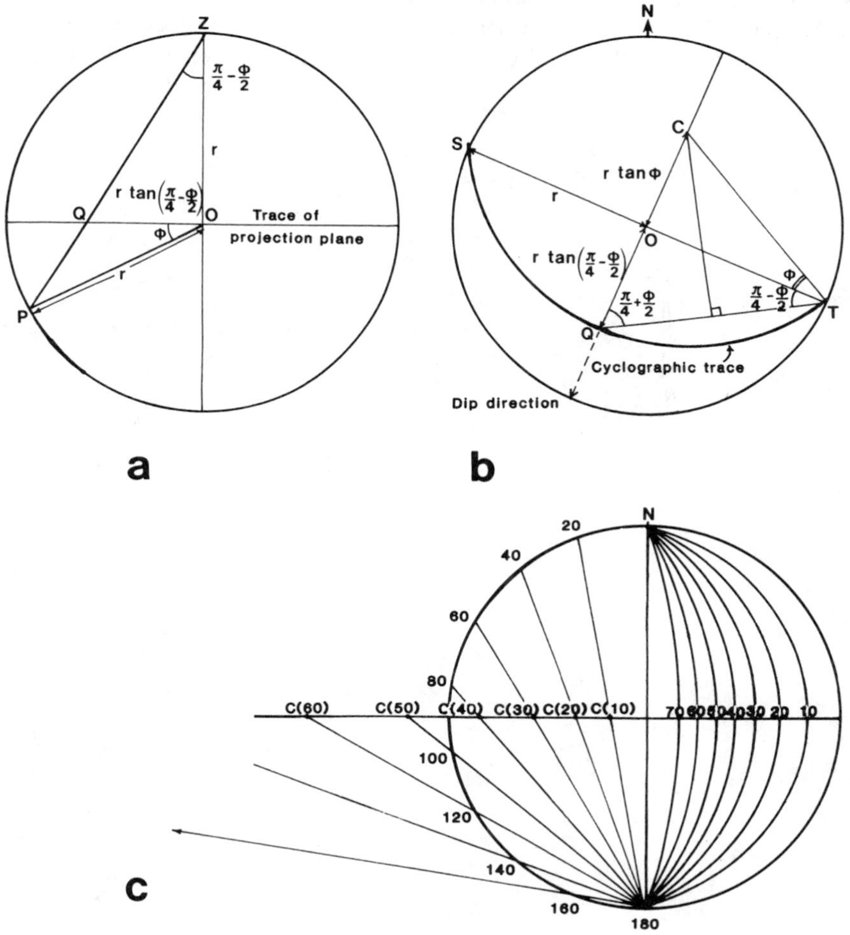

Figure 5-10. Construction of a great circle on a stereonet grid. (a) Vertical cross section through the projection sphere showing the trace of a plane OP that dips at an angle ø. Q is the stereographic projection of P (adapted from Hobbs et al., 1976); (b) stereographic projection of the plane dipping at an angle ø. The cyclographic trace is a circular arc with center at C (adapted from Hobbs et al., 1976); (c) geometric construction to determine the centers of circular arcs representing planes with different dips (adapted from Ragan, 1985).

intersection of the trace of the bed with the primitive and are defined by the strike of the plane. Therefore, once we know the positions of points Q, S, and T, we can define the cyclographic trace of our inclined plane. To locate point C, which is the center of the circular arc of which the cyclographic trace is a segment, we draw the perpendicular bisector of the chord QT and find the point (C) where the bisector of QT intersects the extension of line OQ. The position of point C can also be found analytically by determining the distance d between point C and the center of the stereogram (point O). This distance is given by the equation

$$d = OC = r \tan ø \qquad \text{(Eq. 5-2).}$$

Using the preceding methods we can draw a series of

great-circle curves for planes striking north-south and dipping east or west at various angles (Fig. 5-10c).

A small-circle arc on a stereogram can be constructed graphically as follows. Figure 5-11a shows a vertical cross section through the center of the projection sphere. On this cross section, line MN is the trace of a vertical plane that is perpendicular to the cross section and does not pass through the center of the sphere; it is also the trace of the small-circle intersection between the sphere and the vertical plane. Straight lines joining M and N to the zenith Z intersect the trace of the projection plane at I and J, and these points are the stereographic projections of M and N respectively. On the stereographic projection (Fig. 5-11b) IJ is a diameter of a projected small circle; the center C of the small circle must lie midway between I and J. The chord MN of the sphere subtends an angle of 2β at the center of the sphere (Fig. 5-11a) and hence an angle β at the zenith Z. The diameter of the small circle, IJ, therefore, can also be given by

$$IJ = OJ - OI$$

$$IJ = r \tan(\pi/4 + \beta/2) - r \tan(\pi/4 - \beta/2)$$

$$IJ = 2r \tan \beta \qquad \text{(Eq. 5-3)}.$$

The distance of the center of the small circle from the center of the sphere is line OC, which is given by

$$OC = OI + IC$$

$$OC = r \tan(\pi/4 - \beta/2) + r \tan \beta = r/\cos \beta$$

$$OC = r/\cos \beta \qquad \text{(Eq. 5-4)}.$$

It can also be shown quite easily that on the stereographic projection (Fig. 5-11b), radius vectors drawn from the center of the small circle to the points of intersection of the small circle with the primitive (P and Q) are tangential to

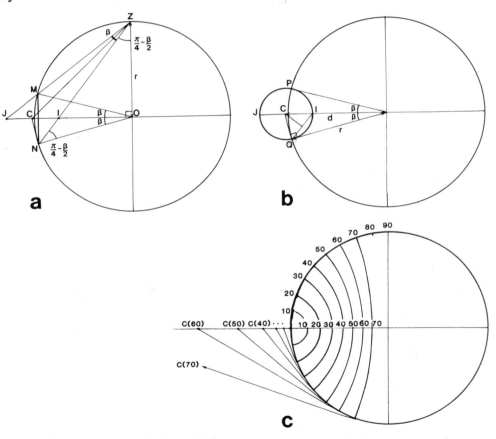

Figure 5-11. Construction of small circles on a stereonet grid. (a) Vertical cross section through the projection sphere showing the trace of a vertical plane MN and its projection JI on the projection plane trace; (b) stereographic projection of the small circle representing the plane. Note that the projection PJQI is also a circle; (c) geometric construction to determine the center of circular arcs representing vertical planes at different distances from the center of the sphere.

the primitive. This relationship permits a quick graphical construction of small-circle arcs for small circles at any particular angle ß to a point on the primitive. Generally, a series of small-circle arcs are drawn to represent small circles at various angles to a horizontal north-south axis (Fig. 5-11c).

Calibration of the Stereonet Grid

As indicated earlier, the perimeter of the stereonet is called the primitive. The poles of the stereonet lie on the primitive and are the points at which the longitudinal great-circle traces converge. The top pole is labeled north and the bottom pole is labeled south. The ends of the horizontal diameter are labeled east (on the right) and west (on the left). The primitive can be divided into four quadrants of 90° each (like a quadrant compass) or can be divided directly into 360° (like an azimuthal compass), with the number of degrees increasing in a clockwise direction. In the azimuthal convention, north has a value of 000° or 360°, east has a value of 090°, south has a value of 180°, and west has a value of 270°. The numbers along the primitive correspond to a strike or trend. Distances measured in from the primitive correspond to dip or plunge angles. Dip or plunge angles are 00° on the primitive circle and increase to 90° at the center of the circle and are always read along a radius that is drawn from the center of the circle to the perimeter. We will see that dips or plunges are usually determined by using the tick marks along the line passing from the east or west mark through the center of the stereonet. The best way to understand how to read a stereonet is to practice plotting lines and planes according to the procedures outlined next.

5-4   PLOTTING TECHNIQUES

In this section we will go through a number of examples, step by step, to familiarize you with plotting various structural elements on a stereonet. The key to successful use of a stereonet is visualization. Before plotting anything on the net, imagine the cereal bowl analogy described earlier, and make sure you have a general image in your mind of where the great circle or point that you will plot will fall on the stereonet.

Preparing the Stereonet and Overlay

Stereographic plotting is usually done on an *overlay* (a piece of tracing paper that is slightly larger than the net) that can be revolved above a fixed stereonet. Before plotting on the overlay, however, it is useful to prepare your stereonet so that it is permanently available for use in both the office and the field. To do this, remove the large

stereonet (Fig. A4-1) from the back of this book and mount it on a piece of stiff cardboard or masonite board. Cover the net with a transparent plastic sheet to prevent it from getting wet or dirty (transparent Contact™ paper serves this purpose very well). Punch a hole in the center of the net with a thumbtack, then insert the thumbtack through this hole from the back of the net. Tape the tack head to the back of the net using clear plastic tape or adhesive tape. This ensures a permanent mount so that the pin does not fall out and also prevents the hole at the center of the net from enlarging and rendering the net unusable. The pin (which should stick out approximately 0.5 cm from the net) serves as an axis of rotation for overlays. When the net is not in use, it is a good idea to cover the pin with a small eraser or a piece of cork to prevent it from ripping paper, clothes, or skin.

To prepare the overlay, place a small piece of clear tape on the back of the overlay at the center. Then, pierce the center of the overlay with the pin at the center of the net so that the overlay is free to revolve above a fixed net, and smooth the overlay against the net. Draw the primitive circle on the overlay, and mark the positions of the north, south, east, and west geographic directions. Use an arrow to indicate the north mark.

Plotting a Plane

Problem 5-1
The attitude of a bed is N80°W,40°S. Plot a stereogram representing the orientation of the bed.

Method 5-1
Step 1: Visualize the bed (Fig. 5-12a), then prepare your overlay, as described above.

Step 2: Count off 80° westward (counterclockwise) from north, and place a tick mark on the primitive to mark the line of strike (Fig. 5-12b). Since the plane intersects the primitive at two points, it is useful to place a tick at the other end of its line of strike (i.e., at S80°E). It is also useful to mark the approximate dip direction at this stage. The dip is in the direction S10°W.

Step 3: Revolve the overlay clockwise to bring the first strike mark to the north position; in this position, the series of great circles now pass through the strike marks (Fig. 5-12c). The dip direction now lies to the left.

Step 4: Count off 40° along the east-west diameter from the primitive on the side of the dip direction (i.e., in from the left side of the net in this case) to locate the great-circle trace representing a plane dipping 40°S (Fig. 5-12c). Draw the great-circle trace connecting the N80°W tick mark to the S80°E tick mark on your overlay. Note that the initial three-dimensional visualization helps in determining whether the dip should be counted in from the left or the right edge of the net. Also note that the straight

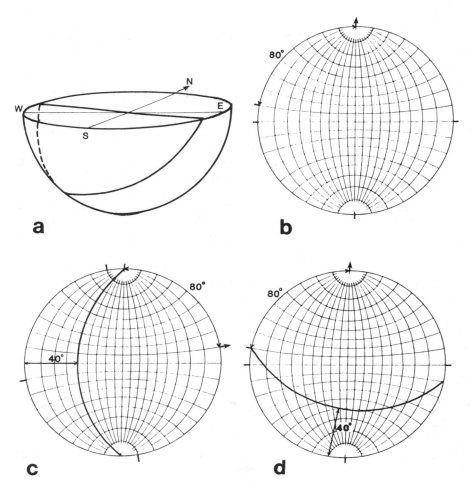

80°

40°

80°

80°

40°

a

b

c

d

Figure 5-12. Procedure for plotting the stereographic projection of a dipping plane.

line connecting the two ends of the great-circle trace and passing through the center of the stereonet is a *strike line* on the plane.

Step 5: Revolve the overlay back to its original position (north-arrow coinciding with north of stereonet), and check the result by visualization (Fig. 5-12d).

Plotting a Line

Problem 5-2

A plunging fold hinge has the orientation 38°,S42°W. Plot the hinge line on a stereogram.

Method 5-2

Step 1: Visualize the problem first (Fig. 5-13a), and prepare the overlay.

Step 2: Count 42° westward (clockwise) from south, and place a tick mark on the primitive to represent the trend of the line (Fig. 5-13b).

Step 3: Revolve the overlay to bring the trend mark to the south position (Fig. 5-13c).

Step 4: Count 38° inward from the primitive toward the center along the south-north diameter. Plot the point to represent the line (Fig. 5-13c).

Step 5: Revolve the overlay back to its original position (north arrow coinciding with north of stereonet) and check the result by visualization (Fig. 5-13d).

It should be noted that in plotting the plunge of the line we used the graduations (along the north-south diameter) marked by the small circles of the stereonet. We could just as easily have revolved the overlay to bring the trend mark to the west position (in step 4) and counted inward along the west-east diameter (in step 5); the resulting point obtained to represent the line would have been exactly the same. Such alternative methods of plotting often help to speed up the plotting process and can be used once you have gained confidence in your ability to use stereographic projections.

Plotting a Line on a Plane

Problem 5-3

Cleavage in an area is oriented 110°,43°SW. A mineral lineation on the cleavage plane trends 160°. Plot the line representing this lineation on the stereogram of the plane, and determine the plunge of the lineation.

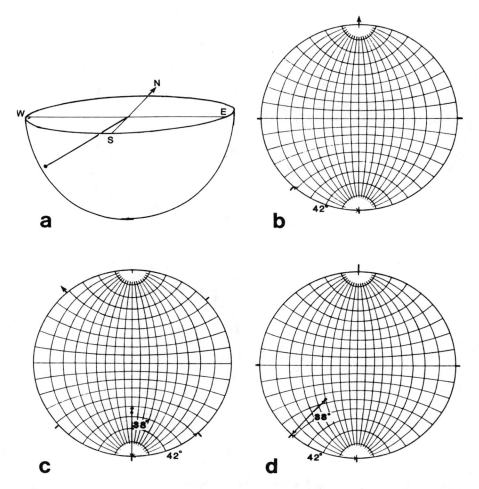

Figure 5-13. Procedure for plotting the stereographic projection of a plunging line.

Method 5-3

Step 1: Visualize the problem (Fig. 5-14a), and prepare your overlay.

Step 2: Note that in this problem we are using an azimuthal convention to represent orientations of planes and lines. Count 110° clockwise from north, and mark the strike orientation on the primitive. Note that your dip direction should be toward the southwest.

Step 3: Revolve the strike mark to south. In this position, the dip direction is to your left. Count 43° inward from the left end of the primitive along the west-east diameter. Draw the great circle to represent the cleavage (Fig. 5-14b).

Step 4: Revolve the overlay back to its original position. Count 160° clockwise from north and mark the trend of the lineation on the primitive (Fig. 5-14c).

Step 5: Revolve the trend mark to south (Fig. 5-14d). With the overlay in this position the lineation must lie on the north-south diameter as well as on the great circle representing cleavage. Therefore, the point of intersection between the cyclographic trace of the plane and the north-south diameter is the point that represents the lineation. The plunge of the lineation can be read off directly by counting inward from the primitive to the lineation along the north-south diameter. The orientation of the lineation is 34°,160°.

Step 6: Revolve the overlay back to its original position, and check the result by visualization (Fig. 5-14e).

Note that if the line were not to plot along the great-circle representation of the plane, then it could not lie in the plane. If your measured attitudes of a plane and a line in the plane do not permit their respective stereographic representations to coincide, then one or both of your measurements are wrong.

Determination of the Rake of a Line

Given the trend and plunge of a line and the strike and dip of the plane on which it occurs, it is very easy to determine the rake of the line in the plane.

Problem 5-4

Cleavage in an area is oriented 110°,43°S. A mineral lineation on the cleavage plane is oriented 34°,160°. What is the rake of this lineation on the cleavage plane?

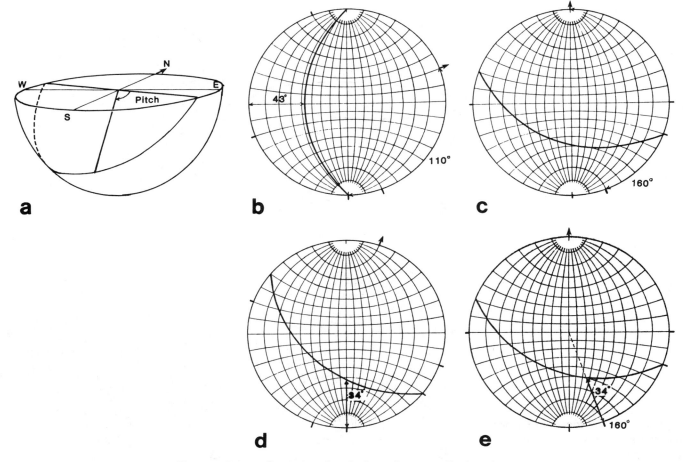

Figure 5-14. Procedure for plotting a line on a dipping plane.

Method 5-4

Step 1: The attitudes of the line and the plane are the same as in Problem 5-3. Follow the procedures outlined in Method 5-3 for plotting the line and the plane (Fig. 5-14).

Step 2: Revolve the overlay so that the strike of the cleavage is oriented north-south (Fig. 5-15a). Count inward from the primitive along the great circle (representing cleavage) to the lineation, keeping in mind that the rake angle must be an acute angle (≤90°). In this case we find that the rake acute angle opens toward the east mark on the overlay. Thus, the orientation of the lineation is given by a rake of 58°E on cleavage oriented 110°,43°S.

Step 3: Revolve back to the original position and check the result by visualization (Fig. 5-15b).

Determination of Plunge and Bearing from Rake

Most of the steps of Methods 5-3 and 5-4 can be reversed if you are given the rake of a line on a plane and the orientation of the plane and are asked to find the plunge and bearing of the line.

Problem 5-5

A cleavage plane is oriented 110°,43°S. A mineral lineation on the plane has a rake of 58°E. What is the bearing and plunge of this lineation?

Method 5-5

Step 1: Visualize the problem (Fig. 5-14a) and prepare your overlay.

Step 2: Mark the strike of the cleavage plane on the primitive. Revolve the strike mark to south, count 43° inward from the left end of the primitive, and draw the great circle to represent the cleavage (Fig. 5-16a).

Step 3: Count 58° inward along the great circle from the correct end (in this case, the south end) of the stereonet. Plot the point representing the mineral lineation (Fig. 5-16a).

Step 4: Revolve the overlay so that the lineation point lies on the north-south diameter. Count the number of degrees between the point and the south end of the

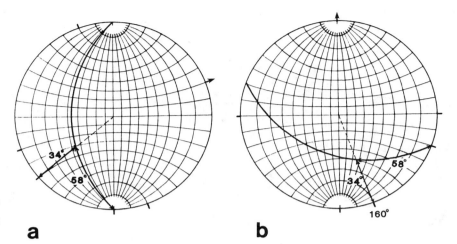

Figure 5-15. Procedure for determining the rake of a line given its plunge.

a   **b**

primitive to get the plunge (Fig. 5-16b). Make a tick mark on the primitive at the south end.

Step 5: Revolve the overlay so that the north arrow overlies the north mark on the stereonet. Read off the trend of the lineation (tick mark) on the primitive (Fig. 5-16c). The lineation has an orientation of 34°,160°.

True Dip and Apparent Dip Problems

It was pointed out earlier (Section 3-4) that the apparent dip of a plane in a given direction is the same as the plunge of a line lying on the plane and having the same bearing as the apparent dip direction. Thus, Method 5-3 can also be used to determine the apparent dip of a plane in any given direction.

Problem 5-6 (Determination of apparent dip angle)

The true attitude of a dike is N25°E,65°SE. The dike is exposed on a vertical quarry wall that trends N10°W.

What is the apparent dip of the dike as viewed in the quarry wall?

Method 5-6

Step 1: Visualize the problem and prepare an overlay (Fig. 5-17a).

Step 2: Mark the strike (N25°E) on the primitive, revolve the mark to north, count 65° in from the right end of the primitive, and draw the great circle to represent the dike (Fig. 5-17b).

Step 3: Revolve the overlay so that the north arrow is over the north mark on the stereonet (Fig. 5-17c). Count off 10° in the direction counterclockwise from north and make a tick to indicate the bearing of the quarry wall and therefore the desired apparent dip direction. The diameter of the stereonet that passes through this point is the great-circle representation of the vertical quarry wall. The point of intersection (P) between this straight line and the great circle that represents the dike is the stereographic projection of the line of intersection between the quarry wall and the dike.

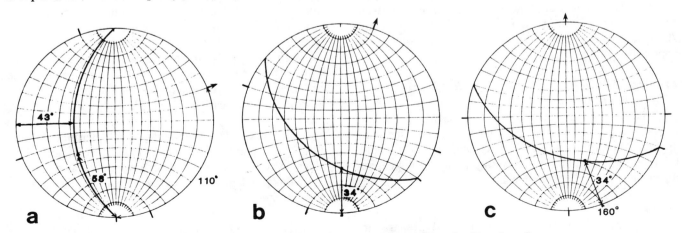

a   **b**   **c**

Figure 5-16. Procedure for determining the plunge of a line given its rake on a dipping plane.

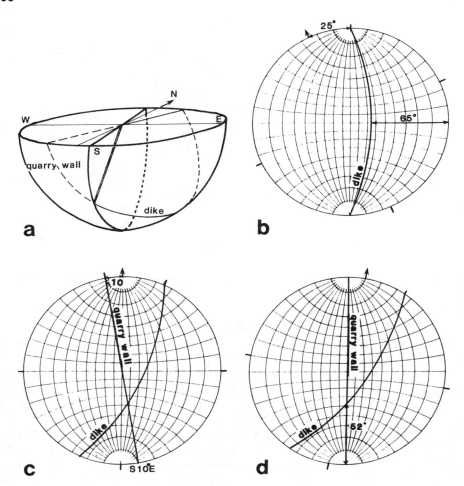

Figure 5-17. Procedure for determining the apparent dip (in a specified direction) of a plane whose true dip is known.

Step 4: Revolve the overlay so that the tick mark is on the north-south diameter (Fig. 5-17d). Count in from the primitive along the the north-south diameter to point P to get the apparent dip (52°). The direction of apparent dip is the same as the bearing of the vertical quarry wall (S10ºE).

Determination of the direction of apparent dip given the apparent dip angle is a somewhat more tricky problem. After plotting the plane, we must resort to a trial-and-error approach.

Problem 5-7 (Determination of the bearing of an apparent dip)

The true attitude of a dike is N25ºE,65ºSE. What are the bearings of the two vertical quarry walls on which the apparent dip of the dike is 40º? Remember that it would be impossible to find a quarry wall on which an apparent dip greater than 65º could be measured.

Method 5-7

Step 1: Visualize the problem and prepare an overlay (Fig. 5-18a).

Step 2: Plot the great-circle trace of the dike on the overlay, following the procedure in Method 5-1 (Fig. 5-18b).

Step 3: Revolve the overlay until the great circle representing the plane intersects the north-south diameter at a distance equal to the apparent dip angle (40º) from the primitive (Fig. 5-18c,d). Two such points will be found.

Step 4: For each of the two points, draw a radius from the center of the stereonet through the point to the primitive, and make a tick mark where each radius intersects the primitive.

Step 5: Revolve the overlay so that the north arrow is over the north mark on the stereonet, and read off the bearings of the two tick marks on the primitive (Fig. 5-18e). The two bearings are N48ºE and S02ºW.

Now, it should be easy for you to determine the true attitude of a plane if you are given the apparent dips of the plane in two directions. Note how much faster it is to solve this problem with a stereonet than to use descriptive geometry.

Problem 5-8 (True dip from two apparent dips)

A distinctive bed is exposed on a highway road cut. At one end of the cut, the apparent dip of the bed is

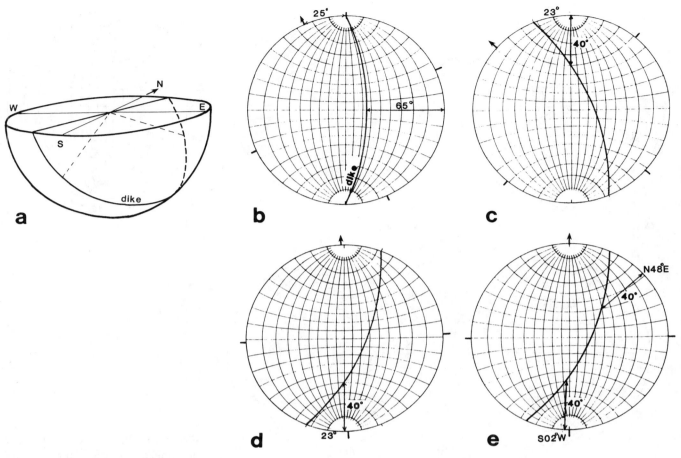

Figure 5-18. Procedure for determining the direction of a specified apparent dip if the true dip of the plane is known.

32º,256º. Around a bend in the road, the apparent dip of the same bed is 27º,125º. Determine the true dip of the bed.

Method 5-8

Step 1. Visualize the two apparent dips as lineations that must be contained on the dipping plane (Fig. 5-19a). Prepare the overlay.

Step 2: Mark the positions of the apparent dip trends (125º and 256º) on the primitive.

Step 3: Bring each trend mark to the north-south line and count the appropriate angles (27º and 32º) inward from the primitive to obtain the points for the apparent dips (Fig. 5-19b, c).

Step 4: Revolve the overlay (by trial and error) until both points lie on the same great circle (Fig. 5-19d). Draw in the great circle. Determine its dip by counting inward from the primitive to the great circle along the east-west diameter (Fig. 5-19d).

Step 5: Revolve back to the original position. Determine the strike of the great circle by counting clockwise from north to the strike mark along the primitive (Fig. 5-19e). The true dip of the bed is 102º,54ºS.

The same procedure can be used when you are given the attitudes of two lineations on a surface and are asked to calculate the attitude of the surface. The true dip of a plane can also be calculated using a similar procedure if you are given the strike of a plane and an apparent dip of a line on the plane or the attitude of a single lineation on the plane.

Determining the Intersection of Two Planes

Remember that the intersection of two planes is a line. In Chapter 3, we presented the geometric method for determining the orientation of this line. Here we present the stereographic method. Again, note how quickly the answer can be obtained with a stereonet.

Problem 5-9

At an outcrop, bedding orientation is S80ºE,20ºS, and cleavage orientation is N30ºE,70ºE. Determine the orientation of the cleavage/bedding intersection lineation.

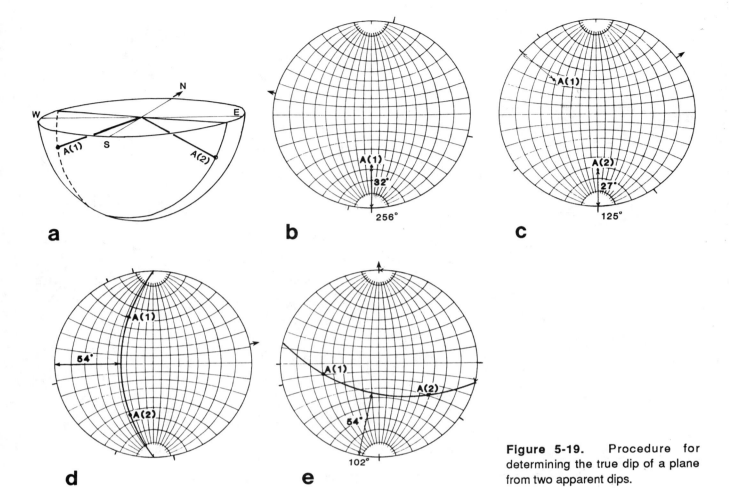

a   b   c

d   e

Figure 5-19. Procedure for determining the true dip of a plane from two apparent dips.

Method 5-9

Step 1: Visualize the problem as two inclined planes intersecting along a line (Fig. 5-20a). Prepare your overlay.

Step 2: Mark the strike of bedding (S80ºE) on the primitive, and revolve the overlay to bring this mark to the south point. Count 20º inward from the left end of the net along the west-east diameter. Draw the great circle representing bedding (Fig. 5-20b).

Step 3: Revolve the overlay back to its original position. Mark the strike of cleavage (N30ºE), and bring this mark to the north point. Count 70º inward from the right end of the net along the east-west diameter. Draw the great circle representing cleavage (Fig. 5-20c). The two

great circles intersect at a point (L) which represents the cleavage-bedding intersection lineation.

Step 4: Revolve the overlay to bring the lineation (L) to the north-south diameter. Count inward from the primitive along the north-south diameter to obtain the plunge angle (Fig. 5-20d). Mark the position where the radial line from L intersects the primitive. This tick mark will give the trend of the lineation.

Step 5: Revolve the overlay back to its original position. Read off the trend of the lineation from the position of the trend mark (Fig. 5-20e). The cleavage-bedding intersection lineation has a plunge of 20º toward S24ºW.

EXERCISES

1. Plot each of the following bed orientations as great circles on the same tracing paper overlay. Label each plane. Be sure to visualize each plane as you plot it.

   (a) N25ºE,44ºNW   (b) N14ºW,85ºSW
   (c) N83ºW,43ºNE   (d) 072º,06ºSE

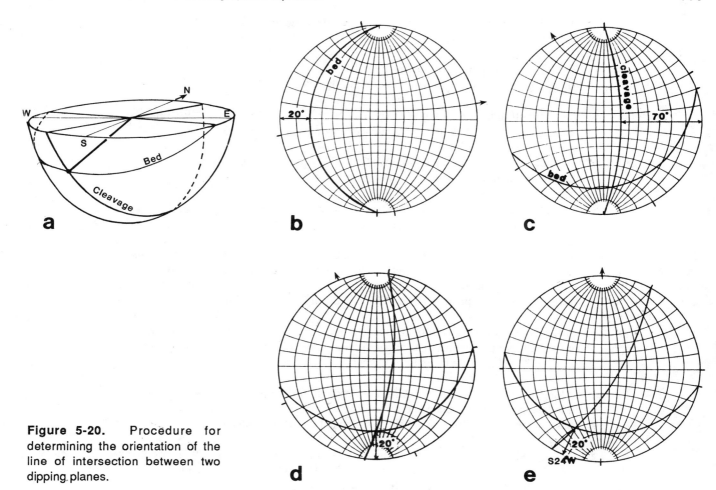

Figure 5-20. Procedure for determining the orientation of the line of intersection between two dipping planes.

(e) 234°,18°NW   (f) 090°,38°N
(g) 047°,01°NW   (h) 180°,90°E

2. Plot the following lineations as points on the same tracing paper overlay. Label each lineation. Be sure to visualize each line as you plot it.

(a) 32°,087°   (b) 43°,217°
(c) 12°,N12°E   (d) 88°,092°
(e) 86°,270°   (f) 59°,N60°E
(g) 59°,S60°E   (h) 59°,N60°W

3. A polydeformed metamorphic rock contains two mineral lineations that lie on the same foliation plane. The first is oriented 14°,N10°E. The second is oriented 58°,S58°E.

   (a) What is the attitude of the foliation plane on which these lineations occur?
   (b) What is the rake of each lineation in the foliation plane?
   (c) What is the angle between the lineations as measured in the plane of foliation?

4. A slip lineation on a fault plane has a rake of 68°NE. The fault is oriented N52°E,83°SE. What is the plunge and bearing of this lineation?

5. A slickenside surface dips N10°E,80°W. Fibrous slickenlines on the surface trend N60°W.

    (a) What is the plunge of the lineation?
    (b) What is the rake of the lineation on the slickenside surface?

6. The true attitude of bedding at Flagstone, New York, is N41°W,65°NE. What is the apparent dip of bedding on a vertical quarry face that trends N20°E?

7. A shale bed has the attitude N65°W,42°SW. What is the apparent dip of the bed in the direction S85°W?

8. The apparent dip of bedding on a quarry wall is 23°,S48°W. On a second quarry wall, it is 34°,N55°W. What is the true strike and dip of the bed?

9. In the Granite Wash Mountains of Arizona, bedding in a flysch sequence is oriented N47°E,34°NW. A spaced cleavage in this unit is oriented N22°E,68°SE. The intersection of bedding with cleavage produces a pronounced lineation that is visible on bedding-plane surfaces. What is the attitude of this lineation?

10. A sandstone bed strikes 140° across a stream. The stream flows down a narrow gorge with vertical walls. The apparent dip of the bed on the walls of the gorge is 25°,095°. What is the true dip of the bed?

11. Along a curving road cut, two different apparent dips on the same contact between a limestone bed and a shale bed were measured. The first was 42°,N30°W and the second was 58°,S70°W. What is the strike and dip of the contact?

12. A fault plane displays two different sets of slip lineations. One set is oriented 22°,325°, and the other set is oriented 49°,041°. What is the orientation of the fault plane that contains the two lineations?

6

STEREOGRAPHIC POLES AND ROTATIONS

6-1 INTRODUCTION

At this point you should feel comfortable with the basic concept of a stereographic projection and should have no trouble visualizing and plotting lines and planes on a stereonet. We now introduce the concept of a *pole to a plane*, describe how it is plotted on a stereonet, and show how it can be used to calculate angles between structures. Once you are adept at plotting poles, we can show how a structure can easily be *rotated* from one orientation to another on a stereonet. Finally, we will apply our ability to plot poles and rotate structures to several practical problems in structural geology. The techniques introduced in this chapter demonstrate the incredible versatility of the stereonet.

6-2 POLE TO A PLANE

A stereogram showing the great-circle traces of many planes is difficult to read because the traces of different planes cross one another and become hard to separate and identify. Fortunately, it is possible to represent the orientation of any plane by specifying the orientation of the *normal* to the plane (Fig. 6-1a). Remember that a normal is the line that is perpendicular to a plane; while a plane projects as a circular arc on the stereographic projection, the normal to the plane is a line and hence projects as a point. The point on a stereogram representing the normal to a plane is called the *pole to the plane* and, by definition, lies 90° from the center of the great circle trace representing the plane (Fig. 6-1b). You can use your hands to help visualize a pole. Hold a pencil between the fingers of your hand so that it is perpendicular to your hand; if your hand represents the plane, the pencil represents the pole to the plane (Fig. 6-1c). The distance of the pole from the center of the primitive is r tan(∂/2), where ∂ is the dip of the plane, and r is the radius of the stereogram (Fig. 6-1d). You might think of a plot of the cyclographic trace of a plane or of a line as a *direct plot*, whereas a plot of a pole is a *reciprocal plot*. In summary, every plane has a unique normal to it, which plots as a unique point (pole) on the stereographic projection. Therefore, we may represent the orientation of any plane by its pole. Diagrams representing poles to surfaces (S-planes) are referred to as *π-diagrams* or *S-pole diagrams*.

Method of Plotting a Pole

Problem 6-1
A bed has an orientation of N30°W,50°SW. Plot the stereographic projection of the bed and its pole.

Method 6-1
Step 1: Visualize the problem. The normal to the plane pierces the projection sphere at a point 90° from the great circle intersection of the plane and the projection sphere (Fig. 6-2a). Prepare your overlay as described in Chapter 5.
Step 2: Mark the strike of the bed on the primitive,

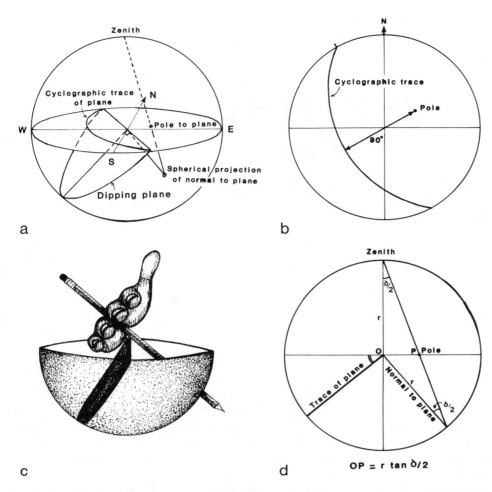

a

b

c

d

$OP = r \tan \delta/2$

Figure 6-1. Relationship between a plane and its pole. (a) Projection sphere showing a dipping plane and its normal, and their projections (cyclographic trace and pole) on the horizontal equatorial plane; (b) stereographic projection showing the cyclographic trace of a dipping plane and its pole; (c) drawing to help visualize the meaning of a pole; (d) vertical cross section through the projection sphere showing the trace of a dipping plane and its normal. The pole (P) is the stereographic projection of the normal. (Adapted from Hobbs et al., 1976.)

and revolve the overlay to bring the strike mark to the north-south axis. Count off the dip (50°) of the plane from the left edge of the primitive inward along the west-east diameter. Draw the great-circle trace for the plane.

Step 3: From the great-circle trace, continue to count 90° inward toward the center and then beyond the center along the west-east diameter. Plot the point P on the west-east diameter to represent the pole; its orientation can be determined just like that of a line (see Chapter 5). The pole (P) has a plunge of 40°,N60°E (Fig. 6-2c).

Note that the plunge of the pole is the complement of the bed dip (40°+50°=90°), and the trend of the pole is the complement of the bed strike (30°+60°=90°). Thus, after bringing the strike mark to the north-south axis (Step 2), we could also locate the pole by counting the dip angle from the center outward in the direction opposite to the dip direction.

Determination of an Intersection Lineation

In Chapter 5 we saw that the intersection of the great circle traces of two planes on a stereogram is a point that

represents the line of intersection between the two planes. The line of intersection can also be determined in terms of the poles to the planes.

Problem 6-2

One limb (labeled A) of a chevron fold is oriented 020°,60°SE and the other limb (labeled B) is oriented 060°,40°NW. What is the plunge and bearing of the fold hinge? Assume that the fold hinge is the line of intersection between the two limbs.

Method 6-2

Step 1: Visualize the problem. The fold hinge is the pole to the plane that contains the normals to the two limbs (Fig. 6-3a). Prepare your overlay.

Step 2: Plot the great-circle traces of the two limbs. Then, plot point P_A, which is the pole to the first limb (limb A), and point P_B, which is the pole to the second limb (limb B; Fig. 6-3b,c).

Step 3: Revolve the overlay so that P_A and P_B lie over the same great circle on the stereonet, and trace this great circle (Fig. 6-3d). This great circle defines the plane that contains both P_A and P_B. Count 90° to the right along the east-west diameter, passing through the center of

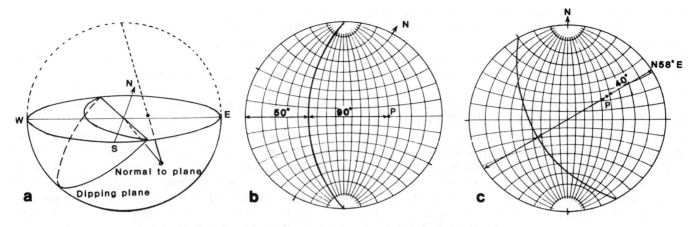

Figure 6-2.   Procedure for plotting the pole to a dipping plane on a stereographic projection.

the stereonet, and locate the pole (point F) to this plane (Fig. 6-3d).

Step 4: Determine the plunge and bearing of point F. Point F represents the line of intersection between the two limbs and thus is the hinge line of the fold. The fold hinge is oriented 22°,035°. Note that the same fold hinge could have been obtained from the point of intersection of the cyclographic traces representing the two limbs.

6-3   ANGLES BETWEEN LINES AND PLANES

Dihedral Angle between Two Planes

A *dihedral angle* is the angle between two planes measured in a third plane that is perpendicular to both planes. The dihedral angle is easily determined by measuring the angle

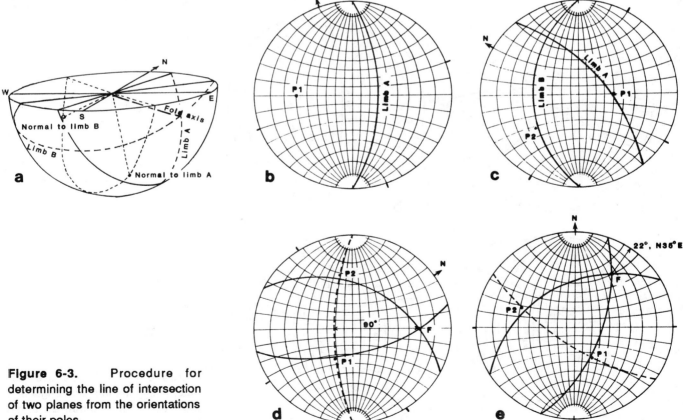

Figure 6-3.   Procedure for determining the line of intersection of two planes from the orientations of their poles.

between the poles to the planes on a stereogram. Note that the poles are lines, and that the angle between two lines can be measured only on a plane containing the two lines. Thus, on a stereogram, the angle between the two poles is measured along the great circle that passes through both poles.

In many cases (e.g., crossing joints) the dihedral angle is usually specified as an acute angle. Therefore, if the measured angle between poles is acute, it may be used directly as the dihedral angle, but if the measured angle between poles is obtuse, its supplement is recorded as the dihedral angle. Specification of the *interlimb angle* of a fold, however, is a bit more tricky. The interlimb angle is always the supplement of the angle between the poles and can be either acute or obtuse. The problem is that it is not always immediately obvious whether to use the acute angle between the poles or the obtuse angle between the poles, as measured on the stereogram, in your calculation of the supplement. In order to make the proper choice, you must visualize the fold and perhaps make a profile sketch (see Problem 6-3).

Problem 6-3

One limb (limb A) of an antiform is oriented N20°E,70°NW. The other limb (limb B) is oriented N30°W,65°NE. Determine the interlimb angle of the fold.

Method 6-3

Step 1: Visualize the problem (Fig. 6-4a), and prepare your overlay.

Step 2: Plot the great circle representing limb A. Locate the pole (P_A) to this great circle (Fig. 6-4b).

Step 3: Repeat the procedure in step 2 and locate the pole (P_B) that represents the normal to limb B (Fig. 6-4c).

Step 4: Revolve the overlay till the two poles lie on the same great circle. This great circle is perpendicular to the line of intersection of the two limbs. The angle between the two poles measured along this great circle is an obtuse angle of 114° (Fig. 6-4d). If you visualize the fold again, you realize that the interlimb angle is an acute angle; it is, therefore, given by the supplement of 114° namely, 66°.

Angle between a Line and a Plane

The angle between a line and a plane is measured by the angle between the line and its orthographic projection on the plane. This angle must be measured in a second plane that both contains the line and is perpendicular to the first plane; in other words, the perpendicular plane must contain the normal to the first plane. On a stereographic projection the angle between a line and a plane is measured along the great circle containing the line and the pole to the plane.

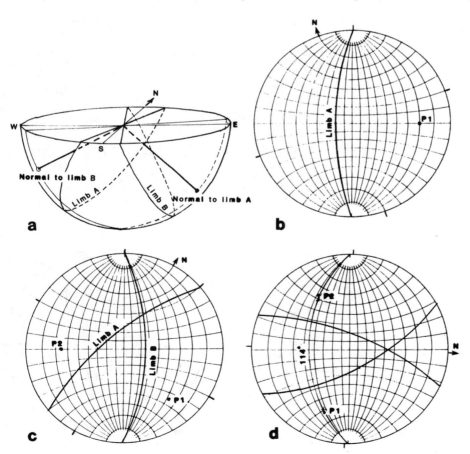

Figure 6-4. Procedure for determining the interlimb angle of a fold using the dihedral angle between the poles to the fold limbs.

Problem 6-4

A fold hinge plunges 30°,260°. A cross section of the fold is exposed on a joint surface whose attitude is 150°,60°E. Determine the angle between the fold hinge and the joint surface.

Method 6-4

Step 1: Visualize the problem (Fig. 6-5a). The angle between the line and the plane must be measured on a specific plane, which is shaded in the drawing. Prepare your overlay.

Step 2: Plot the fold hinge (H) on the stereogram (Fig. 6-5b).

Step 3: Plot the great circle representing the joint surface on the stereogram. Count over 90° along the east-west diameter to find the pole (P) to this plane (Fig. 6-5c).

Step 4: Revolve the overlay until the hinge (H) and the pole (P) lie along a common great circle. The great-circle trace through H and P intersects the great-circle trace representing the joint at point X (i.e., point X represents the line of intersection between the two planes). Measure the angle between H and X along the great circle. The angle between the hinge line and the joint face is 75° (Fig. 6-5d).

Note that the joint face is not perpendicular to the hinge line. Therefore, the exposure of the cross section of the fold on the joint face is not a profile view of the fold.

Bisecting the Angle between Two Planes

A plane that bisects the angle between two intersecting planes must contain the line of intersection between the two planes and the line that bisects the dihedral angle between the planes. For most chevron folds and kink folds it is reasonable to assume that the plane that bisects the angle between the two limbs of the fold and contains the fold hinge is the axial plane of the fold.

Problem 6-5

The two limbs of a chevron fold (limb A and limb B) are oriented N60°E,40°SE and N80°W,50°NE, respectively. Determine the attitude of the axial plane of this fold.

Method 6-5

Step 1: Visualize the problem. We will make the assumption that the axial plane bisects the angle between the two limbs (Fig. 6-6a). Prepare your overlay.

Step 2: Draw the great circle representing limb A,

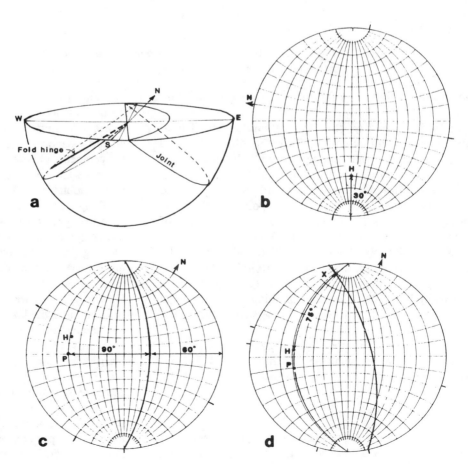

Figure 6-5. Procedure for determining the angle between a line and a plane.

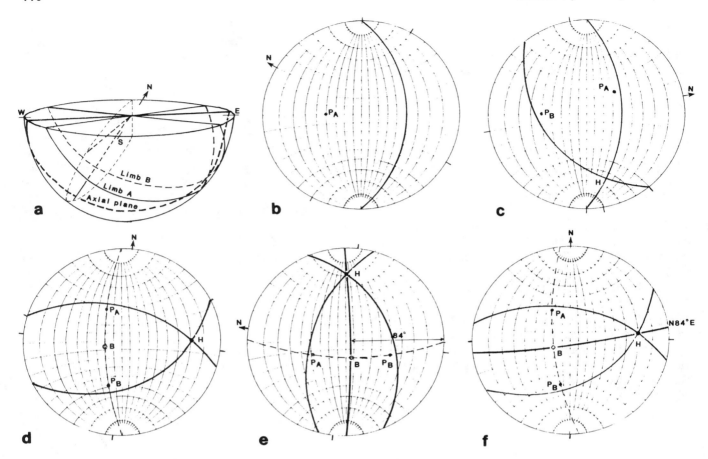

Figure 6-6. Procedure for determining the plane bisecting the angle between two intersecting planes.

and locate its pole (P_A) 90° from the great circle (Fig. 6-6b).

Step 3: Similarly, plot the great-circle trace of limb B, then locate and plot its pole (P_B) (Fig. 6-6c). The two great-circle traces (representing the two limbs) intersect at point H, which represents the hinge line of the fold.

Step 4: Revolve the overlay until the two poles lie on the same great circle, and find the point (B) along the great circle that has the same angular distance from each of the two poles to the fold limbs (Fig. 6-6d).

Step 5: Revolve the overlay till B and H lie on the same great circle, and trace the great circle. This great circle represents the plane that bisects the angle between the two limbs and contains the fold hinge and thus is the axial plane of the fold. Its dip, from the stereogram, is 84° (Fig. 6-6e).

Step 6: Revolve the overlay back to its original position. The axial plane has a strike of N84°E (Fig. 6-6f). Thus, the attitude of the axial plane is N84°E,84°S.

6-4   ROTATION

In order to solve certain problems in structural geology it is necessary to simulate the physical rotation of a structural

element around an axis in space. For example, *rotation* around an axis may be needed in order to undo the reorientation effects of late-stage tilting on the attitude of an early-stage fold. In order to accomplish this we rotate the geometric elements of the fold through a specified angle around a specified axis and then replot the elements in a new orientation. This process is quite different from anything we have done so far. In all the plotting procedures described previously, we have revolved the overlay about the center of the net for convenience in measuring and plotting data. The overlay always carried with it a fixed north mark, so the orientations of planes and lines never changed with respect to the stereogram reference frame. However, when we rotate a line or plane in space, the orientation of the line or plane does change with respect to the stereogram (overlay) reference frame. Note that we use the words "revolve" and "rotate" differently in this book for clarification purposes; revolve refers to physically moving the overlay with respect to the stereonet, and rotate refers to actual reorientation of structural elements.

There are two basic procedures that can be used to achieve any rotation: (1) rotation about a *vertical axis* (the axis has a plunge of 90°) and (2) rotation about a *horizontal axis* (the axis has a plunge of 00°). Rotation around any *inclined axis* (where the plunge of the axis lies between 00° and 90°) is most easily performed by a combination of rotations around horizontal and/or vertical axes.

Rotation about a Vertical Axis

Rotation about a vertical axis is the simplest type of rotation to perform. The axis for such a rotation is the center point of the stereonet. *Rotation of a line* is achieved by moving the point that represents the line along a small circle that is coaxial with the primitive; when the rotation is complete, you simply replot the line. Note that small circles that are coaxial with the primitive are *not* the small circles inscribed on the stereonet grid. To visualize why the small circle track makes sense, imagine an inclined line with one end fixed on the vertical rotation axis; if this line is rotated around the axis, its free end describes a right-circular cone that intersects the projection sphere in a small circle (Fig. 6-7a). *Rotation of a plane* is achieved by moving the strike line of the great-circle trace around the primitive; when the rotation is complete, the plane is replotted. The same result is achieved if the pole to the plane is rotated around a small circle that is coaxial with the primitive.

Once a rotation around a vertical axis has been accomplished, the line or plane has a new orientation with respect to the stereogram reference frame. For a line the plunge direction changes, but the plunge magnitude does not, and for a plane the strike changes, but the dip magnitude does not. In order to uniquely specify the sense of rotation, you must indicate your viewing direction with respect to the axis. A clockwise rotation viewed looking down a vertical axis is the same as a counterclockwise rotation viewed looking up along the same axis.

Problem 6-6
The attitude of a line is 40°,S50°E. What is the orientation of the line after it has been rotated by 40° clockwise around a vertical axis (viewed looking down the axis)?

Method 6-6
Step 1: Visualize the problem. As the line rotates, it describes a segment of the surface of a right-circular cone whose axis passes through the center of the stereonet (Fig. 6-7a). Prepare the overlay.
Step 2: Plot the point (L) representing the line on the overlay (Fig. 6-7b).
Step 3: Draw a tick mark on the primitive that

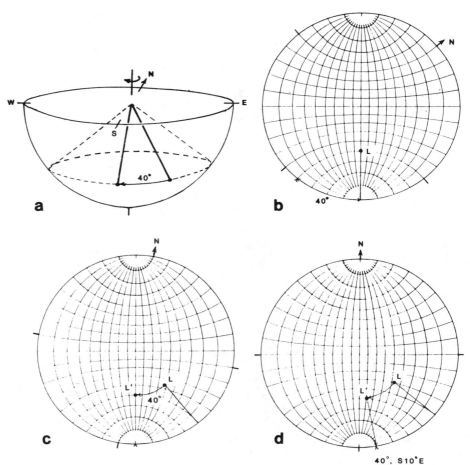

Figure 6-7. Procedure for rotating a plunging line around a vertical axis.

indicates the bearing of the line. Without moving the overlay, count 40° clockwise along the primitive, and make a second tick at S10°E (Fig. 6-7b).

Step 4: Revolve the overlay so that the second tick lies over the north-south diameter. Count in by 40° from the primitive and mark point L', which represents the rotated position of the line in the stereogram reference frame (Fig. 6-7c). The rotated line is oriented 40°,S10°E (Fig. 6-7d). Note that the bearing of the line changed, but not the plunge of the line.

Problem 6-7

A plane is oriented N30°W,40°NE. What is the orientation of the plane after it has been rotated by 70° counterclockwise around a vertical axis (viewed looking down the axis)?

Method 6-7

Step 1: Visualize the problem (Fig. 6-8a), and prepare your overlay.

Step 2: Plot the great-circle trace representing the plane, and plot its pole P (Fig. 6-8b).

Step 3: Locate the N30°W strike mark on the primitive. Count 70° counterclockwise along the primitive

and make a tick. This tick is at S80°W (Fig. 6-8b) and represents the new strike of the plane.

Step 4: Bring the S80°W tick mark to the north-south axis, count in 40° along the east-west axis, plot the new great-circle trace (Fig. 6-8c), and plot the new pole P'.

Step 5: Revolve the overlay so that the north arrow overlies the north pole of the stereonet. Now you can read the strike of the rotated plane. It is S80°W (or N80°E). The dip angle of the plane does not change during the rotation. The orientation of the plane after rotation is, therefore, N80°E,40°SE (Fig. 6-8d). Note that P' lies 70° counterclockwise from P along a small circle that is concentric with the primitive.

Rotation about a Horizontal Axis

A line representing a horizontal rotation axis plots as a point along the primitive of a stereonet. In order to simulate a rotation around a horizontal axis, we must first revolve the overlay so as to bring the rotation axis to the north-south diameter of the net. The rotation of any line around this axis is then accomplished by moving the point representing the line along a small-circle grid line on the

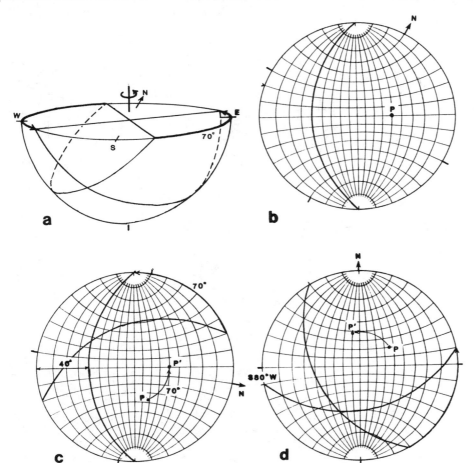

Figure 6-8. Procedure for rotating a dipping plane around a vertical axis.

stereonet. (*Note:* These small circles are *not* coaxial with the primitive). Rotation of a line past the primitive will cause the line to reappear on the diametrically opposite side of the stereogram.   Rotation of a plane can be accomplished by moving individual points of the great-circle trace that represents the plane along small-circle paths through equal angular distances. After the rotation, all the points will lie on a new great-circle trace that represents the orientation of the rotated plane (Fig. 6-9a,b). This method works both when rotating a plane around its own strike and when rotating a plane around a horizontal axis that is at an angle to the strike. If a plane is rotated around its own strike, only the dip of the plane will change. If the horizontal rotation axis is not parallel to the strike of the plane, the rotation will result in a change in both the dip and the strike of the plane.

During a rotation the pole to a plane moves through the same angular distance and in the same direction as the points on the great circle (Fig. 6-9a,b), so the rotation of a plane can also be accomplished merely by moving the point that represents the pole along a small circle on the stereonet. Once the rotated position of the pole is found, the great-circle trace of the new plane that it represents can be drawn.   Any linear structure (L) that has a fixed orientation with respect to the plane will also move through the same angular distance in the same direction along a small-circle path to a new position L' as the plane rotates.   A lineation that lies in a plane before rotation must lie in the rotated plane after rotation. Movement along small circles during rotation around a horizontal axis

makes sense, for remember that each small circle of the stereonet represents the intersection of a right-circular cone, whose axis is horizontal and trends north-south, with the surface of the projection sphere.

In performing rotations, it is very critical to visualize the process at every step, particularly to ensure that the rotation is in the correct direction.   Imagine a plane oriented north-south, 60°E.   If the plane is rotated counterclockwise, viewed looking north, its dip will decrease (Fig. 6-9a,b).   If it is rotated counterclockwise by 60°, it will become horizontal (and its pole will become vertical), and if it is rotated by more than 60° it will begin to dip to the west. If the plane is rotated clockwise, it will steepen (Fig. 6-9a, c).   If it is rotated clockwise by 30°, it will become vertical (and its pole will become horizontal). If it is rotated by more than 30°, it will go past vertical and begin to dip to the west (and the pole to the plane will jump to the diametrically opposite side of the stereogram). Rotation of an overturned plane back up to a horizontal position is sometimes difficult to visualize; remember that an overturned plane must pass through the vertical position before being rotated up to the horizontal position.   The following examples will help you to visualize rotations.

Problem 6-8 (Rotation of a line around a horizontal axis)

A mineral stretching lineation plunges 30° toward 220°. What is the orientation of this lineation after it has been rotated 30° counterclockwise (as viewed looking toward N10°W) around a N10°W-trending horizontal axis?

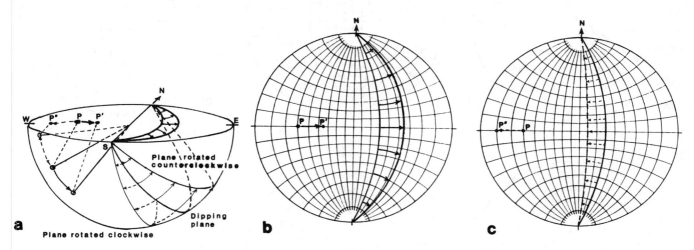

Figure 6-9. Rotation about a horizontal axis.   (a)   Spherical projection showing effects of clockwise and counterclockwise rotations on a dipping plane, its normal, and a lineation on the plane;   (b) stereographic projection showing counterclockwise rotation (about a horizontal north-south axis) of a great-circle trace, its pole, and a lineation;   (c) stereographic projection showing clockwise rotation of the same great-circle trace and its pole, about the same axis.

Method 6-8

Step 1: Visualize the problem (Fig. 6-10a). Note that the axis is oblique to the the bearing of the line. The rotation should increase the plunge of the line and change its bearing. Prepare your overlay.

Step 2: Plot the point (L) that represents the lineation on the overlay (Fig. 6-10b). Draw a tick mark at N10°W on the primitive; this mark represents the horizontal rotation axis (Fig. 6-10b).

Step 3: Revolve the overlay 10° clockwise so that the rotation axis coincides with the north-south axis of the stereonet. Point L has also moved and lies over a new small circle on the stereonet. Count off 30° to the right along this small circle, and make a new dot (L') on the overlay. L' represents the new position of the line after rotation (Fig. 6-10c).

Step 4: Revolve the overlay so that L' lies over the north-south diameter of the stereonet (Fig. 6-10d). Measure the plunge of the line represented by L'; the plunge is 51°. Make a tick on the primitive.

Step 5: Revolve the overlay so that the north arrow of the overlay is over the north mark of the stereonet (Fig. 6-10e). Determine the bearing of L'. L' is a line oriented 51°,202°.

Problem 6-9 (Rotation of a plane around strike)

The attitude of a plane is N20°E,20°SE. (a) What is the attitude of the plane after it has been rotated by 30° clockwise (as viewed looking northeast) around an axis parallel to strike? (b) What is the attitude of the plane after it has been rotated 30° counterclockwise (as viewed looking northeast) around an axis parallel to strike.

Method 6-9

Step 1: Visualize the problem (Fig. 6-11a). In case a, the clockwise rotation will cause the dip of the plane to increase by 30°. In case b, the counterclockwise rotation will cause the bed dip to decrease. A 20° counterclockwise rotation will take the bed to horizontal, so a 30° counterclockwise rotation will take the bed past horizontal so that it dips to the northwest. Note that since the rotation axis is the line of strike, in neither case does the rotation cause the strike of the plane to change. Prepare your overlay.

Step 2: Plot the great circle trace and the pole P of the original plane on the overlay. Rotate the overlay counterclockwise by 20° so that the intersection of the great-circle trace with the primitive lies over the north mark of the stereonet (Fig. 6-11b).

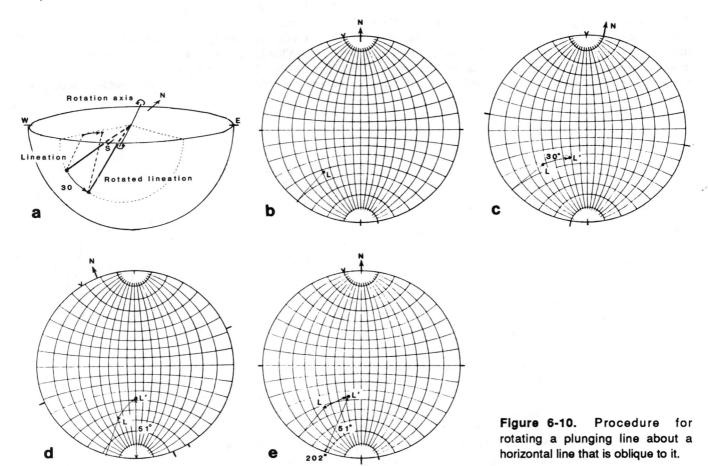

Figure 6-10. Procedure for rotating a plunging line about a horizontal line that is oblique to it.

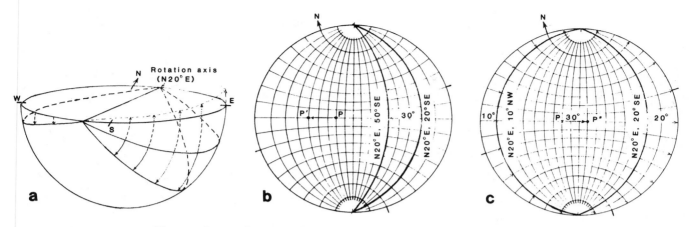

Figure 6-11. Procedure for rotating a dipping plane about its strike.

Step 3: To rotate by 30° clockwise, count over 30° to the left along the east-west diameter, and draw a new great-circle trace (Fig. 6-11b). This trace represents the rotated plane oriented N20°E,50°SE. The pole to the plane also moves 30° to the left (to P'), away from the center of the stereogram (Fig. 6-11b).

Step 4: To rotate 30° counterclockwise, start counting to the right from the original trace of the plane (Fig. 6-11c). Note that when you have counted 20° you have reached the primitive, and the plane is horizontal. Continued rotation causes the plane to begin to dip northwestward, and you must count an additional 10° inward from the diametrically opposite end of the primitive (Fig. 6-11c). Draw the new great circle trace. This trace represents the rotated plane, and it is oriented N20°E, 10°NW. During the rotation, the pole to the plane moves 30° to the right (to P"); the pole moves through the vertical position in the process (Fig. 6-11c).

Problem 6-10 (Rotation of a plane around an axis oblique to strike)

A dipping bed is oriented N20°E,70°SE. What is the orientation of the bed after it has been rotated 30° counterclockwise (viewed from the south) around a horizontal axis that trends N10°W?

Method 6-10

Step 1: Visualize the problem (Fig. 6-12a). Note that the rotation axis is oblique to the strike of the plane. Thus, rotation around the axis will not only change the dip of the plane, but will also change its strike. Prepare your overlay.

Step 2: Plot the great-circle trace of the plane and the pole to the plane (P) (Fig. 6-12b). Next, make a tick indicating N10°W on the primitive to indicate the trend of the rotation axis.

Step 3: Revolve the overlay clockwise so that the rotation axis lies over the north mark of the stereonet.

Point P now lies over a new small circle on the stereonet (Fig. 6-12c). Count 30° to the right along the small circle and plot point P', which is the pole to the rotated plane.

Step 4: Revolve the overlay so that P' lies over the east-west diameter of the stereonet. Count 90° along the east-west diameter and draw the great-circle trace of the rotated plane (Fig. 6-12d).

Step 5: Revolve the overlay so that its north arrow is over the north mark of the stereonet (Fig. 6-12e). The rotated plane is oriented N30°E,45°SE.

Problem 6-11 (Rotation of a lineation in a plane)

An overturned bed is oriented N30°W,40°SW. A current lineation on the bedding plane has a rake of 30°NW. What was the orientation of the lineation when the bedding was horizontal?

Method 6-11

Step 1: Visualize the problem (Fig. 6-13a). In order for the bed to rotate back to horizontal, it must be rotated around its line of strike by 50° counterclockwise (viewed looking northwest) to bring it to the vertical position and then an additional 90° counterclockwise to bring it to horizontal. Prepare your overlay.

Step 2: Plot the great-circle trace representing the plane and plot the lineation (L) on the plane, following methods described earlier. Orient the overlay so that the strike line of the bed lies over the north-south diameter of the stereonet (Fig. 6-13b).

Step 3: From the great-circle trace, count 140° to the right along the east-west diameter and trace the new great circle that coincides with the primitive (Fig. 6-13b). This great-circle trace represents the horizontal right-side-up plane. Note that in order to reach this position, the plane passed through the vertical position.

Step 4: At the end of step 2, point L lies on a small circle. Count 140° to the right along this small circle and

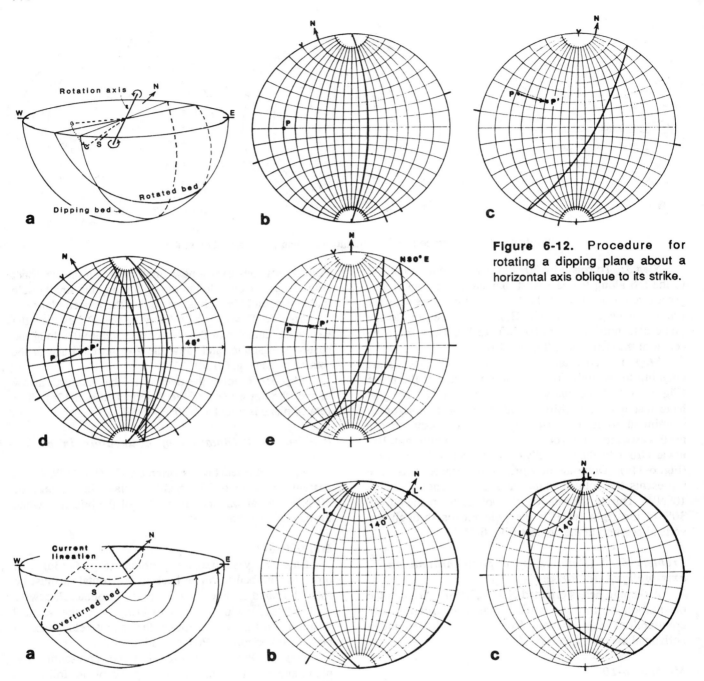

Figure 6-12. Procedure for rotating a dipping plane about a horizontal axis oblique to its strike.

Figure 6-13. Procedure for rotating a lineation on a plane about the strike of the plane.

plot point L'. Point L' represents the horizontal position of the lineation after rotation. Note that L' lies on the trace of the great circle representing the rotated plane and that it still has the original rake. Note also that the bearing of the lineation is quite different from the bearing that it had before rotation. The rotation changes the plunge and bearing of the lineation, but not its rake.

Step 5: Revolve the overlay so that the north arrow of the overlay coincides with the north mark of the

stereonet. The original trend of the lineation (its trend before the bed was tilted) can now be determined directly; it is due north and is horizontal (00°,000°) (Fig. 6-13c).

Rotation about an Inclined Axis

It is possible to perform rotation around an inclined axis directly, but the method is cumbersome. It is far easier to accomplish such a rotation by a three-stage process. First, the inclined axis is rotated to horizontal (or vertical) around a second axis that is horizontal and is orthogonal to the inclined axis. The structural elements to be rotated go through the same angular movements (along small-circle paths) as did the axis during this step. Second, the necessary rotations are performed around the rotated axis, following the methods described earlier. Third, the axis is rotated back to its original inclined position. Once again, the rotated structure elements pass through the same angular movements along small-circle paths.

Problem 6-12 (Rotation around an inclined axis)

The bedding on one limb of a fold is oriented N60°W,40°SW. What is the orientation of the bedding after it has been rotated by 40° counterclockwise (viewed looking down plunge) around the fold axis, which plunges 30°,S80°W?

Method 6-12

Step 1: Visualize the problem and prepare your overlay (Fig. 6-14a). In this problem we first rotate the fold axis to horizontal, then we rotate the bedding by the required amount, and finally, we rotate the fold axis back to its original position.

Step 2: Plot the position of the rotation axis (fold axis), R, and the pole to the fold limb, P (Fig. 6-14b).

Step 3: Rotate the overlay counterclockwise so that R lies over the east-west diameter of the stereonet. Rotate R to horizontal (Fig. 6-14c) by counting 30° along the east-west diameter to the left until you reach the primitive, and plot R' at this position on the primitive. R' indicates the position of the axis after it has been rotated to horizontal. Move P by the same amount and in the same direction along the small circle that underlies it, and plot point P'. Point P' represents the position of the pole after it was carried along passively during the operation that brought the fold axis to horizontal (Fig. 6-14c).

Step 4: Revolve the overlay so that R' lies over the north mark of the stereonet. P' is carried along on the overlay. Rotate P' counterclockwise around R' by counting 40° to the right along the small circle that underlies P'. Plot point P" at this locality (Fig. 6-14d). R' does not move during this stage; it is the rotation axis.

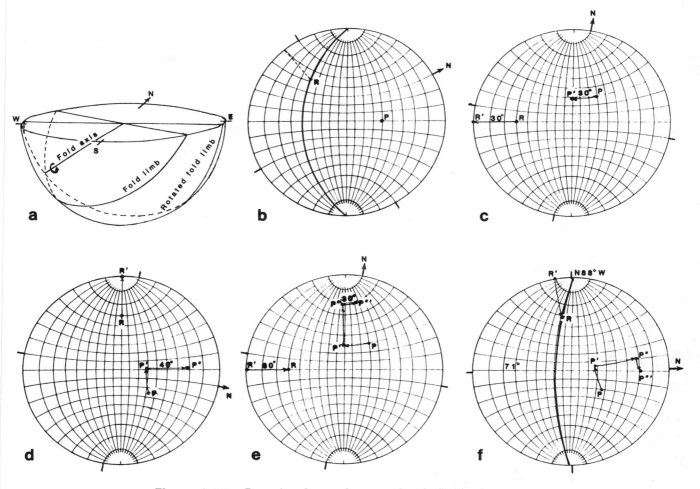

Figure 6-14. Procedure for rotation around an inclined axis.

Step 5: Revolve the overlay so that R' is again over the east-west diameter. P" is passively carried along. Count in by 30° along the east-west diameter, so that you return R' to R (Fig. 6-14e). Move P" along the small circle that underlies it by 30° to the right and plot a new point P'''. P''' is the final rotated position of the pole (Fig. 6-14e).

Step 7: Revolve the overlay so that P''' lies over the east-west diameter. Count off 90° along the east-west diameter and trace the great circle that represents the rotated plane (Fig. 6-14f). Its orientation is N88°W,71°S.

6-5 APPLICATIONS OF STEREOGRAPHIC ROTATIONS

In this section we will work through a few typical examples that illustrate how stereographic rotations can be used to solve geological problems. You need not learn any new procedures; all that is necessary is to reword each problem so that you can define an axis and an amount of rotation.

Pretilt Orientation of Structures below an Unconformity

Problem 6-13

The beds above an angular unconformity are presently oriented N10°E,50°W. The beds below the unconformity are presently oriented N40°E,80°E. What was the orientation of the subunconformity beds prior to tilting of the unconformity?

Method 6-13

Step 1: Visualize the problem. Assume that the beds above the unconformity have the same dip as the unconformity surface and that the unconformity surface was originally horizontal. Removal of the effects of postunconformity tilting can be accomplished by rotating the unconformity to horizontal by an amount equal to the dip of the unconformity around an axis parallel to strike of the unconformity (Fig. 6-15a). Prepare your overlay.

Step 2: Plot the great-circle trace of the unconformity, and mark an X on the primitive where this trace intersects the primitive at N10°E. Point X represents

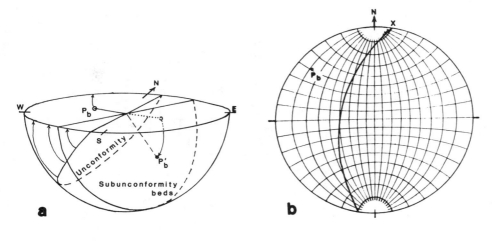

Figure 6-15. Procedure for determining the pretilt orientation of structures below an unconformity.

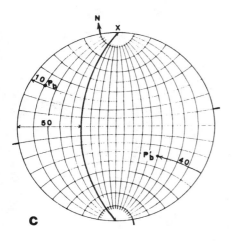
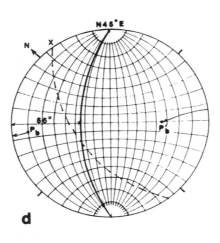

the horizontal rotation axis.   Plot the pole (P_b) representing the attitude of the subunconformity beds. (Fig. 6-15b).

Step 3:   Revolve the overlay counterclockwise by 10° so that point X lies over the north-south axis of the stereonet (Fig. 6-15c).   Rotate the great-circle trace representing the unconformity out by 50° to the left so that it coincides with the primitive.   This operation brings the unconformity to horizontal.

Step 4:   With the overlay in this position, move point P_b to the left along the underlying small circle by 50°, and plot P_b' at this locality.   Note that after 10° rotation along the small-circle path, the pole becomes horizontal; to complete the rotation, move to the diametrically opposite point on the stereonet, and continue for an additional 40° of rotation. P_b' is the pole to the subunconformity beds after the unconformity has been rotated to horizontal.

Step 5:   Revolve the overlay so that P_b' lies over the east-west diameter of the stereonet, count 90° along the east-west diameter, and draw the great-circle trace (Fig. 6-15d).   Revolve the overlay back to its original position, so that the north arrow of the overlay aligns with the north mark of the stereonet.   The new great-circle trace now represents the pretilting orientation of subunconformity beds.   Prior to tilting, the subunconformity beds were oriented N45^{o}E,55°NW.

Unfolding and Refolding Folds

If a lineation occurs on the limb of a nonplunging fold, calculation of the pretilt orientation of the lineation is relatively easy. The plane on which the lineation occurs is returned to horizontal by rotating it through an angle equal to its dip around an axis parallel to its strike, following the procedure described in Method 6-11. The lineation is rotated passively along with the plane. If the fold is plunging, the procedure for unfolding the fold is more difficult. Usually, a two-stage procedure is used similar to that described in Method 6-12. First, the plunging axis of the fold is rotated to horizontal; then, the limbs are rotated around this new horizontal axis so that they become horizontal. Lineations on the limbs are passively rotated along with the limbs.

Problem 6-14 (Unfolding and refolding a plunging fold)

The axis of an anticline is oriented 30°,N10^{o}E. The east limb is oriented N19^{o}W,50°E and contains sole marks that trend due east.   The west limb is oriented N50^{o}E,40°NW. (a) Determine the orientation of the sedimentary lineation before folding. (b) What is the present orientation of the sole marks on the west limb?

Method 6-14

Step 1:   Visualize the problem (Fig. 6-16a,b). First, we must determine the prefolding orientation of the sole marks; then, we must refold the beds to determine the orientation of the lineation on the west limb of the fold. Prepare your overlay.

Step 2:   Plot the great circles representing the two limbs and locate their poles (P1 and P2, respectively). The great circles intersect at the fold axis (F). Also plot the orientation of the lineation (L) on the east limb (Fig. 6-16c).

Step 3:   Revolve the overlay so that Point F lies on the east-west diameter of the stereonet. Rotate F around a horizontal axis up to horizontal by counting along the east-west diameter to the primitive.   Plot Point F' on the primitive to represent the rotated axis (Fig. 6-16d). The points representing the other structures move concurrently along small circles: lineation L moves to L', pole P1 moves to P1', and pole P2 moves to P2' (Fig. 6-16d).

Step 4:   Revolve the overlay so that the F' lies on the north-south diameter of the stereonet; it now serves as a horizontal rotation axis (Fig. 6-16e). Rotate the east limb to the horizontal position by moving P1' to the center of the stereonet and replotting it as P1". The rotated lineation L' must be horizontal when the bed is horizontal; thus, it moves along a small circle to L" on the primitive (Fig. 6-16e). The position of L" gives the original orientation of the sedimentary lineation; it trends N72^{o}E. This is the answer to part (a) of the problem.

Step 5:   Now that we know the trend of the lineation on unfolded bedding, we can refold the bed and determine the present orientation of the lineation on the west limb. First, rotate the west limb to the horizontal position; its pole (P2') moves to the center of the stereogram (P2") (Fig. 6-16f). Remember that F' is the rotation axis. The point diametrically opposite to L" now gives the original orientation of the sedimentary lineation on the west limb (L*).

Step 6:   Rotate the west limb back to its folded position with respect to a horizontal axis; its pole (P2") moves back to its original position (P2'), and the lineation (L*) moves along a small-circle path to a new position (L**) (Fig. 6-16f). We have not yet accommodated for the plunge of the fold.

Step 7:   Revolve the overlay to bring the fold axis (F') back to the east-west diameter of the stereonet, and rotate the fold axis (F') back to its original orientation (F) by counting inward along the east-west diameter (Fig. 6-16g).   At the same time all other points move along small-circle paths by the same amount and in the same direction. P2' moves back to P2, and L** moves to L*** (Fig. 6-16g).

Step 8:   Revolve the overlay back to its original orientation.   Now L*** gives the orientation of the

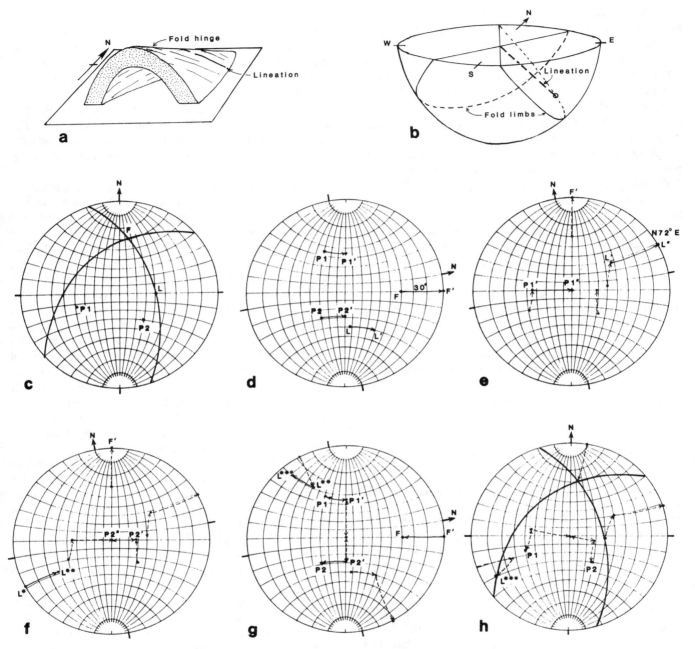

Figure 6-16. Procedure for unfolding and folding a plunging fold and determining the orientation of a prefolding lineation.

lineation on the west limb of the fold (Fig. 6-16h). The lineation on the west limb is oriented 09°,S61°W.

Determining Net Slip on a Fault

In Chapter 4, we saw how to calculate net slip on a vertical fault from map data entirely through the use of descriptive geometry. The construction is considerably simplified if the angular relationships between various planes and lines are obtained stereographically and the rotations are

performed stereographically. The information is then transferred to the orthographic construction. The following example is the most general case of net slip determination (i.e., an inclined fault plane). Special cases of this problem (e.g., a vertical fault plane) can, of course, be solved by modifying this method.

Problem 6-15

A fault (FF') is oriented N90°E,40°S. A dike dipping N30°W,35°E is exposed at A on the south side of the fault, and at A' on the north side of the fault. A bed dipping N30°E,60°W is exposed at B on the south side of the fault and at B' on the north side of the fault. The relative positions of A, A', B, B' are shown on the map (Fig. 6-17d). Determine the azimuth of the horizontal projection of net slip, the plunge of net slip, the rake of net slip on the fault plane, and the amount and relative movement direction of net slip.

Method 6-15

Step 1: Visualize the problem (Fig. 6-17a,b). Complete the map showing the outcrop pattern of the fault, and the bed and the dike north and south of the fault (Fig. 6-17d). Prepare a stereographic overlay.

Step 2: Plot the attitude of the fault and the dike on the stereogram (Fig. 6-17c); the point of intersection of these two great circles gives the attitude of the trace of the dike on the fault plane. Plot the attitude of the bed on the stereogram (Fig. 6-17c). The point of intersection of this

great circle and the fault great circle gives the attitude of the trace of the bed on the fault plane.

Step 3: To determine the magnitude of net slip we need to construct a section parallel to the fault plane. In this section the orientations of all lines (such as bed traces) on the fault plane can be represented by their rakes with respect to the fault strike. The rake of the dike and of the bedding traces on the fault are measured directly off the stereogram (Fig. 6-17c) by bringing the fault strike to the north-south axis of the stereonet and counting along the great-circle trace of the fault.

Step 4: Draw a section (Fig. 6-17e) parallel to the fault plane, with the dike and bed traces in their proper orientations (using the measured rakes) and their proper relative positions (passing through A, A', B, B'). The traces through A and B intersect at S' on the south side of the fault. The traces through A' and B' intersect at N' on the north side of the fault. The line N'S' gives the magnitude of net slip; it is 25 m (use the same scale as the

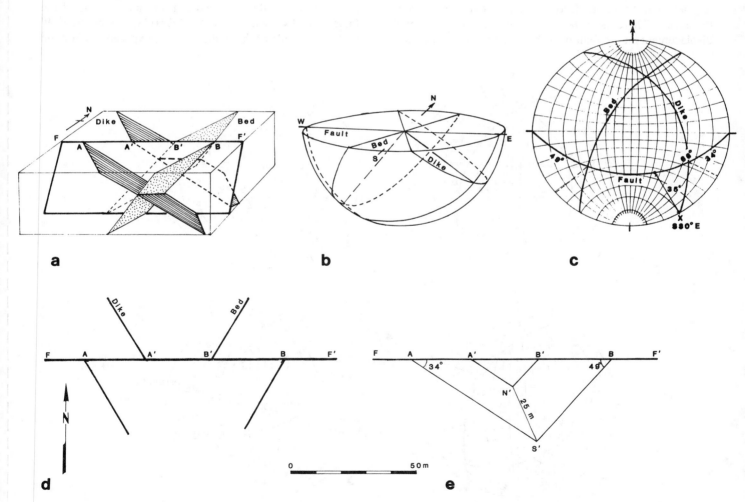

Figure 6-17. Procedure for determining net slip on a fault, using a stereographic projection to determine angular relationships.

map). It is also evident from this diagram that N' has moved up with respect to S'; therefore, because the fault dips south, it is a normal fault.

Step 5: Measure the rake of the net slip on the fault plane (Fig. 6-17e) with a protractor, by measuring the acute angle between the fault strike line and the line of net slip.

Step 6: Return to the stereogram, and plot a line with the same rake as the net slip so that it plots as a point (X) on the great-circle trace of the fault (Fig. 6-17c). The plunge and bearing of X gives the attitude of the line representing the net slip. The plunge of this line can be read directly off the stereogram by bringing this point to the NS axis of the net. The plunge of net slip is 35°. The bearing of net slip is S30°E.

Determining Displacement for Rotational Faulting

Fault tips (where faults end) are areas in which the displacement on a fault decreases to zero over a very short distance. There is usually a component of rotational displacement on the fault over this short segment. For any fault with rotational motion, the magnitude of displacement as a function of position along the length of the fault can be determined if the rotation pole is known. If no gaps or overlaps are to develop across the fault, the rotation pole must be perpendicular to the fault plane. Alternatively, if offsets of markers at several localities along the trace of the fault are known, the rotational pole can be determined. In either case, working out the angular relationships on a stereogram before proceeding to an orthographic construction makes the solution of such a problem much simpler. The main value of the stereonet, once again, is the ease with which rotations about the horizontal and vertical axes can be performed.

Problem 6-16

A fault oriented N30°W,40°E disrupts a bed and a dike. (a) The bed is oriented N20°E,30°W on the west side of the fault. If the fault has 40° counterclockwise rotational movement, determine the attitude of the bed on the east side of the fault. (b) The bed is exposed at W on the west side of the fault and at E on the east side of the fault. The dike, which dips N80°E,60°S, is exposed at W' on the west side of the fault; it is also exposed at E' on the

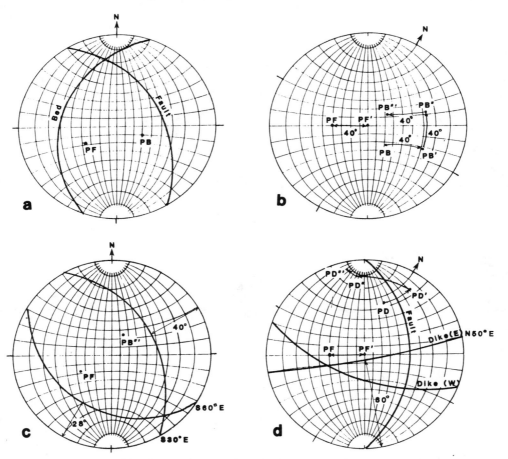

Figure 6-18. Procedure for determining the reorientation of planes due to rotational faulting.

east side of the fault. What are the orientation and magnitude of net slip and the position of the pole of rotation for faulting?

Method 6-16

The problem is presented in two parts to allow the two major parts of the solution to be illustrated separately.

Step 1: Prepare your overlay. Plot the great-circle traces for the fault and the bed on the west side of the fault (Fig. 6-18a). Also plot the poles to the fault (PF) and the bed (PB). Remember that rotational motion on a fault takes place around an axis perpendicular to the fault plane.

Step 2: Revolve the overlay to bring the fault strike to the north-south position. Rotate the fault through 40° to the horizontal; the fault pole moves to PF'. The bedding pole moves along a small circle path to PB' (Fig. 6-18b).

Step 3: The fault plane is now horizontal, so the pole to the fault plane is vertical, and we can apply the rotational fault motion directly about the vertical axis. The bedding pole moves 40° counterclockwise to PB" (Fig. 6-18b).

Step 4: Rotate the fault about the horizontal north-south axis to bring it back to the actual 40° dip

position of the fault. The bedding pole moves through 40° along a small circle path to a new position PB''' (Fig. 6-18b). At the same time the fault pole moves back to PF.

Step 5: Bring the pole PB''' to the east-west axis and draw in the great circle for bedding by counting 90° from the pole along the east-west axis. Revolve the overlay so that the fault great circle returns to its original position (S30°E,40°E). The bedding great circle shows the orientation of bedding on the east side of the fault; the beds dip S60°E,28°SW (Fig. 6-18c).

Step 6: The same method can be followed to determine the attitude of the dike on the east side of the fault (Fig. 6-18d). The dike dips N50°E,80°S.

Step 7: To solve the second part of the problem (part b), draw a map showing the relative positions of the dike and the bed on either side of the fault trace (Fig. 6-19a).

Step 8: The rakes of the traces of the bed and dike on both the east and west sides of the fault can be determined from the stereogram (Fig. 6-19b). Transfer this information to a projection drawn parallel to the fault plane, showing the units in their appropriate positions (Fig. 6-19c). The line (XY) joining the intersection of the

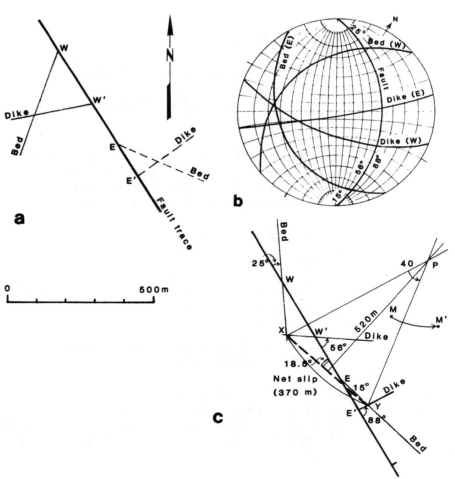

Figure 6-19. Procedure for determining the net slip on a rotational fault and the position of the pole of rotation.

bed and dike traces on the east and west sides of the fault gives the net slip; the length of the line gives the magnitude of net slip to be 370 m. The orientation of net slip is given by the rake of the line on the fault plane, which is 18.5°S (Fig. 6-19c).

Step 9: The pole of rotation on the fault must lie on the perpendicular bisector of the line of net slip, and the net slip subtends the rotation angle (40°) at the pole. The other two angles of the triangle (angle PXY and angle PYX) are equal and must each be 70°. By constructing these angles at X and Y, we can find the position of the pole (P). The pole lies at a distance of 540 m along the perpendicular bisector of the net slip line.

Now, for example, if the position of a mineral vein (M) is known on one side of the fault, its position on the other side (M') can be determined easily, by drawing a circular arc around P with radius PM and then subtending an angle of 40° at P.

In the preceding problem, the amount and direction of rotation were given. If the rotation has to be determined, two possible positions for the rotation poles generally exist. Additional information, such as slickenlines or additional displaced markers, must then be used to determine which of these rotation poles is the correct one.

EXERCISES

1. Plot the following planes as poles on a stereogram. Put all poles on the same overlay, and label each.

    (a) 025°,44°NW   (b) N14°W,85°SW   (c) N83°W,43°NE
    (d) 072°,06°SE   (e) 234°,18°NW   (f) 040°,90°SE
    (g) 090°,38°N

2. One limb of a chevron fold is oriented N23°E,57°SE, and the other limb is oriented N12°W,71°SW.

    (a) What is the plunge and bearing of the fold hinge?
    (b) What is the interlimb angle?
    (c) Assume the axial plane bisects the angle between the two fold limbs. What is the orientation of the axial plane?

3. The following measurements are the attitudes of opposing limbs of nonplunging folds. Determine the interlimb angle for each fold. In order to choose the proper angle in your calculation of the supplement, be sure to visualize the fold and note the orientation of the axial plane (i.e., draw a profile sketch).

Limb A	Limb B	Axial plane
(a) 360°,50°W	360°,30°E	360°,10°W
(b) 360°,50°W	360°,30°E	360°,70°E
(c) 360°,10°W	360°,50°W	360°,30°W
(d) 360°,10°W	360°,70°E	360°,60°W

4. Measurements of dip on opposite limbs of an anticline in southern Illinois are N15°E,32°SE and N10°W,72°NW.

    (a) What is the orientation of the fold hinge?
    (b) What is the attitude of the axial plane?
    (c) What is the plunge and bearing of the axial-plane trace on horizontal gound?
    (d) What is the plunge and bearing of the axial-plane trace on a slope whose surface is oriented N10°W,30°SW?

5. In a metamorphic terrane, regional foliation is oriented 085°,65°S. A slip lineation on a fault plane that cuts the foliation is oriented 80°,255°. What is the angle between the foliation and the slip lineation?

6. In eastern New York, there is an important unconformity called the Taconic unconformity. It separates Mid-Ordovician flysch from Devonian limestone. At a locality near the town of Catskill, the limestone (and the unconformity) is oriented N15°E,44°NW. An anticline occurs in the underlying flysch. One limb is presently oriented N60°E,73°NW, and the other limb is presently oriented N20°E,41°SE. Flute casts occur in the Ordovician strata and have a pitch of 55°NE on the northwest-dipping limb.

    (a) What was the orientation of each of the two fold limbs before tilting of the unconformity?
    (b) What was the orientation of the fold axis prior to tilting?
    (c) What was the trend of the current direction responsible for the formation of the flute casts in Ordovician time?
    (d) What is the present orientation of the flute casts on the southeast-dipping limb?

7. Mineral lineation occurs on foliation planes of the Lion Den mylonite at Battleship Peak, Arizona. The foliation at this locality is oriented N22°W,73°SW. The rake of the lineations is 20°N. Mapping indicates that the Lion Den mylonite was folded subsequent to formation of the foliation. In fact, at the measurement locality the gneiss is on the limb of a large antiform. Foliation at a locality (South Ridge) on the other limb is oriented N64°W,48°NE.

    (a) Assume that the lineation formed while the mylonite was horizontal. What was the orientation of the lineation prior to folding? (*Hint:* Calculate the plunge and bearing of the fold axis first.)
    (b) Predict the present orientation of the lineation at South Ridge.

8. A fault striking N90°W is exposed on a featureless plain. The fault has a dip of 60°N. The outcrop positions of a dike (N30°W,35°NE) and a coal bed (N30°E,60°NW) are known on both sides of the fault.

Feature	North side	South side
Dike	100 m	0 m
Coal bed	250 m	400 m

    Determine the following:

    (a) Trend and plunge of net slip.
    (b) Amount of net slip.
    (c) Rake of the net slip on the fault plane.
    (d) Relative movement along the fault.

    Determine all necessary angles from a stereographic projection.

9. The map in Figure 6-M1 is a simplified geologic map of part of the east coast of Conanicut Island in Narrangansett Bay, Rhode Island. The bedrock is composed of slate and phyllite that contains recognizable Cambrian trilobites. Regionally, these rocks have been complexly deformed, but the area of the map does not

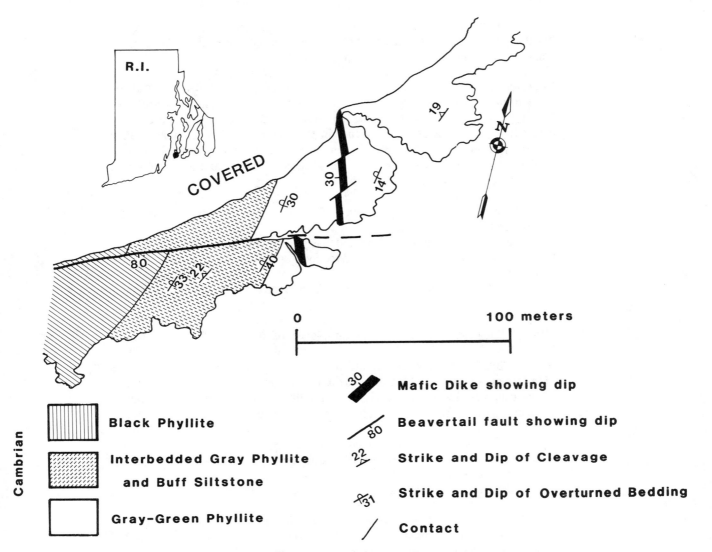

Figure 6-M1. Geologic map of part of the east coast of Conanicut Island, Rhode Island, for exercise 6-9.   Mapping is by students of the Colgate University Geology Field Camp (1983 & 1985).

contain any major fold hinges. The map area includes an exposure of the Beavertail fault. (Problem contributed by A. Goldstein.)

(a) From the geological relationships shown, determine the *sense of separation* of the mafic dike and the *sense of separation* of the bedding contacts.
(b) On the basis of your answers to part (a), make a preliminary guess as to the orientation of the net slip describing the cumulative movement on the fault.
(c) Calculate the sense, direction, and amount of displacement on the Beavertail fault, based on the offset markers shown.
(d) Slip lineations on the fault plane, where it is exposed, rake 85°ENE. Is the orientation of these lineations parallel or oblique to your calculated true separation vector? If not, how can the discrepancy be explained?
(e) How can the age of movement on this fault be constrained?

10.   The map of Figure 6-M2 shows part of the Singatse fault in eastern Nevada. Note that the early Tertiary Conglomerate and the Tertiary Ignimbrite units pinch out

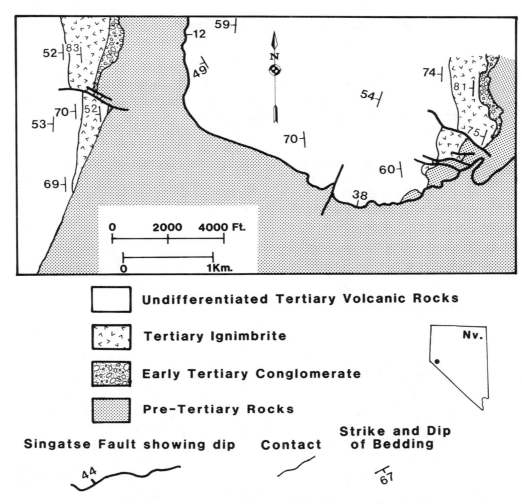

Figure 6-M2. Geologic map of part of the Singatse Range, western Nevada for exercise 6-10. (Modified from Profett, 1977.)

along the unconformity that separates the pre-Tertiary sequence from the Tertiary sequence. The pinchout, which represents the intersection of two planes, appears as a point on the map. The pinchout is exposed at two localities. Assume that the dip of the unconformity at the top of the pre-Tertiary sequence has the same dip as the adjacent Tertiary sequence but a different strike. Also assume that the map area is a horizontal featureless plain. (Problem contributed by A. Goldstein.)

(a) Using the preceding assumptions and the data on the map, determine the orientation and amount of displacement on the Singatse fault.   First, you must find the locations at which the line of pinchout pierces the hanging wall and footwall of the fault. Find the trend and plunge of the line of pinchout using your equal-area net, and draw two cross sections through the two points of pinchout (one for the hanging wall and one for the footwall). Include on these cross sections the line of pinchout and the trace of the Singatse fault. Your instructor will help you do this if you have difficulty. From these cross sections the locations of the two piercing points should be clear. Project these to the surface and put them on the map. Now, draw another cross section containing both piercing points, and project the fault and piercing points onto the cross section. The strike of this cross section is the direction of the displacement, and the amount can be measured from the section.

(b) Classify the Singatse fault (e.g., is it dip-slip, strike-slip, or oblique-slip).

(c) Why does the fault presently have a curving map pattern?

11. Figure 6-M3 is modified from a map by Harwood (1983) and is a more typical map expression of faults than is either of the two previous maps. The area is in the northern Sierra Nevada of California, 35 km northwest of Lake Tahoe. Ordovician through Permian sedimentary and volcanic units dip toward the northeast and are displaced along several vertical fault strands of an unnamed fault zone.

(a) What can be determined about the direction and magnitude of the faults, given only the information on the map?

(b) Assuming that the faults are strike-slip faults, what are the direction, sense and amount of cumulative displacement across the fault zone?

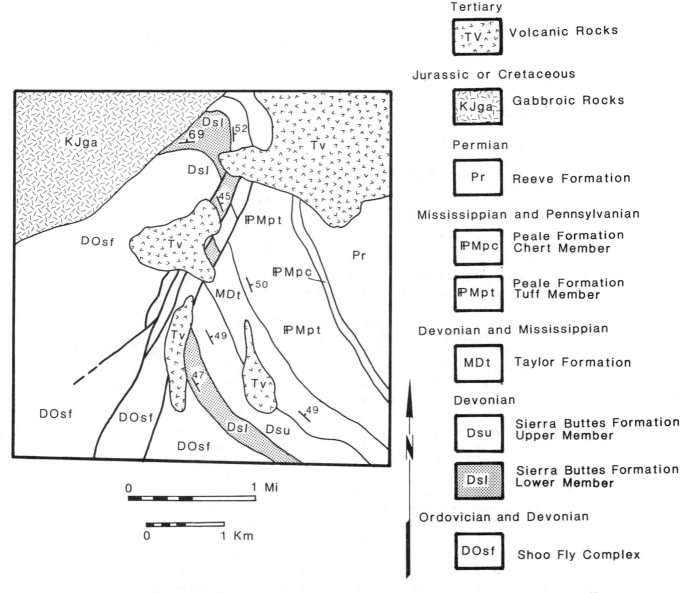

Figure 6-M3. Geologic map of part of the Sierra Nevada Mountains, California for exercise 6-11. (Modified from Harwood, 1983.)

(c) Assuming that the faults are pure dip-slip faults, what are the direction, sense and amount of cumulative displacement across the fault zone?

(d) What was the time period during which the faulting took place?

12. A fault plane in the Wind River Mountains is oriented N20°W,70°SW. The fault is a normal fault; displacement on the fault decreases progressively toward the northwest along the trace of the fault. Strata of the hanging wall are oriented N70°W,35°SW, whereas strata of the footwall are horizontal. At the locality where these measurements were made the displacement on the fault is 100 m.

(a) What is the attitude of the axis of rotation describing movement on the fault?

(b) By how much did the hanging-wall beds rotate (and in what sense) to attain their present attitude?

(c) Approximately how far and in what direction must the geologist walk to find the "hinge" of the fault (the point at which displacement on the fault is 0)?

7

CALCULATION
OF LAYER ATTITUDE
IN DRILL HOLES

7-1 INTRODUCTION

Modern techniques have made the drilling of holes in solid rock a routine, though expensive, task. Drilling is commonly done for the purpose of tapping oil-bearing reservoirs, but often, holes are made simply to obtain important data about subsurface geology. For example, mining companies determine the size and extent of ore bodies by drilling holes at equal intervals on a square grid; such data allow construction of subsurface maps. Similarly, the Deep Sea Drilling Project (DSDP) routinely drills sediments with underlying igneous rocks on the deepest ocean bottoms, thereby providing information on the structure of the crust on ocean floors. Generally, rock in a drill hole is pulverized during the course of drilling and is flushed to the surface in a slurry called drilling mud. If necessary, however, it is possible to keep a drill core intact by using a hollow drill bit.

Information on the orientation of subsurface layers can be obtained from the examination of recovered drill core and from downhole *logging*. Logging refers to the process of sending an electronic instrument, called a *sonde*, down the hole. The instrument records physical properties (e.g., electrical resistivity) as a function of depth. The physical properties are interpreted in terms of rock type and character. Traditional logging techniques indicate only the depth of the intersection between the hole and a known layer, but newer instruments permit direct measurement of layer attitude in holes under appropriate conditions.

In order to use drilling data to provide information on the orientation of subsurface layering, one must understand the geometric limitations of such data. The purpose of this chapter is to demonstrate how drilling data can be used to determine completely the orientation of subsurface layering or at least to constrain the possible ranges of orientation. The constraints on the attitude of a subsurface horizon depend on the number of drill holes in which data are available and on the orientation of the drill holes.

7-2 DATA FROM ONE DRILL HOLE

Use of Unoriented Core

If a drill core could be kept from rotating as it was brought to the surface, it could provide attitude information equivalent to that obtained by measuring a surface outcrop with a compass. Unfortunately, prevention of rotation during extraction of a core is generally not possible. The only usable structural information that we get from a single unoriented drill core is the depth to a particular horizon and the inclination of planar structures (e.g., bedding, foliation, and fractures) with respect to the drill axis. The orientation of a planar structure is uniquely determined only if it is perpendicular to the core axis; the attitude of planes at other orientations with respect to the core axis cannot be uniquely determined from a single unoriented core.

In order to understand the preceding concept, imagine a recovered core that contains a bedding plane inclined to the axis of the core. If you rotate the drill core through 360° (Fig. 7-1a), the range of possible orientations of the bed describes a right-circular cone whose axis is the core axis. Depending on the orientation of the drill hole and the angle that the planar structure makes with the core axis, the cone defining the possible orientations of the planar structure intersects the earth's surface as a circle (Fig. 7-1a), an ellipse (Fig. 7-1b,c), a parabola (Fig. 7-1d), or a hyperbola (Fig. 7-1e).

Use of an Oriented Core

During near-surface drilling, it is sometimes possible to draw an orientation mark on a core so that the orientation of the core with respect to the drill hole is known even if the core is broken and rotated by an unknown amount during drilling. Orientation of a core must be done if the core is to be used for paleomagnetic work or for analysis of rock anisotropy. The orientation of a core is indicated

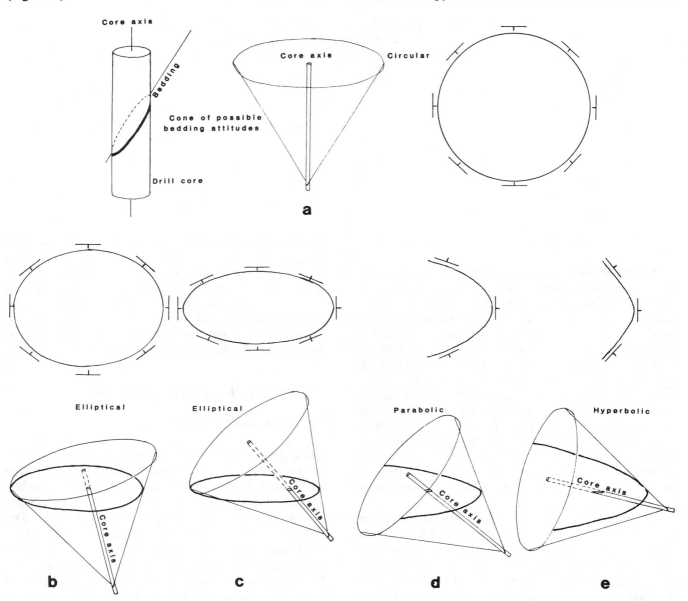

Figure 7-1. Inclined bedding in a drill core may have had a range of possible orientations described by a right-circular cone around the core axis. The inclination of the drill hole determines the conic section (circle, ellipse, parabola, or hyperbola) in which the cone intersects the earth's surface.

either by affixing an oriented marker to the top of the core, if the core is vertical, or by scratching a line along the top side of the core, if the core is inclined (Fig. 7-2).

It is easy to determine the orientation of a planar feature that cuts across a vertical core. The angle between a strike line on the plane and the orientation mark indicates the strike of the plane, and the angle between the plane and the core axis is the true dip of the plane.

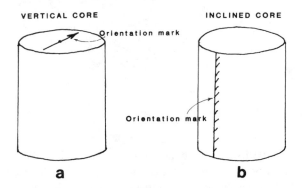

Figure 7-2. Orientation marks used on near-surface drill cores. (a) Vertical core with mark on top; (b) inclined core with line on upper side.

If the core is inclined, the attitude of the plane can be calculated using a stereonet. This calculation requires two stages: first, determination of the orientation of the plane with respect to the core axis as if the core axis were vertical, then rotation of the core axis into the real-world reference frame. In the following example the *apparent strike line* is defined as a line that lies in the plane and is perpendicular to the axis of the core (Fig. 7-3). The angle ø is the angle between the apparent strike line and a reference line inscribed on a plane perpendicular to the core axis. The angle μ is the angle between the plane and the core axis as measured in a plane that contains the core axis and is perpendicular to the apparent strike line.

Problem 7-1

A vein cuts across an oriented inclined core (Fig. 7-4a). The orientation mark (AC) on the side of the core plunges 40°,220°. What is the attitude of the vein in the real-world frame of reference?

Method 7-1

Step 1: The problem as stated does not contain enough information to determine the vein attitude. We must first specify the orientation of the vein in the frame of reference of the core. To do this, cut the core above and below the vein on saw cuts perpendicular to the core axis (Fig. 7-4a). Line AB is a reference line drawn across the top (AB is perpendicular to AC). Lines AB and AC

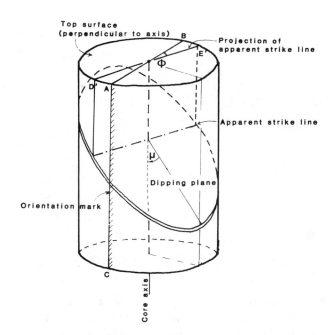

Figure 7-3. Relative orientation of a planar structure with respect to an oriented, inclined core.

together define a plane through the core that contains the core axis.

Step 2: Now we must find the apparent strike. The apparent strike line of the vein is determined by finding two diametrically opposite points (D and E) on the vein that are the same distance from the top surface of the core. Points D' and E' are then marked on the top surface of the core. DD' and EE' are parallel to the core axis. The angle between the apparent strike line and the line AB is the angle between D'E' and AB; we measure it to be 20° clockwise (Fig. 7-4a).

Step 3: Now we must find the angle (μ) between the core axis and the vein. We rotate the core so that we are looking along the apparent strike line and measure μ with a protractor; μ = 30°.

Step 4: We have now specified the vein attitude with respect to the core axis (apparent strike = 20°; μ = 30°). To proceed from this point, it is easiest to visualize the problem by imagining that the core axis is vertical, and that line AB strikes due north. If the core actually had this orientation, the vein would be oriented N20°E,70°SE; we will call this attitude the "relative" attitude of the vein. We must now do two rotations to get the core and the vein into the real-world frame of reference. Prepare a stereonet overlay.

Step 5: Plot the great-circle trace (VV) indicating the relative attitude of the vein and its pole (P) on a stereonet (Fig. 7-4b). Rotate 40° clockwise (as viewed looking down the axis) around a vertical axis to make line

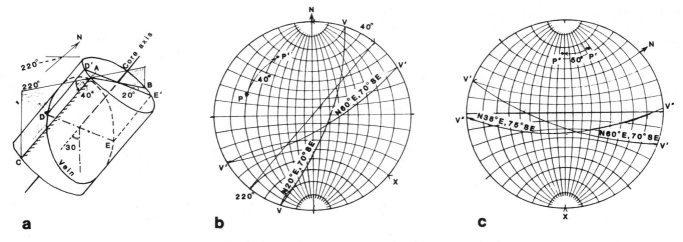

Figure 7-4. Determining the orientation of a planar structure from an oriented core.

AB parallel to the real bearing of the orientation mark. During this operation, the apparent strike line of the vein rotates to N60°E, and the relative dip of the vein remains unchanged. Plot the rotated vein attitude as a new great-circle trace (V'V'), and plot the rotated pole (P').

Step 6: Next, rotate by 50° clockwise (as viewed to the northwest) around a horizontal axis that is perpendicular to plane AB. This operation returns the core to its true inclination. Plot a point (X) on the primitive of the stereogram at 220° - 90°= 130°. Revolve the overlay so that X lies over the south mark on the stereonet. Rotate the pole (P') 50° to the left along a small-circle path (Fig. 7-4c), trace the new great-circle trace, and label it V"V". This trace represents the attitude of the vein in the real-world frame of reference; the vein attitude is N38°E,75°SE.

Note: You must visualize every step of the preceding procedure in order to make sure that the rotations are in the correct sense.

Dipmeter Surveying

In certain situations it is prohibitively expensive to drill many holes. Such is the case, for example, in offshore drilling or in oil exploration in mountainous areas. Under these circumstances a dipmeter survey is generally done to determine bedding attitudes within a single wildcat well.

A *dipmeter* is an instrument consisting of three electrodes that are distributed in azimuth at 120° intervals around the logging sonde (Fig. 7-5a). Each electrode measures and records the microresistivity of bedding units that the instrument passes through as it is lowered down a bore hole. (The prefix "micro" simply means that the instrument has high resolution). Thus, three micro-

resistivity logs, which accurately plot resistivity as a function of depth, are produced for each hole. The log for each electrode shows characteristic "kicks" (sudden changes in resistivity) for key beds. If the bed is inclined to the hole, the depths at which each electrode crosses the bed are different. The elevations at which the three electrodes cross the bed can be used to calculate the attitude of the bed; the method used is the same as the three-point method described in Chapter 3.

In order to calculate the attitude of the bed in the earth reference frame, the orientation of the dipmeter must be known. A dipmeter log typically shows the orientation of one electrode with respect to magnetic north (Fig. 7-5b), and the *drift* (angle of deviation from the vertical) of the bore hole. These measurements are resolved into components on north-south and east-west vertical planes. The following example, done for a vertical hole, illustrates the method for incorporating measurement of dipmeter orientation into the calculation of bed attitude measured by the dipmeter. The method can be modified quite easily for dip determination from inclined drill holes.

Problem 7-2

A dipmeter survey is done on a 9-in. diameter vertical drill hole through uniformly dipping beds. The azimuth of electrode A is N40°E. Electrode A cross a key bed at a depth of 200 ft, electrode B crosses the bed at a depth 4 in. below the 200-ft level, and electrode C crosses the bed at a depth of 1.8 in. below the 200-ft level. Determine the attitude of the bed.

Method 7-2

Step 1: Examine the angular relationship of the bedding plane ABC to the horizontal electrode plane of the dipmeter A'B'C' (Fig. 7-6a). Triangle A'B'C' is an

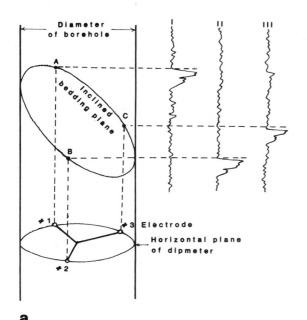

a

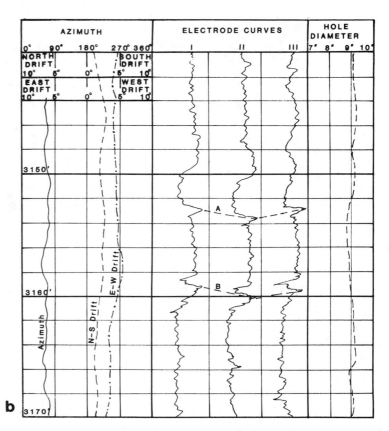

b

CONTINUOUS DIPMETER CURVES

Figure 7-5. (a) The essential features of a dipmeter showing three electrodes 120° apart that record the microresistivity of beds as the instrument is lowered down a bore hole; (b) typical dipmeter log showing microresistivity curves and orientation and "drift" of the instrument. (Adapted from Badgley, 1959.)

equilateral triangle with sides $\mu\sqrt{3}$ (where μ is the radius of the bore hole) and is perpendicular to the axis of the hole. Therefore,

$$A'B' = B'C' = C'A' = 4.5(\sqrt{3}).$$

The slope of line BA, as measured in a vertical plane, is the angle ABA' (ø) and is given by

$$n \tan ø = AA'/AB' = 4/(4.5\sqrt{3}) = 0.5132$$

$$ø = \tan^{-1}(0.5132) = 27°.$$

Similarly, the slope of line CA, measured in a vertical plane, is given by the angle ACA', where

$$ACA' = ∂ = \tan^{-1}(AA'/CA') = \tan^{-1}[1.8/(4.5\sqrt{3})]$$

$$= \tan^{-1}(0.2309) = 13°.$$

Step 2: Prepare a stereographic overlay (Fig. 7-6b). Be sure to use an equal-angle (Wulff) net! The azimuth of A (given by OA') bisects the angle between A'B' and A'C', which give the relative orientations of B and C with respect

to A. If we assume that point A lies at the center of the spherical projection, then lines AB and AC, which have plunges of 27° and 18°, intersect the lower hemisphere at two points whose stereographic projections plot at B* and C* at distances of 27° and 13° from the primitive, respectively. Angle C*AB* is 60° and is bisected by the azimuth of A.

Step 3: Revolve the overlay until B* and C* lie on the same great circle. Trace this great circle, which represents the true dip of bedding and has a dip of 28° (Fig. 7-6c). Revolve the stereogram back to its original position to get the strike of bedding. The bedding attitude is N84°W, 28°S (Fig. 7-6d).

Note: In a dipmeter survey, directions are generally measured with respect to magnetic north rather than true north. Hence, to obtain the true dip direction for the beds we must correct for the local magnetic declination.

Problem 7-3

The drill hole in the preceding problem has drift angles (deviation from the vertical) of 09°N on the north-south plane and 06°E on the east-west plane. Determine the true attitude of bedding.

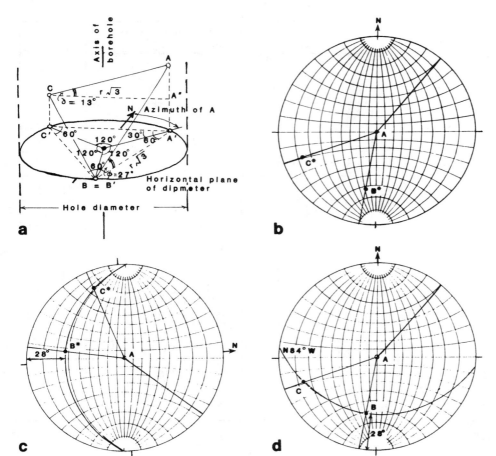

Figure 7-6. Determining the dip of beds from dipmeter data.

Method 7-3

Step 1: The apparent dip of the plane perpendicular to the drill axis (i.e., the plane of the dipmeter) measured in the north-south and east-west vertical planes is the same as the drift angle of the bore hole (measured from the vertical) on these planes (Fig. 7-7a). Thus, the plane of the dipmeter has an apparent dip of 09°S and 06°W. Prepare your stereographic overlay. Plot the two apparent dips on the stereogram, and draw the great circle passing through them (Fig. 7-7b).

Step 2: Also trace the bedding great circle (determined in the previous problem) and plot its pole (P) (Fig. 7-7b).

Step 3: Bring the strike of the dipmeter great circle to the north-south axis, and rotate this plane to the horizontal. This operation has the same effect as rotating the drill hole to the true vertical position. Concurrently, the pole to bedding (P) moves along a small circle to P' (Fig. 7-7c).

Step 4: Draw in the new bedding great circle with P' as pole (Fig. 7-7c); this gives the true orientation of bedding as N85°E,18°S.

Note: This method of correcting for bore-hole drift from the true vertical can also be used in determining

bedding dips from inclined drill holes. Of course, drift from the true inclination of the bore hole must also be accounted for in this case.

7-3   DATA FROM TWO DRILL HOLES

We saw earlier that it is not possible to specify the true attitude of layering in a single unoriented drill core. If the same layering is oberved in two unoriented drill cores, it is possible to considerably reduce the range of possible attitudes that the layering can have. Depending on the inclination of the drill holes and the presence or absence of a marker horizon, a number of different constraints can be determined. If a marker bed is found in both cores, it is usually possible to use a combination of orthographic and stereographic constructions to narrow the possible choice of orientations to one. If no marker bed is present, a stereographic construction will usually yield more than one possible orientation for the layering .

The orientation of a drill hole plots on the stereographic projection as a point. The cone, which defines the possible orientations of bedding around the drill-core axis, intersects the lower hemisphere of the spherical projection in a circle. Any such circle projects on

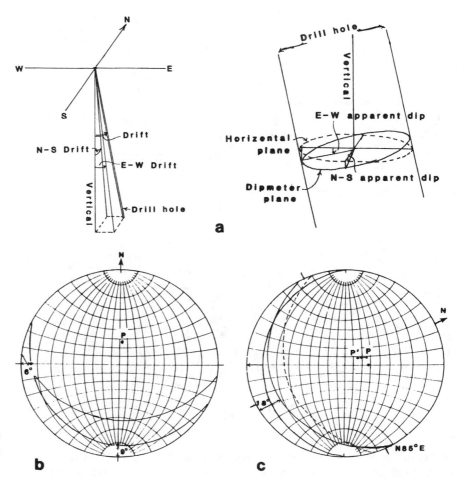

Figure 7-7. Correcting dips determined from dipmeter surveying for drill-hole drift.

the stereographic projection as a circle. We noted this property in Chapter 5 and will now demonstrate that this is a general property of stereographic projections before proceeding with two-drill-hole problems.

Representation of Circles on a Stereonet

Consider a circular cone with its vertex at the center of the projection sphere (Fig. 7-8a). The cone intersects the sphere in a small circle. Figure 7-8b shows a vertical section through the center of the projection sphere that passes through the center of the small circle. Line AB is a diameter of the small circle, and point C is its center. The lines joining points on this circle to the zenith (Z) form a cone with axis CZ, of which AZB is a section. The right section of this latter cone (through BOD) is an ellipse, since an oblique section (AB) is a circle. There should be a second circular section DE symmetrically inclined to the axis (CZ); section DE is called the conjugate section to AB. If we draw a line AF parallel to the projection plane, then angle FAZ = angle ZBA, as they are subtended by equal arcs FZ and ZA. Also, angle ZBA = angle ZDE, since AB and DE are conjugate sections. Therefore, both AF and the plane of projection (PQ) are parallel to the

circular section DE of the cone. Since parallel sections of a cone are similar, the section of the cone (A'B') in the projection plane (PQ) is a circle. The projection of C is C', and it does not lie halfway between A' and B'; thus the center of the small circle on the sphere does not project to the center of the small circle on a stereographic projection.

To convince ourselves of the preceding property we can draw a small circle whose center (A) is at the center of the stereogram and then rotate it out toward the edge of the stereogram by rotation around the north-south axis (Fig. 7-8c). Individual points on the circle move through the same angular distance along small-circle paths, and the new points define a circle, whose center (B) can be obtained by bisecting the east-west diameter. This center does not coincide with the rotated center (B'). If we rotate the circle farther, until its original center (D') lies on the primitive, the circle coincides with the small circle of the net. Its center (D) can be found by drawing a tangent to the primitive at the point where the small circle intersects the primitive; the intersection of this tangent and the east-west axis gives the center (D) of the small circle.

Any small circle can be constructed by drawing the corresponding cone section and projecting its intersection with the sphere onto the plane of the stereonet. To do this,

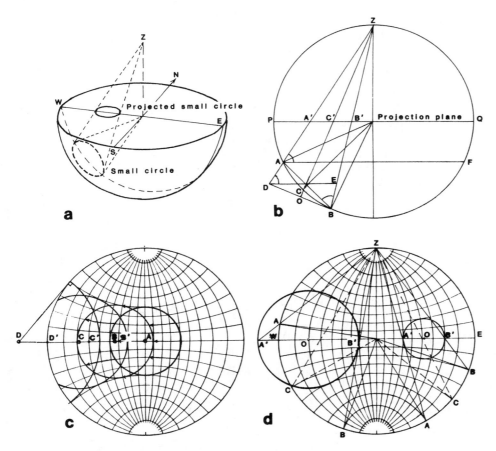

Figure 7-8. Representing small circles on a stereonet. (a) A small circle on the projection sphere projects as a small circle on the stereogram; (b) geometric proof to show that small circles always project as circles on the stereogram; (c) effect of rotating a small circle from the center to the circumference of a stereogram; (d) method for plotting small circles representing cones of possible orientation around a drill core. (Adapted from Phillips, 1971).

the cone is first brought to the east-west diameter, and the net is converted (as a mental exercise) to a vertical section along the east-west diameter (Fig. 7-8d). The plunge of the core axis (C) and the vertex angle of the cone are plotted at the center of the circle (the circumference of the net can be used as a protractor to count off necessary angles). Intersections with the sphere (A_1B_1 or A_2B_2) are then projected ($A_1'B_1'$ or $A_2'B_2'$) onto the horizontal diameter by drawing straight lines from those points to the zenith (Z). The center of the small circle (O) lies halfway betwen its inner and outer limits along the horizontal diameter. The net can now be converted back to a stereogram (horizontal projection) and the small circle drawn on the projection. This method works for all small circles (Fig. 7-8d), including those that lie partly outside the primitive (e.g., A_2B_2, for which the intersection of the cone with the sphere extends into the upper hemisphere).

We can now look at a few examples to see how a combination of orthographic and stereographic projections can help us solve two-drill-hole problems.

Two Vertical Drill Holes with a Marker Bed

Problem 7-4

Two vertical drill holes are 400 m apart on a line trending N70°E. A bed is encountered at a depth of 200 m

in the western hole and 300 m in the eastern hole. Bedding makes an angle of 40° with the core axis in both holes. Determine the possible attitudes of bedding.

Method 7-4

Step 1: Draw a map showing the location of both drill holes (A and B) (Fig. 7-9a).

Step 2: Using AB as a folding line, draw a cross section. Plot the depths at which the bed is encountered in both holes (A' and B'). Line A'B' gives the apparent dip of the bed along the line of cross section (13°,N70°E) (Fig. 7-9b).

Step 3: Draw cross sections of the cones defining possible bedding orientations around each drill hole (Fig. 7-9b). The intersections of these cones with the ground-surface line define the ends of the diameters of circles representing possible bedding orientations.

Step 4: Complete the circular cross sections of the cones on the map (Fig. 7-9a) using the diameters obtained from the cross section. Tangents common to these circles define the possible strike orientations (N59.5°E and N80.5°E).

Step 5: Plot the apparent dip (from Step 2) on a stereogram (Fig. 7-9c). Using the possible strike directions (from Step 4), draw great circles passing through the point marking apparent dip. Each of these great circles

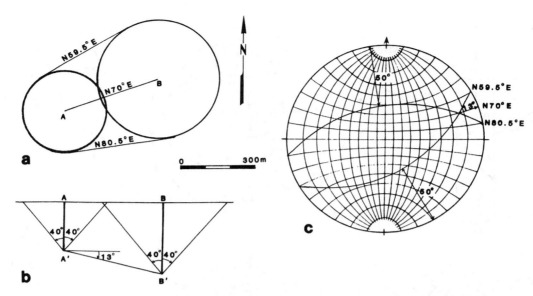

Figure 7-9.   Determining bedding attitudes from two vertical drill holes that encounter a marker bed.

represents a possible true attitude of bedding (N80.5°E,50°N and N59.5°E,50°S).

Note: If there was no marker bed, it would not be possible to determine an apparent dip for bedding. It would also be impossible to determine possible strike orientations for bedding, and hence there would be an infinite number of possible solutions that would satisfy the data.

Two Nonparallel Drill Holes with a Marker Bed

Problem 7-5

Two inclined drill holes are drilled 100 m apart along a line trending S20°E. The first hole is inclined 50° toward N30°E, encounters a marker bed at a depth of 25 m, and has a core-bedding angle of 48°. The second hole is inclined 55° toward S40°W, encounters the same bed at 45 m, and has a core-bedding angle of 54°. Determine the attitude of bedding.

Method 7-5

For an inclined drill hole the cones representing the possible orientations of bedding intersect the projection sphere in a circle. As explained earlier, any circle on the projection sphere projects as a circle on the plane of the stereographic projection. Just like bedding, the possible orientation of the normal to bedding also generates a cone around the core axis, and the vertex angle of this cone is the complement of the vertex angle of the cone representing possible bedding orientations (Fig. 7-10a). In practice, the cone generated by possible orientations of pole to bedding is used to represent possible bedding

orientations; this cone intersects the projection sphere in a circle and projects on the stereogram as a *pole circle*. The pole circles for two inclined drill holes intersect at points that give the possible orientations of the pole to bedding.

Step 1: Plot drill hole 1 (50°,N30°E) on a stereogram (Fig. 7-10b). This represents the orientation of the core axis; note that, except for a vertical drill hole, the core axis does not coincide with the center of the stereogram.

Step 2: The plunge of the bedding pole is given by (90° - bedding dip). Thus, the angle between the pole to bedding and the core axis is 90° - 48° = 42°. Bring the point representing drill hole 1 to the east-west axis and count off 42° in either direction from it to obtain the diameter of the pole circle, and draw this small circle using a compass.

Step 3: Plot drill hole 2 (55°,S40°W) on the stereogram (Fig. 7-10c).

Step 4: The angle between the pole to bedding and the core axis is 90° - 54° = 36°. Using the same procedure as step 2, draw the pole circle (small circle) for drill hole 2.

Step 5: The two small circles intersect at two points, giving the poles (P_1 and P_2) for two possible bedding attitudes: N30°E,18°SE and N40°E,14°NW (Fig. 7-10d). An orthographic construction must be used to decide which one of these gives the true attitude of bedding.

Step 6: Draw a map showing the locations of the drill holes and their trends (Fig. 7-10e). For each hole use the surface trend of the hole as a folding line to draw a vertical cross section that shows the plunge of the drill hole and the distance along the drill hole to the bed. Complete the right-angled triangles to obtain the surface

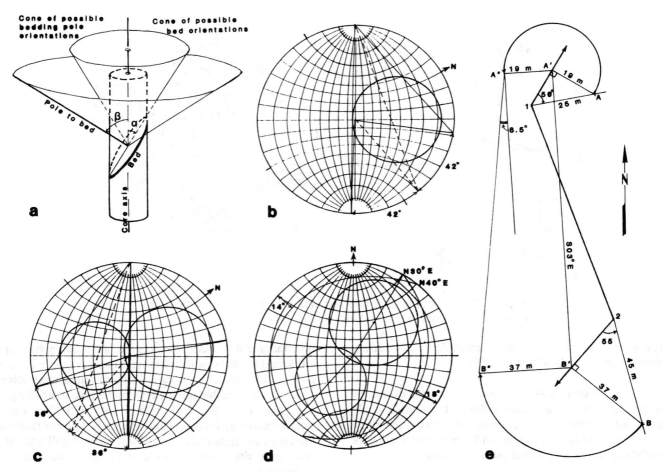

Figure 7-10. Determining bedding attitude from two inclined, nonparallel drill holes that encounter a marker bed.

projections (A' and B') of the drill-hole/bed intersections and depths (19 m and 37 m) at which these intersections take place.

Step 7: Use A'B' as a folding line to draw a vertical cross section, and plot the known depths (19 m at A', 37 m at B') at each end to get points A" and B". The slope of line A"B" gives an inclination of the bed (8.5°) along line A'B ' (S03°E); this gives an apparent dip on the bed.

Step 8: This apparent dip is compatible with only one of the poles determined stereographically, corresponding to a bedding attitude of N30°E,18°SE; therefore, this is the true attitude of the bed.

Note: If there was no marker bed present, it would not be possible to determine an apparent dip for bedding. Thus, it would not be possible to determine a unique orientation for bedding. However, the intersection of the pole circles on the stereogram for the two inclined drill holes would give two possible orientations for bedding. Thus, these drill holes are more useful than vertical drill holes or parallel drill holes, which, as you may remember,

gave an infinite number of possible solutions when no marker beds were present.

7-4   DATA FROM THREE DRILL HOLES

Three nonparallel drill holes allow us to completely determine the attitude of a structural plane, regardless of whether or not any marker horizons are present. As we have seen, if a marker horizon is present, we need only two drill holes to completely determine the orientation of the marker horizon; in a situation where a marker horizon is present, the third drill hole provides superfluous information but can be used to check the accuracy of the two-hole solution. Even without a marker horizon, the point of intersection of the pole circles for three drill holes on a stereographic projection gives a unique pole to the structural plane whose attitude is to be determined.

Problem 7-6

Three nonparallel drill holes have the following orientations and core-bedding angles:

Hole attitude	Core-bed angle
1: 29º,N45ºW	39º
2: 51º,S13ºW	41º
3: 46º,N55ºE	51º

Determine the attitude of bedding.

Method 7-6

Step 1: Plot drill hole 1 (29º,N45ºW) on a stereogram. The angle between the pole to bedding and the core axis is 90º - 39º = 51º. Bring the point representing drill hole 1 to the east-west axis of the stereonet and count off 51º on either side of it (along the east-west axis) to obtain the diameter of the pole circle. Using the midpoint of the diameter as center, draw a small circle using a compass (Fig. 7-11a).

Step 2: Plot drill hole 2 (51º,S13ºW) on the stereogram. The angle between the pole to bedding and the core axis is 90º - 41º = 49º. Use the same method as step 1 to draw the pole circle around drill hole 2 (Fig. 7-11b).

Step 3: Plot drill hole 3 (46º,N05ºE) on the stereogram. The angle between the pole to bedding and the core axis is 90º - 51º = 39º. Use the same method as step 1 to draw the pole circle around drill hole 3 (Fig. 7-11c).

Step 4: The three pole circles (small circles) intersect at a point that represents the pole to bedding. Using this pole, draw the great-circle trace representing bedding. The attitude of bedding is N75ºE,13ºS (Fig. 7-11d).

7-6   USING ROTATION TO SOLVE DRILL-HOLE PROBLEMS

The stereographic constructions for drill holes described so far can be done in an alternative manner that eliminates the need to use a compass for drawing the small (pole) circles. The method involves rotating the cone axis to the horizontal position, so that the small circles can be drawn using the "small-circle" traces of the stereonet. We will illustrate the method here using the same three-drill-hole problem described in Method 7-6.

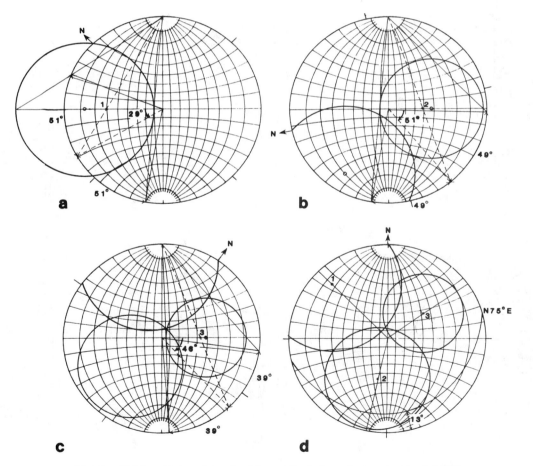

Figure 7-11. Determining bedding attitude from three nonparallel drill holes.

Problem 7-7

Three nonparallel drill holes have the following orientations and core-bedding angles:

Hole attitude	Core-bed angle
1: 29º,N45ºW	39º
2: 51º,S13ºW	41º
3: 46º,N55ºE	51º

Determine the attitude of bedding.

Method 7-7

Step 1: Plot the points representing drill holes 1, 2, and 3 on a stereogram (Fig. 7-12a). Find the great circle passing through points 1 and 2 and rotate it to the horizontal position. The points (1' and 2') now lie on the primitive (Fig. 7-12b).

Step 2: Bring 1' to the north-south axis. Draw the 51º small circle to represent the pole circle for hole 1 (the angle between the pole and the core axis is 90º - 39º = 51º) (Fig. 7-12c).

Step 3: Bring 2' to the north-south axis and draw the 49º small circle to represent the pole circle for hole 2 (the angle between the pole and the core axis is 90º - 41º = 49º) (Fig. 7-12c). The two small circles intersect at four points (A', B', C', and D').

Step 4: Rotate the plane containing the two drill holes back to its original orientation. A', B', C', and D' move along small-circle paths by the same amount to their true orientations A, B, C, and D (Fig. 7-12d).

Step 5: The same procedure can be followed for a second pair of drill holes (2 and 3) to obtain the true orientations of the points of intersection (E and F) of their pole circles (Fig. 7-12e). Note that it is not necessary to draw the entire small-circle traces at this stage, but just enough to find the points of intersection of the small circles; this removes much of the confusion caused by criss-crossing lines on the stereogram.

Step 6: By comparing the stereograms from Steps 4 and 5 (Figs. 7-12d, e), we find that points A and E coincide, giving the position of the point of intersection of the pole circles for all three drill holes. This point gives the pole to bedding, and the great-circle trace for bedding can be drawn (Fig. 7-12f). The attitude of bedding is N75ºE,13ºS.

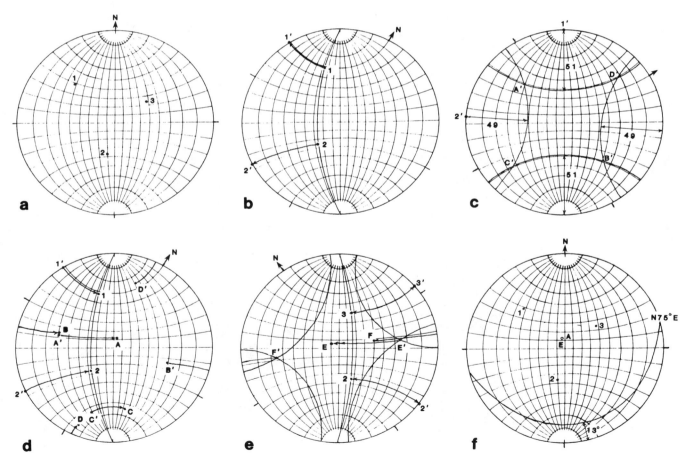

Figure 7-12. Using rotation to determine bedding attitude from three nonparallel drill holes.

EXERCISES

1. The orientation mark on a vertical drill core has an azimuth of 40°. Bedding within the core has an apparent strike of 40° clockwise and a core axis-bedding angle of 48°. What is the attitude of bedding?

2. An inclined drill core has an orientation mark (on its upper side) that has a plunge of 62°,210°. The core cuts a mineral vein whose apparent strike (with respect to the orientation mark) is 30° counterclockwise and which makes an angle of 30° with the core axis. What is the attitude of the vein?

3. A dipmeter survey on a vertical drill hole gives the following data:

   Electrode A crosses a key bed at 300 ft
   Electrode B crosses the bed at 300 ft 9 in.
   Electrode C crosses the bed at 300 ft 5 in.

   (a) If the azimuth of A is N72°E, and B is 120° clockwise from A (looking down the drill hole) determine the attitude of the bed.
   (b) If, in addition, the drill hole has a drift of 7°S on the north-south plane and 4°W on the east-west plane, determine the true attitude of bedding.

4. A mining company drills a series of vertical drill holes on a north-south, east-west grid with a spacing of 100 m. One drill hole encounters a planar ore body at 150 m. The drill hole immediately to the east of the first one encounters the same body at 190 m.

   (a) At what depth should the ore body be encountered in the hole immediately to the east of the second hole? How far west of the first hole should the ore body be expected to be found at the surface?
   (b) If the core-bedding angle in the drill holes is 50°, what are the possible attitudes of the ore body?

5. Two drill holes are 400 m apart on a line trending N70°W. Both drill holes are inclined, plunging 80° toward N70°W. The western hole encounters a coal bed at a depth of 300 m, and the same bed is encountered in the eastern hole at 200 m. If the core-bedding angle is 40° in both holes, determine the possible attitudes of bedding.

6. A drill hole plunging 60°,N35°E encounters a marker horizon at 30 m, and the core-bedding angle is 62°. A vertical drill hole, 50 m west of the first hole, encounters the same horizon at 15 m, and the core-bedding angle is 45°. What is the attitude of the marker horizon?

7. In an area of uniformly dipping siltstone beds with no marker horizons, three drill-holes are drilled to determine the subsurface attitude of bedding. From the drill-hole data (given below), determine the attitude of bedding.

Drill hole	Plunge	Core-bedding angle
A	50°,90°	45°
B	58°,298°	85°
C	40°,200°	50°

CHAPTER

8

EQUAL-AREA
PROJECTIONS
AND STRUCTURAL
ANALYSIS

8-1   INTRODUCTION

In previous chapters we used a specific type of azimuthal projection, called the stereographic projection, to solve a range of geometric problems in structural geology. The stereographic projection has two important properties: (1) The projection preserves angular relationships and is, therefore, often called an *equal-angle projection*. This means that the angle between the tangents to two intersecting great-circle traces at their point of intersection is the same as the angle between the two real planes that the great-circle traces represent (Fig. 8-1a). (2) The stereographic projection does not conserve area. This means that projections of identical circles inscribed on different parts of a projection sphere appear as circles of different sizes on the stereogram (Fig. 8-1b). In fact, the stereographic projection of a circle may vary in area by up to a factor of two, depending on where it is projected; a circle of a given area will appear to be much larger if plotted near the primitive than if it is plotted at the center of the net (Fig. 8-1b). Likewise, a 10° X 10° area at the edge of a Wulff net is much larger than a 10° X 10° area at the center of the net (Fig. 8-1c).

The latter property makes the stereographic projection useless for applications in which the statistical treatment of orientation data is of interest.   Such applications are common in structural analysis. For example, information on the *preferred orientation* (most common orientation) of joints in an area may provide information on paleo-stress

fields. The orientation of the joints can be represented on a *rose diagram* or a *histogram* (see Chapter 12), but these graphs represent orientation only in two dimensions (i.e., they can represent strike or dip, but not both).   An appropriate azimuthal projection can represent a preferred orientation in three dimensions as a cluster of poles, if the concentration of poles per unit area of the projection is proportional to the real concentration of planes of a specific orientation. A stereographic projection, because it distorts area, cannot be used for such representations; equal concentrations of poles at different localities on the surface of a projection sphere appear as unequal concentrations of poles on the plane of a stereographic projection.

In problems for which the statistical distribution of points is important, an alternative form of azimuthal projection called the *Lambert* or *equal-area projection* is used. A grid constructed on an equal-area projection is called a *Schmidt net*, named after a German petrologist (Fig. 8-2). Such a projection does not cause the area of a projected circle to vary with its position, although its shape does change (Fig. 8-3a); thus, the concentration of a cluster of points does not vary with position on the projection. Likewise, on a Schmidt net the size of a 10° X 10° area near the primitive is the same as that at the center (Fig. 8-3b). The purpose of this chapter is to introduce the equal-area projection, show how data distributions on such projections can be represented by contours, and illustrate some applications of the equal-area projection to structural analysis.

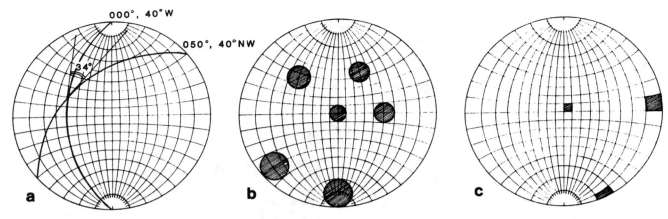

Dihedral angle between two planes = 34°

Figure 8-1.   Properties of a stereographic or equal-angle projection. (a) The angle between two planes is the same as the angle between the tangents to the great-circle traces of the two planes; (b) identical circles on projection sphere project as circles of different sizes; (c) 10° X10° area at edge of projection is larger than at the center.

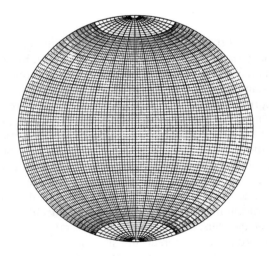

Figure 8-2.   The Schmidt or equal-area net.

8-2 EQUAL-AREA PROJECTIONS AND THE SCHMIDT NET

Construction of an Equal-Area Net

The equal-area projection is simply another form of azimuthal projection that can be used to project a lower-hemisphere spherical projection onto a horizontal plane.   The geometric basis for construction of the equal-area projection is shown in Figure 8-4. This figure shows a vertical cross section through the center of a projection sphere; Z is the zenith of the sphere, C is the center of the sphere, and C' is the base of the sphere. The projection plane is tangent to the sphere at C', which is the center of the azimuthal projection. Any inclined line (CP) that passes through the center of the projection sphere intersects the surface of the sphere at a point P. Point P is

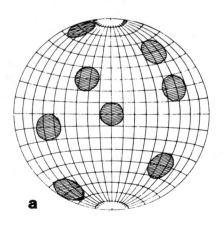

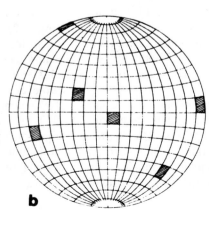

Figure 8-3.   Properties of an equal-area projection. (a) Identical circles on projection sphere project as ellipses with various axial ratios but having the same area; (b) 10° X 10° area at edge of projection is the same size as at the center.

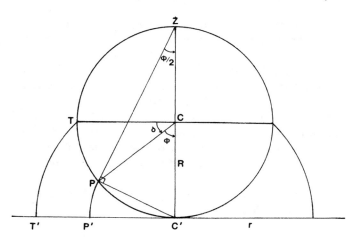

Figure 8-4. The geometric basis for constructing an equal-area projection. (Adapted from Ragan, 1985.)

the spherical projection of line CP. A circular arc, whose center is at C' and which passes through point P, intersects the projection plane at P'. P' is the projection of P on the azimuthal projection plane. The distance of point P' from the center of the azimuthal projection (C') can be calculated as follows:

$$\angle \text{ (C'CP)} = \phi = 90^\circ - \partial,$$

where ∂ is the plunge of line CP, so

$$\angle \text{ (C'ZP)} = \phi/2.$$

Triangle C'ZP is a right triangle, so

$$PC' = 2r[\sin(\phi/2)] \qquad \text{(Eq. 8-1)}$$

$$P'C' = PC' = 2r[\sin(\phi/2)], \qquad \text{(Eq. 8-2)}$$

where r is the radius of the projection sphere. Using a similar method, we can calculate the radius (R = C'T') of the primitive on the projection plane:

$$T'C' = TC' = 2r[\sin(\pi/2)] = 2r/\sqrt{2} = r\sqrt{2} \qquad \text{(Eq. 8-3)}.$$

Remember that in the case of the equal-angle projection, it was easiest to visualize the projection by passing the projection plane through the center of the projection sphere, so that the radius of the primitive equaled the radius of the projection sphere. It is similarly convenient to scale an equal-area projection to be the same radius as the projection sphere. In order to change the scale of the primitive so that it has the same radius as the projection sphere, we make T'C' = r by dividing Equation 8-3 by $\sqrt{2}$, and we determine the position of any point within the scaled projection circle by dividing Equation 8-2 by $\sqrt{2}$:

$$P'C' = \sqrt{2}r[\sin (\phi/2)] \qquad \text{(Eq. 8-4)}.$$

Note that in the projection technique described above, a 2°-wide segment of the surface of the projection sphere will correspond to the same line length on the azimuthal projection, regardless of whether the segment is near the equator or near the pole. Therefore, an azimuthal projection constructed according to the preceding method is an equal-area projection.

Using this projection procedure, it is possible to construct an equal-area net (Fig. 8-2). The net is merely a grid of curves. The suite of curves on this grid that run from the north to south poles represent the equal-area projections of a suite of planes of different dips passing through the north-south horizontal axis of the projection sphere. The second suite of curves represents the equal-area projections of right-circular cones whose vertices are at the center of the projection sphere and whose axes are coaxial with the north-south axis of the projection sphere. Thus, the equal-area grid is analogous to the grid on a Wulff net, and it is used in exactly the same way for plotting lines, planes, and poles. The curves on an equal-area net, in contrast to those on an equal-angle net, however, are elliptical arcs, not segments of circular arcs. Nevertheless, the north-south trending grid lines are usually referred to as great circles, and the other set of grid lines are referred to as small circles. The trace of a plane on an equal-area net is called a great-circle trace.

Which Net Is Which?

There is sometimes confusion about the names assigned to different types of azimuthal projections. A stereographic projection is one type of azimuthal projection. The terms Wulff net or stereonet refer only to grids drawn on a stereographic projection, and a stereogram refers only to a plot of points or curves on a stereographic projection. An equal-area projection is a second type of azimuthal projection. An equal-area projection is not a stereographic projection. The term Schmidt net refers to a grid drawn on an equal-area projection, and it is not the same as a stereonet; formally, the term stereonet should be used only with respect to a grid on a stereographic projection, and the term equal-area plot should be used for points or curves drawn on an equal-area projection. In practice, however, geologists tend to use the term stereonet loosely, to refer to either a Wulff net or a Schmidt net (see Chapter 15).

A question often arises as to when it is appropriate to use a Schmidt net instead of a Wulff net, and vice versa, for plotting data. The Schmidt net must be used in all applications where the concentration of points on the plot is significant; thus, it is particularly applicable for analysis of a large number of measurements. A Wulff net must be used where angles between structures on the net will be measured with a protractor (e.g., problems in Chapter 7). In applications where lines, planes, and poles are to be plotted for geometric calculations without a protractor (i.e., all problems in Chapters 5 and 6), either net can be used; thus, all figures in Chapters 5 and 6 could have been drawn on a Schmidt net. The Schmidt net, therefore, has the most common application for problems in structural geology and is usually the net that geologists carry with them to the field.

We will see that measurements made at many localities around certain structures yield characteristic distribution patterns of poles on a Schmidt net. The distributions of poles shown in Figure 8-5 represent more-or-less ideal patterns. In actual geologic examples the distribution patterns are never quite perfect, and patterns may be difficult to recognize. If the scatter from an ideal pattern is large, the pattern may be unrecognizable unless more data are obtained.

8-3 CONTOURING OF EQUAL-AREA PLOTS

From the experience gained in the exercises of the previous three chapters, you should now be adept at visualizing the orientation of a structure represented on an azimuthal projection. In the process of collecting data on a structure in the field, you will have occasion to make numerous measurements of either planar or linear attitudes. A plot of such data may show clusters of points (poles or lineations) on either a stereoplot or an equal-area plot. A projection that shows only points is called a *scatter diagram* or a *point diagram*. From clusters on a scatter diagram it is often possible to estimate the dominant orientation of a structural element in your study area. But in order to obtain a more precise representation of variations in orientation, you must quantify the number of points per unit area of the projection. Such quantification can be done on an equal-area net and may allow you to recognize subtle variations in the preferred orientations of a structural element as measured in different localities. The most efficient way of representing variations in the concentration of points on an equal-area plot is by *contouring* of point data.

A contour line on an equal-area plot separates zones of the plot in which the densities of point data are different. Densities of point data are usually measured as a *percentage of the total number of points per 1% area of the stereogram*. If the total area of the plot is 100 cm^2, 1% of the plot is 1 cm^2; if there are 100 points plotted on the equal-area net, and 10 points lie in a specific 1-cm^2 area, then the density of points in that area is 10% of total points per 1% of area. Contour lines are drawn on an equal-area plot to separate zones in which the percentage of

total points per 1% area falls within a specified range. For example, if the contour interval is 2, then the lowest contour line is drawn separating the zone in which there are fewer than 2% of total points per 1% of area from the zone in which there are more than 2% of total points per 1% of area. The next contour line is drawn to separate the zone in which there are 2 to 4% of total points per 1% area from the zone in which there are more than 4% of total points per 1% area, and so forth. Admittedly, describing the contour interval on a contoured equal-area plot is a bit of a tongue twister.

Certain general rules can be followed when contouring an equal-area plot:

1. On the basis of the minimum and maximum concentrations, contour intervals should be chosen such that there are no more than 6 contours on the final plot. There should be a constant contour interval.

2. The lowest contour is usually drawn at 1 point per 1% area. The highest contour should be chosen to emphasize and differentiate maxima that are large enough to stand out clearly on the projection.

3. A contour that crosses the primitive has to reappear at the diametrically opposite end of the stereogram.

4. It is easiest to start drawing the contours at the area of greatest concentration and to work outward.

5. After preliminary contouring, it may be useful to go back to the counting net to determine the true maximum. This is done by moving the 1% counting circle (described below) around the net until the largest number of points lie within the counting circle. The center of the counting circle then locates the true maximum.

6. It is also useful to smooth out some contours, after preliminary contouring, or to eliminate contours if the lines are too close to one another.

7. The values of the contours should be indicated in the legend in the form 1-5-9-13% per 1% area, maximum 14%, for example. On the finished plot the area of highest concentration should be blackened. Progressively decreasing densities of stippling are used for areas of lesser concentration. The lowest concentration is left blank. On many plots only the areas within the contours of highest concentration are shaded at all.

8. It is very useful to present the contoured diagram side by side with the scatter diagram (showing points on the stereogram) in order to convey as much objective data as possible.

Once the diagram has been contoured, the mean or dominant orientations of principal structures can readily be determined from the positions on the net where the greatest number of points occur. It is common practice to abstract these data by plotting, on a separate equal-area (or equal-angle) net, the orientations of the principal structural

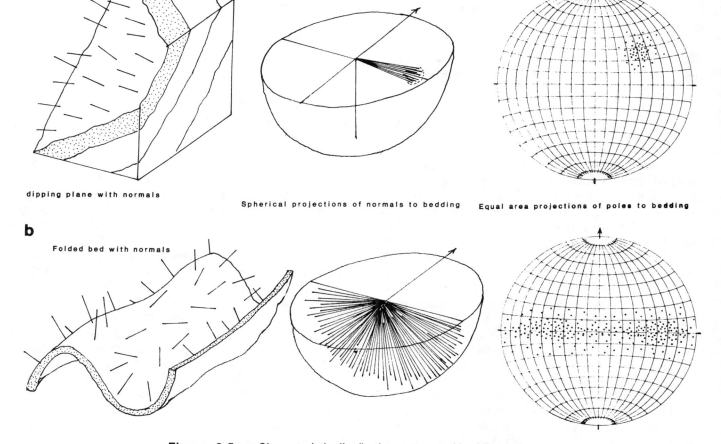

Figure 8-5.   Characteristic distribution patterns of bedding poles on an equal-area projection. (a) Uniformly dipping beds; (b) folded bed (adapted from Ragan, 1985).

elements within a region.   A diagram on which a single great circle or point is used to represent the mean or dominant orientations of several structural elements is called a *synoptic diagram*.

Increasingly, computers are being used to construct contoured equal-area plots, but it still is important to understand contouring principles using graphical methods. There are a number of graphical methods that are used for contouring point data, some of which are very versatile and can easily be used in the field. For most graphical methods it is convenient to use a 20-cm-diameter stereonet for ease of counting, but the large size of this net makes it inconvenient for transporting it to the field. The methods described here refer to a 15-cm-diameter stereonet (Appendix 4 provides a net with this diameter).

Schmidt or Grid Method of Contouring

The most commonly used method of contouring is the *Schmidt method*, which works well for large data sets

(>400) and/or for high concentrations of points. The method employs a square grid, so it is sometimes also called the *grid method*. The steps of this method are outlined next.

Problem 8-1

An equal-area plot of the poles to foliation within the Carthage-Colton mylonite zone in the Adirondacks is given (see Fig. 8-8a). Contour this plot using the Schmidt method.

Method 8-1 (Schmidt method)

Step 1: Construct a square grid with grid points 0.75 cm apart. The dimensions of the grid must exceed that of your equal-area plot.

Step 2: Construct a *Schmidt counter* (Fig. 8-6). A Schmidt counter contains two circular holes at opposite ends of a cardboard strip. Each hole has an area equal to 1% of the total area of your equal-area projection. You will see that two diametrically opposite holes are needed to

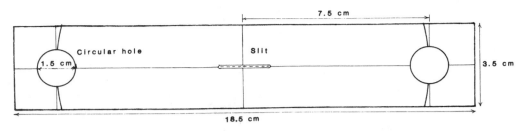

Figure 8-6. Construction of a Schmidt counter for a 15-cm-diameter Schmidt net. The full size counter appears in Appendix 4. (Adapted from Turner and Weiss, 1963.)

count points along the edges of the equal-area plot, while only one circle is needed for counting points in the interior of the plot. The holes to be used with our 15-cm-diameter net should be 1.5 cm in diameter. The counter can be constructed from a strip of poster board that is 18 cm long and 3.5 cm wide (Fig. 8-6). Draw two 1.5-cm-diameter circular holes in the strip, one at each end, such that the centers of the holes are 15 cm apart. In ink, draw a circular arc of 15 cm diameter that passes through the centers of the counting holes, and draw lines through the centers of the holes perpendicular to the line that joins the two centers; these ink marks should be visible on the cardboard borders of the counting holes and will serve as guides during counting (Fig. 8-7). Now, cut out the counting holes, and cut a 2-cm-long thin slot in the middle of the strip, halfway along the inscribed line joining the two centers of the holes. Draw a cross-line to mark the center of the slot.

Step 3: Place an overlay containing the point data of an equal-area plot (Fig. 8-8a), a tracing of the primitive, and a north reference mark, over the square grid. The center of the overlay must coincide with the intersection of two grid lines. Fix the overlay with tape.

Step 4: Place a second overlay, showing only the trace of a 15-cm-diameter circle and a north mark, on top of the first. The circles on the two overlays should be concentric, and the north marks should coincide.

Step 5: Place one end of the counter over the two overlays, such that the center of the circular hole coincides with a grid point; you may use the ink line that passes through the center of the hole as a guide (Fig. 8-7b). The number of points that are visible within the hole represent the number of points per 1% area. Make a dot at the grid point on the second overlay, count the number of points on the first overlay that are visible within the hole, and write it next to the dot on the second overlay. Reposition the counter so that its center is over the adjacent grid point, and repeat the procedure for all the grid points. Leave blank the areas in which there are no points (Fig. 8-8b).

Step 6: In the peripheral zone near the primitive, one counting circle will not fall entirely within the primitive, and you will need to use both circles of the counter. Pierce the center of the grid and the two overlays with a thumbtack, and put the point of the tack through the central slot of the counter. Points within both counting circles at diametrically opposite ends of the projection must be counted together (Fig. 8-7a). Intersections of the grid lines with the primitive are used to center the counting circle where necessary.

Step 7: Once numbers have been written in at each grid point on the second overlay (Fig. 8-8b), convert the number of points (n) next to each dot to a percentage by the equation

$$n(100)/N = \%,$$

where N is the total number of points on the plot. Draw contours at intervals corresponding to the appropriate point densities (Fig. 8-8c).

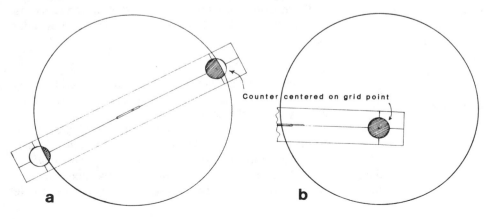

Figure 8-7. Method for counting points for contouring. (a) For points that lie close to the primitive; (b) for points inside the primitive. (Adapted from Turner and Weiss, 1963.)

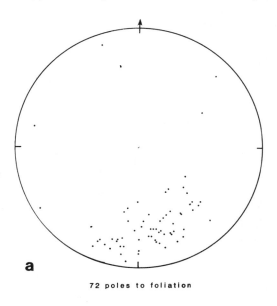

72 poles to foliation

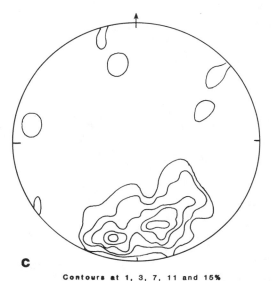

Figure 8-8. Procedure for contouring described in Problem 8-1. (a) Equal-area projection of poles to 72 foliation measurements; (b) point count using grid and Schmidt counter; (c) the final contoured diagram with contours at 1, 3, 7, 11, and 15%. A Schmidt counting grid is available in Appendix 4.

Contours at 1, 3, 7, 11 and 15%

Mellis Method of Contouring

This method is convenient for a small number of points (<100) and for populations of points that do not show local high concentrations. It is particularly useful for determining the contour of minimum density (i.e., usually the one-point contour).

Problem 8-2

Contour the data in Figure 8-9a using the Mellis method.

Method 8-2 (Mellis contouring)

Step 1: Construct an overlay on which only the primitive and the north mark are shown. Construct a counting circle that is 1.5 cm in diameter (i.e., 1% of the total area of your equal-area projection). Place the overlay over your equal-area plot so that it is concentric and aligned with the north arrow. Tape it down.

Step 2: Draw a 1.5-cm-diameter circle around each point in the population (Fig. 8-9a). The overlapping areas of two circles have a 2X concentration of a single circle, overlapping areas of three circles have a 3X concentration of a single circle, and so on. The percentage of the total number of points that is represented by one point can be calculated. The 2X and 3X areas are merely double and triple that percentage respectively.

Step 3: Place a third overlay over the second, and outline the areas of different point concentrations. It is best not to smooth the contour lines in this type of plot (Fig. 8-9b).

The Mellis method is the least subjective and most accurate method for contouring. The results are completely

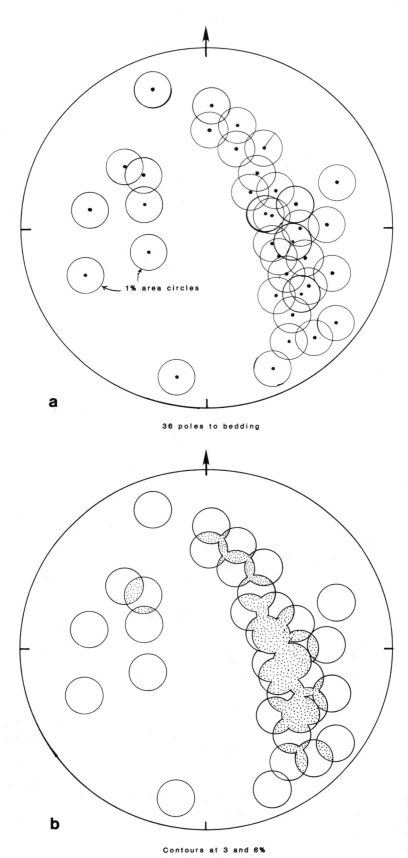

36 poles to bedding

a

Contours at 3 and 6%

b

Figure 8-9. The Mellis method of contouring described in Problem 8-2. (a) Equal-area projection of 36 poles to bedding. 1% area circles drawn around each point; (b) contoured diagram with contours at 3 and 6%.

reproducible, and for the same population of points the results will be identical for any two workers using the method. Its use, however, is limited to small populations and low concentrations; the method would obviously be cumbersome and difficult to use if four or more circles overlapped in a certain area.

Kalsbeek Method of Contouring

The Kalsbeek method (Kalsbeek, 1963) of contouring is quick and easy and thus is particularly appropriate in the field. It can be used with any population of point data on an equal-area plot.

Problem 8-3

The equal-area plot in Figure 8-11a is the same as that used for Problem 8-1. Contour the data using the Kalsbeek method.

Method 8-3 (Kalsbeek contouring)

Step 1: Obtain a Kalsbeek counting net (one is provided in the back of this book). This net is subdivided into small triangles (Fig. 8-10). Each set of six triangles forms a hexagonal area that covers 1% of the total area of the net. The triangles are arranged so that the net has six radial rays. The counting areas at the ends of these rays are semicircles, rather than hexagons.

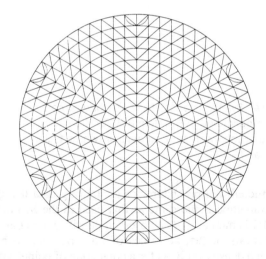

Figure 8-10. Counting net. (From Kalsbeek, 1963.)

Step 2: Place an overlay containing your equal-area scatter plot over the counting net, with the north mark of your plot at the tip of one of the radial rays. Place a second overlay, with only the primitive and the north mark, over the first.

Step 3: Count the number of points that fall within each hexagon, and write the number of points next to a dot

at the center of the hexagon on the second overlay (Fig. 8-11b). Since the hexagons overlap, each point is counted on three occasions. Along the primitive combine the points within each half-hexagon with points from the half-hexagon at the diametrically opposite end of the stereogram. Count the points at the ends of the six radial rays of the net by using the two complementary semicircles at opposite ends of the diameters. As with the Schmidt method, leave areas with no points blank.

Step 4: When you have finished counting, translate the numbers into percentages of the total number of points, and contour the results (Fig. 8-11c) in the same way as you did for the Schmidt method. Note that the contoured plot looks similar to that obtained in Problem 8-1 (Fig. 8-8c), but it is not exactly the same.

Kamb Method

Kamb (1959) proposed a contouring method that permits graphic analysis of the statistical significance of point concentrations on an equal-area plot. In the Schmidt or Mellis methods just described, the area (A) of the counting circle was 1% of the total area of the equal-area projection. We could, alternatively, choose A to be any fraction (of the total area) from 0 to 1.

Consider an equal-area plot on which there are N points. If the distribution of points is statistically uniform, there are (N X A) points within a counting circle of area A and [N X (1 - A)] points outside the counting circle. Call (N X A) the *expected number* of points. If the *actual number* of points (n) that falls within the counting circle is significantly greater than the expected number, then we have a *significant cluster*. The distribution of n values is a *binomial distribution* (see a statistics book for the definition of a binomial distribution), and the mean (μ) and standard deviation (σ) of such a distribution are given by

$$\mu = (NA) \qquad \qquad \text{(Eq. 8-5)}$$

$$\sigma = [N(A)(1 - A)]^{1/2} = NA[(1 - A)/NA]^{1/2}$$

or,

$$\sigma/(NA) = [(1 - A)/NA]^{1/2} \qquad \text{(Eq. 8-6)}.$$

To smooth out wild fluctuations from expected densities, A is chosen such that if the population has no preferred orientation, the number of points (NA) expected to fall within the counting circle is 3σ of the number of points (n) that actually fall within the counting circle under random sampling of the population. Thus, by setting $\sigma/(NA) = 1/3$, we can calculate from Equation 8-2 the appropriate area (A) of the counting circle for a given fabric represented by N points:

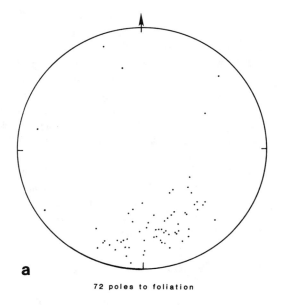

a

72 poles to foliation

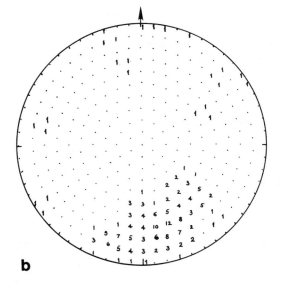

b

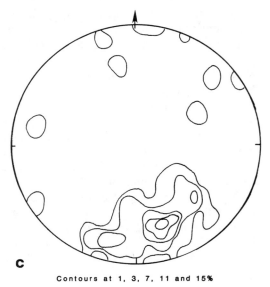

c

Contours at 1, 3, 7, 11 and 15%

Figure 8-11. The Kalsbeek method of contouring described in Problem 8-3, for the same data as Problem 8-1. Contours drawn at 1, 3, 7, 11, and 15%.

$$(1 - A)/A = N/(3^2)$$

$$1/A = (N/9) + 1 = (N + 9)/9$$

$$A = 9/(N + 9) = \pi\, r^2 \qquad \text{(Eq. 8-7)}.$$

Therefore, the radius (r) of the counting circle, expressed as a fraction of the radius of the equal-area projection, should be

$$r = 3/[(N + 9)\pi]^{1/2} \qquad \text{(Eq. 8-8)}.$$

The spacing of grid points on the counting net is also simply r.

Once the diameter of the counting circle has been determined, and an appropriate counting grid has been constructed, the procedure for contouring data using the Kamb method is the same as that used in the Schmidt or the Mellis methods. The observed densities are contoured at intervals of 2σ, at values of 2σ, 4σ, etc., with the expected density (N X A) for a population of points with no preferred orientation being 3σ. Generally, the area of a counting circle used in constructing a Kamb contoured plot is larger than that used for conventional contouring; therefore, the Kamb method smooths out the contour irregularities that are of no statistical significance. However, it does not significantly change the positions of contour lines for large populations. The method is most conveniently applied by using a computer, as the size of the counting circle must be calculated for each data set that is plotted.

Problem 8-4

The equal-area plot of Figure 8-12a shows the same data as that used in Problem 8-1. Determine the appropriate diameter for the counting circle, for the Kamb method of contouring, and contour the data using this method.

Method 8-4

Step 1: Determine the appropriate diameter of the counting circle. Using Equation 8-8 for N = 72, r is found to be 0.188 of the radius of the equal-area projection. For a 15-cm-diameter net, the counting area is a circle of diameter 2.82 cm.

Step 2: Construct a counting grid with a spacing equal to the radius of the counting circle, i.e., 1.41 cm.

Step 3: Follow the same method as in Schmidt contouring to obtain concentrations at the grid points (Fig. 8-12b).

Step 4: Using Equation 8-6, calculate a value for σ (1.57). Contour the concentrations at 2σ, 4σ, 6σ, and 8σ (Fig. 8-12c). Notice that the contoured diagram is significantly different from those obtained by the Schmidt and Kalsbeek methods.

Additional methods of contouring equal-area plots and of statistical analysis of equal-area plots are described by Vistelius (1966).

8-4 PATTERNS OF POINT DATA ON EQUAL-AREA PROJECTIONS

The distribution of points on an equal-area projection graphically expresses the degree of preferred orientation (or lack thereof) of a particular structural element (such as foliation or lineation). The key to interpreting the

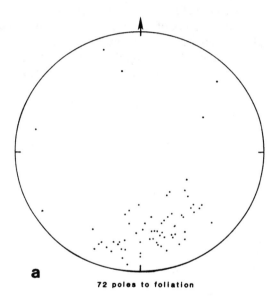

a

72 poles to foliation

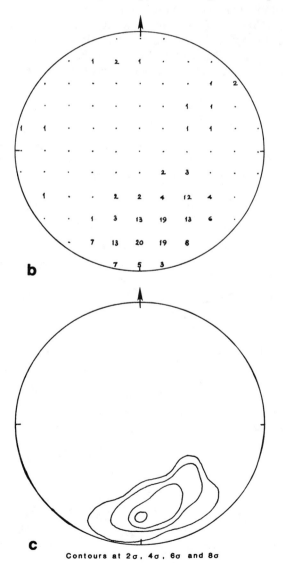

b

c

Contours at 2σ, 4σ, 6σ and 8σ

Figure 8-12. The Kamb method of contouring described in Problem 8-4, for the same data as Problem 8-1. Contours drawn at 2σ, 4σ, 6σ, and 8σ.

projection lies in recognizing the pattern defined by the distribution of points (where "points" refers to either the projection of a line representing a linear structure or the projection of a pole representing a planar structure). Recognition of patterns is often easier to do with a contoured diagram. There are four main patterns that can be recognized:

Uniform Distribution: A statistically random distribution in the orientation of structural elements is expressed by a scatter of points on an equal-area plot in which there are no obvious local concentrations. The lack of any concentration on a plot is called a *uniform distribution* (Fig. 8-13a).

Point Maximum: A preferred orientation of structural elements is represented by a high concentration (significant cluster) of points symmetrically distributed around a single mean orientation (Fig. 8-13b). The center of the cluster is the *point maximum*. A single data set can show more than one point maximum.

Great-Circle Girdle: A concentration of points along an arc approximating a great circle is called a *great-circle girdle* (Fig. 8-13c). The pole of the great-circle girdle is called the *girdle axis*. A girdle may contain one or

more distinct maxima. In some cases, two girdles may intersect, forming a *crossed girdle pattern*. A girdle pattern for linear elements indicates that the lineations all lie in a single plane but are not parallel to one another. In such a case the girdle approximates the attitude of the plane containing the lineations, and the girdle axis is the pole to that plane. A girdle pattern for poles to planar elements indicates that the planes could all intersect along the same line. For example, a girdle pattern is obtained by plotting poles to bedding taken around a cylindrical fold (see below).

Small-Circle Girdle: A *small-circle girdle* is a concentration of points along an arc that approximates a small circle (Fig. 8-13d). Such a girdle may contain one or more distinct maxima. For both linear and planar elements such a girdle indicates a preferred orientation in a cone about a single axis (the girdle axis).

We can describe the point-distribution pattern on an equal-area plot in terms of the type of *symmetry* displayed (e.g., the number of mirror planes that can be drawn on the plot, across which the point clusters are mirror images of one another) by analogy with the description of point groups in crystallography. For example, a fold may be

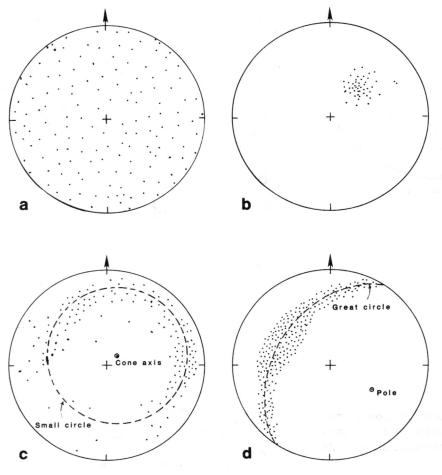

Figure 8-13. Patterns of point data on an equal-area projection. (a) Uniform distribution; (b) point maximum; (c) great-circle girdle; (d) small-circle girdle.

described as orthorhombic or monoclinic, depending on the pattern of clusters displayed on a plot of poles to bedding. For further discussion of this terminology see Turner and Weiss (1963).

8-5   ANALYSIS OF FOLDING WITH AN EQUAL-AREA NET

Geometrically, a fold is merely a curved surface. There are two basic types of folds: (1) *Cylindrical folds* are generated by moving an imaginary straight line parallel to itself in space. The line that generates the fold is called the *fold axis*. (2) *Noncylindrical folds* are generated by a line that moves in a nonparallel manner through space. If one end of the generating line is fixed, the resulting fold form is called a *conical fold*. If the movement of the generating line is nonsystematic, a *complex fold* results. Sometimes complex folds can be subdivided into parts that are approximately cylindrical. The geometry of cylindrical or conical surfaces can be analyzed with either ß-diagrams or π-diagrams on an equal-area projection.

ß-Diagrams of Cylindrical Folds

Every segment of a cylindrically folded surface contains a line segment that is parallel to the fold axis. Any two tangential planes to the folded surface will intersect along a line that is parallel to the fold axis. On an equal-area projection, therefore, the great circles representing the attitudes of the folded surface at different points on the fold should all intersect at a common point representing the fold axis. This point is called the *β-axis*. In practice, however, real folds do not have a perfectly cylindrical form, so strike and dip measurements around the fold produce great circles that do not all intersect at a common point, although the points of intersection do show a point maximum that gives an average orientation for the ß-axis. For n plotted planar attitudes (Fig. 8-14), the number of intersections (x) is given by the arithmetic progression

$$x = 0 + 1 + 2 + \ldots + (n - 1) = n(n - 1)/2 \quad \text{(Eq. 8-9).}$$

Thus, if there are 200 plotted planes, the number of intersections is 19,900. Contouring of the intersection

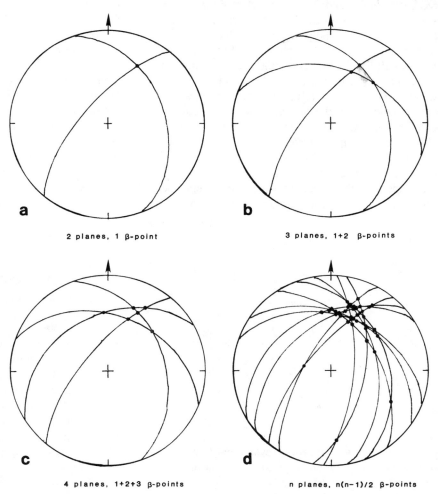

a   2 planes, 1 β-point

b   3 planes, 1+2 β-points

c   4 planes, 1+2+3 β-points

d   n planes, n(n-1)/2 β-points

Figure 8-14. β-diagram of cylindrical fold. The number of intersections of great circles increases rapidly as the number of plotted planes increases. (Adapted from Ramsay, 1967.)

points will emphasize the maximum concentration of intersections.

A plot of ß-axes is *not* generally the best way to represent attitude measurements on a fold, for several reasons. First, the number of points on a ß-axis plot is far greater than the actual number of measurements; thus, such a plot may make you think you have more data than you actually have. Second, if there is any scatter in the original data, there can be concentrations of ß-axes away from the main concentration, leading to an erroneous interpretation. Such errors become unacceptably large if the interlimb angle is very small (<40°), as in tight folds, or very large (>140°), as in open folds. Finally, construction of a ß-diagram is time consuming, because a large number of great circles must be plotted, and the number of intersections can become unmanageably large for even a small data set.

π-Diagrams of Cylindrical Folds

Because of the disadvantages of the ß-axis diagram, a *π-diagram* is the preferred method for representing measurements from a folded surface. A π-diagram is an equal-area plot of the poles to planes that are tangential to the folded surface. Practically, this means that if we have strike and dip measurements from many locations on a fold, we plot the pole for each plane rather than the great-circle trace. On a cylindrical fold, each of the poles is perpendicular to the fold axis; thus, the poles are parallel to a plane perpendicular to the fold axis. On an equal-area plot the poles approximate a great-circle girdle, which is variously called the *S-pole circle*, the *pole circle*, or the *π-circle* (Fig. 8-15). The pole to the π-circle is the π-axis (Fig. 8-15), and it represents the fold axis. The π-axis should coincide with the ß-axis on a plot. For a very open fold, with a very large interlimb angle, the π-diagram will

show an elliptical point maximum. With progressive decrease in the interlimb angle, the pole pattern changes from a point maximum, to an incomplete great-circle girdle, and finally, to a complete great-cricle girdle (Fig. 8-16; e.g., Ragan, 1985).

A π-diagram not only gives information on the orientation of a fold axis but also contains clues to the form of the fold. For example, if a fold has a broad, rounded hinge, the density of points will be uniform within the π-circle girdle, and the two extreme points on the girdle will define the interlimb angle (Fig. 8-17a). The π-circle girdle for a fold with planar limbs and a narrow hinge zone will contain maxima on the girdle corresponding to the two limbs, and these maxima can be used to determine the interlimb angle (Fig. 8-17b). For a chevron fold there is no well-defined girdle, and the π-circle on the projection is defined by two point maxima corresponding to the two limbs (Fig. 8-17c). Most natural folds show patterns that are intermediate between the broad-hinge girdle and the two-maxima (limbs) girdle.

It is generally not possible to say anything conclusive about the symmetry of folds from just a π-diagram, because factors other than dips in the two limbs of the folds determine fold symmetry (Hobbs, Means, and Williams, 1976; Ramsay, 1967). A concentration of points along a girdle may also be a consequence of sampling bias. However, if the spatial distribution of measurements in a train of folds is uniform, there will be fewer readings from the short limbs of asymmetric folds, resulting in an asymmetry in the pole pattern on the π-diagram (Fig. 8-17d). Generally, in order to determine asymmetry of folds, we need additional information such as variation in thickness from limb to limb, orientation of the enveloping surface, and orientation of the axial surface (see Chapter 11).

The orientation of the axial surface (or the axial plane) can be determined if the π-axis is known and if the orientation of the axial trace at a locality can be determined; the great circle passing through these two points gives the orientation of the axial plane (Fig. 8-18a). In the case of chevron folds and kink folds, the axial plane may be defined as a plane containing the bisector of the interlimb angle. The bisector is represented by the point whose angular distance from the two point maxima (measured along the π-circle girdle) is the same. The great circle passing through the bisector and π-axis represents the axial plane (Fig. 8-18b).

The attitudes of the fold axis and the axial plane are, of course, reflected in the position of the π-circle girdle on an equal-area projection. For example, if the fold axis is horizontal, then the π-axis lies on the primitive and the girdle passes through the center of the net, but if the fold axis is plunging, then the π-axis lies inside the primitive, and the girdle follows a curve that does not pass through

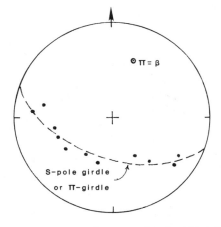

Figure 8-15. π–diagram of cylindrical fold. Poles to planes lie on a great circle girdle whose pole gives the fold axis (π = ß).

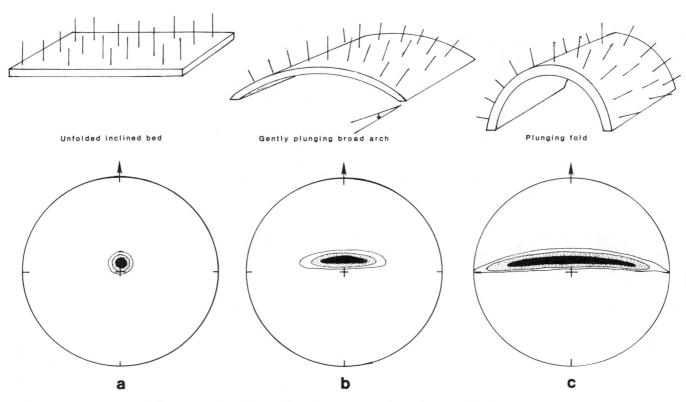

Figure 8-16. Change in pole pattern on the π-diagram with decrease in fold interlimb angle. (Adapted from Ragan, 1985.)

the center of the net. If the axial plane of the fold is vertical, it is represented by a diameter of the equal-angle plot; if it is horizontal, it is represented by the primitive; and if it is inclined, it is represented by some intermediate great circle. A fold with a plunging axis and an inclined axial plane may display a complex pattern on an equal-area net. Figure 8-19 shows several examples of π-diagrams and the folds that they represent.

π-Diagrams of Noncylindrical Folds

If the folded surface is conical, with the cone having an apical angle μ, each pole makes an angle of (90° - μ/2) with respect to the cone axis. In other words, the poles to bedding generate a coaxial cone with an apical angle of (180 - μ). Thus, the poles define a small circle, with its center representing the cone axis (Fig. 8-20). If an approximate small-circle pattern is recognized, it may be worthwhile to replot the poles on a Wulff net, since a small circle projects as a circle on the stereographic projection. A small circle can then be fitted to the plotted points, and the center of the circle (representing the cone axis) can be located. The cone axis can be rotated to the primitive, and the small circles of the net can be used to analyze the angular relationships within the fold.

In nonconical noncylindrical folds, both the axial surface and the fold axis vary in attitude, and construction of a π-diagram will generally give several possible orientations for the π-axis. Commonly, areas of superposed folding exhibit this kind of geometry. To analyze such folds, they must be subdivided, by trial and error if necessary, into domains of plane cylindrical folding (see Problem 8-7). Each domain should have its own constant π-axis orientation. In *plane* noncylindrical folds the axial surface is planar and has a constant orientation, although the orientation of the fold axis (π-axis) may vary. The mean orientation of the axial plane is defined as the great circle passing through the axes of the different cylindrical domains (Fig. 8-21).

8-6 ANALYSIS OF FABRICS WITH AN EQUAL-AREA NET

Types of Fabrics

The internal geometric and spatial configurations of the components of a rock constitute its *fabric*. If a fabric is visible in a rock regardless of the scale of observation, it is said to be *penetrative*. Rocks that have penetrative fabric

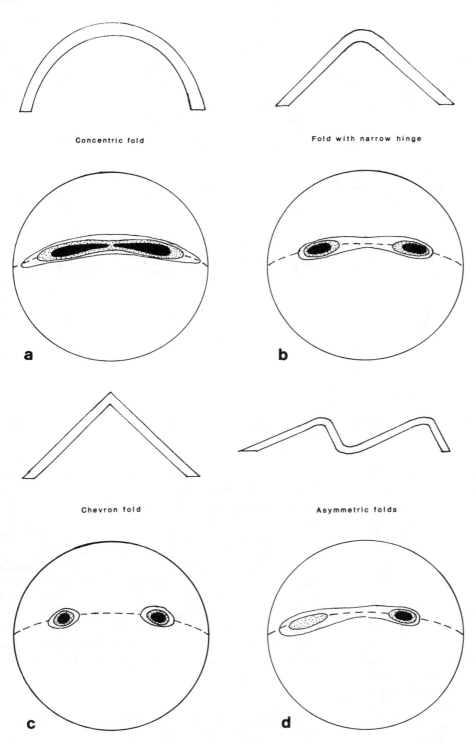

Concentric fold

Fold with narrow hinge

Chevron fold

Asymmetric folds

Figure 8-17. Variation in π–diagrams with change in fold form. (a) Fold with broad, rounded hinge; (b) fold with narrow hinge; (c) chevron fold; (d) asymmetric folds.

resulting from deformation are referred to as *tectonites*. Three major classes of tectonites are recognized, based on whether the fabric can be described as a foliation, a lineation, or both. (1) *S-tectonites* have a strong foliation but no lineation (Fig. 8-22a). The foliation is defined by parallel alignment of platy minerals, lenticular mineral

aggregates, or flattened grains. The letter S is used because of the longstanding convention of referring to foliations as S-surfaces (Turner & Weiss, 1963). (2) *L-tectonites* have a well-developed lineation but no foliation (Fig. 8-22b). The lineation in an L-tectonite is defined by alignment of prismatic minerals or uniaxially elongated grains parallel to

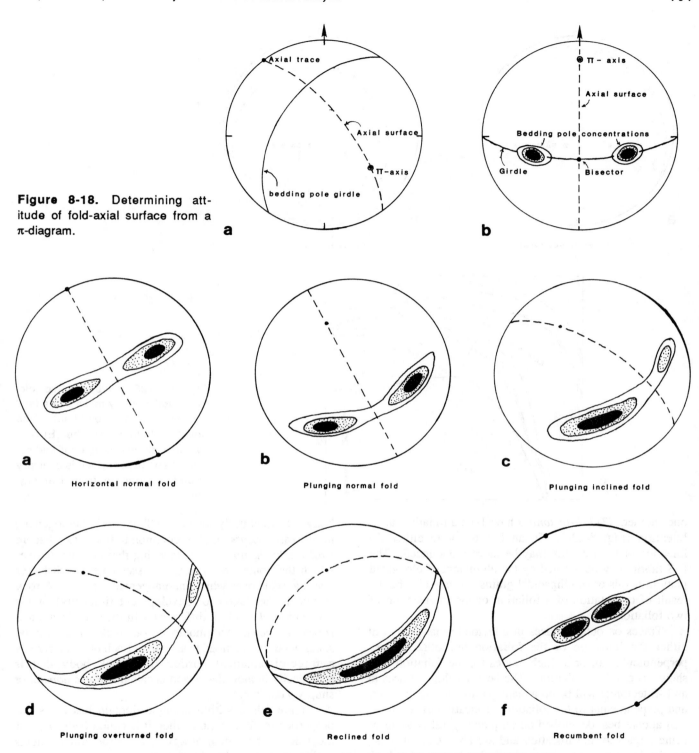

Figure 8-18. Determining attitude of fold-axial surface from a π-diagram.

Figure 8-19. Variations in position of π-circle girdle with changes in attitude of the fold axis and axial plane. (a) Nonplunging upright (normal) fold; (b) plunging normal fold; (c) plunging inclined fold; (d) plunging overturned fold (note the presence of vertical beds indicated by the plotting of some bedding poles on the primitive); (e) reclined fold; (f) recumbent fold; axial plane coincides with the primitive. (See Fig. 11-17.)

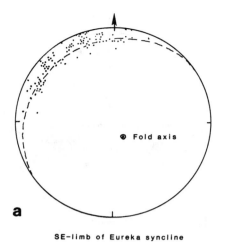

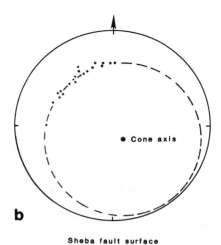

SE-limb of Eureka syncline   Sheba fault surface

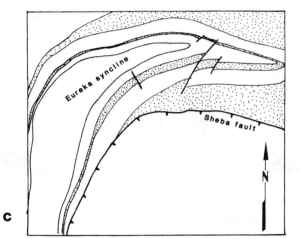

Figure 8-20. Flexural-slip folding of sedimentary rocks and a fault (at low angles to bedding). The southeast limb of the Eureka syncline shows cylindrical folding. The Sheba fault surface is conically folded. (Adapted from Ramsay, 1967.)

one another. (3) *L-S-tectonites* have both a foliation and a lineation (Fig. 8-22c). In an L-S-tectonite either the lineation or the foliation may be more pronounced. The L-S fabric may be defined by the alignment of elongated platy minerals or of ellipsoidal grains. It may also be the result of crenulation of a foliation or the intersection of two foliations.

Traces on outcrop faces in a tectonite may represent either the lineation or the foliation or both. Planes perpendicular to, or at high angles to, the foliation will show traces of the foliation. The best-developed traces in an L-S tectonite will be on a plane parallel to the lineation and perpendicular to the foliation. Lineation traces alone will appear best developed on all planes parallel to, or at acute angles to, the lineation and are absent or, at most, poorly developed on planes perpendicular to, or at high angles to, the lineation. In an L-S-tectonite an outcrop face parallel to the foliation itself shows the true attitude of the lineation.

In order to apply the methods of structural analysis, the fabric in a rock body must be *homogeneous*, meaning that equal volumes of rock from different localities in the body are structurally identical. True fabric homogeneity never really occurs, but it is common to find rocks that are *statistically homogeneous*, meaning that the sample over which the homogeneity is to be assessed is much larger than the scale over which inhomogeneity occurs. A rock in which the degree of development (intensity) or the orientation of a fabric differs as a function of location is *inhomogeneous*. An inhomogeneous rock can usually be subdivided into homogeneous parts. Each of these parts is a three-dimensional portion of a rock body that is statistically homogeneous and is called a *fabric domain* or simply a *domain*.

If the fabric within a single domain has the same properties in all directions, then it is called *isotropic*. In most deformed rocks, however, the structural elements within any domain show some degree of preferred orientation, and the fabric is, therefore, said to be *anisotropic*. Rocks with anisotropic fabrics may be S-, L-, or L-S-tectonites. Equal-area nets are useful in analyzing fabric in two ways. First, they may be used to calculate the true orientation of fabrics, given partial measurements on different planes; second, they may be used to describe

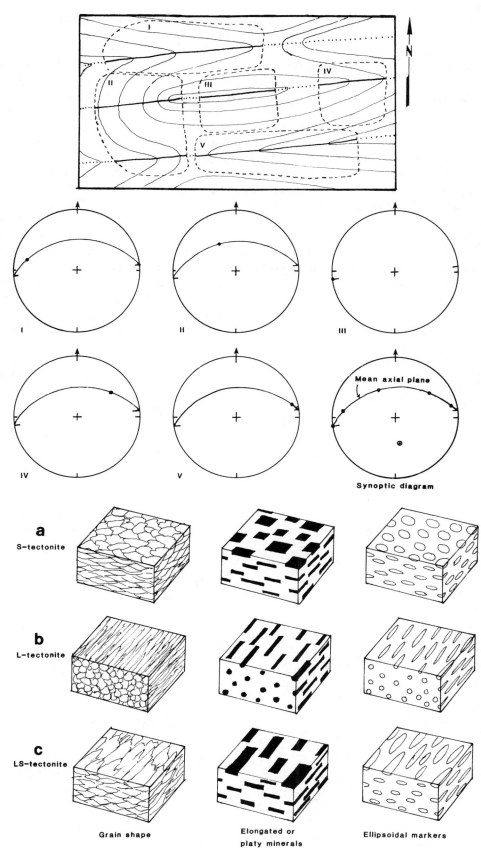

Figure 8-21. Plane noncylindrical folding. The area is subdivided into domains of cylindrical folding, each with its own fold axis, but all the fold axes lie on a common axial plane (as shown on the synoptic equal-area plot). (Adapted from Turner and Weiss, 1963.)

Figure 8-22. The penetrative fabric of a rock defined by overall grain shape, elongated or platy minerals, and ellipsoidal markers. Three different classes are defined. (a) S-tectonite; (b) L-tectonite; (c) LS-tectonite.

variations in the geometry of fabrics that occur between different domains. Finally, in rocks with multiple fabrics an equal-area projection may be the only way of distinguishing various fabric elements.

Calculation of Planar and Linear Attitudes

The trace of a planar structure on any surface is its apparent dip in that plane.   If two or more such apparent dips can be measured, the orientation of the planar structure can be determined on an equal-area projection (Method 5-8). With more than two traces, the plane is defined by the best-fitting great circle that passes through the data points.

Problem 8-5

The trace of a foliation (S) is seen on three nonparallel faces. The attitudes of the faces were measured, and the rake of the foliation trace on each face was measured, with the following results:

Attitude	Rake
N44°E,60°NW	24°SW
S66°E,80°SW	40°NW
S11°E,70°NE	28°NW

Determine the orientation of the foliation (S).

Method 8-5

Step 1:   On an equal-area projection, plot the great-circle trace of each face (Fig. 8-23a, b, c). Measure the rake of the lineation on each face, and plot the point representing the trace of the foliation for each face.

Step 2: Find the best-fitting great circle that passes through the three points representing the foliation traces (Fig. 8-23d). The great circle represents the plane of the foliation, which has an attitude of N30°E,40°NW.

Often, lineations (such as minor fold axes or mineral lineations) can be directly measured on an exposed surface. At some localities, however, linear structures do not lie on a plane of easy breakage and thus cannot be measured directly. The orientation or poorly exposed lineations can be determined by measuring the apparent lineation on two or more differently oriented faces and plotting the data on an equal-area projection.   Imagine a rod-shaped fabric element (e.g., a dowel with a circular cross section). Any section of the rod oblique to its axis will be elliptical (Fig. 8-24). The long axis of the ellipse is an apparent lineation on the plane of exposure; the true linear structure is

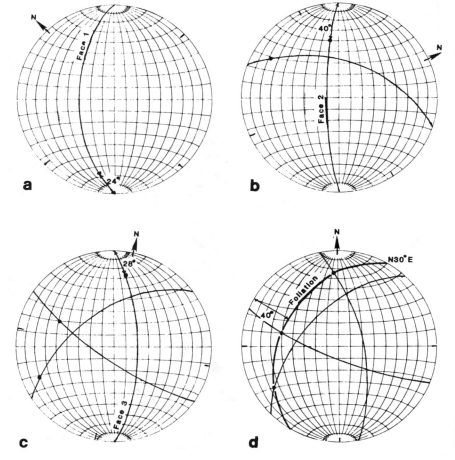

Figure 8-23. Procedure for determining orientation of a foliation described in Problem 8-6.

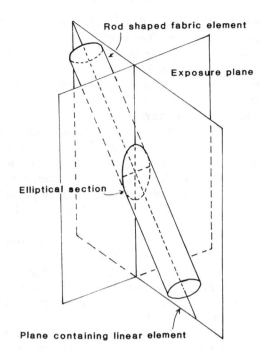

Figure 8-24. Apparent lineation on a planar surface obliquely intersecting a rod-shaped linear structure.

contained in a plane perpendicular to the exposure plane and passing through the long axis of the ellipse. The orientation of the apparent lineation can be measured either by its rake on the plane of exposure or by its plunge and bearing. The measurements are plotted on a stereogram to obtain the true attitude of the lineation. The following problem illustrates the method.

Problem 8-6

Traces of a lineation were measured on three nonparallel faces. The orientations of the three faces and the rake of the apparent lineation in each face are as follows:

Orientation	Rake
S56°E,52°NW	15°NW
N82°E,30°S	84°W
N09°E,70°W	22°S

Determine the true attitude of the lineation.

Method 8-6

Step 1: On an equal-area projection, plot each face, its pole, and the trace of the lineation on the face (Fig. 8-25a, b, c).

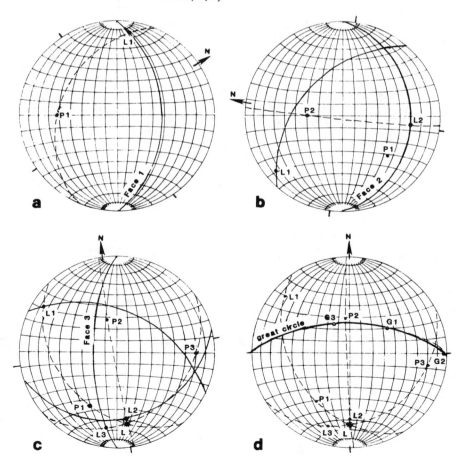

Figure 8-25. Procedure for determining orientation of a lineation described in Problem 8-7.

Step 2: For each face draw the great circle passing through the pole and the lineation trace (Fig. 8-25a, b, c). The three great circles intersect (ideally) in one point defining the attitude of the lineation (Fig. 8-25c) as 24°,S2°W. This is referred to as Lowe's method (Lowe, 1946).

Step 3: Alternatively, after step 1, find the poles (G_1, G_2, G_3) to the great circles that pass through the pole to the face and the lineation trace. All these "new" poles lie on a great circle that represents the plane perpendicular to the lineation (Fig. 8-25d). Thus, the pole to this great circle gives the attitude of the lineation (Fig. 8-25d). This method, which was devised by Cruden (1971), avoids the problem of trying to find a single point of intersection to define the attitude of the lineation by allowing a best-fit great circle to be drawn.

Analysis of Fabric Geometry

Patterns of variation in the attitude of fabrics around folds may help determine the chronology of fabric development with respect to the fold. A number of patterns are possible, depending on the nature of the fabric, the timing

of fabric development with respect to folding, and the mechanism of folding. The patterns are discussed next individually. A complete discussion of fabric types is beyond the scope of this book; our goal here is solely to show how fabric geometry can be practically analyzed with the equal-area net.

Foliation Postdating Folding: If a plane cylindrical fold that folds S_1 and formed coevally with S_2 is cut by a later planar foliation (S_3), the intersection lineation between S_3 and S_1 will vary around the fold (Fig. 8-26a). However, all the attitudes of the lineation lie on a single plane (S_3) and fall along a great-circle girdle (Fig. 8-26b). Also, the angle between the lineation and the F_2 fold axis varies as a function of S_1 attitude (Fig. 8-26b).

Flexural-Slip Folding of a Lineation: Development of *flexural-slip folds* is accommodated by layer-parallel slip with minimal internal distortion of layers. Thus, to a first approximation, the movement of a layer during folding can be described as a rotation, and the angle (μ) between the fold axis and the preexisting lineation remains constant everywhere on the folded surface (Ramsay, 1967; Fig. 8-27a). On an equal-area projection, the points representing the lineation, therefore, lie on a

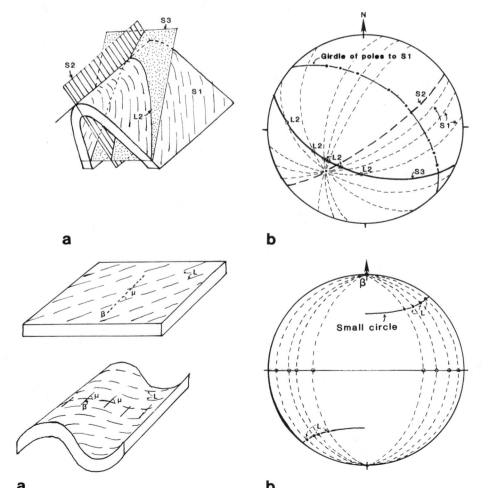

Figure 8-26. Intersection lineation produced by a later planar foliation (S_3) cutting an earlier folded foliation (S_1). (Adapted from Turner and Weiss, 1963.)

Figure 8-27. Flexural-slip folding of a preexisting lineation. Lineation points lie on a small circle centered on the fold axis. Lineation that was perpendicular to the fold axis (open circles on equal-area plot) lies on a great circle after folding. (Adapted from Ramsay, 1967.)

small circle centered on the fold axis (ß) (Fig. 8-27b), unless the original lineation is perpendicular to the fold axis in which case the folded lineation lies on a great circle (Fig. 8-27b). Remember that the rotation of a line around an axis inscribes a small circle (see Chapter 7).

In reality, the discounting of material distortion of layers is not correct. Because individual layers in a flexural-slip fold are buckled, each folded layer has a neutral surface that shows ideal concentric geometry, but the outer arc is extended and the inner arc shortened (Fig. 8-28a). Therefore, on the outer arc, the angle between the lineation and the fold axis is slightly increased ($\mu' > \mu$), and the lineation points lie on an arc that is broader than the small-circle arc but is still centered on the fold axis. (Fig. 8-28b). Similarly, the angle between the fold axis and the lineation is slightly decreased ($\mu'' < \mu$) on the inner arc, and the lineation points lie on an arc that is narrower than the small-circle arc (Fig. 8-28b).

Passive Folding of a Lineation: Development of a *passive fold* is geometrically analogous to the passive reorientation of a marker layer by shearing on a set of close-spaced planes that are oblique to the foliation. In reality, discrete slip planes need not exist. The axial plane of the fold is parallel to the hypothetical shear planes, and the fold axis is parallel to the shear plane-marker layer intersection lineation. Points along an original linear feature on the marker layer are transported variable distances along parallel lines (in the slip direction) and are positioned on the surface of the folded layer so that the folded lineation is contained in a plane defined by the original lineation and the slip direction (Fig. 8-29a). Thus, on an equal-area projection the points representing the folded lineation lie on a great circle that is oblique to the fold axis (Fig. 8-29b). This geometry is the same as that for an intersection lineation due to a foliation superposed on a preexisting fold, except that in this case, there may be no foliation developed parallel to the plane containing the lineation.

Complex Refolding of Lineations: Many natural folds show complex patterns for refolded lineation; the lineations often lie in arcs intermediate between a small circle and a great circle. Such modifications may result from layer-parallel shortening prior to folding, homogeneous flattening after folding, or some form of

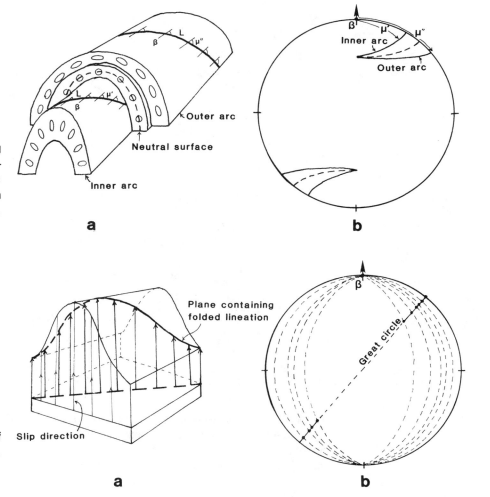

Figure 8-28. Effect of buckling of individual layers during flexural-slip folding. The small-circle arc pattern of lineations is modified in the outer and inner arcs of the fold. (Adapted from Ramsay, 1967.)

a

b

Figure 8-29. Passive folding of a lineation. Lineation points lie on a great circle oblique to the fold axis. (Adapted from Ramsay, 1967.)

a

b

longitudinal layer-parallel strain accompanying folding. Details of these various possibilities are discussed in Turner and Weiss (1963) and Ramsay (1967).

Flexural-Slip Folding of Obliquely Inclined Surfaces: Folding of rocks containing two preexisting foliations (S_1 and S_2) that are oblique to one another results in simultaneous folding of both foliations (Ramsay, 1967). The foliations could, for example, be bedding and cleavage, or two preexisting cleavages, or even cross bedding and its enclosing master bedding. The geometric patterns resulting from such folding are readily analyzed on an equal-area net. During flexural-slip folding, if the S_1/S_2 intersection lineation is parallel to the fold axis, then both surfaces are folded into cylindrical folds that are coaxial (Fig. 8-30a). If, on the other hand, the S_1/S_2 intersection lineation is perpendicular to the fold axis, and one surface (S_2) is folded into a cylindrical fold, the other surface (S_2) maintains its dihedral angle with respect to S_1 and is folded in conical form, with the cone axis parallel to the fold axis for S_1 (Fig. 8-30b). Oblique intersections between S_1 and S_2 give rise to more complex patterns, with the S_1/S_2 intersection lineation falling on a

small-circle arc centered on the fold axis for S_1, and the dihedral angle between S_1 and S_2 varying continuously around the fold (Ramsay, 1967) (Fig. 8-30c).

Passive Folding of Obliquely Inclined Surfaces: During passive folding the S_1/S_2 intersection lineation is folded (Fig. 8-31a) but still lies in a plane defined by its original orientation and the slip direction on the hypothetical shear planes. Thus, after folding, the intersection lineations lie on a great circle (Fig. 8-31b). Both S_1 and S_2 are folded into cylindrical folds with a common axial plane (S_3) parallel to the shear planes. The two folded surfaces have different fold axes (ß1 and ß2) determined by their lines of intersection with the shear planes (Fig. 8-31b, c). The dihedral angle between S_1 and S_2 generally varies across the fold (Ramsay, 1967) (Fig. 8-31c).

π-Diagram Analysis of Superposed Folds

Superposed folding refers to the overprint of a later generation of folds over an earlier one. Depending on their orientation, the later generation of folding can cause

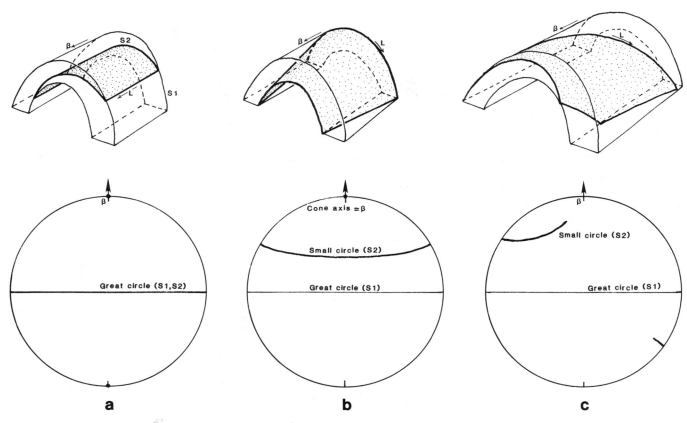

Figure 8-30. Flexural-slip folding of obliquely inclined surfaces. (a) S_1/S_2 intersection lineation parallel to the fold axis; (b) S_1/S_2 intersection lineation perpendicular to the fold axis; (c) S_1/S_2 intersection lineation oblique to the fold axis. (Adapted from Ramsay, 1967.)

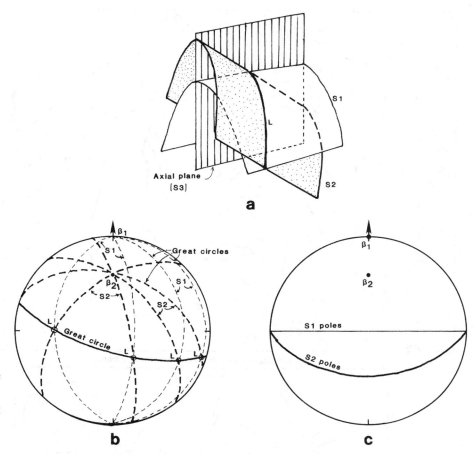

Figure 8-31. Passive folding of obliquely inclined surfaces. S_1/S_2 intersection lineation is folded but lies on a plane (great circle) after folding.

reorientation of the earlier folds. Typically, in areas of superposed folding there are multiple generations of folds and multiple sets of foliations. Sometimes, each foliation set can be shown to be in an axial-planar orientation with respect to a particular generation of folds. In analyzing an area of superposed folding, the first step is to recognize and define domains of plane cylindrical folding of any foliation. The foliation that is analyzed may be different in different domains. The earliest foliation possible is bedding and is usually labeled S_0. Successive later foliations are labeled $S_1, S_2, \ldots$, etc. Next, we illustrate how an area of superposed folding can be analyzed with the aid of an equal-area net. Additional examples are provided in Chapter 15.

Problem 8-7

The example shown in Figure 8-32 is taken from Turner and Weiss (1963). The map shows folded foliation (S_1). There are two kinds of axial traces: first, broken lines (F_2), which are folded; and second, solid lines (F_3), which do not show any consistent folding. With the aid of an equal-area net, analyze the structures shown in the map of Figure 8-32. Identify structural domains, and determine the generation of fold responsible for the orientation of foliation in each domain.

Method 8-7

Step 1: Divide the map into domains of plane cylindrical folding by choosing areas with straight axial traces. Plot poles to S_1 foliation within each domain on a separate equal-area plot. By trial and error, adjust domain boundaries so that the plot from each domain displays a single π-axis; π-axes for different domains will be different. The axial trace, determined from the map, and π-axis permit calculation of the orientation of the axial plane for each domain.

Step 2: Group the domains based on orientation of the axial plane defined by S_1 foliation attitudes within the domain. In this example, we can do this by inspection: Domains I to VII have S_3 as the axial plane, while domains VIII to XIV have S_2 as the axial plane.

Step 3: Draw *synoptic diagrams* for each group of domains: (a) For domains I to VII the fold axes (determined from S_1 poles) lie on a great circle and are almost coplanar to S_3 planes; (b) for domains VIII to XIV the axes (from S_1 poles) also lie on a great circle. The S_2 planes intersect to define an axis that lies on the same great circle as the axes from domains I to VII. Generally, in natural examples the older folds do not show such a regular pattern because of inhomogeneities that develop during refolding.

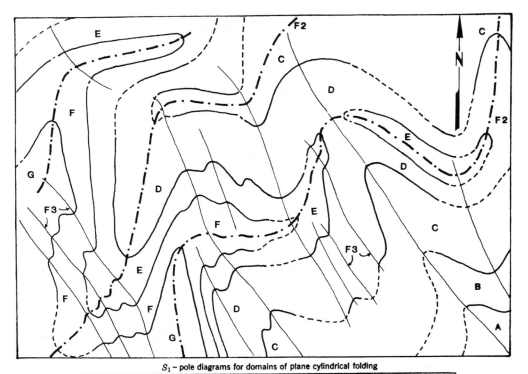

S_1 – pole diagrams for domains of plane cylindrical folding

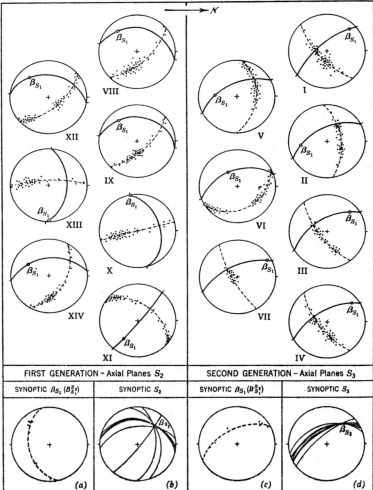

Figure 8-32. π-diagram analysis of superposed folding described in Problem 8-5. (Adapted from Turner and Weiss, 1963.)

Step 4: If S_2 and S_3 cannot be distinguished by inspection, then we can attempt to group the domains either by trial and error or by using minor structures (folds, axial-plane foliations, intersection lineations) to establish age relationships among structures. The latter is usually the more fruitful approach, since any macroscopic analysis of complex folding is unsatisfactory without the information provided by relationships among various minor structures, and among minor and major structures.

EXERCISES

1 . The following series of attitude measurements were obtained from the limbs of a fold:

106°,36°SW	N40°E,60°SE
150°,45°SW	S03°E,65°SW
079°,40°SE	N53°E,50°SE

(a) Plot a ß-diagram using these data.
(b) Plot a π-diagram using these data.
(c) Describe the structural significance of the ß-axis, the π-axis, and the π-circle girdle. Give the orientation of each. Which diagram (π or ß) is easier for you to interpret?

2 . A geologist measured a prominent mineral lineation on foliation surfaces of the Hightower Gneiss exposed on a horizontal pediment near Tarantula Gulch, Arizona. A simplified version of the map of this area is presented as Figure 8-M1. The geologist measured the pitch of the lineations. This problem emphasizes the fact that geometric calculations can readily be done with an equal-area net.

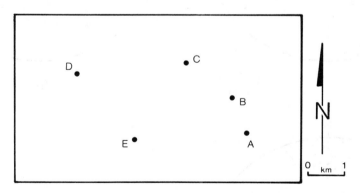

Figure 8-M1. Map of the Tarantula Gulch area for exercise 2.

(a) Complete the following table.

	Attitude of foliation	Rake of lineation	Plunge and bearing of lineation
A	N37°E,30°SE	42°SW	
B	N-S,40°E	11°S	
C	N23°W,60°NE	04°NW	
D	N55°W,70°SW	27°SE	
E	N85°W,40°S	75°E	

(b) Using appropriate structural symbols, complete Figure 8-M1 by plotting the foliation and lineation attitudes at the appropriate stations. Indicate the axial trace of the fold.

(c) Calculate the attitude of the fold axis, and using your mapped axial trace, calculate the attitude of the fold-axial plane.

(d) Calculate the angle between lineation and the fold axis at each station. Based on the results of this calculation, do you think the fold at Tarantula Gulch formed by a flexural-slip mechanism or a passive mechanism?

3. Below are attitudes of poles to bedding planes measured around a fold. From these measurements determine whether the fold is cylindrical or conical.

16°,N10°E	38°,N02°W
47°,N16°W	50°,N49°W
37°,N75°W	24°,N82°W

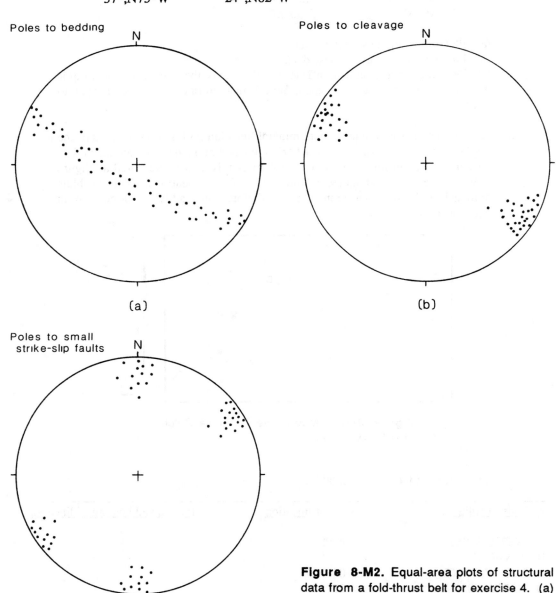

(a)

(b)

(c)

Figure 8-M2. Equal-area plots of structural data from a fold-thrust belt for exercise 4. (a) Poles to bedding; (b) poles to cleavage; (c) poles to small strike-slip faults.

4. Figure 8-M2 provides several equal-area plots of structural data collected in a fold-thrust belt. If you wish, you may reduce a Schmidt net to the appropriate size to determine carefully the attitudes of the structures shown, but you should be able to estimate the attitudes of structures from the equal-area plots.

(a) *Estimate* the attitudes of the structures shown in each plot.
(b) Describe the contour pattern (e.g., a girdle, a point concentration).
(c) With the results of part a, construct a synoptic diagram of the data.
(d) Write a brief interpretation of the structures in the area.

5. The scatter plots of Figure 8-M3 show the attitudes of two types of cleavage that were measured on an upright syncline with a horizontal hinge. Cleavage A is a slaty cleavage found in beds of fine-grained slate. Cleavage B is a spaced cleavage found in beds of slightly micaceous quartzite.

(a) Describe the differences between the two plots.
(b) Which cleavage is more likely to represent the axial-plane attitude of the fold?
(c) If you are familiar with the process of cleavage formation, explain why the two plots are so different from one another.

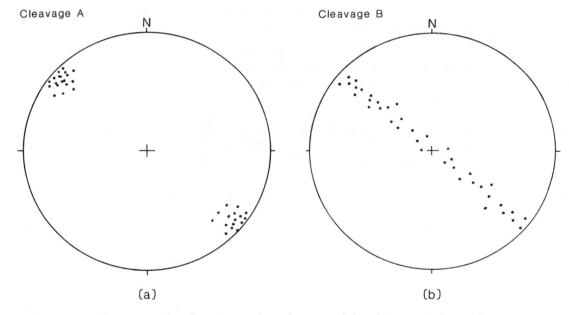

Figure 8-M3. Equal-area plots of structural data from a syncline. (a) Poles to slaty cleavage set A; (b) poles to spaced cleavage set B.

6. Figure 8-M4a shows a contoured equal-area plot of poles to foliation in a Precambrian augen gneiss that occurs in the Buckskin Mountains "metamorphic core complex" of Arizona. The plot was contoured by the Schmidt method, and the contours are at 2%-4%-6%-8% per 1% area. The area within the 8% contour is blackened. Figure 8-4Mb shows a contoured equal-area plot of lineations in shear zones that cut the foliation of the mylonites. Interpret these plots by answering the following questions.

(a) Is the augen-gneiss foliation folded, and if so what is the approximate attitude of the axes associated with the folds?

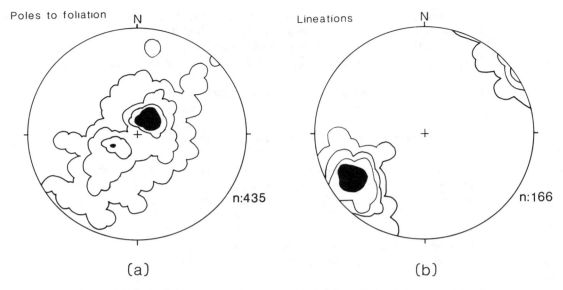

Figure 8-M4. Contoured equal-area plots of structural data from the Buckskin Mountains. The small 'n' signifies the number of measurements represented on the plot. Contours are at 2%-4%-6%-8% per 1% area. (a) Poles to foliation planes in augen gneiss; (b) Lineations in shear-zone rocks.

(b) What is the geometric relationship between the structure displayed by the augen gneiss and the shear zone lineation direction?

(c) Assume that the lineation in the shear zones is down-dip. Estimate the orientation of the shear zones.

7. Your instructor will provide structural data plotted on an equal-area plot that is the same size as the plots provided in Appendix 4. Use one or more of the contouring methods described in this chapter to construct a contoured plot of the data.

II

SPECIAL TOPICS

The purpose of this part is to provide an introduction to the specialized techniques that are used in subdisciplines of structural geology. Included in this part are discussions of how to: interpret geologic maps, interpret the results of rock-deformation experiments, describe natural fracture arrays, describe mesoscopic and microscopic structures (such as folds, shear zones, and foliations), construct cross sections of folds, balance cross sections, measure two-dimensional strain in rocks, and analyze polydeformed terranes. Each chapter is self contained, so a few subjects are discussed in more than one chapter. The chapters have been edited so as to conform to the overall style of the book.

9

ASPECTS
OF GEOLOGIC
MAP INTERPRETATION

Lucian B. Platt

9-1   INTRODUCTION

Map making is a fine art. To appreciate fully and to understand a geologic map takes thoughtful and careful analysis. The purpose of this chapter is to explain what geologic maps show and do not show and to provide illustrations of the process of map analysis. A geologic map is a subtle combination of observed facts and interpretations. The skillful map reader can sort out some of these, but how? We consider first how information is displayed on a map, and then, using techniques developed in earlier chapters as well as some additional methods, we consider what one might anticipate learning from a map.

A geologic map shows the distribution of different kinds of rocks at the surface of the earth. If you want to know the location of a particular kind of rock, say granodiorite in southern Montana, apparently all you have to do is find that unit in the map explanation and then locate its color or pattern or other identifying symbol on the map, right? The location of the granodiorite symbol on the map corresponds to the location of the rock on the ground, and some people think there is nothing more to a geologic map than this. On the contrary, a geologic map is a treatise on the geology of the area depicted that can provide information on the geometry of structures and on the geologic history of the mapped region. In the case of our granodiorite, its contacts with adjacent units might demonstrate when it was intruded, whether it has been cut by faults or whether it cuts faults itself, and whether it had been exposed to erosion prior to the present erosion cycle.

To read a geologic map on a two-dimensional piece of paper and to grasp the four-dimensional aspects (including time) of the geology portrayed requires practice and attention both to details and to the overall map pattern. This chapter will help the map reader sort out map data and detect inconsistencies, contradictions, and errors in the way that structures are expressed on a map. It is not intended to be an all-inclusive monograph on geologic maps but rather to provide an introduction to the thought process involved in map interpretation.

9-2   LIMITATIONS OF MAP SCALE

Geologic maps are prepared and published at many scales, from 1:5,000,000 for North America (Goddard, 1965) to 1:5,000 or even larger for some mines. Regardless of the scale, good maps are almost cluttered with details; that is, they have a similar amount of information per square centimeter of map. The scale of the map, however, limits the scale of the information it can display. For example, 1 mm on a map at a scale of 1:1,000,000 represents 1 km on the ground; no ink line is thin enough to show the details of structure on a hillside on such a map. Obviously, enlarging the 1:1,000,000 map does not produce a map with greater detail.

At 1:25,000, a common scale for current geologic maps, a fine line that is 0.25 mm on the map covers a swath of 6 or 7 m (about 20 ft) on the ground. This line can show a feature the size of a hillside, but it cannot show the details of folds in an outcrop. Of course, the contour lines on a topographic map have a finite width too, and the accuracy with which they are drawn limits the details a geologist can show. Clearly, there is a minimum size of feature that can be illustrated with lines on any particular map.

9-3 OBSERVING LOCAL DETAILS ON A MAP

A geologic map is a collection of details that should fit together to form a meaningful pattern. In looking at the geology of a relatively small area (a few square kilometers) of stratified rocks the sequence of which is known, it is useful to keep in mind the following propositions:

1. Mappable units (formations or members that can be indicated on the map at the scale of the map) or key beds (marker horizons that can be followed through the map area) are laterally continuous, at least locally. Thus, where a unit or key bed is offset by displacement on a fault, it does not vanish into oblivion but rather should appear somewhere on the other side of the fault.

2. A mappable unit has a finite thickness, so it has a definite top surface and a definite bottom surface, which are not the same plane. Rock layers are not Möbius strips, so regardless of whether you know which side of the unit is stratigraphically younger, the top contact at one locality on the map cannot magically become the bottom contact at another locality.

3. In general, different rock types produce different topographic expressions, because of their different erodability.

Next, we illustrate how these propositions can be used to recognize local geological details on a map.

Example 9-1

Scrutinizing small particulars of topography can allow you to identify pertinent particulars of geology. Figure 9-1 shows contour lines on a hill. Every contour line has two sharp corners. If you connect these corners on adjacent contour lines, you trace out on the map the edge of a hard rock layer up the northwest side of the hill and down the northeast side. If you connect two corners on the same contour line, you trace a strike line on the resistant bed. The strike lines are progressively higher southward, so the beds dip north. As a useful exercise, choose an arbitrary scale and contour interval for this figure and then reckon the implied angle of dip. For example, if the contour interval is 10 m and the strike lines are 50 m apart on the

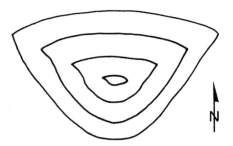

Figure 9-1. Sketch map showing a pattern of contour lines. The pattern contains clues to the attitude of strata on the hill slope. See text for explanation.

map, the layer has a grade of 20%, equal to about 10° of dip. A calculator would give the dip to several decimal places, but such precision does not match the quality of the data and thus is unwarranted. Clearly, a geologic map, by its very nature, cannot be squeezed to disgorge data to four significant figures, but it can produce hard data nonetheless.

Example 9-2

As another example of how to pay attention to local details on a map, we will examine how the displacement of contacts by a fault might be misinterpreted. The map in Figure 9-2 does not have any contour lines, but it is reasonable to assume that the little intermittent creeks flow down little valleys. The northern contact between units A and B dips south at some moderate angle, but at Wet Creek the contact does not continue up the east bank. The same contact between units A and B is shown dipping north less than a kilometer to the south, so units A and B are in a little syncline. The north flank of this syncline appears to have a steeper dip, because the V of the contact is shorter on that side of the fold.

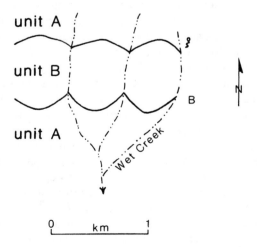

Figure 9-2. Sketch map showing the contacts between units A and B near Wet Creek.

Because the contacts on each limb do not continue directly across Wet Creek, the creek probably runs along a fault, but we do not yet know the offset on this fault. If the fault is down on the east side of the creek, the contacts should be wider apart on that side. If the fault is up on the east side of the creek, the contacts on the limbs should be closer together. Strike-slip movement on the fault would displace both contacts in the same direction and by about the same amount. It is clear that the contacts must be located with care if the fault is to be properly understood. The position of the outcrop of unit B alone on the east bank of Wet Creek does not adequately constrain the sense of movement on the fault; we need to know the positions of the contacts themselves.

In Figure 9-3 the carefully located contacts on the east side of Wet Creek are shown. The implication of the offsets of these mapped lines is that the east side of the fault went down relative to the west side. Even though the fault surface itself is not visible, its existence is thereby established.

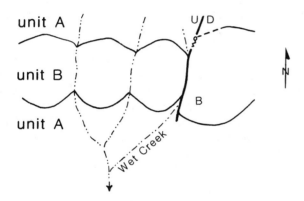

Figure 9-3. Sketch map showing contacts between units A and B near Wet Creek. Trace of the Wet Creek fault is also shown.

By giving attention to details, we obtained a lot of information from Figure 9-3 about the geometry of structures in the area of Wet Creek, even though no contours and no dips are shown. We were not able, however, to estimate the dip of the fault, but we could if the fault continues north over the hill and down the other side. This example has been an application of the first proposition - that units continue and their contacts continue.

9-4   UNDERSTANDING INTERSECTIONS OF GEOLOGIC CONTACTS

A line on a geologic map is the trace on the land surface of the contact between two mappable units regardless of whether that contact is a thrust fault separating upper plate

from lower plate rocks or whether it is the conformable surface between successive formations. You should focus special attention on how and where lines join or intersect on a map, because such intersections define the chronology of geologic events in the map area. What kinds of intersections are possible? The following example discusses relationships shown in the sketch maps of Figure 9-4a-c. Notice that the line pattern, by itself, is nearly the same in all three cases, yet the geologic relationships depicted in each case are quite different.

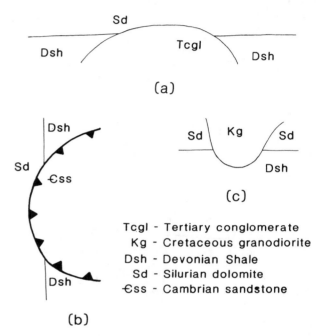

Tcgl - Tertiary conglomerate
Kg - Cretaceous granodiorite
Dsh - Devonian Shale
Sd - Silurian dolomite
Css - Cambrian sandstone

Figure 9-4. Types of cross-cutting relationships. (a) Unconformity; (b) thrust fault; (c) intrusive contact.

Example 9-3

In Figure 9-4a Tertiary conglomerate overlaps the contact between two Paleozoic formations. From this pattern the unconformity at the base of the Tertiary unit is clear. The Paleozoic rocks were exposed to erosion one or more times, including the last day before deposition of the Tertiary conglomerate. We cannot determine from the map whether Upper Paleozoic and/or Mesozoic strata were deposited in the area; if they were, they were completely removed by erosion. Figure 9-4b shows a similar pattern of lines, but the geologic relations are quite different. In this figure the edge of the Cambrian rocks lies across the contact between Silurian and Devonian rocks, and the map symbol shows the edge of the Cambrian to be a thrust fault placing the older rocks on top of the younger. The younger rocks presumably strike north, parallel to their mutual boundary. Figure 9-4c shows that a Cretaceous granodiorite intruded into Paleozoic sedimentary rocks. In each of the three cases (Fig. 9-4a-c) the curved contact is the youngest contact shown; in Figure 9-4b the oldest rocks are inside the curve, whereas in Figure 9-4a and c the oldest rocks are outside the curve.

The intersections of lines in Figure 9-4 are easy to interpret once the symbols for the units and the contacts are recognized.   Combinations of these maps require systematic and stepwise appraisal. Such an analysis is carried out for the map in Figure 9-5a, starting with the youngest rocks.

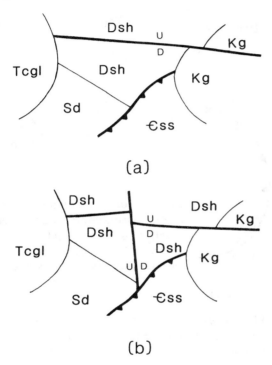

Figure 9-5. Figures for interpretation of cross-cutting relations. The symbols on these maps are the same as those used on Figure 9-4.

Example 9-4

In the map of Figure 9-5a, Tertiary conglomerate lies unconformably upon Silurian dolomite and Devonian shale. Unit Tcgl also overlies and is younger than the fault striking east from its edge. Because the thrust fault is not covered by the Tertiary conglomerate, it might seem, at first glance, that either feature might be younger. Let us turn our attention to the next older unit, the Cretaceous granodiorite. The granodiorite is older than the east-striking fault but is younger than the thrust. Thus, the map relations demonstrate the following sequence of events: (1) deposition of the Paleozoic rocks, (2) thrusting

of the Cambrian rocks over the Middle Paleozoic rocks some time before the intrusion of the Cretaceous granodiorite body, (3) movement on the normal fault, and (4) deposition of the Tertiary conglomerate.

The map explanation indicates the ages of the units. The way that the contacts between the units intersect and offset one another proves that the thrust fault is older than the granodiorite, which in turn is older than the east-striking fault, which in turn is older than the Tertiary conglomerate. Therefore, the Tertiary conglomerate must be younger than the thrust fault, even though the fault and the conglomerate do not touch on the map.

The map in Figure 9-5a was drawn in 1924. Recently, a geologist studied the area again, with the aid of aerial photos and an improved topographic base map, and produced the new map shown in Figure 9-5b. The new map is quite similar to the old map, but on the new map there is a south-striking fault that offsets the Silurian/ Devonian contact before disappearing beneath the thrust plate of Cambrian rocks. The new map has revealed something new and interesting about the geologic history of the area, namely, that there was some normal faulting prior to the thrusting. Not only does the south-striking fault offset the Silurian/Devonian contact, but it also offsets the east-striking fault to the north. The east-striking fault displaces the border of the Cretaceous granodiorite body, which, in turn, cuts off the thrust. Now, note that the depiction of the newly discovered south-striking fault yields an interesting relationship! This fault, as shown, is both older than the thrust and younger than the (younger) granodiorite. Either there is an error in the mapping or there was a recurrence of movement on the northern part of the fault.

Analysis of the intersection of lines on Figure 9-5b has brought to light a problem in the depiction of geology on the map. The sixth part of this chapter will deal further with problems on maps, but first we will review two more considerations useful in evaluating maps.

9-5   IMPLICATIONS OF UNIT THICKNESS

We have studied how the outcrop width of a unit varies with its thickness, its dip, and the slope of the ground (see Chapters 4 and 5). In this section we will see how examination of the variations in outcrop width can help locate hidden structure.

Example 9-5

Figure 9-6 is a cross section of a conformable stack of stratified rocks that dips 30° to the east. If the map explanation indicates that each unit is 500 m thick, then

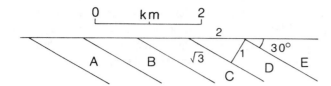

Figure 9-6. Cross section illustrating width of outcrop belt as a function of unit thickness.

each unit will form a band that is 1 km wide on the ground. This is illustrated by the triangle in unit D; this 30°/60°/90° triangle has sides whose lengths are in the ratio 2:1:√3. A similar conclusion about the expected width of a formation can be made in any situation.

Figure 9-7 is a cross section sketched on the graph paper in a field notebook to the scale of the geologic map being examined. These are the same units shown in Figure 9-6 and have the same thickness (500 m each), but here they dip 70°. Unit D is exposed on a hill sloping west at 20° with unit C beneath it. Across the alluvium-filled valley, the contact between units A and B is also dipping east at 70° on a slope of 25° east. Again, as in Figure 9-6, the five units are apparently in order across the valley, although the contact between units B and C is covered by alluvium. Nevertheless, a glance at Figure 9-7 indicates that something is not quite satisfactory; the valley is too wide to be underlain by only 1000 m of 70°-dipping strata (500 m of unit B and 500 m of unit C). The cross section does not provide a hint of what to do about this anomaly; the rest of the map should give an indication of a possible solution (e.g., a fold or a thrust fault that thickens the section, a normal fault that extends the section, or a big sill). Obviously, one should look at the ends of the alluvium to see what emerges from beneath it.

So far in this chapter we have made several suggestions concerning how to read relations on a map. Even the simple fact of continuity - that there are no open gaps in the earth - can force you to question a map, and questioning will lead to further discovery. Thus, in Figure 9-7 something mysterious (an unknown structure) is under the alluvium, because there is simply too much space for the alluvium to be underlain by only units B and C. It follows that, whatever the mysterious feature is, the position of the contact between units B and C is unknown.

9-6 STEPS IN ANALYSIS OF GEOLOGIC MAPS

Previous sections of this chapter considered certain particulars on geologic maps and what such details imply. Section 9-2 pointed out that the map scale limits the size of visible features, and Section 9-3 brought out the fact that little local pictures must fit together to make a big picture. One kind of little picture was emphasized in Section 9-4, namely how contacts and faults meet and offset or overlap one another. In this section we use these considerations to assess how the separate parts of a geologic map work together to give an image of the area depicted.

Checking the Map Explanation

How do you start? The map explanation is a good place. After checking the scale and the map orientation, the same questions may occur to you as are directed at any piece of science: Who, what, when, where, why, and how?

1. *Who did the mapping?* Different geologists have different points of view. Interpretive bias can strongly control the way in which structures are depicted.

2. *What does the map purport to show?* Ask if it is a sketch map or an outcrop map (i.e., what level of interpretation is depicted and if the contacts shown are intended to show the exact or the general location of a feature).

3. *Where is the map area?* What is the regional context in which the map area occurs? Such information can help you predict the suite of structures (e.g., styles of folds and faults) that may occur in the map area. In addition, consideration of the location of the mapped area can affect your confidence in the map; map relations shown on a map of a jungle swamp are probably less tightly constrained than relations shown on a map of a desert range.

4. *When was the mapping done?* Remember that maps are, at some level, interpretations, and that interpretations change as geology advances. If the map is good, geologic relations should be depicted correctly

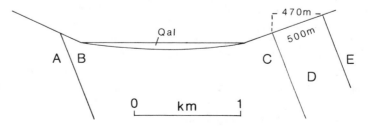

Figure 9-7. Cross section illustrating a problem in the representation of unit thickness.

regardless of whether the map is old or new. Be particularly careful to check whether the stratigraphy of the map area was still in doubt at the time the map was made.

5. *Why was the map made?* Some maps are made solely for the purpose of depicting the general geology of an area, whereas others are made to solve a specific geologic problem. For example, if the mapper was interested in locating gem-bearing pegmatites, the map may not show folds in the country rock even if they are present; if you are interested in the folds, you are out of luck. On the other hand, if the mapper was looking for gold placer deposits, and you are interested in tracing an unconformity, you may find that the map does show what you want--the position of postunconformity gravel.

6. *How was the map made?* Was the map the result of a reconnaissance project or a 9-year detailed analysis? Did the geologist making the map have an accurate topographic base on which to plot the geology? The detail of the mapping partly determines the confidence that you should have in the map.

Scanning the Map

Giving the map itself a "once-over" is time well spent. Such a scan of a geologic map can help you assess the quality of the map and give you a feel for the style of geology in the map area. When scanning a map, you may ask the following questions (in no particular order):

1. *Does the map seem orderly or is it just a tangle of lines and blobs?* Some maps show no systematic arrangement of different kinds of rocks; rather, they show a disorderly collection of patches of the units identified in the map explanation. The map may have this appearance because of (a) poor exposure in the map area (as may occur in country with glacial deposits); (b) slumping in the map area; or (c) breakup of the mapped area by nonsystematic faulting and folding.

Even if exposure is poor and/or units truly lack continuity, the map should show the geologist's interpretation of how discrete exposures relate to one another. One way to do this is to display actual outcrop in more vivid color than the inferred extensions of units. Another way is to specify in the explanation that one of the mapping units consists of "blocks in shale." If no such effort has been made, and the map looks like a "blob job" of this and that here and there, it is not a very useful map. In areas of terrible exposure the mapper may feel reluctant to offer an interpretation that systematizes what little is known. In a sense this reluctance is an avoidance of responsibility, because who is better qualified to interpret a map area than the person who walked around it?

Keep in mind, however, that not all map areas are equally interpretable; in some localities, the exposure may be so bad that many interpretations are viable, and all interpretations must be viewed with skepticism.

2. *How well documented are contact traces?* In some regions it is possible to put your finger on a contact, but in many localities it is not. You can get a sense of the constraints on contact positions in stratified or foliated rock by seeing how close attitude measurements in adjacent units are to a contact. If they are close, the position of the contact is probably well constrained.

3. *Are contacts among units orderly and is there a systematic relationship between suites of structures?* There should be a systematic arrangement of rock types, and their contacts should also be orderly. For example, does one set of faults systematically offset a second set of faults with a different strike, or is the fault pattern a mosaic in which faults with one strike both cut and are cut by faults with other orientations. The four faults in Figure 9-8 form a tangle, a sort of mosaic. Which fault is younger than what? Another anomaly in this map is that no fault continues across any other. The pattern of Figure 9-8 is a peculiar map pattern of interlocking and nonsequential faults. If you find such a pattern on a map, you should suspect the quality of the map. Although there are inherent difficulties in determining relative offsets, places

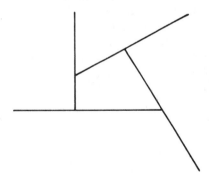

Figure 9-8.   Sketch map of a fault mosaic.

where one fault intersects another deserve special attention, because it is necessary to know which fault is the older in order to understand the geologic history of the area.

Let us clarify what we mean by "inherent difficulties" in the previous sentence. Consider two formations separated by a fault oblique to their outcrop belts on the map. In terrane of reasonable exposure, outcrops may be found anywhere in either belt that can provide opportunities to determine the local attitude of the formations. In contrast, most faults are rather thin sheets and consist of crushed material that is more easily eroded than the uncrushed formations away from the fault. If a fault surface is not exposed, one cannot measure its attitude directly. Hence, to measure offsets along faults, the field geologist needs to make an extra effort to establish locations of attitudes and formation contacts as close as possible to the fault. Thus, a question that you can legitimately ask while evaluating a map is whether it shows attitudes close to faults, especially where they cross. Again, remember to check whether the fault offsets are systematic or tangled.

4. *Do the interactions between contacts and topography make sense?* If layer attitudes are indicated on the map, make sure that the contacts cut across topography in an appropriate manner (see Chapter 5). If layer attitudes are not shown, be wary if the V-direction of a contact as it crosses adjacent valleys is not consistent.

5. *Do displacements on faults make sense?* As you follow the trace of a fault, contacts that are cut by the fault should be displaced. You may have seen a geologic map on which two faults cross without any offset of one by the other. Although this is possible, just as it is possible for a fault to cross a stratigraphic contact without displacing it, it is only barely possible (Redmond, 1972). Unless the displacement on a fault is so small that it cannot be portrayed at the scale of the map, a fault that does not displace a contact or any other kind of surface must have slipped exactly parallel to the trace of the contact on the fault. Possible slip directions on a fault range over 180°, but only one direction of slip can produce slip without offsetting any specific other plane. The intersection of a fault with a contact is rarely a perfect X, and the intersection of two faults is probably never a perfect X. Places where faults intersect deserve special attention. Getting an attitude measurement on a fault is nice work if you can get it, but because of the erodability of faults, such measurements are often impossible. In order to document a fault, a geologist needs to establish attitudes and contact positions as close as possible to the fault.

6. *Do fold patterns make sense?* Make sure that the hinges, crestlines, or axial-plane traces of folds in the map area are indicated by a map symbol. If a dip reversal occurs and no fold symbol is present, there is a problem. Remember that two adjacent antiforms must be separated by a synform or by a fault. Check that the type fold indicated is compatible with the progression of strata from core to limb (e.g., on a simple upright anticline, the oldest rock is in the core and the youngest rock on the limbs).

Example 9-6

Keeping in mind the preceding questions, we will scan a map and see what turns up. The map, Figure 9-9, is adapted from a map of part of the Great Basin in the western United States. This puts things in a regional context. We know that the Great Basin is a region that was affected first by thrust faulting and later by normal faulting, so we will anticipate seeing examples of both types of structures. We also know that the Great Basin is an arid region and that carbonate units should form steep slopes. At the scale of the map and at the contour interval shown there is nearly 2 km of relief in a distance of less than 8 km from the southeast part of the map to the northwest part. The ground is therefore steep, so considering the arid climate, good exposures are virtually assured (in fact, cliffs and ledges might limit where a geologist can walk around).

The map explanation identifies seven formations by the symbols Sl (for Lower Silurian) through M (for Mississippian). The brief descriptions of the map suggest that the units can be identified with confidence by anyone who tries. The lack of talus or any other Quaternary map unit is notable (it may imply good exposure of bedrock or maybe just lack interest on the part of the mapper concerning Quaternary units).

A single glance at the map itself reveals several problems. Overall the map shows too few attitude measurements for an area of 70 km² of probably good outcrop. Many of the faults are drawn with solid lines, suggesting that the map's author considered that their position was well constrained, yet few attitude symbols are shown near the faults, and no direct measurements of fault attitudes are indicated. We count eight fault intersections without any offset (see Question 3 above). All the faults except the thrust were drawn straight regardless of topography; to find this many vertical faults in an area is unusual. Not one of the faults lies in a gully, although one might expect shattered carbonates along the faults. At three localities northwest-trending faults are truncated at their junction with the northeast-trending faults, but their continuations cannot be found. Again, we should be suspicious, even though it is possible that the continuations of the truncated faults are outside the map area.

In the middle of the map there is a belt of Middle Silurian sandstone dipping homoclinally northeast at about 20°. The outcrop belt between the two northwest-trending faults is almost 3 km wide. This outcrop pattern suggests that the unit is about 1200 m thick (note the 1400-m contour line at the southwest and northeast edges of the patch and note that sine 20° = 0.34), but the map explanation claims that it is only 800 m thick. Either there is a contact missing, or the unit is not continuous and homoclinal within this fault block. The inaccuracy in the thickness representation makes us suspect that the map is incorrect or incomplete.

Another potential problem is in the northwestern part of the map where Mississippian rocks (dipping 23° to the

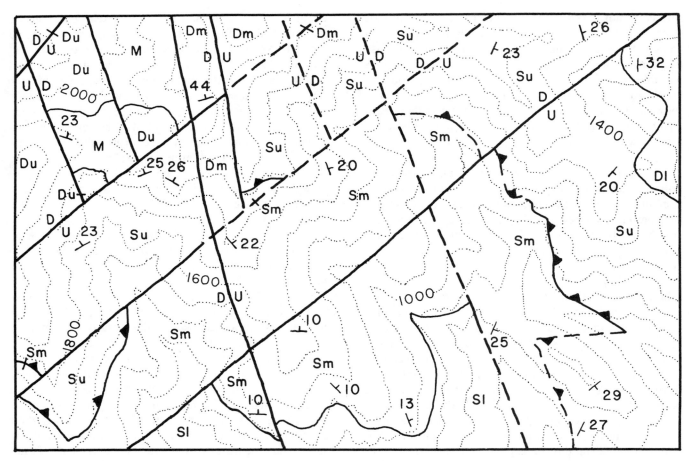

Explanation

M	Limestone, massive ledges, abundant corals, about 400 m thick
Du	Silty limestone, distinctive tan weathering, about 500 m thick
Dm	Limestone, white chert nodules in thicker ledges, 350 m thick
Dl	Dolomite, interbedded light and dark gray, fine-grained, 200 m thick
Su	Limestone, thin-bedded silty, gray, 700 - 800 m thick
Sm	Sandstone, calcareous, outcrops rare, weathers to blocks, 800 m thick
Sl	Limestone, gray fresh and weathered, high ledges, about 600 m thick

0   km   3

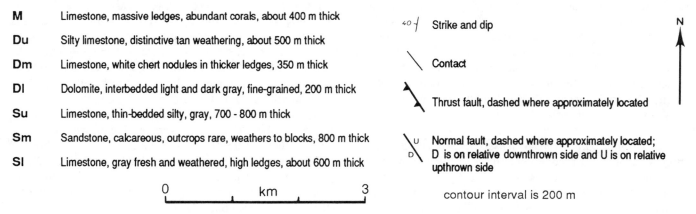

Strike and dip

Contact

Thrust fault, dashed where approximately located

Normal fault, dashed where approximately located;
D is on relative downthrown side and U is on relative
upthrown side

contour interval is 200 m

Figure 9-9. Geologic map of a portion of the Great Basin. Several
errors exist on the map.

northeast) are exposed on a topographic saddle. On the
south side of the saddle, Upper Devonian silty limestone
dips under the Mississippian unit, but the younger unit
seems to dip beneath the Devonian unit on the north side
of the saddle.

No fold hinge traces are shown on the map, despite the

fact that bed dip direction in some localities is northwest,
whereas in other localities it is southeast. The discordance
in dip directions could be a consequence of rotation of fault
blocks. We need to look more closely to see if dip
directions within a fault block are uniform. In most of the
blocks they are, but in the west-central part of the map

there is a block where strike changes from northeast to northwest over a very short interval, and it is likely that a fold should be indicated.

Our initial scan of the map has turned up so many anomalous situations that we might conclude that the map is not dependable. If we sift carefully, however, we can derive useful information from it. First, because the units are distinctive, we infer that the units are correctly labeled as to dip location and direction. Unit contacts seem reasonable; shallowly dipping contacts closely follow contour lines, and moderately dipping contacts appear to show the appropriate V-pattern. Because the geologist probably walked along the ridges where the attitude symbols are marked, contacts and faults may well cross the ridge where they are shown. The map suggests that there may be two sets of faults, one striking northeast and another striking north-northwest. A single thrust fault occurs in the area, but the trace of this fault has been broken up by later faulting. The thrust fault is shallowly dipping; in the eastern half of the map, the fault clearly dips to the east. The thrust fault seems to have the same attitude as adjacent bedding, and it always lies between units Sm and Su; it does not cut across stratigraphy in the map area. We should wonder, however, about the significance of the fault; the units juxtaposed by the fault are in their proper stratigraphic sequence, with the younger unit above the fault and the lower unit below. Is the lack of stratigraphic repetition reasonable for a *thrust* fault? If it is, then the fault is a bedding-plane "flat" (see Chapter 14). From the map data, however, we cannot completely rule out that the fault is a low-angle normal fault.

In summary, our examination of the map and its explanation has unearthed several severe problems, but we can realistically infer that middle Paleozoic sedimentary rocks in the area dip gently to moderately northeast. Faults are numerous, have uncertain dips, and intersect in unknown ways. In view of the numerous problems we must admit that we cannot significantly improve this map with one single change or one simplifying hypothesis. In this area, one day invested climbing up to find out how one fault intersection actually maps out in detail would not clarify the situation much. Some problems with particular maps do lend themselves to the consideration of alternative

hypotheses that significantly improve the map. The next section of this chapter deals with two such examples.

9-7   THE METHOD OF MULTIPLE WORKING HYPOTHESES

Look again at Figures 9-2 and 9-3. When these were discussed early in this chapter, they illuminated the proposition that contacts continue, although offsets across faults are interesting. In discussing these examples, we automatically considered alternative interpretations: What if geometry X? What are the possibilities if I know geometry Y? How do I efficiently test these alternatives - that is, what data can I collect to discriminate among them?

Example 9-7

Figure 9-2 presents an outcrop of unit B directly across the creek from the contact between units A and B, so at first glance it appears that the east side of Wet Creek fault must have gone down; a more careful look offers other possibilities. The location of the outcrop at B, 100 m east and a bit above the creek, actually does not preclude any displacements on Wet Creek fault except east side up by enough to eliminate unit B on the east side, because the contact downhill from that outcrop is not located. Figure 9-10 provides the local map pattern resulting from each of three possible interpretations of the geology in this area: In Figure 9-10a the fault is up on the east, or is left-lateral; in Figure 9-10b the fault is down on the east, or is right-lateral; in Figure 9-10c no fault is necessary to fit the local exposures. Corroborating evidence for and against these possibilities is necessary to test them. Such evidence was provided by exposures of the north side of the syncline farther up Wet Creek.

An important concept is illustrated by this simple example. If the geologist mapping this area had already decided what kind of fault the Wet Creek fault was - had arrived at Wet Creek with a bias in favor of strike-slip faults, for example - he or she might not have visualized the several other possible interpretations of the area and

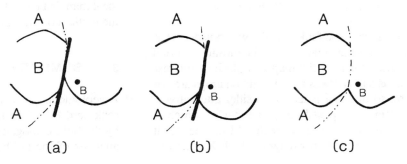

Figure 9-10. Geologic sketch maps showing details of fault relationships along Wet Creek.

(a)   (b)   (c)

thus might not have paid sufficient attention to the offset of the northern contact between units A and B shown in Figure 9-3. This method of considering alternatives as one goes along has been known since the late nineteenth century as the *method of multiple working hypotheses* (Chamberlin, 1890; 1897). It is an essential method of thinking about geologic maps and is useful in many other endeavors. Indeed, it is so important in all kinds of science that the American Association for the Advancement of Science maintains a stock of reprints of the 1890 article.

Example 9-8

Analysis of Figure 9-11 puts the method of multiple working hypotheses to work. This map could be from a low wooded island in a region with known Jurassic and Cretaceous stratigraphy and complex structural history, including thrusts and normal faults. The map has a fault mosaic somewhat like the mosaic in Figure 9-8, but in Figure 9-11, fault Y is offset by fault Z; both are shown with solid lines, and dips are recorded near the intersection, so we can have confidence in this offset relationship. The contacts and dips shown will aid us in interpreting this map, which we could not do with Figure 9-8.

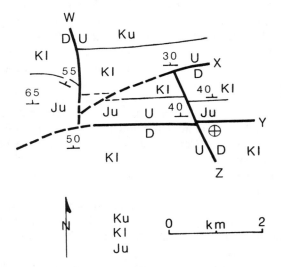

Figure 9-11. Geologic sketch map showing a fault array with an impossible sequence of faulting. Faults are indicated by heavier lines.

Having recognized the fault mozaic, we consider what might be done to alter the map so that the mosaic is sorted out. The southwest part of the map is a place of interest because the faults and contacts are dashed; outcrops are not abundant here, so this is a place to consider alternatives. Figure 9-12 offers two possibilities, neither of which is much of a change, although each shifts the fault intersections a bit. In the upper diagram, fault X is clearly

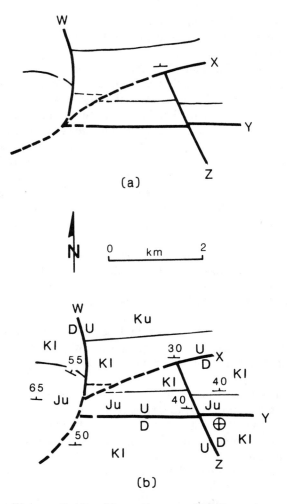

Figure 9-12. Alternative modifications to the fault array shown in Figure 9-11.

the youngest, and fault Y appears to be older than X or Z. This partly sorts out the mosaic, but fault W is left out of the sequence. In Figure 9-12b W is the youngest fault, and Y is the oldest, but X and Z terminate against other faults. We also note that the Ju/Kl contact displays right-lateral separation, which is opposite to the sense expected if the fault is down on the west, as indicated by the D and U symbols. Perhaps we should reverse the sense of displacement. Neither of these modifications yields an ideal map, but each demonstrates the thought process called the method of multiple working hypotheses.

9-8 SUMMARY

A geologic map is a subtle combination of creative art and rock-hard fact. The factual part is placed on the map first by the field geologist and should get first attention in your analysis of the product, but at least a bit of creativity is included in the map and deserves your attention when reading it.

This chapter began by discussing how the scale of a map restricts what it can include; thus, hill-scale folds cannot be shown on a map of the whole state of Colorado, and regional trends may not be apparent on a single quadrangle. In the next three sections we used little maps of stratified rock to demonstrate how to recognize various structural relationships on a map. We could not discuss all possible features, but the geometric aspects of Figures 9-4, 9-5, and 9-7 can be carried over to situations not illustrated. Throughout the chapter the map explanation was emphasized as an integral part of a geologic map, and a part that you must examine carefully; it is a specialized dictionary in which the formations and symbols on the map are defined. We also discussed the difficult procedure of evaluating a map and discovering its inconsistencies. Critical map makers and users rely on the method of multiple working hypotheses to help in construction or interpretation of a map.

EXERCISES

GROUP A

This group of questions is designed to get you to think through explicit map relations. They are arranged in order of increasing difficulty. Answers to these questions should be quite specific.

1. Figure 9-M1 shows units V, W, and X of unknown age.

(a) What is the dip of the conformable units W and X?
(b) What is the relationship of W and X to unit T?

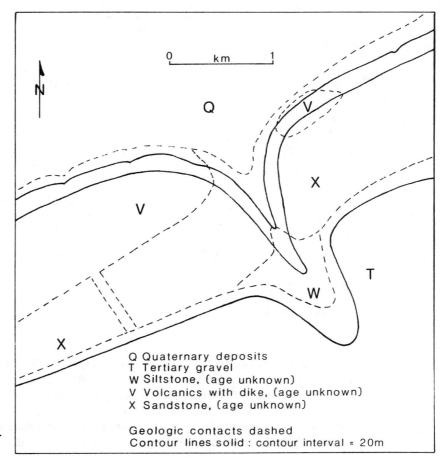

Q Quaternary deposits
T Tertiary gravel
W Siltstone, (age unknown)
V Volcanics with dike, (age unknown)
X Sandstone, (age unknown)

Geologic contacts dashed
Contour lines solid : contour interval = 20m

Figure 9-M1. For use with exercise 1.

(c) What is the dip of unit V? What is its age relation to T?

(d) Give the geologic history illustrated by the map relations on Figure 9-M1.

2. In Figure 9-M2 contour lines are dotted, and the contour interval is 20 m. Contacts between stratigraphic units A, B, C, and D are solid. Units B and C are each 40 m thick, but the thicknesses of units A and D are unknown. The map scale is 1:4000.

(a) What is the dip of the fault?

(b) What is the direction of displacement on the fault?

(c) Draw geologic cross sections XX' and YY'.

(d) What is the amount and direction of stratigraphic throw of the fault at YY'? What is the amount and direction of stratigraphic throw on XX'? Explain the basis of your answers.

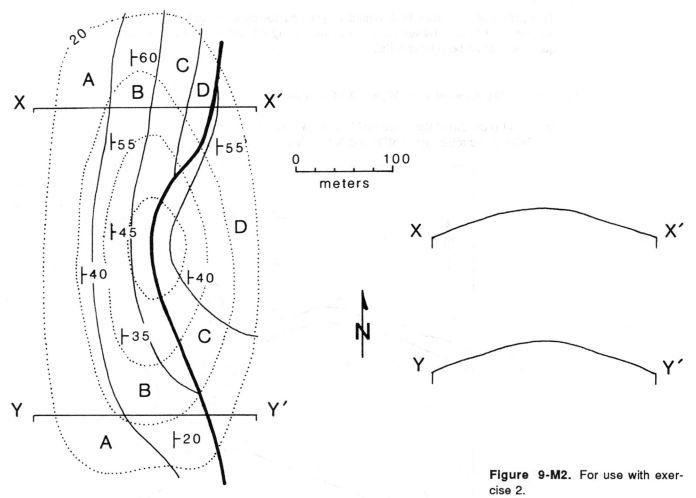

Figure 9-M2. For use with exercise 2.

3. Obtain Map #4 from the Williams & Heinz Map portfolio No. 1 (Williams and Heinz Company, 8119 Central Avenue, Capitol Heights, MD 20743). This map shows the geology of central Pennsylvania (taken from the State geologic map). The map has no contour lines, but the general lay of the land can be inferred from various features, especially the stream pattern. For the western part of the map, topography at the same scale can be seen on the U.S.G.S. Pittsburgh 2° sheet.

(a) Based on the V's of contacts in stream valleys, what is the attitude of the beds in the northwestern part of the map?

(b) What is the dip of the Silurian units along strike for a few miles northeast of Altoona?

(c) Are the Cambrian rocks 3 mi southeast of Tyrone part of an anticline or a syncline?

(d) How far along strike does this structure extend in each direction from Tyrone?

(e) In the southeast part of the map note the line between Mifflin and Huntingdon counties. Examine the structural geology along the Juniata River from the county line northwest to Petersburg. Describe the geology in a paragraph, or draw a diagrammatic geologic cross section (at a suitable scale) along the river, assuming flat ground.

Group B

In this group of questions not only must you read the map, but you must find and evaluate difficulties and then offer possible solutions, or at least alternatives. For any one of these problems there may not be a single perfect answer; in this way they represent real geologic problems.

4. For the map of Figure 9-M3 we do not know either the sequence or attitudes of the units or the dip of the fault. You have a choice in interpreting the situation. Make a suitable choice and follow the directions. Then make another reasonable choice and follow the directions a second time.

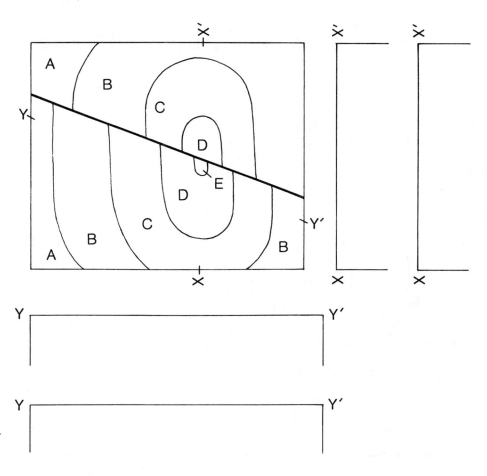

Figure 9-M3. For use with exercise 4.

(a) Complete cross sections XX' and YY'. It is easier to do YY' first.
(b) Give the sequence of rock units in the order of deposition, oldest first.
(c) Indicate the direction of movement of the fault.
(d) Name the kind of fault you have shown in your cross section XX'.
(e) Repeat a-d with a different view of the structure.

5. In Figure 9-M4 contacts between units A, B, C, and D and Quaternary alluvium
    are shown as solid lines. Something of the topography can be inferred from the
    intermittent streams. Complete cross section YY' and explain the relations
    beneath the alluvium.

6. The four faults in Figure 9-M5 offset each other. Thus, in the upper left corner of
    the figure, fault B offsets fault A, and in the lower right, fault A offsets fault D.

    (a) Of the six intersections now forming a tangle with no logical time sequence,
    is there an intersection that can be reversed to form an orderly sequence? For
    example, in the lower right corner, if the intersection were changed to make D
    offset A, would this alter the picture satisfactorily?
    (b) Is there more than one such intersection?
    (c) State the sequence of faulting for each solution that you propose.

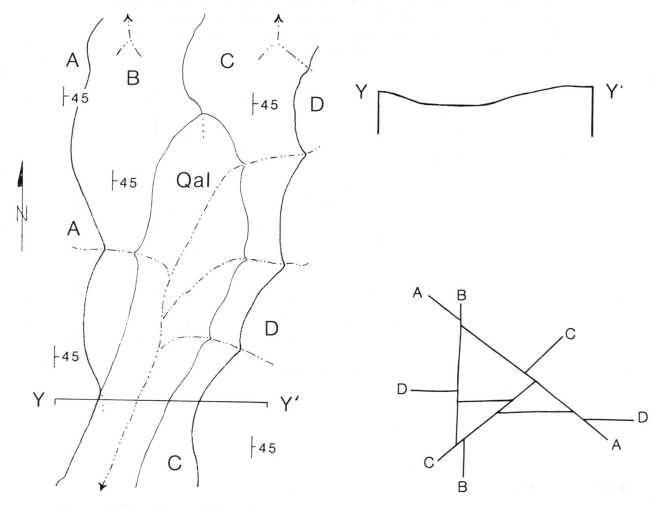

Figure 9-M4.   For use with exercise 5.   **Figure 9-M5.**   For use with exercise 6.

7 . Look again at Figure 9-12b. Assume this map is of essentially flat ground.

(a)   Draw a north-south geological cross section east of fault Z, interpreting fault Y as a normal fault dipping steeply south.

(b)   Draw another north-south cross section east of fault Z, this time interpreting fault Y as a thrust dipping steeply north.

(c)   Discuss the possible implications of each of these cross sections for the geometry of bedding west of fault Z and for displacement on fault W.

(d)   Discuss the possible implication of discovering that the 50º dip is on overturned rocks.

10

ANALYSIS OF DATA FROM ROCK-DEFORMATION EXPERIMENTS

Terry Engelder

Stephen Marshak

10-1 INTRODUCTION

Experimental rock deformation refers to the laboratory study of the mechanical characteristics of rocks. These characteristics include *rheology* (the response of rock to stress), *strength* (the maximum stress that can be sustained by a rock before it fails), and *friction* (the resistance to sliding on a fracture surface in the rock). In such work, specimens of rocks (usually machined cubes or cylinders) are subjected to measured stresses under varying conditions of temperature, confining pressure, strain rate, pore pressure, and chemical environment. Rock-deformation experiments can be used to model both *brittle deformation*, meaning deformation that involves formation of and movement on discrete fractures in a rock, and *ductile deformation*, meaning deformation that occurs without loss of cohesion across a plane. Structural geologists should be able to read and interpret descriptions of experiments, because experiments simulate deformation in the earth and therefore provide insight into the physical processes by which common structures, such as faults and folds, form.

Ideally, structural-geology students should have the opportunity to observe or participate in experimental studies. Unfortunately, most schools do not have the equipment to demonstrate such studies. The purpose of this chapter is to describe rock-deformation experiments and provide an opportunity to work with methods for representing and interpreting experimental results. The format of this chapter differs from that of previous chapters in that you are asked to work through the interpretations of

the experiments that are outlined in the text. The discussion in this chapter is limited to certain types of rock-deformation experiments; we consider experiments concerning rock strength under brittle, brittle-ductile, and ductile conditions, and we consider experiments concerning rock friction. Ultimately, these experiments give information about the state of stress in the upper crust. To introduce the terminology of experimental rock deformation, we begin by describing some of the equipment used in rock-deformation experiments and by describing the diagrams commonly used to represent rock-deformation data.

10-2 THE ROCK-DEFORMATION EXPERIMENT

Experimental Apparatus

Many rock-deformation experiments are carried out on a *triaxial load machine* (Fig. 10-1). This device is a sophisticated press that is able to exert a stress on a rock cylinder or cube. Stress (force per unit area) is commonly measured in pascals (1 Pa = 1 X 10^{-5} kg m^{-1} s^{-2}), where 1 MPa = 10 bars = 145 psi (psi means "pounds per square inch"). The length/diameter ratio of rock cylinders used in experiments is generally around 2:1, and cylinders range in size from 5-mm diameter (small), through 2.5-cm diameter (standard), up to 30-cm diameter (large). Work with different size cylinders is important because strength is scale dependent. Larger cylinders are more likely to contain

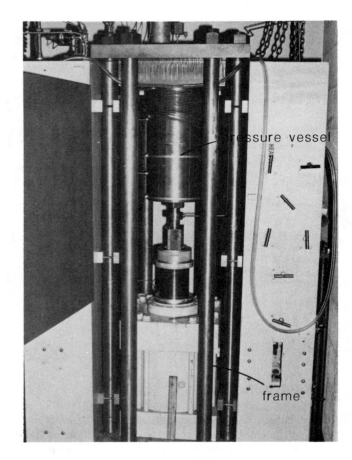

Pore pressure
fluid access

Confining
pressure
fluid access

S = sample
V = pressure vessel
P = piston
• = O-ring seal

(a)   (b)

(c)

Figure 10-1. Triaxial load machine. (a) Diagrammatic cross-sectional sketch of a portion of a machine showing the pressure vessel, sample, and piston; (b) photograph of a machine; (c) photograph of two deformed samples. The left sample contains an induced fracture. The right sample contains a saw-cut for friction experiments. There is a 5 mm-thick layer of gouge along the cut. Samples are 3.5" long and 2" in diameter.

flaws such as microfractures that will cause local stress concentrations leading to failure (see Means, 1976, Jaeger and Cook, 1979, and Kulander et al., 1979 for further discussion of stress concentration).

During an experiment a rock cylinder is squeezed by displacement of a piston in the machine (Fig. 10-1). In response to the displacement of the piston, the axial length of the cylinder changes. This change, the *axial strain* of

the cylinder, is measured by an electrical transducer attached to the cylinder. The axial stress felt by the cylinder during an experiment is measured by a load cell aligned with the piston and the cylinder.

Typically, a triaxial experiment is designed so that physical conditions can be varied during the experiment. For example, it is possible to control the following parameters:

Strain rate (the time rate of change of the rock-cylinder length), by specifying the rate at which the piston moves during the experiment.

Temperature, by heating the cylinder during an experiment.

Confining pressure (the pressure exerted on the sides of the cylinder), by placing the cylinder in a pressure vessel containing a confining medium (usually kerosene or argon). The pressure of the confining medium is controlled during the experiment. The cylinder is usually jacketed in copper, lead, or plastic to isolate it from the confining medium. A solid medium such as talc is sometimes used for very high pressure experiments.

Pore pressure (the pressure of the fluid that fills pores in the rock cylinder), by allowing fluid to reach the cylinder via small conduits in the machine (Fig. 10-1). The pressure of the fluid in the cylinder is regulated independently of the axial stress or of the pressure in the confining fluid.

The term triaxial refers to the fact that all three principal stresses (σ_1, σ_2, and σ_3) can be manipulated during an experiment, so that none is necessarily equal to atmospheric pressure in the laboratory. The subscript 1 signifies the maximum principal compressive stress, the subscript 2 signifies intermediate principal compressive stress, and the subscript 3 signifies the minimum principal compressive stress. If a rock cylinder is used, the axial stress is σ_1 in compression, and the confining pressure is σ_2 and σ_3 (σ_2 and σ_3 are equal because of the sample geometry and the experimental configuration). The *differential stress*, which is defined as the axial stress minus the confining pressure is,

$$\Delta\sigma = \sigma_1 - \sigma_3 \qquad \text{(Eq. 10-1)}.$$

If the pore pressure (P_p) is not equal to 0, then the preceding equation may be rewritten in terms of *effective stress* (*σ_1):

$$*\sigma_1 \; (\textit{effective axial stress}) = \sigma_1 - P_p \qquad \text{(Eq. 10-2)}$$

$$*\sigma_3 \; (\textit{effective confining pressure}) = \sigma_3 - P_p$$
$$\text{(Eq. 10-3)}$$

and thus,

$$\Delta\sigma = *\sigma_1 - *\sigma_3 = \sigma_1 - \sigma_3 \qquad \text{(Eq. 10-4)}.$$

Two basic experiments are possible with a triaxial load machine. The first type is called a *constant strain-rate experiment*. During such an experiment, as the name suggests, the piston of the machine moves at the same rate throughout the experiment. The second type is called a

constant stress experiment (also called a *creep test*). During a creep test the stress is held constant, and a variation in strain rate is measured as a function of time. Creep tests are useful in constraining the *constitutive equations* (quite simply, equations that specify strain rate as a function of stress) associated with different deformation mechanisms. This chapter is restricted to descriptions of constant strain-rate experiments.

Representation of Data on Stress-Strain Plots

There are a number of ways to represent the results of rock-deformation experiments. Commonly, results are displayed on a *stress-strain plot*. A stress-strain plot is merely a graph, constructed in Cartesian coordinates, that plots differential stress (measured in bars, pascals, or other valid units) on the vertical axis against strain in a specified direction on the horizontal axis (Fig. 10-2). The direction of strain is usually parallel to the axis of the test cylinder. Remember that *strain* is defined as the ratio of change of length of line over the initial length; it is a dimensionless quantity usually represented by a percentage.

On a stress-strain plot the results of a constant strain-rate experiment involving *elastic* deformation plot as a straight line (portion of the curve labeled "elastic deformation" on Fig. 10-2a). The slope of this line is a rock property called *Young's modulus*. This straight-line relationship indicates that, for an elastically deforming material, strain is directly proportional to stress. The stress-strain relationship can be represented by the equation $\sigma = E\varepsilon$, where E is Young's modulus and ε is the strain.

It is commonly observed in rock-deformation experiments that a rock behaves elastically until a certain strain is achieved; then the rock either *fails* brittlely (i.e., it fractures, suddenly loses cohesion, and can no longer support the stress, so the stress drops; Fig. 10-2b, curve A), *yields* by fracture (rock does not necessarily lose all cohesion when it fractures), or deforms plastically. If the curve on a stress-strain plot still has a positive slope after yielding has occurred, the rock is said to exhibit *strain hardening* during deformation (Fig. 10-2b, curves F). If the curve has a negative slope after yielding, the rock is said to exhibit *strain softening* (Fig. 10-2b, curve D). A sample that yielded by fracture may fail after a certain amount of strain hardening or strain softening (Fig. 10-2b, curves B and C). *Perfectly plastic* behavior on a stress-strain plot is represented by a horizontal straight line once the rock yields (Fig. 10-2b, curve E). Such a plot means that, for a plastically deforming material, strain will continue to increase as long as the stress is held at or above the ultimate strength of the rock.

Elastic strain is *recoverable*, in that if stress is

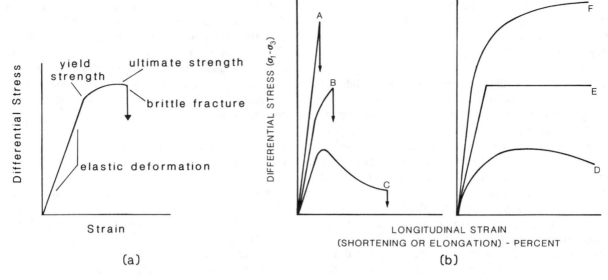

Figure 10-2. Stress-strain plots. (a) General stress-strain plot for an experiment in which a sample first deforms elastically, then yields and deforms plastically before failing. (b) Representative stress-strain plots. A = elastic deformation followed by brittle failure; note that failure is indicated by a sudden stress drop. B = elastic deformation followed by yielding, plastic deformation, then brittle failure. C = elastic deformation followed by yielding, strain softening, then brittle failure. D = strain hardening followed by strain softening. E = elastic deformation followed by yielding, then plastic deformation. F = elastic deformation followed by yielding, then strain hardening (adapted from Handin, 1966).

removed while the rock is behaving elastically, the strain returns to 0. Plastic strain is *nonrecoverable* or *permanent*, in the sense that if stress is removed after plastic strain has developed, the strain does not disappear.

A number of terms are useful in describing the behavior of a rock as displayed on a stress-strain plot. Note that in these definitions the term *strength* is measured in units of stress.

Yield strength: The value of stress ($\Delta\sigma$) at the bend in the stress-strain curve that marks the onset of permanent plastic strain.

Fracture strength: The value of stress ($\Delta\sigma$) at which a rock fails by brittle fracture. This is represented by a sudden drop in stress on the stress-strain curve.

Ultimate strength: The maximum stress ($\Delta\sigma$) that a rock sustains during an experiment (i.e., the maximum ordinate of the stress-strain curve).

Ductility: The total percent permanent strain before failure by fracture, as indicated by a marked stress drop.

The results of a constant strain-rate experiment can be be plotted not only on a stress-strain plot (Fig. 10-3a), but can also be plotted on a strain-time plot, which shows how

strain changes as a function of time (Fig. 10-3b). On such a plot the horizontal axis is time, and the vertical axis is strain. Obviously, the plot for a constant strain-rate experiment must be a straight line.

The results of creep tests can be depicted by curves on a stress-strain plot (Fig. 10-3c), but the curves do not give an indication of the change in strain as a function of time. Therefore, it is often preferable to plot results of a creep test on a strain-time plot (Fig. 10-3d). The plot of results for a creep test is usually not a straight line, because the strain rate changes as strain hardening or strain softening occurs.

Representation of State of Stress on Mohr Diagrams

The stress across a specified plane can be represented by a stress vector. The *stress vector*, **P**, acting across a randomly oriented plane is inclined to the plane (Fig. 10-4) and thus can be resolved into σ_n, a *normal stress component* (perpendicular to the plane), and τ, a *shear stress component* (parallel to the plane). The state of stress at a point cannot be represented by a stress vector. Stress at a point can, however, be represented by a *stress ellipsoid*

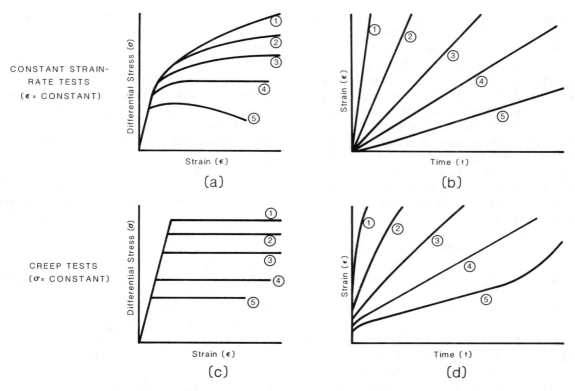

Figure 10-3. Correspondence between stress-versus-strain plots and strain-versus-time plots (adapted from Heard, 1963). (a) Stress-strain curves for constant strain-rate tests. Curve 5 is the slowest strain rate, and curve 1 is the fastest strain rate; (b) constant strain-rate experiments plotted on strain-time coordinates for the same experiments as in (a); (c) creep experiments plotted on stress-strain coordinates. Curve 1 is the fastest strain rate, and curve 5 is the slowest strain rate; (d) creep experiments plotted on strain-time coordinates for the same experiments as in (c).

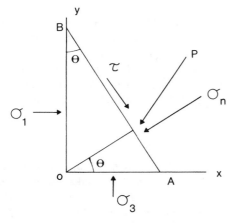

Figure 10-4. The general orientation of the principal stresses, σ_1 and σ_3, and the stress vector, **P**, resolved into σ_n and τ. The angle θ is between the normal to the plane AB and the σ_1 ('x') direction.

or a *stress tensor* (see Means, 1976, for a complete definition of these important terms). For a given stress state, there are three unique planes on which the magnitude of the shear stress component is zero. These planes are called the *principal planes*, and the stress vectors acting across them are called the *principal stresses*, where σ_1 equals the maximum principal stress at the point in question, and σ_3 equals the minimum principal stress. Principal stresses have no shear component and correspond to the principal axes of the stress ellipsoid.

If we are given the principal stresses representing the stress state at a point, we can calculate the stress vector acting across any plane of a specified orientation that passes through that point. Such calculations can be done analytically with force-balance equations or with the stress tensor (see Means, 1976, or Suppe, 1985). Fortunately, there is an easy-to-use graphic device that allows us quickly to determine the relative magnitudes of the normal and shear stress vectors in two dimensions acting across a specified plane without going through any calculations. This device is called a *Mohr circle*. A plot containing a Mohr circle is called a *Mohr diagram*.

A Mohr circle is plotted on Cartesian axes; the x-axis represents values of normal stress (σ_n), and the y-axis represents values of shear stress (τ). The coordinates of each point on the circle represent the values of the normal and shear stress components acting across a plane of an orientation specified by the angle θ, which is measured between the pole to the plane and the maximum principal stress σ_1 (Fig. 10-4). The reason that the relative values for normal and shear stress plot along the circumference of a Mohr circle is that the circle is the locus of all points that satisfy the equations

$$\tau = \frac{1}{2}(\sigma_1 - \sigma_3)\sin 2\theta \qquad \text{(Eq. 10-5)}$$
and
$$\sigma_n = \frac{1}{2}(\sigma_1 + \sigma_3) + \frac{1}{2}(\sigma_1 - \sigma_3)\cos 2\theta$$

(Eq. 10-6).

For a derivation of these equations using the balance of forces method, see Suppe (1985) or Means (1976).

Examination of Equations 10-5 and 10-6 indicates that the right-hand intersection of the Mohr circle with the x-axis is σ_1, and the left-hand intersection is σ_3. Therefore, to plot a circle, simply plot a point representing σ_1 and another point representing σ_3 on the x-axis (remember that principal stresses have no shear component). Place the anchor needle of a compass midway between these two points, and place the pencil of the compass on either one of the points; then simply trace out a circle with the pencil. The length of the diameter of the circle is equal to *differential stress* ($\sigma_1 - \sigma_3$), and the x-coordinate of the center of the circle is the *mean stress* [$(\sigma_1 + \sigma_3)/2$]. The values of σ_n and τ at a point on the circle represent the normal and shear components, respectively, of the stress vector acting on a plane oriented at and angle of 2θ measured counterclockwise from σ_1. The sign convention for Mohr diagrams used in geology is as follows: positive stresses on the x-axis are compressive, negative stresses on the x-axis are tensile, positive stresses on the y-axis represent a left-lateral shear couple, and negative stresses on the y-axis represent a right-lateral shear couple.

From the Mohr diagram the relative magnitudes of τ and σ_n on a plane inclined to the principal stresses can easily be visualized. For example, a plane inclined at 80° to σ_1 is being subjected to a greater normal stress than a plane oriented at 40° to σ_1. The plane on which shear stress is greatest is the plane oriented at 45° to σ_1.

Problem 10-1

The mean stress at a point in a rock is 40 MPa, and the differential stress is 20 MPa. (a) What are the values of σ_1 and σ_3 in the rock? (b) What are the magnitudes of τ and σ_n acting on a plane that is inclined at 30° to σ_1?

Method 10-1

Step 1: Construct coordinate axes calibrated in MPa. The horizontal axis represents σ_n, and the vertical axis represents τ (Fig. 10-5).

Step 2: Construct a Mohr circle representing the specified state of stress. The diameter of the circle is 20 MPa, and the center is positioned along the x-axis at 40 MPa. The values of σ_1 and σ_3 (Problem 10-1a) can now be read directly: σ_1 is 50 MPa, and σ_3 is 30 MPa.

Step 3: Draw a line from the center of the circle, point N, to a point, O, on the circle so that line ON makes a counterclockwise angle of 2θ (=120°) with respect to the x-axis. Remember that θ is the angle between the *normal* to the specified plane and the direction of σ_1.

Step 4: The values of τ and σ_n acting on the plane are specified by the coordinates of point O.

Representation of Failure Criteria on Mohr Diagrams

As discussed earlier, failure of a rock is manifested by a sudden stress drop. Failure under brittle conditions can be indicative of either (1) development of a new fracture surface in an intact rock or (2) slip on a preexisting fracture in a previously broken rock. A *failure criterion* is a specification of the stress state at which failure occurs. A

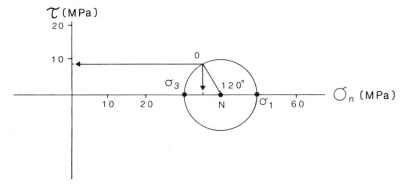

Figure 10-5. A Mohr diagram. Point N marks the value of mean stress, which is 40 MPa. $2\theta = 120^\circ$, $\sigma_1 = 50$ MPa, $\sigma_3 = 30$ MPa. The normal stress on a plane oriented such that the angle between σ_1 and the pole to the plane is 60° is specified by the coordinates of point O.

Mohr diagram can be used to represent certain types of failure criteria. On a Mohr diagram, a failure criterion is represented by a curve that separates a field in which the state of stress is such that a rock remains stable from a field in which the state of stress is such that a rock is unstable and either ruptures or deforms by slip on a preexisting fracture. The curve is called a *failure envelope*. A failure envelope on a Mohr diagram is empirical, in that it is drawn based on laboratory experiments, not on theoretical calculations.

One of the most widely known failure criterion is the *Coulomb-Mohr failure criterion*, which is described by the equation

$$\tau = C + (\mu^*)\sigma_n \qquad \text{(Eq. 10-7).}$$

where τ is the shear stress at failure, C is a constant called the *cohesion* (the y-intercept), and μ^* is a constant called the *coefficient of internal friction* (the slope of the line). As indicated by the equation, the failure envelope defined by the Coulomb-Mohr criterion is a straight line whose slope is μ^* and whose y-intercept is the cohesion (Fig. 10-6a). The envelope cuts the abscissa at a point representing the tensile strength ($T = -\sigma_3$).

To understand what is meant by a failure envelope, consider a stress state defined by a Mohr circle that falls below the envelope and does not touch it. Such a stress state is stable, in that the rock subjected to the stress state

does not fail (Fig. 10-6b). If the differential stress is increased, the diameter of the Mohr circle increases. If the differential stress is increased to a sufficiently large value, the Mohr circle becomes tangent to the envelope (Fig. 10-6c). At this instant the rock fails by formation of a fracture. If the differential stress is constant, but the mean stress decreases, the Mohr circle moves to the left along the x-axis of the Mohr diagram and eventually may touch the envelope. Again, the instant that a stress state is achieved such that the Mohr circle defining the stress state touches the envelope, the rock fails. As a consequence, it is impossible to have stress states defined by Mohr circles that extend beyond the envelope (Fig. 10-6d), because the rock will fail before such a stress state can be achieved.

Failure envelopes on a Mohr diagram are empirical, in that they are determined experimentally rather than by means of theoretical calculations. In the experiment described next, we see how a failure envelope is determined. You will find that real envelopes are not always perfectly straight lines.

10-3   ANALYSIS OF ROCK STRENGTH AND FAILURE CRITERIA

The study of rock strength at elevated temperature and pressure was greatly advanced in the 1950s by John Handin and his colleagues at the Shell Development Company.

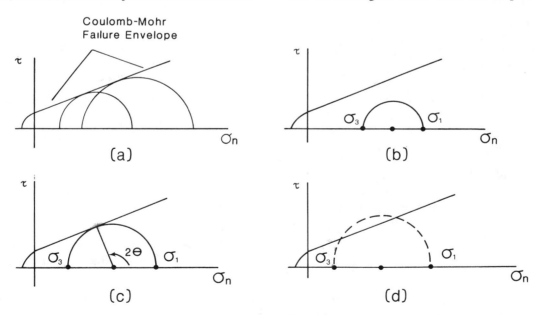

Figure 10-6. Coulomb-Mohr failure envelope. (a) Failure envelope showing the position of the y-intercept; (b) Mohr circle that is not tangent to the failure envelope and therefore represents a stable stress state; (c) Mohr circle that is tangent to the envelope and represents the stress state at the instant of failure; (d) Mohr circle representing an unstable and impossible stress state.

Handin's group systematically characterized the brittle strength of various types of rock (e.g., Handin and Hager, 1957). Experiment 1 shows how results obtained at the Shell Lab during this period can be used to determine a failure envelope for a brittle rock.

Determination of the Failure Envelope for Brittle Fracture

Experiment 10-1 (Oil Creek Sandstone)

The purpose of this experiment was to determine the failure envelope that characterizes the strength of Oil Creek Sandstone (a massive, very fine-grained, well-sorted, well-cemented Ordovician sandstone from Grayson County, Texas). To obtain each experimental data point a jacketed cylinder of sandstone was placed in a triaxial rock-deformation machine. Then the confining pressure was set at a specified value, and the axial stress (σ_1) was increased until the specimen failed. Note that by increasing the value of σ_1 while σ_3 was held constant, we increased the differential stress. This increase was represented by an increase in the diameter of the Mohr circle defining the stress state in the sample. Individual experiments differed from one another in the value of the confining pressure (σ_3) set at the beginning of the experiment. All experiments were carried out at room temperature and at a strain rate of about 10^{-3} per second.

Results 10-1

The raw data of each experiment (run) were replotted on a graph in which the x-axis was axial strain and the y-axis was differential stress. In each case, when the sample achieved its ultimate strength, it failed brittlely by formation of a discrete shear fracture. In Figure 10-7 three stress-strain curves obtained during the experiment are shown (from Handin and Hager, 1957). The confining pressure for each run is indicated on the graph. The strength of the sample under the specified confining pressure is the maximum stress achieved before the sudden stress drop occurred. The stress drop is indicated by the small arrow at the end of the curve. Figure 10-8 provides a sketch of a ruptured specimen after completion of the experiment, showing the orientation of the shear fracture (the measured $\theta = 67°$).

Interpretation 10-1 (to be completed by the student)

(a) Using the stress-strain plots of Figure 10-7, determine the differential stress and the mean stress at the point of brittle fracture for each run.

(b) Draw the coordinate axes of a Mohr diagram at an appropriate scale. On the diagram draw the Mohr circle showing the stress state at failure for each run (i.e., you should draw three nonconcentric circles centered at the mean stress for the respective experiment). Note that you must

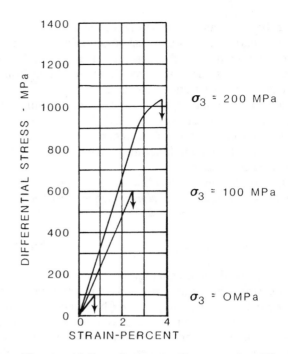

Figure 10-7. Stress-strain curves for Oil Creek Sandstone (adapted from Handin and Hager, 1957). The confining pressure is indicated next to the curve.

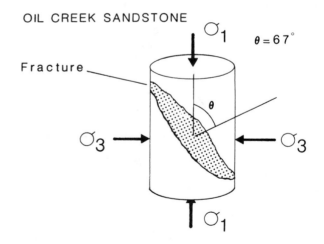

Figure 10-8. A drawing of fractured Oil Creek Sandstone. The stippled plane is the fracture surface.

first calculate σ_1 from knowledge of the differential stress and σ_3.

(c) Draw the two curves that are tangent to all three circles. One curve will lie above the x-axis, and one curve will lie below the x-axis. These curves define the failure envelope. Are the curves straight or bent?

(d) Calculate the equation that approximately describes the curve that lies above the x-axis (i.e., determine appropriate values for the constants in Equation 10-7). This equation is the Coulomb-Mohr failure criterion for Oil Creek Sandstone.

(e) The orientation of the failure plane is determined by the obtuse angle between the x-axis and the line connecting the center of a particular circle and the point of tangency with the envelope. Is the orientation of the failure plane indicated by each circle the same?

(f) Is the orientation of the observed fracture plane the same as the fracture orientation suggested by the Coulomb-Mohr failure criterion?

(g) How does confining pressure affect the fracture strength of Oil Creek Sandstone?

(h) From your answers to parts a-f, complete Table 10-1. Write a paragraph outlining the conclusions that can be drawn from the experiments on Oil Creek Sandstone. In particular, suggest what applications the results might have toward predicting the stability of boreholes?

Study of the Brittle-Ductile Transition

Experiment 10-2   (Berea Sandstone)

Handin et al. (1963) conducted a set of experiments on Berea Sandstone (medium-grained, poorly cemented Mississippian sandstone from Ohio), using the procedures outlined in Experiment 10-1. The Berea Sandstone is weaker than the Oil Creek Sandstone. In these experiments the specimens contained pore fluid under pressure. Handin et al. ran five experiments. In each experiment, the confining pressure was the same (200 MPa), but the pore pressure varied (0, 50, 100, 150, and 200 MPa, respectively). The purpose of these experiments was to determine the failure envelope that characterized the strength of Berea Sandstone under conditions of elevated pore pressure. (The data used here have been modified slightly from the original experiments in order to make the results more obvious.)

Results 10-2

Figure 10-9 provides the stress-strain plots for five runs (adapted from Handin et al., 1963). The effective confining pressure associated with each curve is indicated next to the curve. It is evident from the curves in Figure 10-9 that Berea Sandstone does not lose complete strength after yielding; there was not a sudden loss of cohesion accompanied by a catastrophic stress drop, as was the case for the Oil Creek Sandstone. At higher confining pressures it was observed that the sandstone specimens continued to strain without a stress drop even after fractures had developed. Note that the ultimate strength continued to increase with increasing confining pressure.

Table 10-1
Oil Creek Sandstone

Run number	Confining pressure	Differential stress at failure	σ_1 at failure	Fracture angle
1				
2				
3				

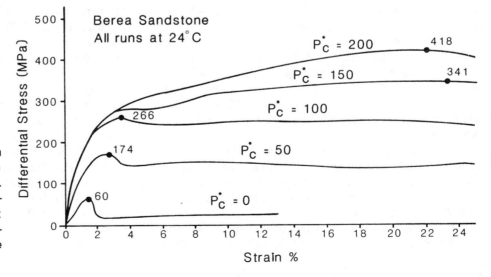

Figure 10-9. Stress-strain curves for Berea Sandstone (adapted from Handin et al., 1963). P_C^*, the effective confining pressure, for each run is indicated next to the curve. The value of differential stress at the ultimate strength is also indicated.

Interpretation 10-2 (to be completed by the student)

(*a*) Look at the curves shown in Figure 10-9. Note that the curves for runs at lower effective confining pressure are different in shape from the curves at higher effective confining pressure. Describe the difference in curve shape. (*Hint:* Compare these curves with those shown in Figs. 10-2 and 10-3a. Explain how the series of curves in Figure 10-9 shows the transition from brittle behavior to ductile behavior). Is the yield strength the same as ultimate strength for all curves? Under what conditions does strain hardening occur?

(*b*) After examining Figure 10-9, indicate how the strain (in percent) at which ultimate strength is reached is related to effective confining pressure. Remembering the definition of ductility (ductility = total percent strain before fracture), explain how ductility is related to effective confining pressure.

(*c*) Construct a Mohr diagram, and plot the Mohr circle corresponding to each experiment. Use the ultimate strength to define the differential stress at failure, and consider the effective confining pressure to be σ_3. Draw the failure envelope so that it is tangent to each circle. How does the failure envelope constructed from this experiment differ from that constructed in Experiment 10-1 for Oil Creek Sandstone?

(*d*) From the Mohr diagram, determine how the fracture angle changes as a function of confining pressure. Is there a systematic change? Try to explain why this change occurs. The measured fracture angles (angle between σ_1 and the plane of the fracture) were 26° at 0 MPa, 27° at 50 MPa, 34° at 100 MPa, 36° at 150 MPa, and 38° at 200 MPa confining pressure, respectively. Note that these angles are somewhat different from those derived from the Mohr diagram, probably reflecting the qualitative nature of the Coulomb-Mohr failure envelope.

(*e*) Based on the observations in this experiment, do you estimate that wet Berea Sandstone subjected to the stress conditions at a depth of 1 km in the earth will lose strength by brittle fracture or maintain strength by ductile-like behavior? What about dry Oil Creek Sandstone? Assume that differential stress in the upper crust is $2/3(\rho gh)$, where ρ is the average density of rock (2.7 g/cm^3), g is the gravitational constant (980 cm/s^2), and h is the depth measured in centimeters. Means (1976) explains why the differential stress in the upper crust is approximately $2/3(\rho gh)$.

(*f*) Note that in this experiment we have used the term "effective" confining pressure to emphasize that the samples contained pore fluid under pressure. Will a change in the pore pressure affect the equation of the failure envelope for a given rock?

(*g*) The confining pressure on a sample of Berea Sandstone is set at 100 MPa, and the axial stress is set at 210 MPa. Based on the failure envelope you determined above, will the sample fail if the pore pressure is 50 MPa? Will a sample fail if the pore pressure is 60 MPa? How does an increase in the pore pressure of 75 MPa affect the position of the Mohr circle? How does an increase in pore pressure of 60 MPa affect the values of mean stress and differential stress?

10-4 ANALYSIS OF DUCTILE DEFORMATION

During brittle deformation only a small elastic strain (< 3%) is achieved before failure occurs by formation of a brittle fracture. During *ductile deformation* of rock, fracture does not occur even after large strain is achieved (> 25%). In other words, during ductile deformation large strains develop without loss of cohesion. Rock that fails by brittle fracture when the strain is between 3% and 25% is said to exhibit brittle-ductile behavior. We observed the transition from brittle to ductile behavior in the experiment on Berea Sandstone. It is obvious that ductile deformation is common in nature, for there are many geologic settings in which folding occurs, without the aid of brittle fracture. Next, we consider triaxial loading experiments in which environmental conditions are manipulated so that ductile deformation can occur.

Experiment 10-3 (Solenhofen Limestone, Carrara Marble, Yule Marble)

The three environmental parameters that are most important in determining whether rock behaves brittlely or ductilely are temperature, confining pressure, and strain rate. Heard (1960) and Edmond and Paterson (1972), among others, have examined how these variables affect the ductility of rock. In the experiments a cylinder of rock was placed in a triaxial loading machine. Confining pressure was exerted by increasing the pressure of argon in the pressure chamber, and temperature was increased by an electrical furnace. The strain rate was varied by changing the rate at which the piston moved. In some experiments the sample was stretched rather than shortened. We describe four sets of experiments.

(*a*) *Variable Confining Pressure:* Edmond and Paterson (1972) deformed cylinders of Carrara Marble at room temperature and at a strain rate of 4 X 10^{-4} s^{-1} They repeated the experiment six times, each time with a new rock cylinder and under a different confining pressure. Differential stress versus axial strain curves for the experiments are shown in Figure 10-10a.

(*b*) *Variable Confining Pressure:* Heard (1960) deformed cylinders of Solenhofen Limestone under compression at room temperature. He repeated the experiment seven times, each time at a different confining

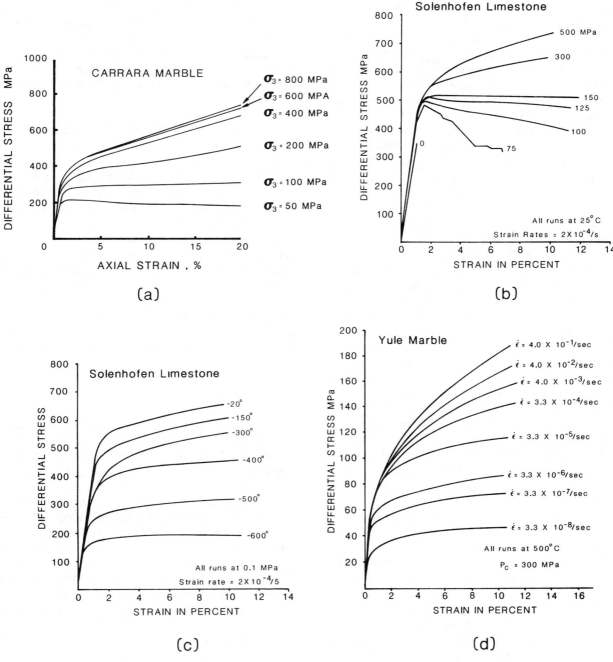

Figure 10-10. Effect of environment on ductile behavior.   (a) Differential stress versus axial strain for triaxial compression tests on Carrara Marble under variable conditions of confining pressure (adapted from Edmond and Paterson, 1972); (b) differential stress versus axial strain for triaxial compression tests on Solenhofen Limestone under variable conditions of confining pressure at a constant strain rate (adapted from Heard, 1960); (c) differential stress versus axial strain for triaxial compression tests of Solenhofen Limestone under variable conditions of temperature (adapted from Heard, 1960); (d) differential stress versus strain for   triaxial extension tests on Yule Marble at 500°C and 500 MPa (adapted from Heard, 1963).

pressure. The results of the experiments and the confining pressures are indicated on Figure 10-10b.

(c) Variable Temperature: Heard (1960) deformed cylinders of Solenhofen Limestone under compression at a strain rate of 2×10^{-4} s^{-1} and a confining pressure of 300 MPa. He repeated the experiment six times, each time with a new rock cylinder. The results and the temperatures are indicated in Figure 10-10c.

(d) Variable Strain Rate: Heard (1963) deformed Yule Marble by extension at a temperature of 500°C. He repeated the experiment eight times, each time with a new rock cylinder and at a different strain rate. The results and the strain rates are indicated in Figure 10-10d.

Interpretation 10-3 (to be completed by the student)

(a) For each set of experiments describe how the yield strength varies as a function of the environmental condition. To emphasize your result make a plot of yield strength (in terms of differential stress) as a function of the environmental parameter (i.e., confining pressure, temperature, strain rate).

(b) Considering the results of these experiments, provide generalizations concerning the relationship of ductility to deformational environment. Under what conditions will ductile deformation be more likely?

(c) Compare the results of the variable confining pressure experiments for Solenhofen Limestone (Experiment 10-3b) with those of the experiments for Carrara Marble (Experiment 10-3a). Solenhofen Limestone is a very fine grained carbonate, whereas Carrara Marble is a relatively coarse grained carbonate. Is there a dependence of ductility on grain size?

(d) Carbonates are relatively ductile compared to granite. What types of stress-strain curves would you expect for granite under the conditions of Experiments 10-3a and 10-3b? (*Hint:* Look again at the stress-strain plot for Oil Creek Sandstone, and remember that some sandstones, like granite, are stronger than carbonates).

(e) Considering the state of stress in the crust, at what depth would you expect ductile behavior to become dominant over brittle behavior (the *brittle-ductile transition*) for Solenhofen Limestone, assuming that temperature does not change with depth? Assume that differential stress in the upper crust is 2/3($\rho g h$), where ρ is the average density of rock (2.7 g/cm^3), g is the gravitational constant (980 cm/s^2), and h is the depth measured in centimeters.

(f) In reality the geothermal gradient in the crust is about 30°C/km (i.e., at 1-km depth, the temperature is 30°C greater than at the surface). Keeping this in mind, at approximately what depth would you expect the brittle-ductile transition for Yule Marble to occur, neglecting the effect of confining pressure?

(g) Folds, which are manifestations of ductile deformation, are known to develop in the upper crust above the brittle-ductile transition. Furthermore, in many deformational settings, folding and brittle faulting occur during the same period of time. Considering the results of the experiments described above, explain this paradox.

10-5 ANALYSIS OF ROCK FRICTION

Experiments 10-1 and 10-2 concerned the initiation of a fracture in an intact rock. Once a rock contain fractures, deformation may continue by additional slip on these fractures. *Friction* is the resistance to sliding on a fracture surface. To initiate sliding on a surface, the component of shear stress parallel to the surface must exceed a critical value called the *frictional strength*. In general, frictional strength depends on the magnitude of normal stress across the surface; as the normal stress increases, it becomes progressively harder for sliding to take place, and thus the shear stress necessary to initiate sliding must increase. The *coefficient of friction* (μ) is the ratio between the shear stress necessary to initiate sliding and the normal stress across the surface:

$$\mu = \tau/\sigma_n \qquad \text{(Eq. 10-8)}.$$

The value of μ can be determined from a single experiment in which the value of τ at a given σ_n is measured. Experimental work suggests that μ is not constant but depends on the value of σ_n. If a series of tests is conducted, each at a different σ_n, and the results are plotted using Cartesian axes (x-axis is σ_n, and y-axis is τ), we can define another coefficient of friction (μ'), which is the slope of the line passing through the data points. This line is a failure envelope that may be used in the same manner as the Coulomb-Mohr envelope to predict frictional sliding on favorably oriented fractures. For measurements made under high pressure, the sloping line intercepts the τ-axis above zero. Therefore, the equation of the line is

$$\tau = S_0 + \mu'\sigma_n \qquad \text{(Eq. 10-9)},$$

where S_0 is the intercept between the friction envelope and the τ-axis. S_0 represents the shear stress necessary to initiate sliding under conditions such that σ_n is 0 (i.e., it is the cohesive strength of the fracture). We can solve for μ' and derive the equation

$$\mu' = (\tau - S_0)/\sigma_n \qquad \text{(Eq. 10-10)}.$$

The coefficient of friction (μ') can also be defined in terms of the angle (ϕ) between the friction envelope and the horizontal. This angle is called the *angle of friction,* and

$$\mu' = \tan \phi \qquad \text{(Eq. 10-11)}.$$

In the following experiments, we see how to determine the value of μ'.

Analysis of Failure Envelopes for Frictional Sliding

Experiment 10-4   (Tennessee Sandstone)

John Handin at Shell Development Company measured the frictional properties of Tennessee Sandstone. For each experimental run he used a 1.9 x 5.0 cm cylinder of sandstone containing one through-going saw cut inclined at 45° to the cylinder axis. Each saw-cut surface was polished to allow a good fit when mated to reform a cylinder. The samples were jacketed in lead and placed in a triaxial rock deformation machine. The stress at which sliding occurred (i.e., the *frictional strength*) was measured by subjecting the specimen to an axial load and observing when a displacement occurred on the saw-cut surface. The experiment was repeated for a range of confining pressures as listed in Table 10-2.

Results 10-4

Table 10-2 lists the μ (the ratio of τ/σ_n) for individual experiments (Handin, 1969). Note that the normal and shear stress terms were calculated by resolving the axial stress on a plane oriented at 45° to the axial stress. All stresses were measured in MPa. Note that the coefficient of friction is dimensionless.

Interpretation 10-4 (to be completed by student)

(a) Derive μ' based on the slope of the line of τ versus σ_n.

(b) Write the general equation for the frictional sliding of Tennessee Sandstone. Note that μ' in this equation is by definition independent of confining pressure.

(c) How does the coefficient of friction (μ) for individual tests depend on confining pressure? This relationship is best illustrated by plotting a graph of μ against confining pressure.

Determination of the Preference for Fracture over Frictional Sliding

Experiment 10-5 (Blair Dolomite, Solenhofen Limestone, Leuders Limestone, Tennessee Sandstone)

The shear stress necessary to initiate sliding on a preexisting fracture depends on the normal stress across the fracture, as can be illustrated by comparing shear stress and normal stress in Table 10-2. The normal stress across a fracture plane depends on its orientation relative to σ_1. If a fracture is oriented such that the normal stress across the fracture is high (i.e., 2θ is small), the rock may fail by formation of a new fracture before the preexisting fracture can slip. It is possible to determine conditions under which frictional sliding precedes fracture for a given rock by comparing the envelope for frictional sliding with the Coulomb-Mohr failure envelope.

Handin (1969) described a series of experiments designed to investigate the preference for new fracturing before slip on preexisting rock fractures. He obtained 1.9 X 5.0-cm cylinders of several rock types (Blair Dolomite, Solenhofen Limestone, Leuders Limestone, and Tennessee Sandstone) and made a cut at a specified angle in each sample. These cuts represented preexisting fractures in the test samples. The opposing surfaces of each cut were lightly polished so that they closed tightly. The samples were jacketed in lead and placed in a triaxial rock-deformation machine. The stress at which sliding occurred (i.e., the *frictional strength*) was measured by subjecting the specimen to an axial load and observing when a displacement occurred on the fracture. The

---

Table 10-2
The Frictional Properties of Tennessee Sandstone

Confining pressure	Shear stress τ	Normal stress σ_n	Coefficient of friction μ
25	76	100	0.76
50	130	180	0.72
75	181	255	0.71
100	231	330	0.70
125	287	410	0.70
150	331	480	0.69
175	386	560	0.69
200	420	620	0.68

experiments were repeated for the same rock type, for different fracture orientations in each rock type, and for a range of confining pressures.

Results 10-5

The results of the work on preference for fracturing or sliding are presented in the form of fracture and friction envelopes in Table 10-3. In some specimens deformation was accommodated entirely by movement on the preexisting fracture, in some specimens deformation was accommodated by formation of a new fracture in addition to sliding on a preexisting fracture, and in some specimens deformation was accommodated only by initiation of a new fracture (Fig. 10-11).

Interpretation 10-5 (to be completed by student)

(a) On a Mohr diagram construct both the Coulomb-Mohr failure envelope and the sliding friction envelope for the four lithologies listed in Table 10-3. Note that the saw cuts are cohesionless, so the frictional failure line passes through the origin.

(b) For the runs identified in Table 10-3 compare the coefficients of sliding friction and the coefficients of internal friction. Consider an experiment in which the confining pressure is 100 MPa, and the differential stress is gradually increased. From the data in Table 10-3 determine whether any of the rocks will fail first by slip on a preexisting fracture inclined at 45° to the axial load or whether they will fail by formation of a new fracture.

(c) Consider an experiment run at a mean stress of 400 MPa. For Tennessee Sandstone what are the angles of the preexisting fracture for which new fracture will take place while the preexisting fracture remains stable (i.e., does not slide)?

(d) The crust of the Earth contains fractures in many orientations. Using your answers to the preceding questions, and Tables 10-2 and 10-3, explain why differential stress magnitude in the upper crust is controlled by frictional sliding criteria rather than by failure criteria.

(e) If the axial stress is 207.6 MPa and the confining pressure is 92.4 MPa, determine the shear stress on each of the two planes shown in the cylinder of rock (Fig. 10-12) used for a deformation experiment. Assuming that both discontinuities exist within the rock cylinder, on which plane will frictional slip be favored? (Draw the Mohr diagram very carefully.)

Table 10-3
Faulting and Friction Envelopes

Rock type	Coulomb-Mohr envelope	Friction envelope
Blair Dolomite	$\tau = 45 + \sigma_n \tan 45°$	$\tau = \sigma_n \tan 21°$
Tennessee Sandstone	$\tau = 50 + \sigma_n \tan 40°$	$\tau = \sigma_n \tan 35°$
Solenhofen Limestone	$\tau = 105 + \sigma_n \tan 28°$	$\tau = \sigma_n \tan 32°$
Leuders Limestone	$\tau = 15 + \sigma_n \tan 28°$	$\tau = \sigma_n \tan 31°$

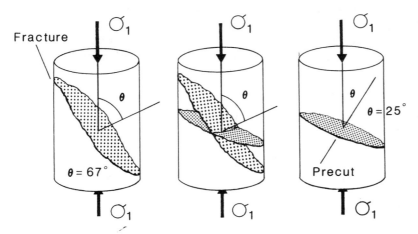

Figure 10-11. Three samples of Blair Dolomite: one with a fracture; one with a saw cut, and one with both. The lightly stippled plane is a fracture that formed in intact rock. The darker shaded plane is a saw cut.

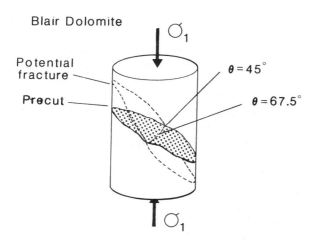

Blair Dolomite

Potential fracture

Precut

$\theta = 45°$

$\theta = 67.5°$

Figure 10-12. A cylinder of Blair Dolomite with two fracture surfaces (for Experiment 10-5).

Friction and an Explanation for Stress in the Earth's Crust

In the late 1970s a large number of *in situ* stress measurements were made in outcrops, drill holes, and mines. Comparison of the results from in situ measurements with laboratory measurements indicated that ambient stresses in the upper crust of the earth are generally too low to initiate fractures in intact rock. It was suggested that the relatively low ambient stress state of the earth's upper crust did not, therefore, reflect the strength of intact rock but rather reflected the magnitude of shear stress necessary to cause sliding on preexisting fractures. The crust is pervaded with joints and fractures; it is intuitively reasonable to assume that slip will occur on one of these fractures long before the magnitude of differential stress becomes high enough to cause rupture of intact rock between the fractures. In order to better understand stresses in the earth's upper crust it is therefore necessary to

understand the conditions under which sliding along natural fractures can initiate.

The data for Tennessee Sandstone from Table 10-2 are plotted on a Mohr diagram (Fig. 10-13). This is the answer to a question in Experiment 10-5. To a first approximation these data appear to follow a linear trend defining a Mohr-like envelope called the *friction envelope*. Byerlee (1978) compiled a large quantity of friction data for a great variety of rock types. After plotting his data in the same format as that of Figure 10-13, Byerlee (1978) observed that most of the friction data for rocks followed a general trend divided into two linear segments. Those data for experiments with a mean stress of less than 200 MPa followed the friction equation

$$\tau = 0.85(\sigma_n) \qquad \text{(Eq. 10-12)},$$

whereas data from experiments with mean stress > 200 Mpa followed the friction equation:

$$\tau = 50 + 0.6(\sigma_n) \qquad \text{(Eq. 10-13)}.$$

Experiment 10-6   (Barre Granite)

In this experiment we wish to demonstrate that Coulomb-Mohr analysis can be used to determine which fracture orientation is most favorable for slip. To do so, we create an experimental rock cylinder of Barre Granite with three fractures (saw cuts that have been slightly polished) as shown in Figure 10-14. Normals to the fractures make angles of 45°, 60°, and 75° to σ_1, respectively. We place the rock in a triaxial load machine, set the confining pressure at 150 MPa, and gradually increase the axial load until failure by sliding on a fracture occurs.

Results   10-6

The granite sample fails by sliding on one of the fractures when the value of σ_1 reaches a sufficiently high

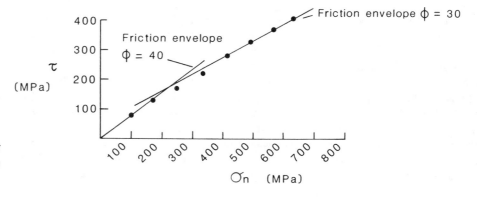

Tennessee Sandstone - Sliding on 45° Saw Cuts

Figure 10-13. Mohr diagram for frictional sliding of Tennessee Sandstone.

Existing Fractures

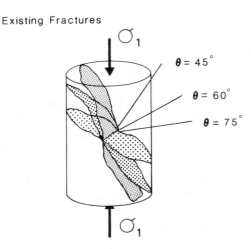

Figure 10-14. Three preexisting fractures in a cylinder of Barre Granite (for Experiment 10-6). The angle θ is the angle between the pole to the fracture and the σ_1 direction. The fractures all intersect along a line that is perpendicular to the cylinder axis.

value. Because the sample is jacketed, we cannot determine which fracture slipped until we remove the sample from the machine and strip off its jacket. In order to prove that the Coulomb-Mohr analysis correctly predicts the orientation of the slipped fracture, we first complete the following steps of interpretation (i.e., predict the differential stress at failure and predict which fracture failed first).

Interpretation 10-6

(a) First, we attempt to predict the differential stress at the time of failure. We know that σ_3 is 150 MPa, so we guess that the mean stress is at least 300 MPa. If this guess is correct, we can use the general friction equation (Eq. 10-13) to define the failure envelope for frictional sliding. We construct a Mohr diagram showing this frictional sliding envelope. The friction equation states that the coefficient of friction is 0.60, that the frictional sliding envelope has a slope of 31°, and that the rock behaves as if it has a "cohesive strength" of about 50 MPa.

(b) We plot the specified σ_3 on the x-axis. We know that the rock failed by sliding, so by trial and error we use a compass to find the Mohr circle that passes through the 150-MPa mark and is tangent to the failure envelope. Then, with a compass, we draw a circle centered at this point and tangent to the sliding envelope. Now that we have the Mohr circle, we determine the mean stress and the differential stress at the time of failure. Considering the mean stress that we determined, was it reasonable to use Equation 10-13 rather than Equation 10-12 to specify the failure envelope?

(c) We draw a line from the center of the Mohr circle to the point of tangency between the circle and the failure envelope and measure the angle 2θ. Given this angle, we can determine the orientation of the fracture that probably slid first. What is it? Note that this plane is not the plane on which shear stress was greatest. Why did it slide first? Why do you expect the other fractures to be stable under these stress conditions?

(d) Consider a hypothetical state of stress in the upper crust of the earth. Vertical stress is σ_1 and is due to the weight of overlying rock. The magnitude of σ_1 is ρgh. In the absence of tectonic stress, horizontal stress in the upper elastic crust is due to lateral expansion of the rock in response to the vertical load. Assume the value of horizontal stress to be $\sigma_2 = \sigma_3 = \frac{1}{3}(\rho gh)$. Determine the mean stress, differential stress, and depth in the crust at which $\sigma_2 = \sigma_3 = 150$ MPa ($\rho = 2.7$ g/cm^3, and g = 9.8 m/s^2). Assuming that the predictions made above are correct, how do these stresses compare with the stresses at the time of failure in the experiment? Do you expect that at a depth of 10 km in the earth that one or both fracture sets will slide in dry rock ? If not, by how much can pore fluid pressure increase before slip on a fracture set will occur.

A Mohr diagram constructed from the data in Table 10-3 for Blair Dolomite is shown in Figure 10-15a. In this diagram the Mohr circle represents the stress state at which a shear fracture develops under a confining pressure of 100 MPa. The Coulomb-Mohr failure envelope is tangent to the Mohr circle at Point O. The radius ON makes a 2θ angle of 135° measured counterclockwise from the x-axis. Therefore, the Mohr diagram indicates that a fracture developed in intact rock that fails under a confining pressure of 100 MPa should be oriented at an angle of 22.5° to σ_1.

Using data from Handin's friction experiments (described above), we find that the friction envelope for Blair Dolomite has a relatively shallow slope (θ = 21°). This envelope is not the same as that specified by Equation 10-12 and, therefore, does not fit the general friction equations given above. This discrepancy may be a consequence of the experimental conditions used by Handin. In his experiments the saw cuts were quite smooth, so there may have been little interlocking across the saw cut; therefore, resistance to shear was less than expected.

Examination of Figure 10-15a allows us to predict the range of possible orientations for which slip on a preexisting fracture is favored over fracture through intact Blair Dolomite. To do this, we locate the intersections between the envelope for frictional sliding and the Mohr circle and label the two points of intersection A and B. Then we draw the two radii of the Mohr circle that

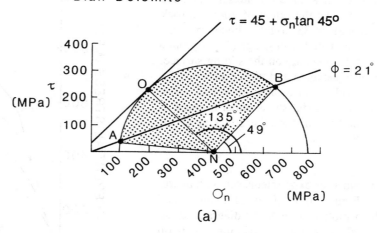

(a)

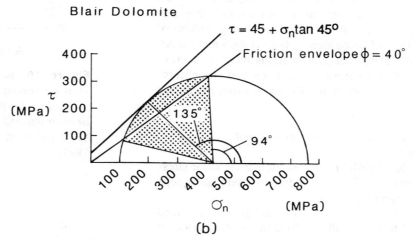

(b)

Figure 10-15. Mohr diagrams for Blair Dolomite. Only fractures represented by lines that fall in the stippled intervals on the diagrams can slip. If there are no favorably oriented fractures in the rock, then the rock fails when the circle touches the Coulomb-Mohr envelope. (a) Diagram showing both the Coulomb-Mohr failure envelope and the envelope for frictional sliding determined by Handin (1969); (b) diagram showing both the Coulomb-Mohr failure envelope and the general envelope for frictional sliding proposed by Byerlee (1978).

terminate at A and B. These radii are lines NA and NB. The wedge of the circle that lies between NA and NB (shaded in Fig. 10-15a) represents the range of 2θ values for which slip on a preexisting fracture will occur at stresses lower than those necessary to cause a new fracture to form. If we had used the general friction equations to represent the frictional strength of Blair Dolomite, then a smaller range of fracture orientations would favor slip over fracture initiation. Figure 10-15b shows that if the general friction equations are applied to Blair Dolomite, an existing fracture inclined at 45° to σ_1 is not favored over development of a new fracture at 22.5° to σ_1. However, a fracture inclined at 43° to σ_1 is equally likely to slip as a new fracture is to form.

10-6 FRICTIONAL PROPERTIES OF FAULT GOUGE

So far the frictional properties of fractures and joints have been examined by using experiments where intact rock is

sliding on intact rock. If slip continues on these breaks in rock, a layer of fault gouge will build up by the grinding and milling of the rock in contact with the slip surface. Before much slip the fracture will become a fault zone with a layer of fault gouge between intact rock. The gouge will act to change the frictional properties of the rock depending on the strength of the fault gouge. The effect of gouge on the frictional properties of rock has been investigated in the laboratory.

Experiment 10-7 (quartz gouge and halite gouge)

Many real faults are not planes along which two clean rock surfaces are juxtaposed. During faulting, fault gouge composed of finely ground rock, may accumulate along the fault. In some circumstances, a particularly ductile material, such as halite, may occur along a fault. Halite can be incorporated along faults that pass through evaporite sequences. In such circumstances the frictional strength of the fault is affected. A number of experiments have been conducted to study how the frictional strength of a fault is affected by the presence of a ductile material along the fault. In experiments to test the effect of gouge, a rock cylinder was cut at an angle of 35° to the cylinder axis, and a 2-mm-thick layer of simulated fault gouge (finely ground quartz or halite) was spread evenly on the sliding surface. The precut cylinder was jacketed and placed between the pistons of a triaxial load rig in a pressure vessel. The sample was subjected to an axial stress under a range of confining pressures, and the differential stress at which sliding initiated was determined.

Results 10-7

The stress-strain plots for experiments run with quartz gouge are shown in Figure 10-16 (from Shimamoto and Logan, 1981). When plotted on a Mohr diagram, these curves suggest that shear stress for frictional slip on gouge-coated surfaces increases with normal stress, as predicted by the general equations for sliding friction of rock. The effect of halite along a fault is shown in Figure 10-17. Under experimental conditions halite is ductile, which means that it deforms with a constant shear stress regardless of the magnitude of normal stress across the plane of shear.

Interpretation 10-7 (to be completed by the student)

(a) Using the Mohr diagram, derive the shear stress and normal stress for frictional sliding of quartz fault gouge at 8% axial shortening.

(b) Plot the shear stress and normal stress determined in part a to determine a sliding friction equation. Does it agree with the general equations for friction of rocks without gouge?

(c) Repeat part a for frictional sliding on halite.

(d) What is the coefficient of sliding friction for halite gouge?

(e) Above a confining pressure of 200 MPa frictional sliding on halite gouge requires the same shear stress as at 200 MPa. This means that the shear stress is independent of normal stress. What then is the coefficient of sliding friction for halite gouge above 200 MPa? Represent this friction criterion for halite on a Mohr diagram.

(f) Referring to Figure 10-18, if σ_3 is 300 MPa, determine the differential stress required for frictional sliding on salt. What would have been the differential stress required for slip on a plane 45° to σ_1 if the general friction equations had applied to salt?

Comments on Experiment 10-7

The Mohr circle for halite has the same diameter regardless of the confining pressure at pressures above 200 MPa. This type of behavior is modeled by the

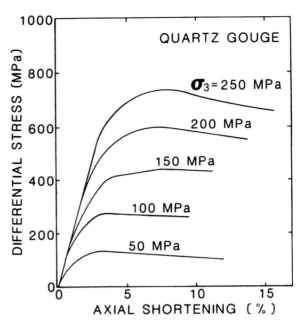

Figure 10-16. Differential stress versus axial shortening curves for the frictional sliding of quartz fault gouge as determined by Shimamoto and Logan (1981) (for Experiment 10-7) (adapted from Shimamoto and Logan, 1981).

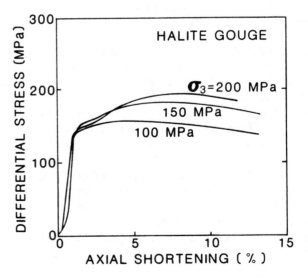

Figure 10-17. Differential stress versus axial shortening curves for the frictional sliding of halite fault gouge as determined by Shimamoto and Logan (1981) (for Experiment 10-7) (adapted from Shimamoto and Logan, 1981).

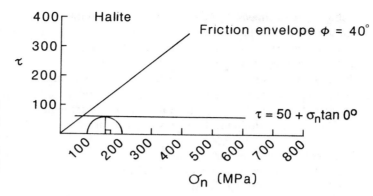

Figure 10-18. A Mohr diagram for the general fictional sliding curve for halite that is acting as a perfectly plastic material along the sliding surface (for Experiment 10-7).

Coulomb-Mohr failure criterion using a constant S_o and $\theta = 0°$. The friction equation for salt is then

$$\tau = S_o + \sigma_n \tan 0°$$

$$S_o = 50 \text{ MPa} \qquad \text{(Eq. 10-14).}$$

The Mohr diagram describing frictional sliding on salt is presented in Figure 10-18. Such a diagram would also represent the behavior of an intact sample of rock that deforms in a perfectly plastic manner. The rock will strain at the same differential stress regardless of the confining pressure.

The significance of Equation 10-14 is that it says that for fault zones deep within the upper crust, frictional resistance to sliding is very low if the fault gouge is salt. Many thrust belts of the world, such as the Appalachian foreland, the Jura of Switzerland, and the Zagros of Iran, are wide because salt "gouge" acts to reduce frictional resistance (Davis and Engelder, 1985).

10-7   ACKNOWLEDGMENTS

We thank John Logan for a helpful review of the manuscript.

ADDITIONAL EXERCISES

1. There are several general classes of stress. All can be drawn in two dimensions using a Mohr circle. Possible stress states in the earth include:

    (a) *Tension and compression:* One principal stress is tensile and the other is compressive. This is the most likely situation for the generation of most tensile fractures (joints) within the earth's crust.

    (b) *Pure shear stress:* A special case of tension and compression, in which $\sigma_1 = -\sigma_2$, so that the planes of maximum shear stress are also planes of pure shear stress (i.e., the normal stress component is zero on these planes). This is a very unusual state of stress for the earth's crust.

    (c) *General compression:* Both principal stresses are compressive. In three dimensions this state of stress in the earth is called triaxial compression. This is the usual state of stress within the crust.

    (d) *Hydrostatic compression:* The stress across all planes is compressive and equal. Pore water within a rock can exert a state of hydrostatic compression provided that the water in the pores can communicate directly with the surface.

    (e) *Lithostatic compression:* The stress across all planes is compressive and equal to the weight of the rock on top of the point at which the measurement is made.

    Draw each of these states of stress considering the nature and relative magnitudes of the principal stresses. Note the differences in location of the center and length of the diameter of the Mohr circle for each state of stress.

2. Using the data shown in Figure 10-14 for Barre Granite, determine the shear and normal stress on the three planes shown in the cylinder of rock used for a deformation experiment if σ_1 is 100 MPa and σ_3 is 50 MPa. Calculate the answer using the appropriate equations, and derive the answer using a Mohr diagram. In this example it is important to appreciate that slip occurs on the plane whose ratio of τ to σ_n is maximum. The plane (at $\theta = 45°$) with the maximum τ did not slip because σ_n was so large that the ratio was not maximum. Likewise, the plane ($\theta = 75°$) with the minimum σ_n did not slip because τ was small.

3. Leuders Limestone is fairly weak lithology. At a confining pressure of 100 MPa, a differential stress of just over 200 MPa is required to fracture intact samples of Leuders Limestone (Fig. 10-M1a).

(a) What is the orientation of the fractures that form in Leuders Limestone?

(b) The friction envelope for Leuders Limestone is also plotted on Figure 10-M1a. Note that it is very close to the Coulomb-Mohr failure envelope. What does this relationship imply?

(c) Is there a preference for slip on an existing fracture at 34° to σ_1 over development of a new fracture at 31° to σ_1?

(d) In the paper in which Byerlee (1978) compiled friction data for many lithologies several rocks showed frictional behavior that deviated from the general friction envelope. Leuders Limestone is one of those lithologies. In Figure 10-M1b the frictional envelope defined by Equation 10-12 is plotted next to the Coulomb-Mohr failure envelope for Leuders Limestone. In this case the envelope for frictional sliding plots well above the Coulomb-Mohr failure envelope. If the general friction equations are applicable, is it ever possible for Leuders Limestone to slide on an existing fracture? If not, how will Leuders Limestone respond to elevated differential stress?

(e) Leuders Limestone is known to fail by sliding on favorably oriented preexisting fractures. What are the implications of this observation for the the acceptance of the envelope defined by Equation 10-12 (i.e., does the equation apply to all cases)?

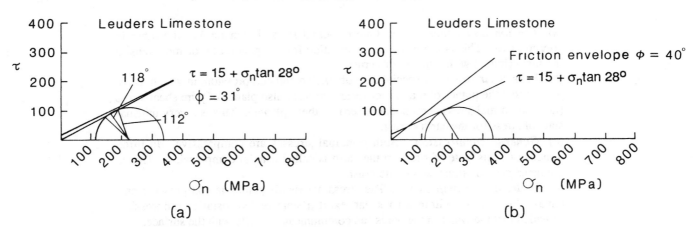

Figure 10-M1. Mohr diagrams for Leuders Limestone (for Exercise 3). (a) Diagram showing the Coulomb-Mohr failure envelope and the envelope for frictional sliding determined by Handin (1969). Circle is tangent to the Coulomb-Mohr envelope; (b) The Coulomb-Mohr failure envelope and the friction envelope defined by Equation 10-12.

11

DESCRIPTION OF MESOSCOPIC STRUCTURES

Gautam Mitra

Stephen Marshak

11-1 INTRODUCTION

The study of *mesoscopic* structures (visible at the scale of outcrops and hand samples) and *microscopic* structures (visible under the microscope) relies on a keen eye and a willingness to observe patiently and carefully. Study of such structures helps in strain analysis, provides information that can help in the interpretation of regional (*macroscopic*) structural relations, and can provide clues about deformation conditions and sense of movement during deformation. The determination of sense of movement of particles of rock during deformation is often called *kinematic analysis* (see Davis, 1984).

The purpose of this chapter is simply to outline terminology and procedures used for describing mesoscopic and microscopic structures and to draw your attention to features of structures that should be observed and measured during structural analysis. Each of the subjects addressed in this chapter could be the subject of a whole book, so we are forced to be brief and we present much of the description in figures or lists. Our intention is to provide a concise synopsis that can be taken to the field or lab for reference. We do not have the space to elaborate on how mesoscopic and microscopic structures can be used as interpretive tools, nor can we provide detailed discussion of the mechanisms of formation. Excellent descriptions and interpretations of mesoscopic and microscopic structures are provided in standard structural geology textbooks.

We hope that this chapter will be used in conjunction with field and laboratory analysis of natural structures. For

that reason, we do not provide exercises at the end of the chapter.

11-2 FOLDS

Folds are the most familiar manifestation of ductile deformation in rocks. They form under a variety of conditions in igneous, sedimentary, and metamorphic rocks. Folding can be a consequence of *primary deformation* or a consequence of *tectonic deformation* during *orogenesis*. Primary folds develop during the formation of the rock. Examples include slump folds in sedimentary rocks (Kuenen, 1953) and flow folds in lava flows. Tectonic folds develop in response to applied stress associated with plate movement and the formation of mountain belts (see King, 1977). Fold geometry is variable and reflects the rheology of the rock, the conditions of deformation, and the rate of deformation.

In order for a fold to be visible in a rock, the rock must possess layering; a fold simply cannot be seen in an isotropic rock, even if the mineral grains in the rock have moved with respect to one another during deformation. The layering that defines a fold can be bedding (in a sedimentary rock), flow banding (in an igneous rock), or metamorphic foliation. Folds can also be defined by the boundaries of a sheet intrusion or by streaks of impurities in an otherwise homogeneous rock.

A complete description of a fold should include a description of its *shape* and its *orientation*. The shape can be described in outcrop or in hand sample and may provide clues about the mechanism of fold formation irrespective of the fold orientation (e.g., Hudleston, 1973). The orientation, on the other hand, can be described only with respect to an external reference frame (i.e., a horizontal plane) and is, therefore, best described at the outcrop. In the context of fold description, the term "shape" refers to four parameters: (1) the form of a single surface, such as a bedding plane, within the fold, (2) the form of a sequence of folded layers, (3) the form of a series of folds involving a single surface, and (4) the shape of a single folded layer in profile.

Shape of a Folded Surface

Figure 11-1 provides an illustration of a pair of folds involving a single surface. A number of geometric elements, useful for describing the shape of a folded surface, are labeled on this figure (Fleuty, 1964; Fleuty, 1987; Ramsay, 1967; Ramsay and Huber, 1987). These are:

Hinge line: The line of maximum curvature (smallest radius of curvature) on a folded surface.

Hinge zone: The area on a folded surface adjacent to the hinge line where the surface has a relatively small radius of curvature.

Crest line: This is the line on the surface of the fold at which the dip changes direction with respect to the horizontal-surface reference frame. The dip directions on either side of this line point away from each other (e.g., the dip changes from southeast to northwest). If a fold is inclined such that dips change at a hinge but are in the same direction (Fig. 11-2), then the crest line or trough line cannot be defined.

Trough line: Same as a crest line except that the dip directions point toward one another.

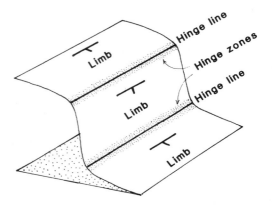

Figure 11-2. A fold in which beds of both limbs dip in the same direction. This fold has no crest or trough line.

Fold axis: Many natural folds have approximately cylindrical shapes or are made up of approximately cylindrical parts (Fig. 11-3). These cylinders are usually statistically defined using equal-area projections (Chapter 8). The cylindrical shape of a folded surface can be generated by moving a line parallel to itself in space. This imaginary line, which is parallel to the axis of the cylinder or cylinders approximating the fold, is called the *fold axis*.

Fold limb: This is the area of the folded surface between hinge zones where the surface has a large radius of curvature. Each convex fold shares a limb with an adjacent concave fold.

Inflection line: This is the line along the folded surface at which the surface changes from convex to concave. An inflection line, therefore, is the boundary between two adjoining folds, although it may not lie halfway between the two adjoining hinges.

Figure 11-4 shows additional geometric elements of a fold that involves a sequence of surfaces. Terms defined by this figure are:

Hinge surface: This is the surface containing the hinge lines of successive folded surfaces within a single

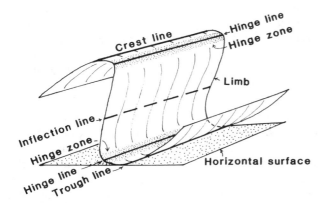

Figure 11-1. The principal geometric elements used to describe a single folded surface.

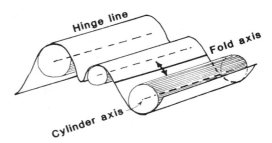

Figure 11-3. Concept of a cylindrically folded surface. The folds are generated by a line (*fold axis*) moving parallel to itself; this line is also parallel to the axes of the cylinders.

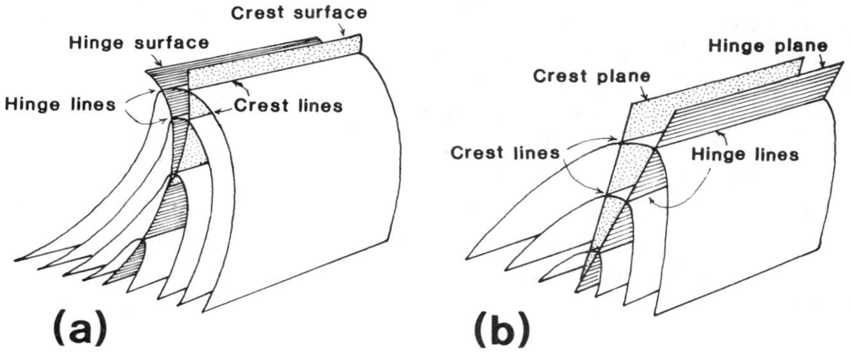

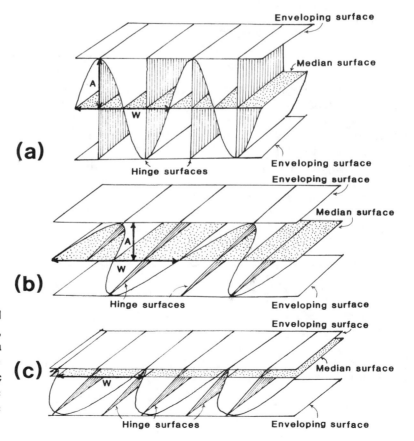

Figure 11-4. Geometric elements of a fold involving several folded surfaces. (a) Fold with curved hinge and crest surfaces; (b) fold with hinge plane and crest plane.

(a)

(b)

fold. The hinge surface sometimes is also called the *axial surface*. In general, the hinge surface is curved (Fig. 11-4a). If the surface is planar, it is referred to as a *hinge plane* or *axial plane* (Fig. 11-4b).

Crest surface: This is the surface containing the crest lines of successive folded surfaces within a single fold. Like the hinge surface, the crest surface can be curved or planar (*crest plane*).

Trough surface: Like the crest surface, this surface contains the trough lines of successive folded surfaces within a single fold.

Terms used to describe a series of folds involving a single surface (Fig. 11-5a-c) include:

Fold system: A group of folds that are spatially and genetically related is called a fold system. A fold system can consist of a single *fold train*, in which the linked folds are comparable in dimension, or it can consist of related *sets* of folds. In such a case, each *set* is composed of folds of comparable dimensions.

Median surface: The surface passing through inflection lines between successive folds in a fold train.

Enveloping surface: The two limiting surfaces between which a fold train oscillates are called enveloping surfaces. These surfaces can be defined by drawing the tangent planes to successive fold-hinge zones. The median surface for a fold train may or may not lie halfway between the two enveloping surfaces.

(a)

(b)

(c)

Figure 11-5. Terminology used to describe geometric elements, amplitude, and wavelength of a series of folds involving a single surface. (a) Periodic symmetric waves; (b) periodic asymmetric waves; (c) periodic asymmetric waves.

Closure direction: The direction in which a fold closes is used to classify it as one of three basic fold types (Fig. 11-6a-c). A fold that closes upward is called an *antiform*. A fold that closes downward is called a *synform*. A fold that closes sideways (left or right) is called a *neutral fold*. The terms anticline and syncline have more restricted usage and depend on the *facing* (see below) of the folds. An *anticline* has the oldest unit of the folded sequence in its core, while a *syncline* has the youngest unit in the folded sequence in its core.

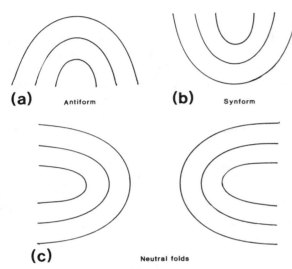

Figure 11-6. Classification of folds based on closure direction. (a) Antiform; (b) synform; (c) neutral folds.

Fold order: In a fold system, smaller folds commonly occur on the limbs of the larger folds. The largest folds in the system are called *first-order folds*, the next largest are called *second-order folds*, and so forth. The enveloping surfaces associated with lower-order folds are folded around the next-higher-order folds (Fig. 11-7). However, all the folds of a system have similar orientations for axial planes and share a common fold axis. Second-order folds formed in response to shear on the limbs of a first-order fold are sometimes called *parasitic folds*.

Description of Folded Surfaces in Profile

A profile of a cylindrical fold is the projection of a fold on a plane drawn perpendicular to the fold axis (Fig. 11-8; see Chapter 13). The terms used to describe the shape of a fold in profile are defined next.

Tightness: The tightness of a fold is a measure of the angle between the limbs of the fold (interlimb angle = α) as shown in Figure 11-9a. The angle α is measured between tangents to the folded surface drawn at the inflection points. The adjectives used to describe tightness are listed below (Table 11-1; after Fleuty, 1964) and are illustrated in Figure 11-9b.

Table 11-1
Terminology for Describing Fold Tightness

Interlimb angle	Adjective
$120°$-$180°$	Gentle
$70°$-$120°$	Open
$30°$-$70°$	Close
$10°$-$30°$	Tight
$0°$-$10°$	Isoclinal
$<0°$	Elasticas

Obviously, the transition between various tightness groups is gradational, so it may be difficult to place a natural fold precisely within a single group, especially if there are variations in geometry from bed to bed. In practice, geologists assign a fold to a certain group by visually estimating the angle between the limbs.

Hinge/limb shape: The hinge of a fold may be *rounded, angular,* or *very angular* (Fig. 11-10a-d). All these terms are self-descriptive. Folds with very angular hinges are called *chevron folds* (Ramsay, 1974) if they have limbs of equal length (Fig. 11-10c), and *kink folds* if they

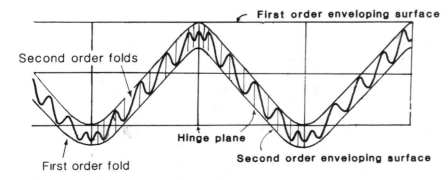

Figure 11-7. First- and second-order folds showing enveloping surfaces in different orientations.

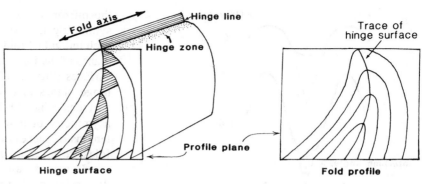

Figure 11-8. Cylindrical fold showing the orientation of the *profile plane* perpendicular to the fold axis. The trace of the fold on the profile plane is the *fold profile*.

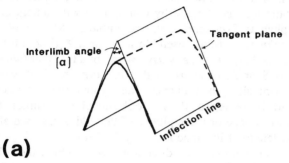

(a)

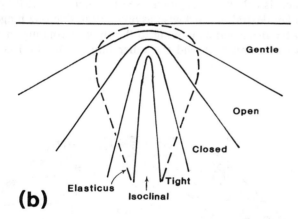

(b)

Figure 11-9. Interlimb angle. (a) Interlimb angle α measured between tangent planes to the folded surface drawn at the inflection lines; (b) classification of fold tightness based on interlimb angles.

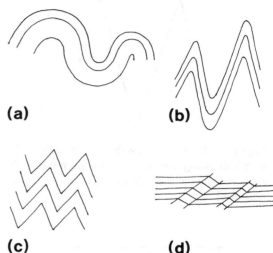

Figure 11-10. Fold profiles showing different hinge and limb geometries. (a) Rounded hinge with curved limbs; (b) angular hinge with planar limbs; (c) very angular hinges (chevron folds); (d) very angular hinges (kink folds).

The categories of folds based on hinge/limb ratio are listed in Table 11-2.

Fold symmetry: A single fold is *symmetric* if its hinge plane is a plane of symmetry and half of the fold is have limbs of unequal length (Fig. 11-10d). The limbs of a fold may be *planar* or *curved;* these terms are also self-explanatory.

Hinge/limb ratio: The relative lengths (in profile) of the hinge zone to the limbs affect the appearance of the fold (Fig. 11-11). Ramsay (1967) proposed the parameter P to describe this characteristic, where

$$P = \frac{\text{length of limb projection on the median-surface trace}}{\text{length of hinge-zone projection on the median-surface trace}}$$

Table 11-2
Description of Hinge/Limb Ratio

Name	Hinge description	P value
Chevron/kink	Very narrow hinge zone	20 - ∞
Angular	Narrow hinge zone	5 - 20
Subrounded	Broad hinge zone	1 - 5
Rounded	Very broad hinge zone	0* - 1

*indeterminate

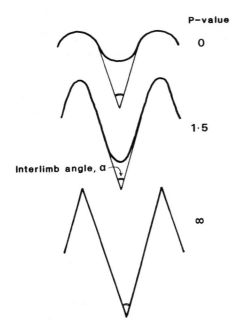

P-value

0

1·5

Interlimb angle, α

∞

Figure 11-11. Folds with the same interlimb angle showing variation of the parameter P (i.e., variation in the relative sizes of the hinge and limb).

the mirror image of the other half. A fold that is not symmetric is *asymmetric*. A fold train is symmetric if each individual fold in the train is symmetric and the hinge planes of the fold train are perpendicular to the enveloping surfaces (Fig. 11-12a). A fold train is asymmetric if the hinge planes of the folds are oblique to the enveloping surfaces (Fig. 11-12b). In general, the symmetry or asymmetry of a series of folds depends on the relative lengths of successive limbs; the two limbs of asymmetric folds are not equal in length.

If a fold pair is viewed down its plunge (see the discussion of down-plunge viewing in Chapter 13), its shape may be defined by comparison with letters of the alphabet: *Symmetric* folds resemble an M or a W, and *asymmetric* folds resemble an S or a Z (Fig. 11-12b). Remember, the *sense of asymmetry* of a fold pair (whether it is S or Z) depends on the direction in which we are looking along the fold axis. The asymmetry of a particular fold is reversed if we look up-plunge rather than down-plunge or if the plunge of the fold is reversed along the length of its hinge line (Fig. 11-12c).

Vergence: The direction toward which the fold is turned is called its *vergence* (see Chapter 16). Z-folds display dextral or clockwise asymmetry in that one hinge can be visualized as moving to the right or rotating in a clockwise sense around the other hinge (Fig. 11-13a).

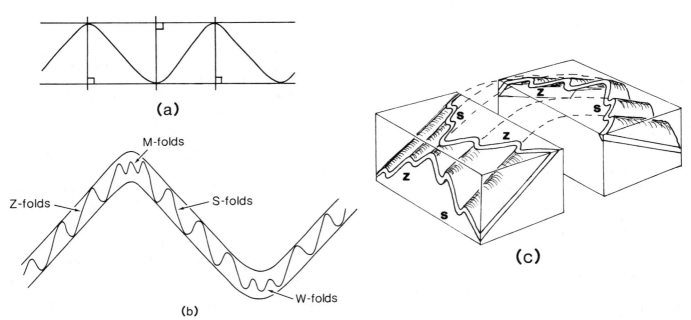

(a)

M-folds

Z-folds

S-folds

W-folds

(b)

(c)

Figure 11-12. Symmetric and asymmetric folds. (a) Profile of a symmetric fold train with hinge planes perpendicular to the enveloping surfaces; (b) profile of a fold train showing that second-order folds may be symmetric (M- or W-) or asymmetric (S- or Z) depending on their position within the larger folds; (c) block diagram illustrating that the asymmetry of a fold depends on the direction from which the fold is observed.

Z-folds are said to have clockwise *vergence*. Likewise, S-folds display sinistral, or counter-clockwise, asymmetry, and are said to have counter-clockwise vergence (Fig. 11-13a).

Generally, lower-order folds in a fold system show a systematic variation in asymmetry across the next-larger-order fold; Z-folds are found in one limb of a larger fold, and S-folds in the other limb, while M- or W-folds occur in the hinge zone (Figs. 11-7; 11-12b). This systematic variation in asymmetry is also referred to as vergence (Hobbs et al., 1976). Note that the minor folds in the two limbs of major folds verge toward the antiformal axial surfaces and away from the synformal axial surfaces (Roberts, 1982). This means that the axial traces of major folds can be located by studying the changes in vergence shown by the minor folds (see Chapter 16).

The vergence of an asymmetric fold can be used as an indication of the sense of transport associated with the development of the fold. Imagine two points, A and B, on an unfolded line (Fig. 11-13b). As an asymmetric fold

develops, point A (which will end up on the higher limb) is displaced with respect to point B (which will end up on the lower limb). In the example of Figure 11-13b, point A moves to the east relative to point B, so we say that this fold *verges* to the east (see Bell, 1981).

Facing: The facing of a fold is often confused with the vergence of a fold, but in fact it refers to quite a different aspect of the fold geometry (see Chapter 16). The facing of a fold refers to the direction in which strata get younger along the axial surface. Figure 11-14a-d shows (a) an east-facing anticline, (b) an east-facing syncline, (c) a west-facing antiformal syncline, and (d) a west-facing synformal anticline. A fold that has a vertical axial plane can be upward or downward facing. To describe the facing of a fold, you must know the relative ages of the units.

Dimensions: If a fold train shows regular, periodic repetition, and if the median surface lies halfway between the enveloping surfaces, we can describe the *dimensions* of the folds in terms of their *amplitude* and *wavelength*. The amplitude (A) is half the perpendicular distance between the enveloping surfaces (Fig. 11-5a,b). The wavelength (W) is the distance from a point on one fold to the equivalent point on the next fold in the train (Fig. 11-5a,b). If the median surface does not lie halfway between the two enveloping surfaces, we can no longer define a single amplitude, although we can still define a wavelength (Fig. 11-5c). If the folds do not show regular periodic repetition, it is not possible to define an amplitude or wavelength for the folds.

Thickness Variation of a Folded Layer

So far, we have described the geometry of folded surfaces. Real folds involve a sequence of layers (e.g., beds) each of which has a finite thickness. Whether or not the thickness

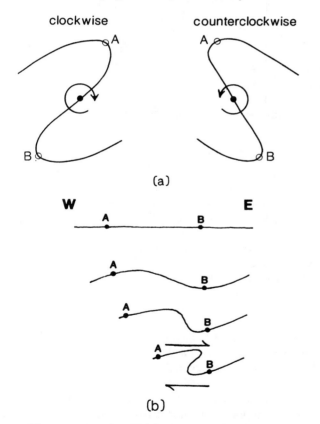

(a)

(b)

Figure 11-13. Fold vergence. (a) Clockwise and counterclockwise vergence of asymmetric folds defined on the basis of relative movement of one hinge with respect to the other; (b) relative movement of two points (A and B) during formation of an asymmetric fold. This fold verges toward the east.

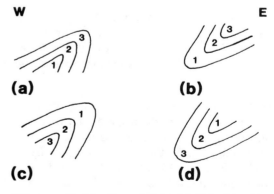

Figure 11-14. Description of fold facing. To describe the facing, the relative ages of the units must be known (1 is oldest, 3 is youngest). (a) an east-facing anticline; (b) an east-facing syncline; (c) a west-facing antiformal syncline; (d) a west-facing synformal anticline.

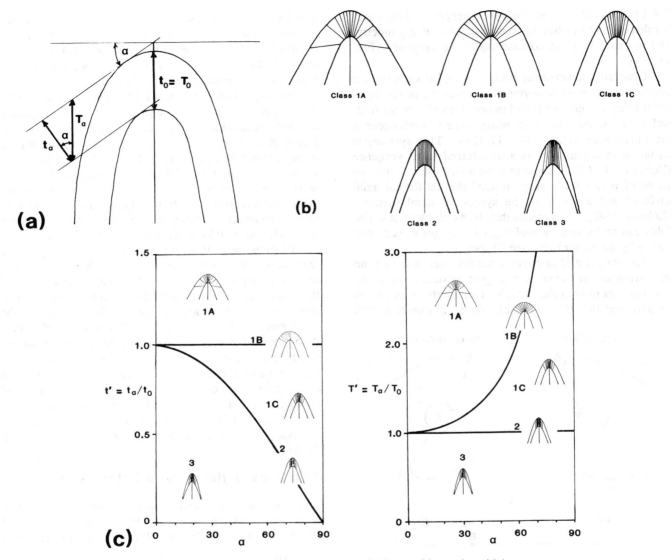

Figure 11-15. Isogonal thickness of folds. (a) Measuring thickness between tangents of equal dip; (b) the dip isogon classification of folds; (c) plots of t' and T' versus dip angle. These plots define fields for different fold classes. (Adapted from Ramsay, 1967.)

of a layer remains constant or changes during folding depends on the *mean ductility* of the rock, which is determined by the pressure, temperature, and strain rate during folding, and on the *contrast in ductility* between successive layers, which depends on the contrast in lithology between layers (Donath and Parker, 1964). Folds which display constant bedding thickness around the fold developed by the process of *flexural slip* (meaning that adjacent layers slid past one another as bending occurred). Flexural-slip folds formed under conditions such that there was a low mean ductility and a large ductility contrast between successive layers. In contrast, formation of folds

in which limb and hinge areas are not the same thickness involved plastic flow mechanisms, and form where rocks have relatively high mean ductility.

Because the thickness variation of a folded layer provides information on the mechanism of fold formation, it is an important characteristic to describe. Ramsay (1967) developed a morphological classification of folds based on the profile of a folded layer. The geometry of a folded layer can be described in terms of the following:

1. The relative curvature of the two bounding surfaces of the layer, which can most easily be described by

constructing *dip isogons*. As described in Chapter 13, a dip isogon is a line that connects points of equal dip on adjacent surfaces in the fold.

2. The distance between the two bounding surfaces of the layer. This distance can be specified in three ways: (a) The *orthogonal thickness* t is the perpendicular distance between two tangents of equal dip (Fig. 11-15a); (b) the *thickness parallel to the axial surface* T is the distance between tangents of equal dip measured parallel to the axial surface (Fig. 11-15a); and (c) the *isogonal thickness* is the thickness measured parallel to dip isogons (see Chapters 4 and 13). Note that if you were to inscribe lines parallel to the axial plane within the limbs of the fold, they might not connect points of equal dip.

If the fold profile is positioned so that the tangent at the hinge is horizontal, then $t = T$ at the hinge, and everywhere else on the fold $t = T(\cos \alpha)$, where α is the dip of the tangent to the folded surface.

Based on the pattern of isogons in the profile of a folded layer, Ramsay (1967) defined five classes of folds (Fig. 11-15b). Table 11-3 provides additional information about these fold classes. Statements about convergence or divergence of dip isogons are made with reference to the inner arc (concave side) of the fold (e.g., *convergent isogons* merge toward the inner arc of the fold).

If the normalized orthogonal thickness (t/t_0) or normalized thickness parallel to the axial surface (T/T_0) is plotted against the dip angle α (Fig. 11-15c), the different fold classes plot in different fields. Thus, by constructing dip isogons for any natural fold and plotting the results graphically, we can determine its fold class.

Procedure 11-1 (Construction of dip isogons)

To construct dip isogons for a particular fold, follow these simple steps (Fig. 11-16a,b; Ragan, 1985).

Step 1: Draw a profile section of the fold. This can be either a down-plunge projection from a map or a tracing from a hand sample cut perpendicular to the fold axis.

Step 2: Draw a series of lines tangent to two successive folded surfaces either with a drafting machine or with a protractor and a triangle as shown in Figure 11-16a. Using a convenient horizontal datum, place the protractor at a certain dip angle and draw successive tangents by sliding the triangle up and down the straight edge of the protractor.

Step 3: Repeat the process for other dip angles, usually at 10° intervals. A smaller dip interval may be needed in some situations if more detail is required.

Step 4: Connect points of equal dip on the surfaces with straight lines. These lines are the dip isogons (Fig. 11-16b).

Table 11-3
Classes of Folds

Fold Class	Characteristics
Class 1A	Strongly convergent dip isogons Curvature of the outer arc < curvature of the inner arc The smallest distance (t and T) between two surfaces is at the hinge
Class 1B	Moderately convergent dip isogons Curvature of outer arc < curvature of the inner arc Isogons are perpendicular to the outer and inner arcs t remains constant throughout the fold T is a minimum at the hinge These folds are often called *parallel folds* or *concentric folds*
Class 1C	Weakly convergent dip isogons Curvature of outer arc < curvature of inner arc t is a maximum at the hinge T is a minimum at the hinge
Class 2	Parallel dip isogons Curvature of outer arc = curvature of inner arc Isogons are parallel to the axial-surface trace t is a maximum at the hinge T remains constant around the fold Such folds are often called *similar folds*
Class 3	Divergent dip isogons Curvature of the outer arc > curvature of the inner arc Largest distance (t and T) between two surfaces is at the hinge

Orientation of Folds

All the fold features we have described so far apply to folds in hand samples as well as in outcrop. However, a fold in outcrop is in its true natural orientation and presents us with some additional geometric information that can be described.

To quantitatively describe the orientation of a fold we

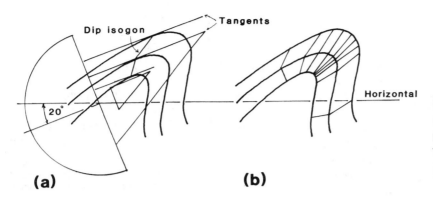

(a)   **(b)**

Figure 11-16. Method for determining points of equal dip on a fold profile. (a) Use of a protractor and triangle; (b) dip isogons for this fold profile. (Adapted from Ragan, 1985.)

must subdivide the fold into parts with straight hinge lines and planar axial surfaces. The orientation of a fold with a straight hinge line and a planar axial surface can be described in terms of the attitude of the hinge line and the attitude of the axial plane. Based on the orientation of these features, a fold can be classified into one of several groups, which are defined on Fig. 11-17. In many folds with inclined axial planes, and in reclined folds, both fold limbs dip in the same direction as the axial plane. This indicates that one of the limbs of the fold rotated through more than 90° from its prefolding subhorizontal position and is now *overturned*. Folds with one overturned limb are

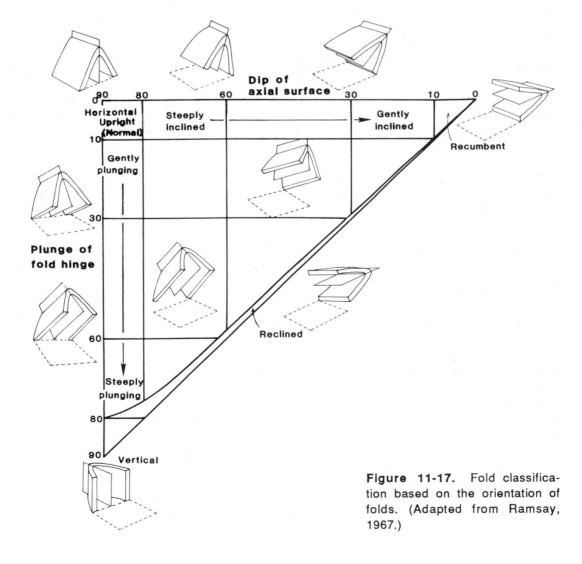

Figure 11-17. Fold classification based on the orientation of folds. (Adapted from Ramsay, 1967.)

called *overturned folds*. In areas of overturned folds it is useful to have primary structures to aid in determining *younging direction* or facing of beds.

The orientation of the *crestal line* and *crestal surface* (and *trough line* and *trough surface*) is usually different from the hinge line except in upright folds (Fig. 11-4). The hinge line of a fold is not always straight, and the plunge of the hinge line may change along its length. If the plunge of the hinge line changes direction along the length of the fold, the fold is *doubly plunging* (Fig. 11-18). If the hinge line (for an antiform, synform, or neutral fold) plunges away from a high point, it produces a *culmination;* similarly, if it plunges towards a low point it produces a *depression* (Fig. 11-18).

The axial surface of a fold may be curved. Thus, for example, a fold may change from an antiform to a neutral fold either along its length (i.e., along the length of its hinge line) or up- and down-section.

Procedure 11-2 (Measurement of the orientation of a fold hinge)

The orientation of a fold hinge cannot be measured if the fold is exposed on a smooth two-dimensional plane.

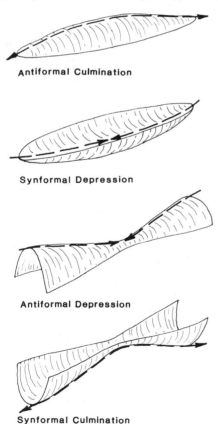

Antiformal Culmination

Synformal Depression

Antiformal Depression

Synformal Culmination

Figure 11-18. Terminology for description of doubly plunging folds.

Fortunately, there is usually local relief on the surface of an outcrop because of the difference in resistance to erosion of various layers. If such relief is available, emphasize the hinge by placing a pencil on it or, if that is not possible, by aligning the pencil with the hinge (Fig. 11-19). Then, remember that a fold hinge is a line, and apply the procedure for measuring lines described in Chapter 1 to measure the attitude of the pencil.

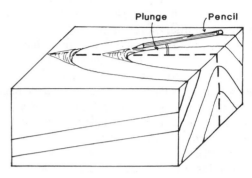

Figure 11-19. Illustration showing how to use a pencil to emphasize the orientation of a fold hinge.

Procedure 11-3 (Measurement of the orientation of a fold axis)

The fold axis is calculated, using the equal-angle or equal-area nets as described in Chapter 8, from attitude measurements at many locations on the fold surface. Collection of data for such a calculation from a mesoscopic fold may be difficult if the fold is very small; it may help to use a compass plate (Chapter 1).

Procedure 11-4 (Measurement of the attitude of a hinge plane)

Remember that the hinge plane is defined by two lines. One of these is a hinge line, and the other is the line composed of the points defined by the intersection of hinge lines on successive layers with the outcrop face. Locate these two lines (you may wish to emphasize them with a pencil); then align your compass plate so that it contains both lines, and measure the attitude of the compass plate (Chapter 1).

Representation of Folds on a Map

Map-scale folds are defined on a map by the outcrop pattern of folded layers (Appendix 1). If both limbs of the fold are present, a traverse across the map will cross the same stratigraphy twice. If the hinge zone of a plunging fold is present in the map area, the map will display a *fold closure* where the outcrop belts of the two limbs join and the outcrop pattern has the maximum curvature. The *fold trace* is the line on a map connecting closures in the outcrop pattern of successive layers or units. In general, the fold trace does *not* coincide exactly with either the *hinge-surface trace* or the *crest-surface trace*. If the fold has a vertical hinge plane, the hinge-plane trace will coincide with the fold trace only in areas of negligible relief (Ragan, 1985). For folds with inclined hinge planes, the real hinge-plane trace is always different from the fold trace (see Chapter 13). The degree of discrepancy between the two depends on the geometry of the fold (e.g., whether the fold is concentric or similar) and on the topography (Ragan, 1985). Thus, the fold trace on a map has no real significance in terms of the fold geometry but is simply a convenient way of representing a fold on a map.

Symbols are usually placed on the fold trace to convey additional information about the fold. Arrows pointing away from the trace represent an antiform, and arrows pointing toward the trace represent a synform (Fig. 11-20a). Similarly, symbols can also be used to represent overturned folds (Fig. 11-20a). If the fold is plunging, an arrow along the fold trace is used to represent direction and amount of plunge (Fig. 11-20b). A strike-and-dip symbol placed on the fold trace can be used to specify the strike and dip of the hinge surface.

Folds that are too small to be represented on the map by their outcrop patterns are usually represented simply by an arrow indicating direction and amount of plunge. It is useful to draw at the tail of the arrow a profile sketch of the fold as viewed looking down-plunge (Fig. 11-20c). Asymmetric (S- and Z-) minor folds are often represented in this way.

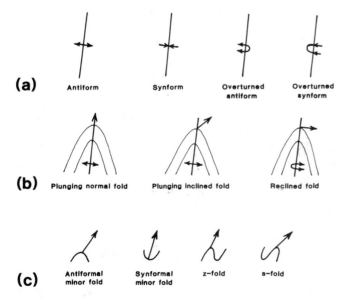

(a)   Antiform   Synform   Overturned antiform   Overturned synform

(b)   Plunging normal fold   Plunging inclined fold   Reclined fold

(c)   Antiformal minor fold   Synformal minor fold   z–fold   s–fold

Figure 11-20. Symbols used to represent folds on a map. (a) Overall shape of major folds; (b) plunge of major folds; (c) shape and orientation of minor folds.

Special Fold Types

There are a number of specific fold types that are distinctive enough to warrant special names. These are described below:

Monocline: This is a fold in which a layer changes dip for an interval then returns to its original dip. A monocline, therefore, has two hinges (Fig. 11-21a). In many localities large monoclines occur over high-angle reverse faults. In such settings the monoclines are commonly assumed to have developed as *drape folds* (Stearns, 1978). However, monoclines can also form by buckling or kink-folding at a site of a preexisting flaw or mechanical disturbance (Reches and Johnson, 1978; Davis, 1978; 1984).

Ptygmatic folds: In metamorphic rocks it is not uncommon to find veins that have been folded into an intestinelike squiggle containing elasticus-shaped folds (e.g., Mitra and Datta, 1978; Klein, 1981). The folds typically involve a single layer and have concentric geometry. Such folds are called ptygmatic folds (Fig. 11-21b).

Sheath folds: Sheath folds generally occur in zones of high shearing (Cobbold and Quinquis, 1980; Malavieille, 1987). A sheath fold has this name because the hinge of the fold is itself folded in the axial plane of the fold, so that the folded surface has the form of a sheath, such as would fit around a sword (Fig. 11-21c).

Kink-domain folds: These are folds that do not have smooth curving hinge zones. Rather, the fold is composed of straight-limb segments separated from one another by sharp hinges (see Faill, 1973). A map of a large kink-domain fold would contain bands in which the beds share the same dip; these bands are called *dip domains* (see Chapters 13 and 14).

Kink bands: A *kink band* is a tabular zone in which an earlier foliation has been deflected into a new orientation (e.g., Dennis, 1987). The boundaries of the zone are kink-fold hinge planes (Fig. 11-10d). Kink bands only form in rock that contains a pre-existing well developed foliation (bedding, cleavage, or schistosity) and has uniform layer thickness (see Dewey, 1965; Anderson, 1964; Weiss, 1980). Typically, kink bands die out along their length in profile either by tapering to a point as the amplitude of the kink diminishes (Fig. 11-22a), or by narrowing into a thin shear zone. The geometry of a kink band leads to a relative displacement of the foliation across the kink band. Kink bands often occur in conjugate pairs (Fig. 11-22b) resulting in *conjugate folds*. The intersection of the kink band with the foliation being kinked determines the orientation of the kink-fold hinge or *kink axis* (Fig. 11-22c). Conjugate kink bands intersect along *kink-intersection axes*, which lie on the *kink-intersection surface* (Fig. 11-22c; see Chapter 13).

Figure 11-21. Examples of special fold types: (a) Monocline (adapted from Huntoon, 1974); (b) outcrop photograph of ptygmatic folds; (c) progressive steps in the development of a sheath fold. Inset shows cross sections of "a-type" and "b-type" sheath folds (adapted from Malavieille, 1987).

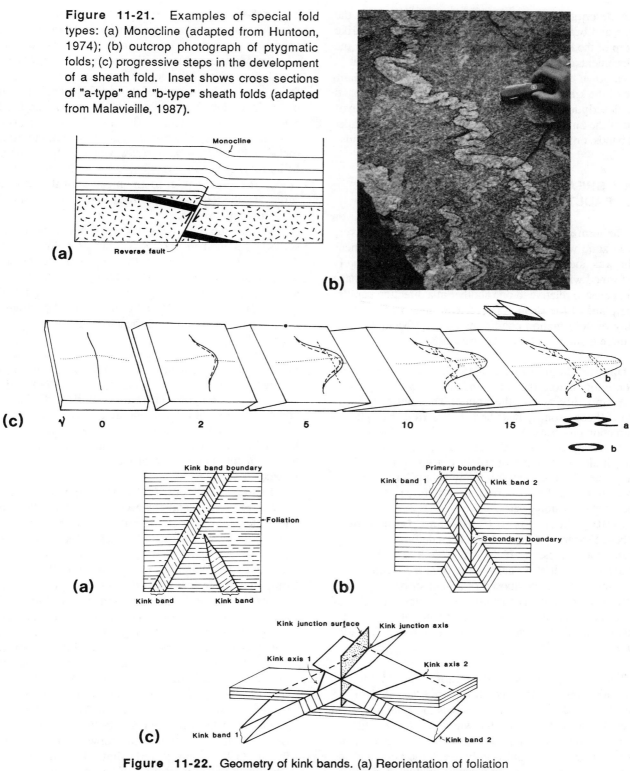

Figure 11-22. Geometry of kink bands. (a) Reorientation of foliation across a parallel-sided kinkband resulting in relative displacement of the foliation. Kink bands die out by tapering to a point (adapted from Anderson, 1964); (b) intersection of conjugate kink bands (adapted from Weiss, 1980); (c) conjugate kink bands showing orientation of kink axes and kink-intersection axes.

A description of a kink band should include: (a) the plunge and bearing of the kink fold hinges; (b) the strike and dip of the kink band; (c) a specification of whether the displacement across the kink band is dextral or sinistral as viewed down the plunge of the kink axis; (d) the interlimb angle of the kink folds; (e) the dimensions of the kink band and a description of its termination; and (f) an indication of whether the kink band is isolated, is part of a set of parallel kink bands, or is part of a conjugate system of kink bands.

11-3   SHEAR ZONES, FAULTS, AND FAULT ZONES

The term *shear zone* is a general term for a relatively narrow zone with subparallel boundaries in which shear strain was localized (see Chapter 15). The relatively undeformed wall rocks on opposite sides of the zone have been displaced relative to one another in a direction parallel to the plane of the shear zone. Shear zones form under a variety of deformation conditions and can be subdivided into three main types (after Ramsay, 1980):

Ductile shear zones: In these zones there is no discontinuity across the zone, and shear strain magnitude varies smoothly across the zone. The fabric of rocks within these zones has been modified by plastic deformation processes.

Brittle-ductile shear zones: There is a discontinuity within the ductilely deformed rock of the shear zone. This discontinuity may be a discrete fracture on which sliding has occurred, or it may be an array of en-echelon extension gashes.

Brittle shear zones (faults and fault zones): In a brittle shear zone the rock has been deformed by brittle deformation processes. If the "zone" is a discrete planar fracture on which slip occurred, it is called a *fault*. If a brittle shear zone is composed of a number of subparallel anastomosing faults separating lens-shaped blocks of undeformed rock, or if it is a tabular band of finite width containing brittlely shattered or pulverized rock, it is called a *fault zone*.

The description of a shear zone should include information on (a) the orientation (strike and dip) of the zone, (b) the relative movement across the zone (direction and amount of net slip), (c) the width of the zone, (d) the style of deformation (brittle or ductile) within the zone, and (e) the nature of the transition between the zone and the wall rocks (is the boundary of the zone gradual or abrupt).

Shear-Zone and Fault-Zone Rocks

The fabric of rocks deformed in shear zones is quite distinctive. Sibson (1977) suggested that shear-zone rocks can be classified based on the proportion of *matrix* (the relatively fine-grained groundmass) formed in the rock as a consequence of shear, on the grain size of the matrix, on whether or not the rock is *cohesive* (a cohesive rock holds together), and on whether or not the rock developed a foliation as a consequence of shear deformation. The three main categories of shear-zone rocks are: (a) the *breccia series*, which includes incohesive nonfoliated rocks, (b) the *cataclasite series*, which includes cohesive random-fabric rocks, and (c) the *mylonite series*, which includes cohesive foliated rocks. In practice, geologists use the term *cataclastic rock* to refer to any member of the cataclasite series and the term *mylonitic rock* to refer to any member of the mylonite series.

Various subcategories (rock types) within each of these series are named in Figure 11-23a (after Sibson, 1977). This chart also indicates the approximate pressure-temperature conditions (metamorphic grade) in which each series forms, assuming a granitic protolith (after Hull et al., 1986). Hand-sample and thin-section photographs of some typical fault-zone and shear-zone rocks are shown in Figure 11-24.

The terms *fault breccia* and *fault gouge* are used for incohesive rock formed by fracturing and crushing in a non-metamorphic (brittle) shear zone. Rocks of the *cataclasite series* form in brittle and brittle-ductile shear zones; they are not foliated but are cohesive. The specific name given to a rock in the cataclasite series (e.g., crush breccia, protocataclasite etc.) depends on the grain size of the fragments and on the proportion of the rock that is matrix.

Brittle shear zone rocks develop when fractures initiate, propagate, and coalesce in the rock (Blenkinsop and Rutter, 1986). Frictional sliding on fractures may give rise to *slickenside* surfaces. (A slickenside is the polished surface of a fracture.) Breccia forms when blocks are surrounded by fractures and separate from the wall rock. Continued shear leads to crushing and grinding and a progressive decrease in grain size of the blocks (Fig. 11-24a). If rock is pulverized under low pressure, gouge ultimately forms, but under high pressure the fine grains are interlocked into cohesive cataclasite (Fig. 11-24b,c).

If sufficient heat is generated during a shear event, some of the rock in the shear zone melts, and the molten material injects into fractures and quickly cools. The resulting glass is called *pseudotachylite*. This rock resembles basaltic volcanic glass (tachylite) and may occur along a fault plane or in vein networks near the fault. Sibson (1975) suggested that the formation of pseudotachylite is indicative of seismic movement on a fault.

Rocks of the mylonite series form by ductile deformation mechanisms (see Bell and Etheridge, 1973; Hobbs et al., 1976; White et al., 1980; Suppe, 1985; Poirier, 1985) and therefore characteristically form in shear

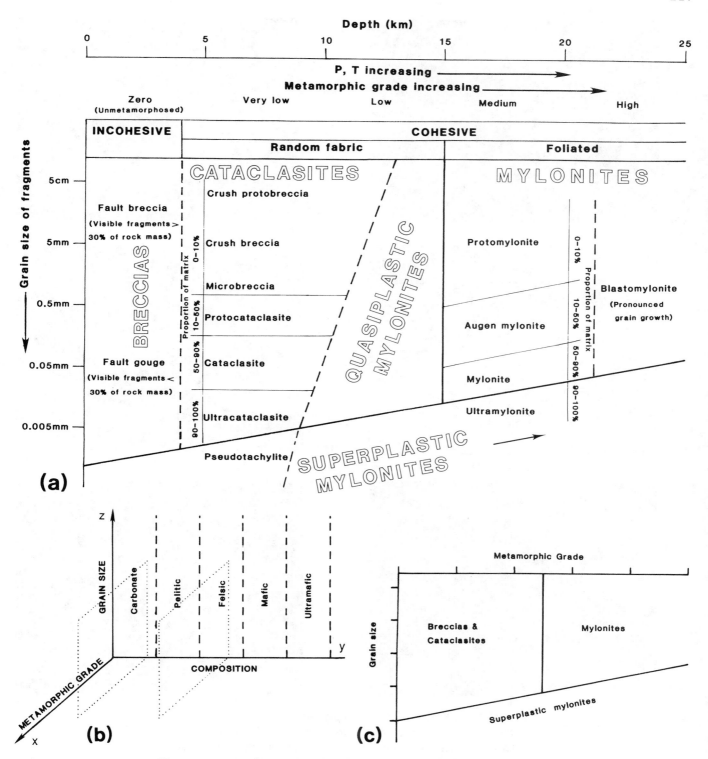

Figure 11-23. Fault-rock terminology. (a) Classification of fault rocks that have been derived from quartzo-feldspathic lithologies (e.g., granite) (adapted from Sibson, 1977); (b) the grain size - metamorphic grade - lithologic composition grid used for classifying fault rocks (after Hull et al., 1986); (c) fault rock diagram for marl showing expanded mylonite and superplastic mylonite fields as compared the those shown on the diagram for granite in part a.

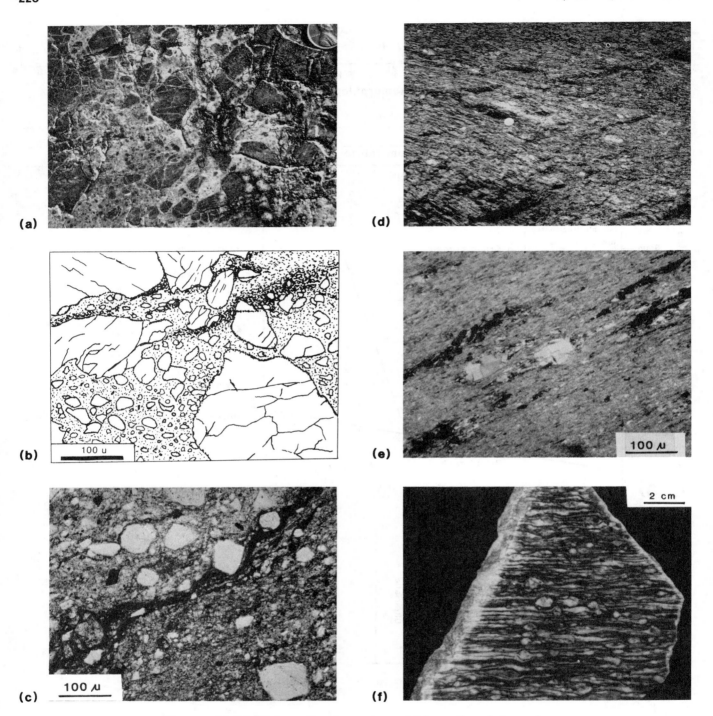

Figure 11-24. Outcrop, hand-sample, and thin-section photographs of some typical fault and shear-zone rocks. (a) Fault breccia from Appalachian thrust, Tennessee; (b) thin section sketch of a cataclasite from the Wind River Mountains, Wyoming; (c) thin section of foliated cataclasite from the White Rock thrust fault, Wind River Mountains, Wyoming; (d) quasiplastic mylonite from the Blue Ridge region, Appalachian Mountains of Virginia; (e) thin section of quasiplastic mylonite. Notice cracking of feldspar porphyroclasts; (f) hand sample of mylonite from the Carthage-Cotton mylonite zone in the Adirondack Mountains of New York (from Lumino, 1987).

zones that were active under medium to high metamorphic grades. The sequence, protomylonite ⇒ augen mylonite ⇒ mylonite ⇒ ultramylonite, represents progressive grain diminution by dynamic recrystallization of original grains.

In a *protomylonite*, most of the rock displays original grain size, though grains are flattened and stretched and display undulose extinction. As ductile shear continues, a once coarse-grained rock evolves into a matrix of extremely small grains. The fine-grained matrix is strongly foliated and lineated; foliation is defined by flow segregation of the matrix into bands, by alignment of fine mica, and by preferred orientation of flattened grains of other minerals. In an *augen mylonite*, the bands of matrix wrap around lenticular coarser grains called *augen* (German for eyes). Some augen are *porphyroclasts*, in that they are composed of relict crystals or clusters of crystals of the parent rock. Porphyroclasts are commonly bordered by pressure-shadow tails formed by solution-reprecipitation, or by tails (often asymmetric) of tiny grains formed by recrystallization. The tails, along with stretched grains and streaks of fine mica, define the characteristic lineation of mylonite-series rocks. A *mylonite* (in the strict sense) has 50-90% matrix; the remainder of the rock is composed of porphyroclasts and ribbons of highly strained relict minerals. In an *ultramylonite*, porphyroclasts are largely absent and the entire rock is composed of extremely fine matrix; ultramylonites are aphanitic. Mylonites or ultramylonites that are composed largely of fine-grained mica are commonly called *phyllonites*. Under high-grade metamorphic conditions, crystals (*porphyroblasts*) grow in the mylonite (e.g., Wintsch and Knipe, 1983) and strained grains are annealed. The resulting rock is called a *blastomylonite*. Eye-shaped porphyroblasts can also be called augen.

The chart of Figure 11-23a shows a field for *quasiplastic mylonites*. A quasiplastic mylonite (Fig. 11-24d) is one that exhibits both ductile and brittle deformation features. In some quasiplastic mylonites of granitic composition, quartz deforms ductilely, while feldspar deforms by fracture and fragmentation (Mitra, 1978; Fig. 11-24e).

The usage of the term *superplastic mylonite* is somewhat controversial. In a broad sense, *superplasticity* simply refers to extremely ductile deformation during which very large strains develop without failure (Schmid, 1983). Thus, any shear-zone rock in which there has been extreme strain without loss of cohesion can be called a superplastic mylonite. Such large ductile strains are typically observed in very fine grained shear-zone rock, irrespective of whether the initial reduction in grain size took place by brittle or ductile processes (Wojtal and Mitra, 1986; Mitra, 1984; Gilotti and Kumpulainen, 1986; Schmid, 1975; Bouillier and Gueguen, 1975). In Figure 11-23a, the term superplastic mylonite is used in the broad sense; the field of superplastic mylonites is shown as

including the extremely fine-grained members of the quasiplastic mylonite series and the mylonite series.

Under medium to high-grade metamorphic conditions, superplastic behavior in rock is thought to be a manifestation of deformation by grain-boundary sliding (Schmid et al., 1977; Schmid, 1983). *Grain-boundary sliding* is a deformation mechanism that involves diffusion and dislocation motion concentrated along grain boundaries, such that grains can switch neighbors without the overall rock losing cohesion. Bouiller and Gueguen (1975) suggested that evidence of grain-boundary sliding in a mylonite is the occurrence of a fine-grained matrix without a crystallographic preferred orientation. If water is present, a form of grain-boundary sliding called particulate flow may occur even under relatively low metamorphic conditions (Borradaile, 1981).

As noted earlier, the boundaries between fields on Figure 11-23a are related to metamorphic grade. Figure 11-23a is drawn for felsic (granitic) rocks and implies that breccia-series rocks form at depths of < 4 km, cataclasite-series rocks form at depths up to 15 km, and mylonite-series rocks form at depths below the brittle/ductile transition (i.e., > 15 km). The depth ranges just listed are very approximate and depend on rock composition, geothermal gradient, and strain rate.

Figure 11-23b schematically illustrates that the boundary between the catalclasite and mylonite fields depends on the original lithology as well. For example, ultramafic rocks stay brittle at relatively high temperatures, whereas carbonate rocks become ductile at relatively low temperatures. Figure 11-23a is a slice parallel to the X-Z plane through the 3-D space of Figure 11-23b. Figure 11-23c shows the X-Z slice for a carbonate rock. Note that the boundary between the cataclasite and mylonite fields is further to the left (i.e., at lower metamorphic grade) for carbonate rocks than it is for granitic rocks.

Sibson (1977) schematically illustrated the relationship between metamorphic conditions (i.e., depth of deformation) and the type of shear-zone rock that forms (Fig. 11-25). Zones formed at shallower levels within the earth's crust are brittle and yield breccia or cataclasite series rocks, whereas zones formed at deeper levels, hence at higher pressures and temperatures, are ductile and yield mylonitic series rocks.

Kinematic Indicators in Fault Zones

A correct interpretation of shear sense on a shear zone may provide critical insight into the tectonic significance of the shear zone or fault. Obviously, the most direct way of determining shear sense is to look for offset markers. As we have seen in Chapters 6, in order to completely specify the net slip on a fault it is necessary to have two nonparallel offset markers or one offset marker and an indication of the direction of slip. In the absence of offset

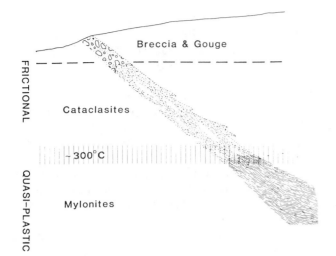

Figure 11-25. Characteristic fault rocks of a shear zone as a function of depth. (After Sibson, 1977.)

(a)

markers, certain structures can assist us in determining shear sense on a fault. These structures are called *kinematic indicators* because they indicate the sense of movement.

Slickenlines or Slip Lineations: Often a slickenside surface is decorated by lineations parallel to the direction of movement on the surface. These lineations are variously referred to as *slickenside lineations*, *slickenlines* (Fleuty, 1975), or *slip lineations*. There are at least three types of slip lineations:

(1) Groove lineations: These are lineations formed by the scratching of one surface against another during movement on the fault or sliding surface (Fig. 11-26a,b). The opposing surfaces of the fault are not perfectly smooth; microscopic *asperities* contact the opposing wall and gouge the opposing wall as the fault moves (Fig. 11-26c). Groove lineations are often ambiguous as indicators of the sense of shear, so their presence can be used only to determine direction of shear.

Figure 11-26. Outcrop, hand-sample, and thin-section photographs of some typical slickenside features. (a) Groove-type slickenlines on slickenside surface from the Appalachian Valley and Ridge province; (b) scanning-electron photomicrograph of a grooved slickenside surface in the Flathead Sandstone, Wind River Mountains, Wyoming (from Mitra & Frost, 1981); (c) scanning-electron photomicrograph showing the cross section of a slickenside surface in granitic rock, Wind River Mountains, Wyoming. The upper wall of the fracture has an asperity (the protrusion indicated by the arrow) that is gouging the lower wall (from Mitra & Frost, 1981).

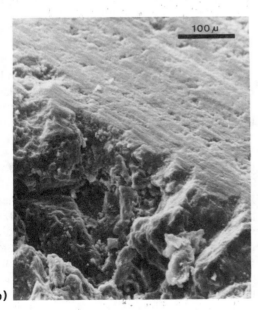

(b)

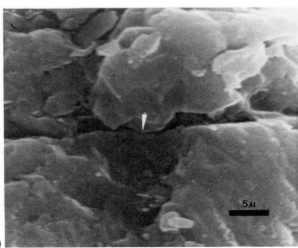

(c)

(2) Fiber lineations: These are lineations that are a consequence of fibrous mineral growth on the surface of the fault. Their presence indicates that as the fault moved, it opened slightly and became the locus of vein-mineral precipitation. The precipitation occurred in increments (see the discussion of *crack-seal deformation* presented later in this chapter) and as a consequence the mineral composing the vein grew in the form of long, thin fibers. When the fault surface is exposed, these fibers appear in imbricate sheets (Fig. 11-27a,b). The long axis of each fiber is parallel to the direction of extension (i.e., movement direction on the fault), and the sense of imbrication of the fibrous sheets gives the sense of movement on the fault (Fig. 11-27c).

Typically, fault-plane veins have a complicated history and may, in cross section, be seen to consist of sheets of blocky spar separated by thin screens of clay residue or wall rock; in such examples, the fibers are visible only on the surface of the vein (Fig. 11-27b). If the fault has had a

complicated movement history, two or three sets of fibers with different orientations may be visible on the fault surface.

(3) Nesting grooves and ridges: During creation of a fault plane and initial sliding on the plane, lineations appear on the fault that are sometimes longer than the displacement on the fault (Means, 1987). Shallow U-shaped grooves on one side of the fault nest with ridges of similar length protruding from the other side. The exact mechanism of formation of these lineations is not clear, although they are known to form under conditions of brittle deformation. They may represent lateral steps in the fault surface. It appears that nesting grooves and ridges can be used to indicate the direction but not the sense of slip.

Steps or Bends on Fault Planes: Fault surfaces are generally not perfectly planar. Rather, the surfaces are locally offset at bends or steps. Because a bend or step in a fault plane is not parallel to the shear direction, compressional or extensional structures may develop at the

(a)

(b)

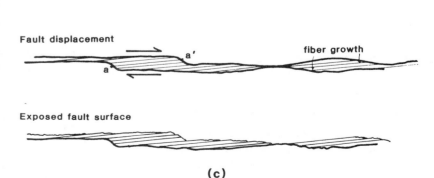

Figure 11-27. Fibrous slip lineations. (a) Partially eroded fault vein of white calcite on dark grey muddy limestone. Fibers are parallel to transport direction and are only on the surface of the vein. The vein interior is composed of calcite spar. Note compass for scale; (b) Imbricate sheets of fibers on the footwall of a fault in the Hudson Valley of New York. The hanging wall (eroded away) moved up with respect to the footwall; (c) formation of imbricate fibrous minerals on a fault surface (adapted from Durney and Ramsay, 1973).

(c)

bend. The shape of the step with respect to the shear sense on the fault determines whether there is compression or extension at the step. To visualize the relationship between step shape and the type of structure that develops at the step, imagine a horizontal fault (Fig. 11-28). *Restraining steps* face opposite to the direction of shear of the hanging wall with respect to the footwall, and *releasing steps* face in the direction of shear.

On mesoscopic faults that have developed under conditions amenable to fiber lineation formation, restraining steps are the site of solution pitting (Fig. 11-28). If the step is perpendicular to the fault plane, the step evolves into a small stylolite, and if it is oblique to the fault plane, it is sometimes called a *slickolite* (a hybrid of the words "slickenline" and "stylolite"); the long axis of a pit on a slickolite is oblique to the plane of the slickolite and is parallel to the fibrous lineations on the fault surface (see Price, 1967; Arthaud and Mattauer, 1969). Releasing steps are the site of vein formation (Fig. 11-28). Recognition of the above structures on a fault are a direct indication of the sense of shear across the fault plane.

Restraining and releasing bends on regional strike-slip faults lead to the development of large structures (Crowell, 1974). Restraining bends are the sites of thrusting and folding, and releasing bends are the sites of normal faulting and pull-apart basin formation.

Hansen Slip-Line Method: If an interval of well-layered rock is deformed in a shear zone, it is likely that a large number of mesoscopic asymmetric folds develop as a consequence of the shear couple. The axial planes of these folds are inclined at a low angle to the boundaries of the shear zone, and the hinges of the folds lie in the plane of the shear zone or at a low angle to it. Because of local variations in the magnitude of shear in the zone, the hinges of the folds are *not* all parallel to one another and are *not* all oriented perpendicular to the slip direction. Rather, the angle between the hinges and the slip direction is quite variable. If the slip direction has been uniform, however, there is a consistent relation between the fold vergence and the orientation of the hinge, and thus it is possible to use mesoscopic folds in a shear

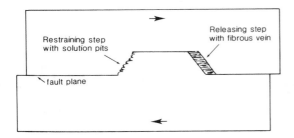

Figure 11-28. Cross section of a jagged fault surface showing the restraining and releasing steps. (Adapted from Marshak et al., 1982.)

zone to determine the shear sense on the fault (Hansen, 1971).

Imagine a thrust fault that dips 30° to the east; the hanging wall moved up to the west (Fig. 11-29a). Small asymmetric folds develop in the zone of shearing. Folds whose hinges plunge in a northerly direction have counterclockwise vergence, whereas folds that plunge in a southerly direction have a clockwise vergence (Fig. 11-29a). This difference reflects the fact that fold vergence must be described as viewed down-plunge. On a stereographic projection, the fold hinges lie on a great circle representing the plane of the fault, with the dextral and sinistral folds in different fields. The gap between the two fields is the separation arc, and the slip direction lies within this arc.

To help visualize why there is a difference in vergence, place a pencil between the palms of your hands so that it is inclined to your fingers (Fig. 11-29b). Shear your hands past one another in the direction parallel to your fingers and watch how the pencil rotates. Repeat the exercise with the pencil plunging in the opposite direction.

Procedure 11-6 (Hansen slip-line method)

Step 1: In the field, measure a number of mesoscopic folds with unambiguous vergence; only clearly defined S- and Z-folds fit the bill (W- and M-folds cannot be used). About 30 folds are needed for a reliable answer. For each fold, measure the bearing and plunge of the fold and the sense of vergence (clockwise or counterclockwise) as viewed down-plunge.

Step 2: On an equal-angle or equal-area net, plot the points representing the fold hinges. Use arrows to indicate the sense of rotation of the clockwise folds and the counterclockwise folds.

Step 3: If the data are appropriate for the method, the fold hinges define a great circle representing the plane of the shear zone. The clockwise folds fall in a different field than the counterclockwise folds. The gap between the two fields is called the *separation arc*.

Step 4: The slip direction lies within the separation arc. To determine the sense of slip (either up-dip or down-dip) you must think through the sense of vergence. Imagine a fault that dips 30° due south. Figure 11-29c shows two possible patterns of fold data; if the data appeared as shown in the stereoplot on the left, then the fault is a thrust, whereas if the data appeared as shown in the stereoplot on the right, then the fault has normal displacement.

Mesoscopic and Microscopic Kinematic Indicators in Ductile Shear Zones

Ductile shear zones and mylonite zones often do not contain offset marker layers for the determination of relative movement sense. However, a number of

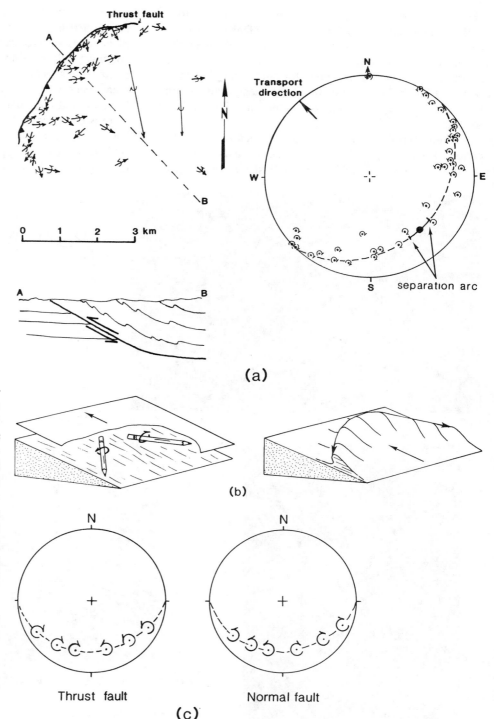

Figure 11-29. The Hansen slip-line method. (a) Map and cross section of a thrust sheet showing the development of asymmetric folds. The stereogram shows the populations of dextral and sinistral folds with the transport direction lying within the separation arc; (b) illustration to help visualize why different folds along the same fault plane can have different vergence; (c) illustration that the sense of slip can be determined by examining the pattern of folds along a great circle. The pattern in the left stereoplot is characteristic of a thrust fault, and the pattern of the right stereoplot is characteristic of a normal fault. Both plots show the plunge and bearing of mesoscopic folds along a fault surface that dips 30° to the south.

small-scale structures (at the hand-sample and thin-section scale) in mylonites can be used to determine relative sense of movement. To obtain the maximum amount of information about sense of shear, samples should be viewed on a plane perpendicular to the foliation and parallel to the stretching or mineral elongation lineation. Here, we describe briefly the main criteria used to determine shear sense in mylonitic rocks (adapted from Simpson, 1986);

more detailed descriptions are given by Simpson and Schmid (1983), Lister and Snoke (1984), Simpson (1986), and Ramsay and Huber (1987).

Sigmoidal foliation: In a relatively narrow shear zone it is possible to observe the variation in orientation of the new foliation (within the zone) with respect to the zone boundaries. The foliation generally shows a smooth change in orientation across the zone, giving rise to a typical sigmoidal pattern (Fig. 11-30a). This pattern can be used to determine the relative sense of movement on the zone, as indicated in the figure. In addition, the angular relationship between the new foliation and the zone boundary can be used to determine the shear strain within the zone and the displacement across the zone (Fig. 11-30b; see Chapter 15).

Shear-band geometry: Some shear-zone rocks contain small, subparallel, evenly spaced (at 1- to 10-cm intervals) shear zones. These small shear zones form

within a larger host shear zone and deflect or cut schistose foliation (Fig. 11-30c,d). The relative age of the small shear zones with respect to the foliation that they deflect is not always clear. Simpson (1986) suggested that the term *shear band* should be used as a general name for the small shear zones when the relative ages of the deflected schistose foliation and the small shear zones are uncertain. When there is evidence that the schistose foliation and the shear bands formed at the same time, the term *S-plane* (S for schistosity) can be applied to the schistose foliation and *C-plane* (C for "cisaillement," the French word for shear)

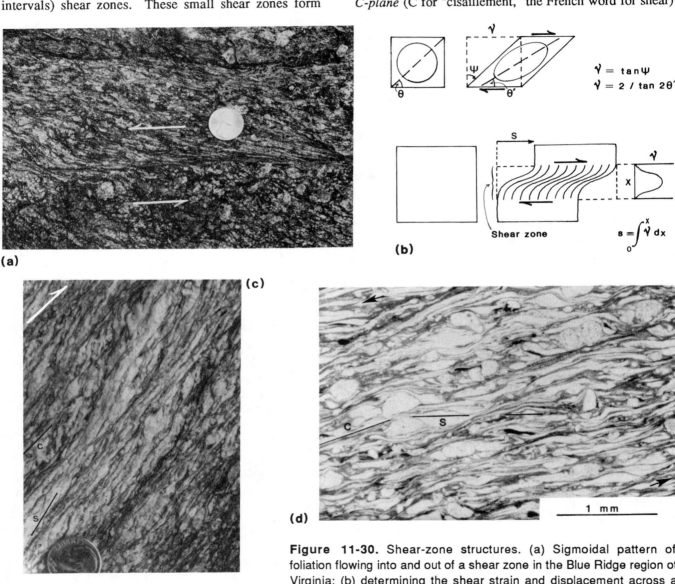

Figure 11-30. Shear-zone structures. (a) Sigmoidal pattern of foliation flowing into and out of a shear zone in the Blue Ridge region of Virginia; (b) determining the shear strain and displacement across a shear zone using the change in orientation of foliation through the zone; (c) small-scale structure within a shear zone showing the development of s-planes (schistosity) and c-planes (shear planes in sheared granodiorite from Palm Canyon, California; (d) photomicrograph of Särv mylonite from the Swedish Caledonides showing well-developed c-planes and s-planes (photo by J. Gilotti); (e) sketch illustrating the use of S-C fabrics as shear indicators.

can be applied to the shear bands (Berthe et al., 1979). Shear-zone rocks with pronounced S-planes and C-planes are called *S-C mylonites* (Lister and Snoke, 1984). They develop best in augen mylonites formed from a micaceous granitic protolith.

The geometric pattern of shear bands with respect to the deflected schistose foliation gives the overall sense of movement across the larger host shear zone (e.g., Simpson, 1986). Shear bands are parallel to the shear-zone boundaries, and the S-planes or the earlier schistose foliation are inclined to the shear-zone boundaries. S-planes or earlier schistose foliation dip away from the direction of shear and curve into shear bands, thereby creating a sigmoidal pattern of foliation (Fig. 11-30e) that directly gives shear sense as described earlier. The line of intersection of shear bands and schistose foliation is approximately perpendicular to the direction of movement.

It is important to emphasize that not all shear bands can be used as shear-sense indicators for the host shear zone. In some cases, shear bands, which closely resemble sigmoidal crenulation cleavage (described later in the chapter), are inclined to the main shear zone boundary and the shear across them can be antithetic or sympathetic to the shear across the larger host shear zone. In addition, shear bands can form conjugate systems that are associated with flattening across the shear zone (Platt and Vissers, 1980; Bell, 1981).

Porphyroclast tails: Porphyroclasts are relict larger grains of relatively rigid minerals that occur in an otherwise fine-grained, foliated mylonite. Porphyroclasts range in size from 0.1 mm to several centimeters. They often have thin mantles and elongated tails of either tiny recrystallized grains of the same composition as the porphyroclast (Fig. 11-31a,b) or of reaction-softened

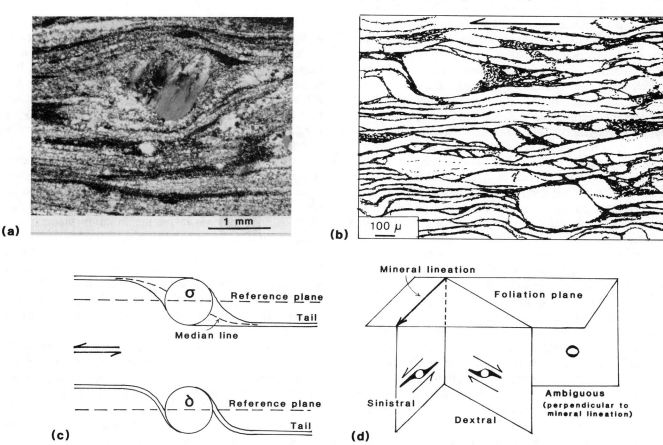

Figure 11-31. Porphyroclast tails in mylonites. (a) Feldspar porphyroclast with dynamically recrystallized asymmetric tails (σ-type) from the Carthage-Colton mylonite zone. The tails indicate clockwise vergence (from Lumino, 1987); (b) photomicrograph sketch of a σ-type tail in Särv mylonite, Sweden, indicating counterclockwise vergence (after a photo by J. Gilotti); (c) σ-type and δ-type tails on porphyroclasts used for determining shear sense in mylonites (adapted from Passchier and Simpson, 1986); (d) change in vergence depending on direction of viewing of the plane of exposure (from Lumino, 1987).

material formed from the porphyroclast (e.g., feldspar altered to white mica and quartz). The asymmetry of the tails on the porphyroclasts can be used to determine the sense of shear within a shear zone (Simpson and Schmid, 1983; Passchier and Simpson, 1986; Simpson, 1986).

To determine the sense of shear from porphyroclast tails we first need a frame of reference. The *median line* of a tail is defined as the line in the cross-sectional plane that runs down the middle of the tail. The *reference plane* is drawn to contain the *porphyroclast axis* (this axis is perpendicular to the shear direction) and the stretching axis (the stretching axis is parallel to the prominent mineral lineation of mylonites); the reference plane should be parallel to the tails at a distance from the porphyroclast and is usually parallel to the foliation of the mylonite.

There are two different geometric types of porphyroclast-tail systems (Passchier and Simpson, 1986). Wedge-shaped tails whose median lines lie on opposite sides of the reference plane at all points are called *σ-type tails*; often the tails themselves are concave on the side toward the reference plane (Fig. 11-31c). Thinner tails whose median lines cross the reference plane next to the porphyroclast are called *δ-type tails*; the tails show embayments of finer matrix adjacent to the porphyroclast and bend into parallelism with the reference plane away from the porphyroclast (Fig. 11-31c). In general, in going from the tail on one side of the porphyroclast to the other, we "step up" across the reference plane either to the right (indicating clockwise rotation or right-lateral shear) or to the left (indicating counterclockwise rotation or left-lateral shear) (Fig. 11-31a,b,c). In the case of δ-type tails, it is important that you "step up" away from the porphyroclast so that you are not misled by the portions of the tails that have moved to opposite sides of the reference plane. Do not confuse the embayments of δ-type tails with the concavities of the σ-type tails. Also, note that in some cases, the tails themselves rotate with the porphyroclast.

In order to obtain the correct sense of shear, you must be careful to note the orientation of the surface that you are examining (remember, the surface should be perpendicular to the foliation and should contain the lineation of the mylonite) and the direction in which you viewed the surface. The same porphyroclast can give the "opposite vergence" when viewed from the "other side" of a plane of exposure (Fig. 11-31d).

Rotated grains: Porphyroclasts of equant crystals (e.g., garnet, albite) tend to rotate when caught in a shear couple and thus can be used to indicate shear sense (e.g., Powell and Vernon, 1979; Rosenfeld, 1970). The sense of rotation is indicated by the pattern of inclusion trails or relict foliation in the grain; the inclusion trails or relict foliation is reoriented with respect to the foliation that has developed in the shear zone. Trails or relict foliation that is planar within the porphyroclast formed prior to

development of the shear zone. Inclusion trails with sigmoid shapes indicate that the porphyroclast grew during development of the shear zone (Fig. 11-32; Zwart, 1962; Vernon, 1976).

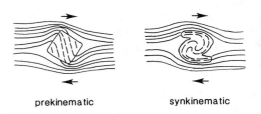

prekinematic   synkinematic

Figure 11-32. Rotated porphyroblast textures useful for determining shear sense and for relating time of development of foliation with time of growth of the porphyroblast. (From Vernon, 1976.)

Mica fish: Large porphyroclasts of white mica in mylonites are commonly referred to as *mica fish* (Lister and Snoke, 1984; Simpson, 1986). The cleavage planes of a majority of the mica grains are oriented at the same low angle to the mylonitic foliation, the normal to the cleavage plane being parallel to the incremental shortening direction (see Lister and Snoke, 1984). The large grains may be bordered by σ-type tails composed of dynamically recrystallized fine mica.

A phenomenon called *fish flash* (S.J. Reynolds, oral commun., 1985; Simpson, 1986) may be used to determine shear sense if the mica fish are large enough to be seen in hand samples and in outcrop. To observe this phenomenon, look down on the foliation plane of the mylonite in a direction parallel to the stretching lineation. Tilt your line of sight (in an outcrop) or the sample (for a hand specimen) until a majority of the mica grains have maximum reflectivity at once (i.e., the fish "flash"). The sight line from your eye to the mica is then parallel to the relative movement vector of the shear zone (Simpson, 1986; Fig. 11-33).

Fractures and displaced grains: Rigid minerals in a deforming ductile matrix often develop cracks along weak planes because they cannot accommodate large strains by crystal plastic mechanisms. With continued shearing the rigid grains rotate, and the fragments of each grain slide past one another along the cracks, allowing the grain to extend in the flow direction. The sense of shearing along individual fractures can be sympathetic with or antithetic to the overall shear sense (Fig. 11-34a).

If the initial fractures are at a low angle to the flow plane, the shear sense on them will be sympathetic with the overall shear sense, and the grain will be extended by motion analogous to motion on low-angle faults (Simpson, 1986; Fig. 11-34b). If the initial fractures are at a high angle (45° to 135°) to the flow plane, the shear

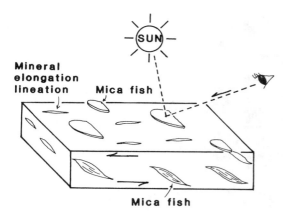

Figure 11-33. The use of "fish flash" as a shear sense indicator. (Adapted from Simpson, 1986.)

sense on them will be antithetic to the overall shear sense, and the grain will be extended by motion analogous to motion on high-angle normal faults or sometimes to motion on high-angle reverse faults (Fig. 11-34b). Rotation may cause initial high-angle fractures to reach a low-angle orientation during progressive deformation, causing the shearing on the fractures to change from antithetic to sympathetic. It is safest to use grains in which fractures are at high angles (50° to 130°) or at very low angles (0° to 20°, 160° to 180°) to the flow plane in a shear-sense determination (Simpson, 1986).

Folded layering: In most shear zones inhomogeneous deformation causes perturbations in the flow foliation of mylonites. These perturbations evolve into asymmetric folds whose vergence is consistent with the shear sense within the zone (Fig. 11-35a). The vergence of such folds can be used as a shear-sense criterion only where mylonitic foliation formed in a shear zone is folded by the late stages of movement of the same

(a)

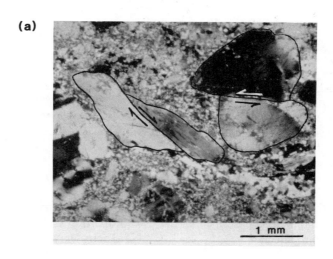

Figure 11-34. Photomicrograph of fractured and sheared grains. (a) Fractured feldspar grains from the Carthage-Colton mylonite zone, New York, showing sympathetic and antithetic shearing. Grains are outlined for emphasis; (b) illustration showing that sympathetic and antithetic shearing depends on initial orientation of fractures with respect to the flow plane. (From Lumino, 1987.)

(b)

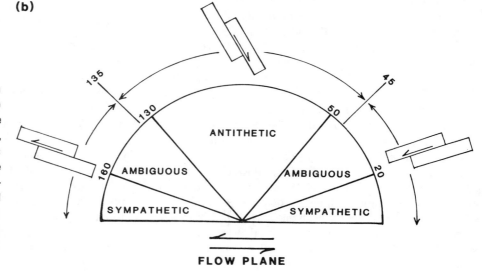

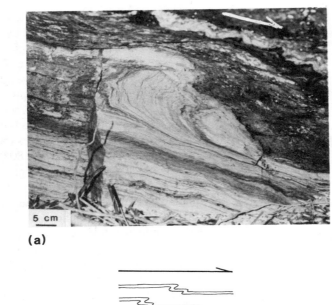

(a)

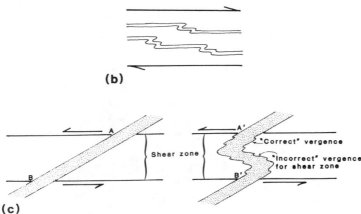

(b)

(c)

Figure 11-35. Use of folded layering as a shear-sense indicator. (a) Isoclinal folding of mylonitic layering indicating top to the west shearing along the Linville Falls thrust, North Carolina; (b) mylonitic foliation in a shear zone folded by later stages of shearing (adapted from Simpson, 1986); (c) folding of pre-existing layering in a shear zone (adapted from Simpson, 1986).

shear event (Simpson, 1986; Fig. 11-35b). When such folds exist, the Hansen slip line method described earlier can be used to determine transport direction.

   If original compositional layers (e.g., bedding, sills or dikes, earlier gneissic layering) are folded in a shear zone, the geometry of the fold depends on the initial orientation of the layer with respect to the shear zone (Ramsay, 1980; Ramsay and Huber, 1987). The small-scale folds formed during this folding may show vergence related to flexural folding of the layer rather than to flow in the shear zone (Fig. 11-35c). Thus, you must be careful about using small-scale folds in shear zones as shear-sense indicators.

   Preferred orientation: The term *preferred orientation* is used with reference both to the *crystallographic* or *lattice preferred orientation* and *grain-shape preferred orientation*. Crystallographic preferred orientation is also called *texture* (Schmid, 1983). Analysis of preferred orientation can permit determination of shear sense, but the methods for such analysis (e.g., Lister and Price, 1978) cannot be discussed here for they require use of somewhat sophisticated techniques of microscopy (see Simpson, 1986 for an introduction). Preferred orientation

also leads to *rock anisotropy,* meaning that physical properties such as magnetic susceptibility and seismic velocity vary with direction in the rock (see Owens and Bamford, 1976; Wood et al., 1976; Kligfield et al., 1981).

11-4   FOLIATIONS

The term *foliation* is a general term that refers to any planar fabric or layering in a rock, with the implication that the rock did not lose cohesion along foliation planes during the formation of the foliation. According to this definition, an array of joints or faults is *not* considered to be a foliation. Foliation can be a consequence of depositional processes (in which case it is called bedding) or a consequence of metamorphic and deformational processes (e.g., Williams, 1977). In this section we are concerned only with foliations created during metamorphism and deformation. We describe three types of foliation (earlier in this chapter, we mentioned a fourth type, mylonitic foliation). The purpose of this discussion is to provide terminology that you can use to describe foliation that you may come across in a field study.

Cleavage

We use the term *cleavage* to refer to foliation formed as a consequence of tectonic deformation in rocks at relatively low metamorphic grades. There are two basic categories of cleavage, namely, disjunctive cleavage and crenulation cleavage (Borradaile et al., 1982).

Disjunctive cleavage: This type of cleavage is defined by planes that cut across earlier foliation such as bedding (Engelder and Marshak, 1985). The cleavage planes, called *cleavage domains*, are zones in which the original rock fabric has been altered by the cleavage-formation process (Fig. 11-36a). Adjacent domains are separated from one another by *microlithons*, which are the bodies of rock in which the original fabric is still present and the earlier foliation is still visible. Earlier foliation in a rock (usually bedding) is not reoriented by formation of a disjunctive cleavage.

The formation of disjunctive-cleavage domains involves preferential solution (by pressure solution or free-face dissolution) of the more soluble minerals in the rock (Engelder and Marshak, 1985). The domain is composed of an accumulation of the less-soluble minerals. For example, in limestone, cleavage domains consist of accumulations of clay and quartz, and in sandstone, domains consist of an accumulation of clay. The clay plates in a domain tend to be packed together parallel to the domain boundaries. The preferred orientation of clay in domains reflects rotation of clay flakes into parallelism as the framework carbonate or quartz grains are removed. Clay in domains may undergo slight recrystallization.

The character of disjunctive cleavage is quite variable and is controlled by the original composition of a rock and by the amount of strain. *Domain spacing* (the distance between adjacent domains) tends to be smaller in rocks that originally had a higher concentration of clay and tends to decrease as strain increases. Therefore, domains are more closely spaced in clay-rich limestone than in clay-poor limestone, and in a given lithology, domains are more closely spaced in a location where there is large strain than in a location where there is low strain. *Domain morphology* refers to the shape of a single domain. Domains tend to be sutured or stylolitic (tooth-like in profile) in clay-poor rocks, and thicker and mesoscopically smoother in clay-rich rocks (Fig. 11-36b). Commonly, individual domains are wavy in cross section, so that a group of domains in an outcrop define an anastomosing or braided pattern (Fig. 11-36b).

The classification of disjunctive cleavage is based on domain spacing and domain morphology. Figure 11-37a provides a scale for specifying terminology to describe cleavage-domain spacing. Note that very widely spaced cleavage domains may be called *stylolites*, and that *slaty cleavage* (Fig. 11-36c; Wood, 1974; Siddans, 1977) refers

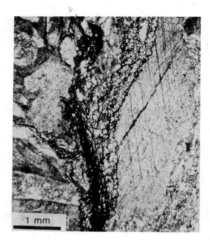

(a)

(b)

(c)

Figure 11-36. Illustrations of disjunctive cleavage. (a) Photomicrograph of a cleavage domain in limestone. The domain is the dark accumulation of clay between the large twinned calcite grains; (b) Non-sutured (smooth) planar cleavage domains in clay-rich limestone in the Canadian Rocky Mountains. The ruler is 30 cm long. See also Figure A1-19a; (c) Vertical slaty cleavage in shallowly dipping mudstone. Notebook is 19 cm long.

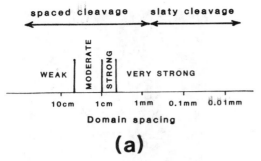

(a)

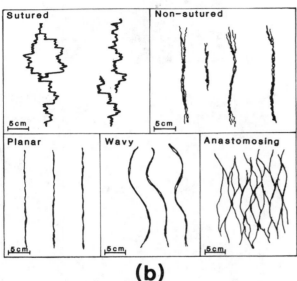

(b)

Figure 11-37. Classification of cleavage.
(a) Scale for specifying cleavage-domain
spacing; (b) chart for describing cleavage-
domain morphology. (Adapted from Engelder
and Marshak, 1985.)

to disjunctive cleavage in which discrete microlithons are
no longer definable in hand specimen. If a preferred
orientation of fine-grained platy minerals is pervasive
throughout the rock, such that there are no unaffected
microlithons, the slaty cleavage can be called *continuous* or
penetrative. Figure 11-37b provides a chart of adjectives to
be used for describing domain morphology.

Crenulation cleavage: This type of cleavage is
characterized by microscale kinking of an earlier fabric.
The earlier fabric may be preexisting slaty cleavage or
finely laminated bedding. There are two main types of
crenulation cleavage (Fig. 11-38; Gray, 1977; Cosgrove,
1976; Gray and Durney, 1979). *Zonal crenulation cleavage*
(Fig. 11-38a) is defined by laminar zones consisting of
microkink fold limbs, in which preexisting foliation is
reoriented. The boundaries of the zones may be distinct
surfaces, or they may be gradational, depending on the
angularity of fold hinges. There are two versions of zonal
crenulation. In *chevron* or *symmetric crenulation cleavage*

the crenulation fabric looks like the bellows of an
accordion (Fig 11-38b) and the boundaries between adjacent
zones are approximate planes of symmetry. In *sigmoidal*
or *asymmetric crenulation cleavage* the relict earlier
foliation within a zone is bent into a sigmoid shape (Fig.
11-38c; see also Platt and Vissers, 1980). Pressure
solution can accompany formation of zonal crenulation
cleavage. In pelitic rock, pressure solution removes quartz
from zones and concentrates it in the hinge areas that form
the boundaries between zones. Thus, there is a
concentration of mica in the zones (Fig. 11-38a).

Discrete crenulation cleavage (Fig. 11-38d) occurs
where redistribution of minerals by pressure solution and
precipitation has been very extensive (Gray, 1977). As a
consequence, the cleavage consists of alternating bands - in
one band, material has been dissolved, and in the adjacent
band, material has been precipitated. There are sharp
discontinuities between bands. In pelitic rocks that contain
this fabric, one band will be composed of quartz, and the
adjacent band will be composed of mica.

Schistosity

Schistosity refers to foliation in metamorphic rocks that
are coarse-grained enough for individual layer-silicate grains
(e.g., mica, chlorite) to be visible to the unaided eye. This
type of foliation is commonly found in medium- and
high-grade metamorphic terranes, where the minerals
comprising the rock formed largely by complete
recrystallization and new mineral growth. The fabric of a
schist (a metamorphic rock with schistosity) is defined by
preferred orientation of the large layer-silicate crystals.
There are three types of schistosity (Borradaile et. al.,
1982): (a) *Domainal schistosity* is characterized by domains
of subparallel mica grains that form films that anastomose
around lenticular domains composed of other minerals (Fig.
11-39a); (b) *Continuous schistosity Type 1* is characterized
by a strong preferred orientation of coarse mica grains and
no mesoscopic lenticular microlithons (Fig. 11-39b);
however, microlithons may be visible if the rock is
examined under a microscope; (c) *Continuous schistosity
Type 2* is characterized by planar fabric elements (layer
silicates or flattened/stretched grains) that have a single
preferred orientation and are distributed throughout the rock
rather than being concentrated in zones. Thus, this type of
schistosity is not defined by a domainal fabric at any scale
(Fig. 11-39c).

Gneissic Layering

Gneissic layering is a foliation developed in high-grade
metamorphic rocks and is defined by compositional
layering that is not primary in origin (Fig. 11-40).
Gneissic layering can form by metamorphic differentiation,

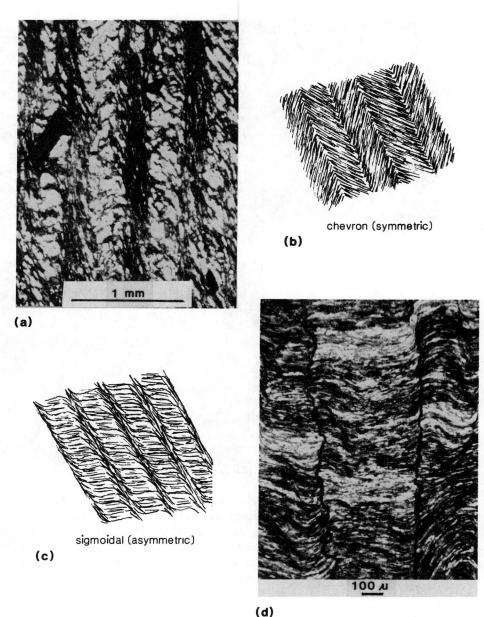

Figure 11-38. Crenulation cleavage. (a) Zonal crenulation cleavage (photo by D. Gray); (b) sketch of symmetric zonal crenulation cleavage; (c) sketch of asymmetric zonal crenulation cleavage; (d) discrete crenulation cleavage (photo by D. Gray).

by transposition and recrystallization of original bedding, by lit-par-lit intrusion, and by partial melting (Spry, 1969).

Description of Foliation

A description of foliation should provide answers to the following questions:

1. What does the foliation look like? Is the foliation best described as bedding, cleavage, schistosity, or gneissic layering? A "flow chart" for foliation description is provided as Figure 11-41. The description should include details concerning the feature that defines the foliation (e.g., is it defined by domains of clay residue probably indicative of solution removal of framework quartz or calcite, by parallel alignment of platy silicates, by flattened crystals of quartz, or by crenulations). It should also include information on whether or not the foliation is domainal or penetrative, on whether or not there has been a redistribution of minerals associated with the foliation (e.g., by pressure solution), and on whether or not the foliation is associated with recrystallization.

2. What is the relationship between foliation and folds in the study area? There are three possible relationships betweeen a foliation (specifically, cleavage or schistosity) and a fold: (a) The foliation is *axial-planar* to the fold (is parallel to the fold's axial plane; see Williams, 1976); (b) The foliation (specifically, cleavage) fans around the fold.

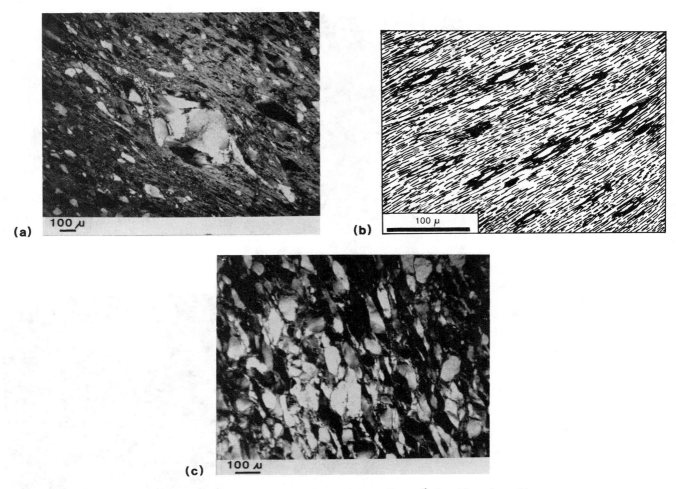

Figure 11-39. Types of schistosity. (a) Domainal schistosity with mica films wrapping around lenticular quartz grains (photo by S. Mitra); (b) photomicrograph sketch of continuous schistosity type 1 defined by strong preferred orientation of mica; (c) continuous schistosity type 2 defined by stretched and flattened grains of quartz in an orthoquartzite (photo by S. Mitra).

Fanning cleavage is cleavage that stays at a high angle to bedding around a fold (e.g., Mitra and Yonkee, 1985); (c) The foliation transects the fold. *Transecting foliation* (specifically, cleavage or schistosity) cuts across the axial plane of a fold (Borradaile, 1978; Gray, 1981).

Cleavage formed in association with folds at low metamorphic grades in clay-rich rocks tends to be axial-planar, whereas cleavage in clay-poor rocks tends to fan around folds (Fig. 11-42). Thus, in the same fold, cleavage in some beds may fan around the fold, and cleavage in other beds may be axial-planar. The change in orientation of cleavage as a function of lithology is called *cleavage refraction*. Fanning cleavage probably initiated prior to development of a fold. The occurrence of a transecting foliation may indicate that there was complex strain history during a single event or that the fold and the foliation developed during separate events (see Chapter 16).

3. Does the domain spacing of the foliation change as a function of structural position? This information may indicate variations in lithology or strain in a region.

4. Is the foliation itself folded? Is the foliation cut or crenulated by a later cleavage? Such information gives clues to the history of polyphase deformation in a region (see Chapter 16). What are the geometric relationships among different foliations, if more than one exist? If a rock has been subjected to more than one phase of deformation, one or more crenulation cleavages may disrupt a slaty cleavage. Be sure to determine if more than one set of folds exist in the rock, and try to associate each foliation with a fold set.

Figure 11-40. Gneissic layering in Precambrian rocks of the Wind River Mountains, Wyoming. The layering is folded and is cut by igneous dikes.

5. Is the foliation related to a shear zone? If it is, it could be used to help determine the sense of shear and perhaps the shear strain of the shear zone.

6. Is the foliation cut by faults or veins or does it cut these structures? In many localities, cleavage or schistosity bends and merges asymptotically with a fault plane.

Answers to the preceding questions will help you gain insight into the conditions under which the foliation formed, will help you to determine the nature of strain resulting from the cleavage, and may help you to place the development of the foliation in the sequence of deformation events that affected the region.

11-5   LINEATION

Many different types of lineation can occur in a rock (Cloos, 1946). Some of the important categories of lineations are listed next.

Slip lineations: These are found on fault surfaces and have been described in detail earlier in this chapter.

Intersection lineations: Any two foliation planes that are oblique to one another intersect along a line. The intersection defines a lineation (Fig. 11-43a). Cleavage/bedding intersection lineations appear as minute parallel lines on the bedding surface; if the cleavage is an axial-plane cleavage, the lineation is parallel to the hinge line of associated folds and may be used to determine the orientation of the hinge. The trace of bedding on the cleavage plane may produce colored stripes referred to as striping (Fig. 11-43a).

Crenulation hinges: The fold hinges of microkink folds associated with crenulation cleavage define a lineation on the early foliation surface (Fig. 11-43b).

Mineral lineations: Metamorphic minerals often grow with a preferred crystallographic and dimensional orientation. Depending on the shape of the crystals and their orientation, one of several lineation patterns is possible (Fig. 11-43). Minerals with one long dimension (e.g., hornblende, tourmaline, sillimanite) arranged with their long dimension parallel to one another will produce an *L-tectonite* (see Chapter 8; Fig. 11-43c). Some mineral lineations are defined by streaks of microcrystalline mica smeared along foliation planes in the direction of shear (Fig. 11-43d,e).

Stretched markers: Spherical markers (e.g., oolites, pebbles) may be deformed into ellipsoids, or crystals can be stretched during plastic deformation to give rise to lineations (e.g., Mosher, 1987). If the strain ellipsoid (Chapter 15) is prolate, the rock is an *L-tectonite* (Fig. 11-43f); if the strain ellipsoid is oblate, the rock is an *S-tectonite* (Fig. 11-43g); and if the strain ellipsoid is triaxial, the rock is an *LS-tectonite* (see Chapter 8; Fig. 11-43h).

Pressure shadows: Material dissolved by pressure solution is often reprecipitated as fibers in the pressure shadow behind rigid grains (Fig. 11-44; e.g., Stromgard, 1973; Beutner and Diegel, 1985). The fibers thus form long tails that define a lineation. The tails in pressure shadows may also form by dynamic recrystallization or reaction softening of rigid grains during the process of mylonitization (see Section 11-3).

Boudinage: Extension of a layered sequence, which

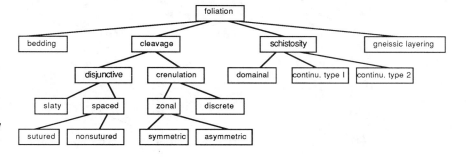

Figure 11-41. Simple "flow chart" for description of foliations.

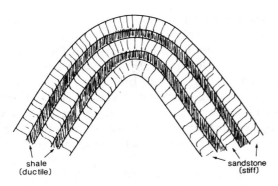

Figure 11-42. Profile sketch showing cleavage refraction. Idealized fanning of cleavage is evident in competent beds and axial-planar cleavage occurs in incompetent beds.

is made up of competent layers interlayered with more ductile units, causes necking and eventual separation of the competent layer into segments, while the ductile material flows into the space between the segments (Fig. 11-45a; e.g., Cloos, 1947; Ramsay, 1967; Paterson and Weiss, 1968; Sanderson, 1974). The segments are called *boudins*, and the overall structure is called *boudinage* (Fig. 11-45a,b). If the competent layer does not separate into pieces, it may still thicken and thin, giving rise to *pinch-and-swell structure* (Fig. 11-45a,c). In the third dimension, both boudins and pinch-and-swell structures extend as elongate bodies that lie in bedding and are parallel to one another, thereby defining a lineation (Fig. 11-45b). In describing boudinage, it is important not only to measure the foliation containing the boudinage and the boudin lineation, but also to note whether the boudins' long dimension in cross section is parallel (symmetric

structure) or oblique (en-echelon structure) to the enclosing foliation, what the boudin terminations look like, and what the relationships are to other major structures (e.g., folds; see DeSitter, 1958). Analysis of boudinage is useful in working out details of the strain history (see Ramsay and Huber, 1983).

Mullions: Shortening of a thin incompetent unit sandwiched between two thick competent units forms cylindrical corrugations on the surfaces of the competent units (Fig. 11-46a). The parallel cylindrical features on the bedding surface define prominent lineations called *mullions* or *antiboudins* (Fig. 11-46a,b; Wilson, 1953; Smith, 1977). Like boudins, these are useful for working out details of the strain history of larger-scale structures.

Pencil Structure: In many localities, weakly deformed shales break into pencil shaped fragments at the surface of weathered outcrops. One explanation for such pencil structure is that the shale breaks along two directions of weakness: one direction is parallel to sedimentary compaction fabric (i.e., original bedding) and the other is parallel to incipient tectonic cleavage (Ramsay and Huber, 1983; Reks and Gray, 1982). The term pencil structure has also been used with reference to shales which break along closely spaced joints and to very well developed metamorphic L-tectonites.

Lineations can be difficult structures to interpret. In some circumstances lineations define the direction of shear, whereas in other circumstances they define the long axis of the strain ellipsoid. A slip lineation is an example of the former, while a stretched marker is an example of the latter. In some localities the long axis of the strain ellipsoid is at right angles to the shear direction, although in most cases the two are at an acute angle to one another. It should be

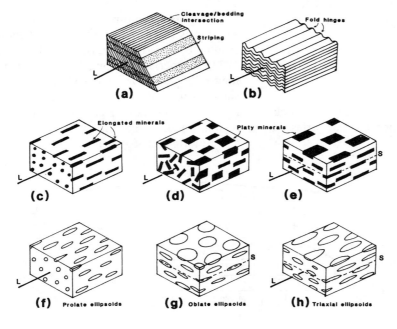

Figure 11-43. Types of tectonic lineation. (a) Intersection lineation; (b) fold hinges of crenulation; (c-e) preferred orientation of elongated and platy minerals; (f-h) deformed markers.

Figure 11-44. Fibrous mineral growth in pressure shadows behind a rigid pyrite grain in Devonian limestone from the Hudson Valley fold-thrust belt (photo by S. Bhagat).

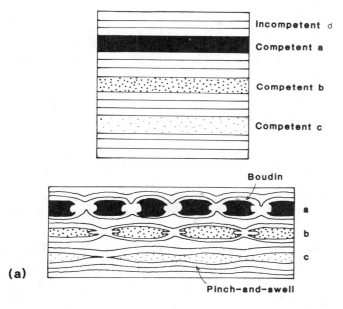

(a)

Incompetent d
Competent a
Competent b
Competent c

Boudin
a
b
c
Pinch-and-swell

(b)

Figure 11-45. Boudinage. (a) Derivation of boudinage as a consequence of extension of a layered sequence in which layers are of different compositions. The nature of boudinage varies according to the competence contrast (a>b>c>d) (adapted from Ramsay, 1967); (b) boudinage in vertical beds of limestones in the Great Valley province of the Appalachians of Maryland. Note hammer for scale; (c) pinch-and-swell structure in quartzite in the Blue Ridge province of the Appalachians of Maryland (from S. Mitra, 1979).

(c)

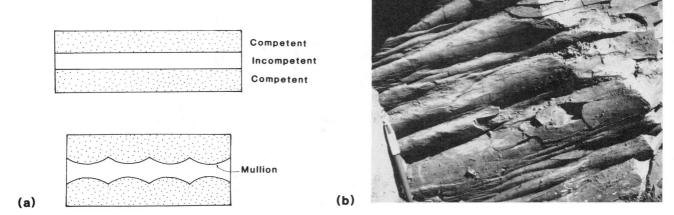

(a)   **(b)**

Figure 11-46. Mullions. (a) Derivation of mullions during the deformation of a layered sequence (adapted from Smith, 1975); (b) Mullions in sandy limestone of the Wyoming fold-thrust belt.

pointed out that in three-dimensional space, the shear direction and the long axis of the strain ellipsoid can never be exactly parallel to one another (see Chapter 15).

11-6 VEINS

Veins are mineral-filled fractures that generally result from precipitation of minerals out of fluids passing through the fractures. The minerals filling the vein can occur in two forms: (a) blocky spar, with fine to coarse grain size (Fig. 11-47a), and (b) crystal fibers (Fig. 11-47b).

Blocky spar generally grows into open cavities. In some localities the veins themselves contain open gaps in which euhedral crystal terminations grow. Generally, such

veins form at shallow levels in the earth's crust, where low lithostatic pressure allows open cavities to form along fractures. At deeper levels in the crust, open fractures can exist only if the hydrostatic pressure of the fluids is sufficiently high to keep the fracture walls open (Secor, 1965).

Fibrous veins form by the *crack-seal process* (Ramsay, 1980). The vein grows incrementally by first cracking and then quickly resealing by precipitation of new mineral at the tips of earlier-formed grains. The fiber grows parallel to the incremental elongation direction and can thus be used to track incremental strain history (Ramsay and Huber, 1983). If the fibers are perpendicular to the walls of the vein, the vein opened normal to the plane of the vein (Fig. 11-48a). If the fibers are inclined to the vein walls, the

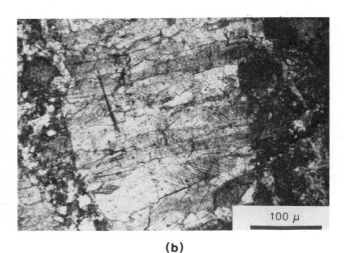

(a)   **(b)**

Figure 11-47. Veins. (a) Outcrop photo of blocky calcite crystals filling a vein. The pen is 15 cm long; (b) photomicrograph of a fibrous calcite vein. The crystals are twinned.

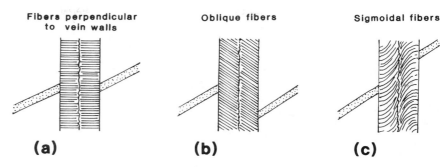

Figure 11-48. Incremental history of an opening fibrous vein. (a) Fibers perpendicular to vein walls; (b) fibers inclined to vein walls; (c) sigmoidal fibers.

direction of vein opening was inclined to the plane of the vein (Fig. 11-48b). Often the fibers are sigmoidal in shape (Figs. 11-47b, 48c), indicating successive noncoaxial stages of vein opening. There are four types of fibrous veins (Durney and Ramsay, 1973; Ramsay and Huber, 1983):

Syntaxial veins: These grow inward from the wall by precipitation at a median line (Fig. 11-49a). Each increment of cracking occurs at the medial line, and the increments of new mineral are precipitated there. In thin section the boundaries between increments are indicated by bands of fluid inclusions. Typically, the wall rock and the vein material are the same composition, so that the vein fibers nucleate on grains of the wall.

Antitaxial veins: These grow outward from the center of the vein (Fig. 11-49b). Each increment of cracking occurs along the two walls of the vein. In thin section, increments can be seen to be separated from one another by thin screens of wall rock. Typically, the wall rock and the vein fibers are composed of different minerals.

Composite veins: These form by a combination of antitaxial and syntaxial growth. Usually, such veins contain two different mineral components (Fig. 11-49c).

Stretched crystal veins: In these veins, each fiber is optically continuous with two halves of a grain that was split during initiation of the vein (Fig. 11-49d).

A complete description of a vein should include the following information: (a) Description of the structural setting of the vein. Is it one member of a systematic set (see Chapter 12)? Is it a member of an en-echelon array? Is it a local structure related to, say, outer-arc extension on a fold? Are there nearby unmineralized joints of the same orientation? (b) Description of the vein character. Is it fibrous or blocky? If the vein is fibrous, define the type of fibrous vein and specify whether the fibers are straight or sigmoidal. (c) Description of the vein fill. What is the composition of the vein fill and of the wall rock? (d) Measurement of the vein orientation. What is the attitude of the vein? If the vein is part of an en-echelon array, be sure to measure the attitude of the vein and the attitude of the enveloping surface.

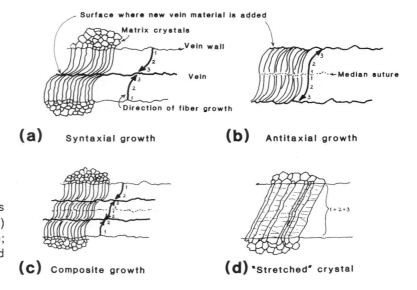

Figure 11-49. Types of fibrous veins. (a) Syntaxial vein; (b) antitaxial vein; (c) composite vein; (d) stretched crystal vein. (Adapted from Durney and Ramsay, 1973.)

12

ANALYSIS
OF FRACTURE ARRAY
GEOMETRY

Arthur Goldstein
Stephen Marshak

12-1   INTRODUCTION

A fracture in rock is a surface on which the rock has lost cohesion. If there is no observable slip on a fracture it is called a *joint*, if there is observable slip it is called a *fault*. Fractures commonly occur in *sets* composed of several subparallel members. A single region may be cut by several sets of fractures; a group of fracture sets is called a *fracture system* or *fracture array*. Information on fracture arrays in a region helps to define the stress and strain history of a region and thus is important for regional tectonic analysis. Information on fractures is also of practical importance, for fractures affect slope stability and foundation strength. In addition, fractures affect a range other phenomena, such as permeability and drainage. This chapter discusses approaches used to describe fracture arrays. We draw attention to morphological and geometric features that should be noted in a survey of fractures. The chapter concludes with a brief introduction to the description of lineaments, for lineaments in some cases are topographic manifestations of fractures.

12-2   CHARACTERISTICS OF JOINTS

Descriptive Terminology for Joints

Joints are fractures that do not display observable displacement parallel to the fracture surface. They are very common structures and occur in a range of sizes in a given

area. *Joint size* refers to the area of the joint surface. Some joints are only a few square centimeters in area, whereas others have areas of several hundred or several thousand square meters. Fractures that affect only a single grain or a few grains in a rock and, therefore, are visible only in thin section are called *microfractures*.

Remember that a joint is a plane and that it does not extend infinitely in all directions; therefore, its orientation is specified by a strike and dip and its dimensions should be specified as an area, not as a line length. The orientation of vertical joints, however, can be represented by strike alone. Also, the full surface area of a joint is usually not exposed; therefore, joint size is usually described in terms of *trace length*, which is the length of the exposed line of intersection between the joint and the surface on which it is exposed.

Joints that have relatively long traces (more than 2 to 5 m) and are parallel or subparallel to many other joints on the outcrop or in the region are called *systematic joints*. A group of parallel or subparallel joints comprise a *joint set*. Short, locally irregular joints that are not part of a set are called *nonsystematic joints* (Hodgson, 1961). The largest, most prominent joints of a set are sometimes called *master joints*. If a large proportion of the joints in an area share the same orientation, they comprise a *prominent joint set*. Recognition of prominent sets often depends on statistical analysis of joint orientation data from a region.

Some joints terminate at the top and bottom of a bed and are called *bedding-contained joints*, whereas others cut across bedding. *Joint spacing* refers to the distance between adjacent joints of the same set measured perpendicular to their surfaces, and is a particularly important parameter in studies of rock permeability. Joint spacing reflects the material properties of the host rock, the thickness of the jointed layer (see Suppe, 1985), and the structural position of the measurement site (joint spacing may be smaller in the hinge of a fold or in the vicinity of a fault; e.g., Wheeler and Dixon, 1980). *Joint intensity* refers to the surface area per unit volume of joints in a rock. *Joint frequency* refers to the number of joint traces cutting a traverse line per unit length of traverse line.

More than one joint set is usually present in any one outcrop. Any group of two or more joint sets is called a *joint system* or *joint array*. Joint sets in a system may intersect at any angle. If two joint sets intersect at nearly 90°, they define an *orthogonal system*. Joint systems in which the dihedral angle between two sets is significantly less than 90° (usually around 60°) have been called *conjugate joint systems*. As we discuss later in this chapter, there is debate as to whether or not true conjugate joint systems actually exist.

Description of Joint Surface Morphology

Many joint surfaces display *plumose structure*. To a first approximation, plumose structure looks like the imprint of a feather (Fig. 12-1). The tiny surface irregularities that comprise the plume develop as a consequence of local variations in fracture propagation velocity and in the stress field, and as a consequence of inhomogeneities in the rock. The main elements of an ideal plumose structure (Fig. 12-1) include the following:

1. The *origin*, which is the point at which the fracture initiated.

2. The *mist*, which is composed of tiny irregularities on the fracture surface, resulting from breaking of bonds in the rock that are not in the plane of the fracture. These irregularities, on some fractures, are arranged in a pattern that looks like the barbs of a feather.

3. The *plume axis*, which is the line from which individual barbs propagate, and which is considered to be parallel to the fracture propagation direction.

4. The *twist hackle*, which is composed of the steps at the edge of a fracture plane. These steps represent a zone in which the fracture has split into a set of smaller en-echelon fractures. The individual members of the en-echelon array are not parallel to the main joint surface.

The entire surface of a joint does not form instantaneously. Rather, a joint initiates at the origin in an intact rock body and propagates outward. The *joint front*, which is the line that separates intact from fractured rock, moves through rock as the joint grows. Some joints display *arrest lines*, which mark a temporary edge of the fracture,

where it stopped before continuing to propagate. Arrest lines are curved, because the joint front at any given time is curved. For a detailed mechanical interpretation of the surface morphology of fractures see Kulander et al. (1979).

12-3 COLLECTION AND REPRESENTATION OF ATTITUDE DATA ON JOINTS

Sampling Schemes

The most commonly asked question in any study of jointing is, what are the orientations of the prominent joint sets in the area? Proper collection of joint-orientation data is crucial to the success of any study of jointing. There are several methods of sampling joint attitudes. These are identified as Steps 2a, 2b, 2c, and 2d below.

Problem 12-1 (Sampling a joint array)
    A number of joints are exposed in a region. Determine the prominent joint sets and their orientation.

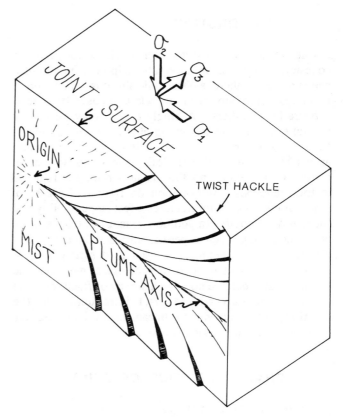

Figure 12-1.   Block diagram illustrating the morphology of a plumose joint. The stress field depicted assumes that the joint is an extension fracture.

Method 12-1

Step 1: Establish *structural domains* in the study region. This means that you must put limits on the area that you are going to sample. There is no point in lumping measurements from unrelated localities or structural settings. The specification of domain limits may be determined on the basis of lithology and/or on the the basis of structural position. For example, you may segregate measurements of joints in a limestone unit from measurements of joints in a granite intrusion cutting the limestone, and you may segregate measurements of joints on one fold limb from measurements of joints on a different fold limb. Once you have established structural domains, select suitable *measurement stations*, which are localities of good exposure where you can measure joints efficiently.

Step 2a (Selection method): In this method (e.g., Nickelsen and Hough, 1967) you visually scan an outcrop and select representatives of the prominent joint sets, then you measure only from four to eight joints of each set. This method is relatively fast, giving you time to make measurements at many stations. It also is quite subjective and works best only where joint patterns are relatively simple. If, however, you are careful to pay attention to such characteristics as size, mineralization, and surface features, even complicated joint patterns can be sampled with this method.

Step 2b (Quantity method): At an appropriate measurement station, measure as many individual joints as possible without regard to size or systematics. Typically, a sample suite collected at one station using this method will consist of 50 to 100 measurements. The concept behind this method is that the prominent joint sets will be obvious on plots of the data. The problem with this method is that large numbers of measurements of small nonsystematic joints can swamp measurements on master joints of systematic sets. The results of quantity measurement, therefore, may not provide insight into the attitudes of systematic sets. Also, because the measurements are made randomly, they cannot be used to specify joint intensity or frequency (Wheeler and Dixon, 1980).

Step 2c (Inventory method): This method (e.g., Davis, 1984) not only provides data that can be used for statistical determination of prominent joint attitudes but also allows determination of joint intensity. To inventory the joints in an area, define a circle up to 10 m in diameter on an outcrop (the size of the circle is determined by the joint density). Measure the attitude and the length of all joints exposed on the surface. If possible, repeat the measurements in the same area on outcrop faces of other orientations to avoid sampling bias (as discussed later).

Step 2d (Traverse method): Lay out a traverse line and measure its orientation; the traverse can be on the ground or along a cliff face or road cut. A reasonable traverse is 20 to 40 m long, but the length will be determined by the exposure quality and joint spacing. Walk along the traverse and measure every joint that crosses the line. Also, indicate which individual joints appear to be prominent in that they are continuous and straight for a long distance (this is a subjective determination). Figure 12-2 shows a form that can be used for recording data along the traverse. This method is time-consuming, but allows quantification of joint-orientation patterns in a region and allows quantification of joint frequency. As we discuss later in this chapter, results of the traverse method in a region are controlled, in part, by the orientation of the traverse, so in order to get better results, two or three traverses at different orientations should be run through the same locality.

Graphical Presentation of Orientation Data

Several types of diagrams can be used to display joint orientation data: the *rose diagram*, the *histogram*, the *running average*, the *equal-area projection*, the *length versus strike (LVS) diagram*, and the *strike versus traverse distance (SVTD) diagram*. Some of these diagrams can be used only with two-dimensional data and thus are most useful in studies of vertical joint sets.

Rose Diagrams: A standard rose diagram is constructed on a grid composed of concentric circles superimposed on a set of radial lines (Fig. 12-3a). The

---

FIELD FORM FOR JOINT DATA

NAME: _____   DATE: _____

EXPOSURE TYPE*: _____   ROCK TYPE: _____

* (A = vertical outcrop face; B = inclined outcrop face; C = horizontal outcrop face;
  1 = natural outcrop; 2 = stream cut; 3 = manmade outcrop)

BEDDING ATTITUDE: _____

STRUCTURAL FEATURES AT THE STATION (folding, faulting, cleavage, etc.):

_____

_____

ATTITUDES OF OTHER STRUCTURES: _____

_____

TYPE OF SAMPLING§: _____   ORIENTATION OF TRAVERSE: _____

   §(traverse, selection, circle inventory, etc.)

#	Joint attitude	Distance along traverse	Joint character
1			
2			
3			
4			

Figure 12-2. Example of a joint-sampling field form.

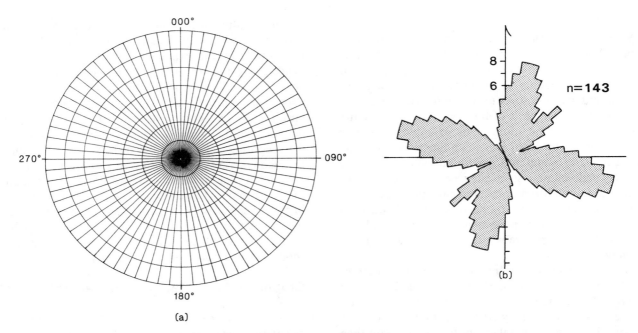

(a)

(b)

Figure 12-3. Rose diagrams. (a) Template that can be used for preparing rose diagrams with a 5° class interval; (b) completed rose diagram of 143 joint measurements. (See Appendix 4 for template.)

radius of each successive circle is one unit greater than the radius of the previous circle. Units of distance measured along a radius represent a quantity of joint measurements. The quantity of measurements represented by a single unit along a radius is chosen so that the total number of joints in the most prominent joint set can be represented on a standard 20-cm-diameter grid. Commonly, each unit represents one or two measurements. The orientation of a radius represents a compass bearing. Radii are normally spaced at 5° intervals. Each 5°-wide section is called a *class interval*. The number (n) of joints whose bearings are in a single class interval is plotted as a filled pie-shaped sector of length n at the scale of the diagram.

The major advantage of a rose diagram is that the data shown are easy to visualize (Fig. 12-3b). The major drawback of a rose diagram is that it is difficult to visually distinguish between two joint sets whose azimuths are less than 15° apart on two different rose diagrams. Also, because the area of a sector representing a measurement increases with increasing distance from the center, the difference between the numbers of joints in two different

sets is visually exaggerated. This second problem can be overcome by plotting data on a Lambert equal-area polar projection (Appendix 4).

Because it does not matter which end of a compass needle is read when representing strike, joint strikes are commonly converted so that all measurements can be placed on one half of the rose diagram.

Histograms: A histogram is similar to a rose diagram in that it uses class intervals (usually 5° wide), but it is plotted on a square grid instead of a circular grid. The x-axis of this grid represents joint azimuth, and the y-axis represents the number of measured joints (Fig. 12-4). Measurements represented on a histogram are not as easy to visualize as those on a rose diagram (compare Figures 12-3b and 12-4, which are plots of the same data set), but peaks with small azimuthal differences are easily discerned, and there is no false exaggeration of peaks.

Running Averages: Rose diagrams or histograms can appear very jagged. Calculation of a *running average* can be used to smooth either type of diagram and thereby make true peaks stand out more clearly. To construct a

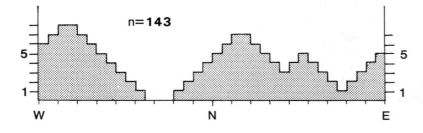

Figure 12-4. Histogram of 143 joint measurements (using the same data set as in Figure 12-3b).

running average, replot the data using a small class interval (say 1º), then average over a range (usually 10º). Shift the area of averaging across the range of orientations in 1º or 2º steps. For example, in Figure 12-5 we show the construction of a 10º running average with a 1º step. If we start at a strike of 30º, we count the total number of measurements that fall within 5º to either side of 30º (from 25º to 35º) and divide by 10. There are 17 measurements, so the average in the 10º around 30º is 1.7. We plot 1.7 at 30º. Then we move to 31º and count all measurements that fall within 5º to either side of 31º (from 26º to 36º) and divide by 10. There are 20 measurements, so we plot 2 at 31º. We repeat the procedure for each successive degree on the x-axis.

Equal-Area Plots: Where the joints in an area are not strictly vertical, the only graphical device that can display both dips and strikes is an equal-area plot of poles to joint planes (see Chapter 8 for method). Be sure to use an equal-area projection so that the point distribution can be contoured. The Kamb method (Chapter 8) is the best method to use to overcome noise in the plot when the number of measurements is large (>200). Remember that the size of the counting circle used in the Kamb method varies with the number of measurements, so you can contour the actual density of points in relation to the density distribution expected for a population without preferred orientation. The size of the counting circle is

$$A = 9/(N + 9) \qquad \text{(Eq. 12-1)}$$

where A is the decimal equivalent area of the counting circle, and N is the number of measurements.

Length versus Strike (LVS) Diagrams: The diagrams just described represent only orientation. Orientation diagrams may present a misleading impression of the nature of jointing in an area. For example, consider a horizontal bed in which there are 100 5-cm-long vertical joints whose traces trend north-south and 50 10-m-long vertical joints whose traces trend east-west. Intuitively, it would seem that the east-west set is tectonically more significant, but on an orientation diagram the north-south set would appear to be more prominent. An LVS diagram accommodates this problem. On an LVS diagram the x-axis represents joint-trace trend, and the y-axis represents cumulative joint length in a specific direction. In the preceding example the total cumulative length of the north-south set is 5 m, whereas the total cumulative length of the east-west set is 500 m. An LVS plot looks like a histogram (Fig. 12-6). Usually, a fairly large class interval (say 10º - 20º) is used in the construction of an LVS diagram. Note that only two-dimensional data can be used for an LVS diagram.

Strike versus Traverse Distance (SVTD) Diagrams: Wise and McCrory (1982) note that, in many cases, joint sets occur in discrete geographic domains. This means that joints in one locality have a different trend than joints in another locality. Such differences can be usually represented by providing separate histograms for separate domains. If measurements from separate domains were lumped together, you would convey the false impression that all joint trends were visible at all localities.

But now consider the possibility of two joint domains that overlap slightly (Fig. 12-7). If you measure joints at one locality in each domain, it is impossible to tell if the region contains one set of joints that change trend between the two stations or two sets of discrete joints each of which occurs only in a limited area, so that they do not overlap. The detection of separate joint domains is possible by sampling along long traverses that span the two regions and then plotting a graph of strike versus distance along the traverse; such a graph is called a strike versus traverse

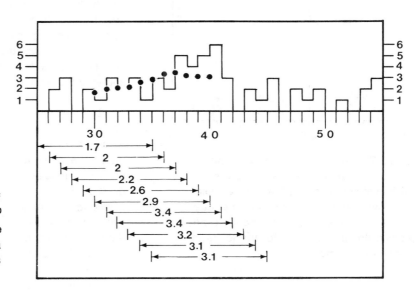

Figure 12-5. Example of smoothing a histogram using a 10º running average with a 1º step. The initial histogram was plotted using a 1º class interval. The dots represent the smoothed values.

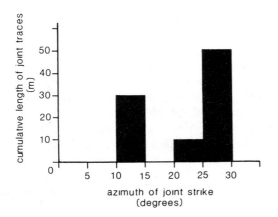

Figure 12-6. Example of a length versus strike diagram, constructed using a 5° class interval.

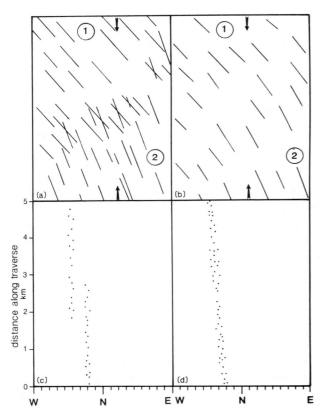

Figure 12-7. Hypothetical application of strike versus traverse distance plots to distinguish joint domains. (a) Two discrete joint sets that overlap only in an intermediate locality; (b) a single joint set whose trend changes with locality; (c) plot for the joints in 'a'; (d) plot for the joints in 'b'.

distance diagram (SVTD diagram). Figure 12-7 shows two possible SVTD diagrams; in case A, two discrete joint sets overlap whereas in case B one smoothly varying joint set is observed. SVTD diagrams can only be applied to two-dimensional data.

Data Corrections for Joint Data Sets

Regardless of the sampling method chosen and of the graphical device used for representing results, there are three factors that make a simple presentation of joint measurements an unrealistic portrayal of the true distribution of joints in an area. The first factor results from the possibility of incorrectly lumping measurements from different structural domains; to avoid this problem, you should define the sampling domains in the region of interest carefully, and if appropriate, plot your data on an SVTD diagram. The second factor is a consequence of the intrinsic bias in collecting orientations along a specified traverse direction; to accommodate this bias, it is necessary to apply a *traverse correction*. The third factor results from tilting of the bedrock containing the joints after the joints formed; to adjust for postformation reorientation of joints, a *tilt correction* must be applied.

Traverse Corrections: Imagine that two vertical joint sets (A and B) cross an east-west-trending 20-m-long traverse line. The joint spacing measured perpendicular to the fractures in each set is 1 m. Set A strikes N80°E and set B strikes north-south. Because of the difference in strike, 20 members of set B cross the traverse line, but only 4 members of set A cross the traverse line (Fig. 12-8). Thus, if the results of a traverse-method sampling of joint attitudes is plotted on a histogram, set A will appear to be less prominent, in terms of quantity, than set B. In other words, the frequency of joint set A appears, incorrectly, to be less than the frequency of joint set B.

A true measurement of joint frequency can be obtained only by sampling at 90° to strike of the joint set. If joints do not strike perpendicular to the trend of the traverse line, the number of measurements used to determine frequency can be adjusted to provide a truer representation by applying the equation

$$N = n/\sin \emptyset \qquad\qquad\qquad \text{(Eq. 12-2)}$$

where N is the corrected number of joints in a class interval, n is the measured number in that interval, and $\emptyset$ is the acute angle between the traverse direction and the strike of the class interval ($\emptyset = 10°$ in Fig. 12-8). After correction, N for joint set A in Figure 12-8 would be 23. The precision of determining joint frequency decreases as the angle $\emptyset$ becomes small (LaPoint and Hudson, 1985). For small $\emptyset$'s, one cannot be confident that the actual

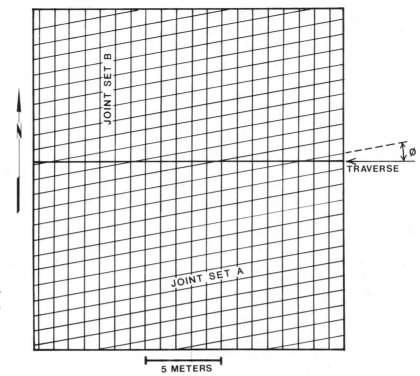

Figure 12-8. Diagram illustrating the rationale for a traverse correction of joint measurements. Ø is the angle between the strike of a joint set and the traverse direction.

numbers of joints measured reflects true joint frequency. To get a truer representation of joint frequency in such a case, it is better to make two mutually perpendicular traverses.

A similar correction must be applied if the dip of the joints is not vertical. For example, if two joint sets strike perpendicular to a horizontal traverse line, you will encounter fewer shallowly dipping than steeply dipping joints. To correct for nonvertical dips, apply the equation

$$N = n/\sin ß \qquad \text{(Eq. 12-3)}$$

where ß is the dip of the joint set. If both strike and dip corrections are necessary, apply the equation

$$N = n/(\sin ø \sin ß) \qquad \text{(Eq. 12-4)}.$$

Traverse corrections can be applied to the distribution of data points on an equal-area net. For a counting circle centered at a particular ø and ß, apply Equation 12-4 to the number of data points that fall in the circle.

Tilt Corrections: If jointing predates the tilting or folding of strata in an area, the present-day orientation of the joints will be different from the original orientation of the joints. In such cases a tilt correction must be applied (see Chapter 6), in which the joints are rotated around the strike of the beds that contain them by an amount equal to the dip of the beds. The rotated orientation of the joints is then plotted graphically. The rotation can be accomplished by computer or graphically on an equal-area net.

Statistical Treatment of Joint Data

Statistical analysis of joint orientation data can be quite complex; the arithmetic mean of several compass measurements cannot be determined by simply adding the strikes together and dividing by the number of measurements. Below we describe a calculation that permits the mean angle of a set of angles (e.g., joint strikes) to be determined (taken from Zar, 1984).

Consider a population of n angles. Each angle is denoted by a_i (Zar, 1984). In order to determine the mean angle (a_m) of the population, we first define the quantities:

$$G = (\Sigma \cos a_i)/n$$

$$H = (\Sigma \sin a_i)/n$$

$$T = \sqrt{(G^2 + H^2)}.$$

Given these quantities, the mean angle (a_m) can be determined by either of the following two equations:

$$a_m = \text{arc cos } (G/T) \qquad \text{(Eq. 12-5)}$$

or

$$a_m = \text{arc sin } (H/T) \qquad \text{(Eq. 12-6)}.$$

Discussion of more advanced statistical treatment of joint orientation data is available in Kohlbeck and Scheidegger (1977), Hay and Abdel-Rahman (1974), Wheeler and Stubbs (1979), and Wheeler and Holland (1978).

12-4   STYLE, AGE ANALYSIS, AND INTERPRETATION OF JOINTS

Style of Jointing

The collection of orientation and spacing data alone does not constitute a complete analysis of jointing in an area. In order to interpret the tectonic significance of jointing (e.g., Engelder and Geiser, 1980; Hancock, 1985) and to classify joints (e.g., Nelson, 1979), it is necessary to characterize the *style* of joints and to determine, if possible, the relative ages of different joint sets. It is also important to record basic characteristics about the geology (e.g., lithology, bedding thickness, existence of other structures) of the outcrop containing the joint.

The style of a joint or joint set (e.g., Wheeler and Stubbs, 1979; Kulander, Barton, and Dean, 1979) includes the following characteristics of the joint:

1. *Surface morphology:* This refers to the appearance of the joint surface. Knowledge of joint surface morphology permits kinematic analysis of the joint (e.g., Arthaud and Mattauer, 1979; Kulander et al., 1979). In particular, it is possible to specify whether a joint formed by extension and has subsequently remained unchanged, or whether the joint surface has been modified by shear or compression. Plumose structure, described earlier, is generally attributed to the tensile fracturing. Slip lineations indicate either that shear has occurred subsequently occurred on a joint surface or that the fracture initiated as a shear surface, and stylolitic pitting indicates the occurrence of compressional stresses across the joint surface (see Price, 1967; Ramsay and Huber, 1987).

2. *Joint size:* This refers to the trace length of a joint or, better, the joint surface area.

3. *Systematics:* This refers to the regularity of joints in a specified orientation. Does a joint appear to be part of a systematic set?

4. *Relation to bedding:* This refers to both the joint attitude with respect to bedding and nature of the interaction between joints and bedding planes. Are joints perpendicular or oblique to bedding? Do joint systematics improve if a tilt correction is applied to the joints in an area? Are joints bedding-confined?

5. *Relation to other structures:* This refers to the geometric relation among joints, folds, and faults. Are joints parallel to, oblique to, or perpendicular to the bearing of fold hinges in the area? Does joint spacing,

morphology, or orientation change in the vicinity of a fault?

6. *Joint intensity:* Calculation of joint intensity is partly a statistical problem and has been approached in many ways. For a discussion of methods for calculating joint intensity, see Karcz and Dickman (1979) and Wheeler and Dixon (1980).

Joints in Folded Regions

In regions of flat-lying rock, it is common for prominent joint sets to be either parallel or perpendicular to the earth's surface (Suppe, 1985). In regions where rocks have been folded, however, a greater range of joint orientations with respect to bedding commonly occur (e.g., Nickelsen, 1979) and it may be useful to define the geometry of the joints in terms of their geometric relation to the folds.

One way of referring to joints in folded regions is to define joints that are parallel to the strike of fold axial planes as *strike joints* and to define joints that cut across the axial plane as *cross-strike joints*. Hancock (1985) described joint orientation with relation to symmetry axes (a, b, and c) of a fold using a procedure similar to the definition of Miller indices for crystals. This convention is described in Figure 12-9. As an example, vertical joints that cut obliquely across an upright fold are called hk0

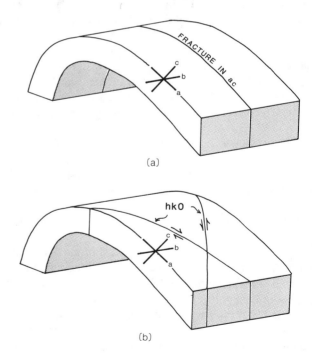

(a)

(b)

Figure 12-9. Terminology for describing joint orientations with respect to folds. (a) Fold showing the orientation of fractures in ac; (b) fold showing the orientations of hk0 fractures.

joints, with the h and k denoting intersection of the joints with both the a axis and the b axis respectively, and the 0 (zero) indicating the lack of intersection with the c axis.

As noted earlier, it is important to apply tilt corrections to joints in folded regions to help determine whether the joints were tilted during folding, and thus may predate folding. While measuring joints in a fold, be sure to define structural domains carefully (i.e., separate measurements near the hinge from measurements on the limbs) and keep track of bedding attitude at each measurement station.

Age Relationships of Joints

There are two methods available for determining age relationships between two different joints. These methods are briefly described next.

Analysis of Joint Intersections: If two non-parallel joints form simultaneously, they will mutually cross-cut one another, and there may be no apparent interaction between the joints along the line of intersection.

If one joint forms before another, the first joint is a free surface (unless it has been recemented) and thus cannot transmit shear stress; therefore, the regional stress field is modified locally in the vicinity of the joint such that principal stresses are parallel or perpendicular to the joint surface. As a second joint grows and approaches the first, its trace bends, so that it intersects the first joint at a right angle. The second joint terminates at the first joint. The bending of the second joint is called *hooking* (Kulander et al., 1979), and the intersection is called a *J-* or *T-intersection*. Where such intersections are observed, the younger joint is always the one that bends and/or terminates at the older joint (Fig. 12-10). The explanation of hooking depends on the assumption that the second joint forms perpendicular to σ_3.

Mineralization on Joints: Joint surfaces can be locally coated with quartz, calcite, or other minerals deposited when water flows along the fracture. Absence of mineralization on one joint set that cuts a different mineralized set indicates either that the unmineralized set postdates the mineralized one or that stress kept the unmineralized set closed to fluid circulation. In favorable circumstances close examination of mineral coatings on joints may help to determine the sequence of jointing. For example, Tillman and Barnes (1983) used fluid-inclusion analysis to deduce the temperature at which minerals were deposited on a joint surface; they found that the mineral coatings of different joint sets on the Appalachian Plateau formed at different temperatures. If it is assumed that the temperature of rocks of the plateau decreased progressively as the area was uplifted, then mineral coatings formed at

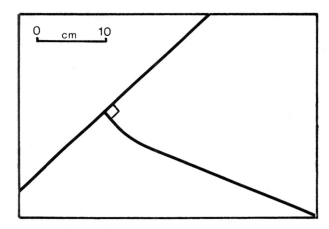

Figure 12-10. Map view of a J-intersection between two joints. The curved joint is younger.

lower temperatures could be assumed to be relatively younger. Presumably, the joints on which the younger coatings formed are younger than joints on which the older coatings formed.

Dynamic Interpretation of Joint Sets

An examination of a Mohr-Coulomb failure envelope (see Chapter 10 for a discussion of failure envelopes) suggests that there are three different types of fractures each of which is associated with a specific stress state (e.g., Suppe, 1985). *Extension* (or *tensile*) fractures form under conditions where the Mohr circle is tangent to the failure envelope at only one point (Fig. 12-11a), and one of the principal stresses is tensile. *Shear fractures* form when the Mohr circle is tangent to the envelope at two points and all principal stresses are compressive (Fig. 12-11b). A third fracture type also forms when the Mohr circle is tangent to the envelope at two points, but under conditions such that one of the principal stresses is tensile (Fig. 12-11c); this third type of fracture has been called either a *transitional-tensile fracture* (Suppe, 1985) or a *hybrid shear fracture* (Hancock, 1985). Note that according to the Mohr-Coulomb diagrams (Fig. 12-11), extension fractures form perpendicular to σ_3 and parallel to σ_1, whereas shear fractures and hybrid shear fractures form at an angle to the principal stresses.

Studies of fracture formation in ceramic materials suggest that there are three ways in which a fracture can propagate through intact rock (Fig. 12-12; Kulander et al., 1979). *Mode I* propagation means that the fracture grows by incremental extension perpendicular to the plane of the fracture at the fracture tip. *Mode II* and *mode III* propagation means that the fracture propagates by incremental shear parallel to the plane of the fracture at the

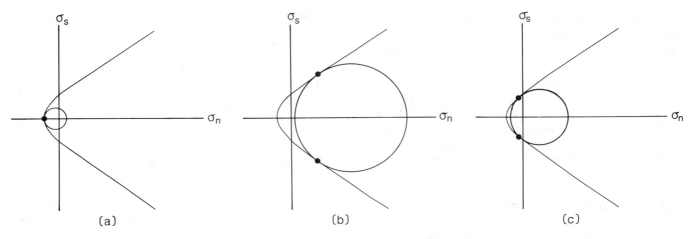

Figure 12-11. Mohr-Coulomb diagrams defining stress state at the time of joint formation. (a) Stress state during formation of an extension fracture; (b) stress state during formation of a shear fracture; (c) stress state during formation of a hybrid shear fracture.

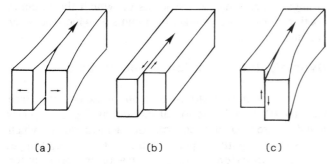

Figure 12-12. Block diagram sketches of fracture opening modes (a) Mode I propagation; (b) mode II propagation; (c) mode III propagation. (Adapted from Kulander et al., 1979.)

fracture tip. Extension fractures form by mode I propagation, shear fractures by mode II or III propagation, and hybrid shear factures by a combination of mode I and modes II and/or III.

The dynamic interpretation of natural joints (i.e., the specification of the stress state at the time of joint formation) remains quite controversial. The controversy revolves around the interpretation of plumose markings on joint surfaces. One school of geologists believe that the presence of a plumose marking on a joint indicates that the joint formed by mode I propagation, that joints are strictly extension fractures, and that joints are a principal plane of stress. A second school believes that some plumose-marked joints are hybrid shear fractures. In particular, this school considers members of conjugate systems of hk0 joints to be hybrid shear fractures even if they have plumose markings (Hancock 1985). In contrast, geologists of the first school believe that that conjugate systems are

merely pairs of extension joint sets each of which formed at a different time under a different stress state. The issue is moot if there is clear evidence of shear on a fracture (crushed grains and slip lineations), in which case the fracture is probably best described as a fault.

Joints which do form strictly by mode I propagation define a principal plane of stress. Based in this fact, several authors have used extension joints as *paleostress-trajectory* indicators (Engelder and Geiser, 1980). A paleostress trajectory is the orientation of a principal stress in the rock at some time in the past. In these studies, the trend of a vertical joint on the ground surface is taken as the trajectory of σ_1 at the time of the joint formation (σ_2 is assumed to be vertical).

Numerous studies have considered the creation of stress states leading to joint formation. In general, it is considered that most joints form in association with uplift of a region (Price, 1966; Suppe, 1985; Engelder, 1987).

12-5   FAULT-ARRAY ANALYSIS

Fault Arrays

In Chapters 4, 6, and 9 we learned how to analyze the sense, direction, amount, and timing of fault displacement resulting from movement on individual faults that measurably offset mappable contacts. In many localities the bedrock is broken by numerous faults. Parallel faults which have parallel directions of motion and similar senses of motion comprise a *fault set*. Two or more fault sets comprise a *fault array* or a *fault system*. The size of faults in fault arrays is quite variable; at some localities individual faults are mappable features on regional maps, and the fault array is a prominent feature of the orogenic belt, whereas at other localities the faults of the array are mesoscopic, in that they have outcrop traces of only a few meters and displacements of only a few centimeters (Price, 1967). There are several types of fault arrays:

1. *Parallel array:* A parallel array is composed of a single set of parallel faults. There are three types of parallel arrays. (a) *Rotational parallel arrays*, are composed of a set of parallel faults, the motion on which leads to rotation of the inter-fault rock bodies around an axis that lies in the fault plane and is perpendicular to the slip direction on the fault (Wernicke and Burchfiel, 1982); (b) *Nonrotational parallel array,* are composed of a set of parallel faults accommodating shear without block rotation; (c) *Relay arrays*, are composed of parallel faults each of which is shorter than the length of the host deformed belt (Fig. 12-13). As one fault dies out along strike, its displacement is transferred to a neighboring fault. As a consequence cumulative displacement across the deformed belt is relatively constant. Relay arrays are common in fold-thrust belts

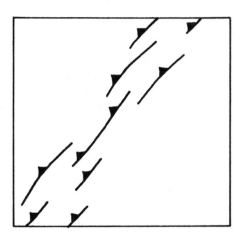

Figure 12-13. Sketch map of a relay array of thrust faults.

2. *Conjugate array:* A conjugate fault array is composed of two fault sets that formed at the same time under the same stress system. The dihedral angle between the two sets is typically between 50° and 70°, but larger angles have been documented (Freund, 1970; Marshak et al., 1982).

3. *Anastomosing array:* In some localities a fault zone contains many intertwined faults that cross one another at a low angle or merge with one another. The sense of movement on all faults is roughly the same.

4. *Complex array:* In a complex array, there are many non-parallel sets of faults. However, the shear sense and shear direction on each fault is such that, overall, the slip on the array results in a single coherent regional movement (e.g., Compton, 1966; Arthaud, 1969; Reches, 1978). Arrays such as these possibly reflect either slip on preexisting surfaces (earlier formed joints, faults, or foliations; Bott, 1959; Compton, 1966), faulting under nonplane strain conditions (Reches, 1978), and noncoaxial strain (Wojtal, 1986). Rhomboid fault patterns (Reches, 1978) are one type of complex array.

5. *Detachment-bounded array:* In both fold-thrust belts and rift zones, sets of faults occur which merge at depth along a subhorizontal detachment fault (see Chapter 14). Such arrays include both planar and listric faults.

5. *Polyphase array.* Some regions are cut by sets of faults each of which formed at a different time and under a different stress field. A group of tectonically unrelated sets can be called a polyphase array.

Description of Fault Arrays

A fault array is a group of planar structures. Therefore, strike and dip data representing a fault array can be treated much like strike and dip data on a joint array. A fault is different from a joint, in that shear occurred parallel to the surface of the fault (see Chapter 11 for a discussion of how to determine direction and sense of slip on a fault). There are several ways to communicate information on slip direction and shear sense on a fault.

Rake Histogram: Data on the rakes of slip lineations on an array of faults can be plotted on a histogram. The vertical axis of the histogram represents number of measurements and the horizontal axis represents rake in class intervals of 5° - 10°. If the majority of slip lineations have low rakes, for example, then the fault array is composed of strike-slip faults.

Annotated Equal-Area Plot: The orientation of a fault on an equal-area plot can be represented by the orientation of the pole to the fault. Additional information can be represented on such a plot if a symbol is plotted at the position of the pole rather than just a dot. Appropriate symbols include, e for extensional faults, c for contractional faults, n for normal faults, r for reverse faults, and s for strike-slip faults. If all the data on a plot refer to strike-slip faults, you may wish to plot open circles for left-lateral faults and filled circles for right-lateral faults. Such plots can also display the orientation of the slip lineations.

Slip-Linear Plot: A slip-linear plot is an equal-area plot on which the symbol for the pole to a fault plane is decorated by a line that indicates direction of slip, or an arrow that indicates the direction and sense of slip. Such diagrams are very useful for representing the kinematics of a fault array. The method of constructing a slip linear plot (Hoeppener, 1955) is described based on a figure from Anastasio (1987).

Problem 12-2 (Construction of a slip-linear plot)

A fault is oriented 000°,60°E. Slip fibers on the fault plunge 50° in a northeasterly direction. The hanging wall of the fault moved relatively up dip. Construct a slip-linear plot representing this fault.

Method 12-2

Step 1: Refer to Figure 12-14a. This figure shows a fault coated with imbricate slip fibers that give the direction and sense of movement. Visualize the problem; the fault is an oblique-slip fault with a component of reverse motion.

Step 2: On an equal-area diagram (Fig. 12-14b), plot the great-circle trace representing the fault plane, the pole to the fault plane, and the point representing the plunge and bearing of the slip fibers.

Step 3: Construct a plane, called the M-plane ('M' stands for 'movement'), which contains the slip fiber lineation and the pole to the fault plane. At the point

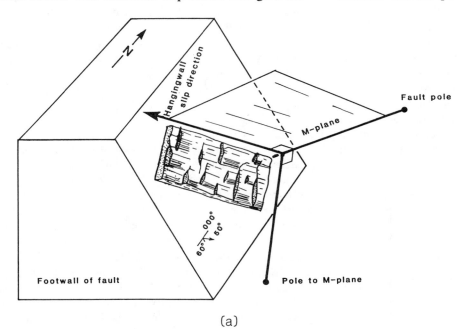

(a)

Figure 12-14. Construction of a slip linear plot. (a) Block diagram illustrating the position of the M-plane with respect to fiber slip lineations; (b) equal-area plot showing the slip linear and the great-circle traces of the fault plane and M-plane; (c) slip linears representing an array of faults in the southern Pyrenees of Spain. (From Anastasio, 1987.)

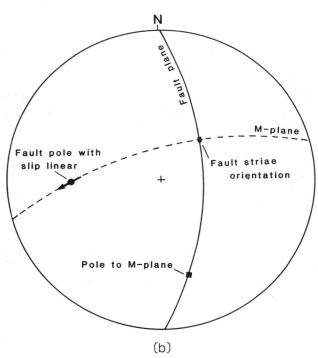

(b)

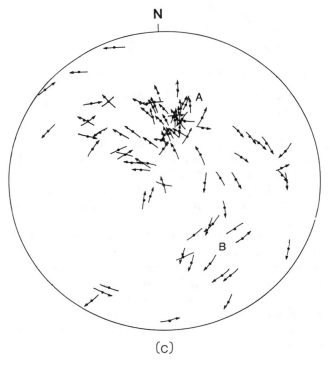

(c)

representing the pole to the slip plane, draw a short line segment along the the great-circle trace of the M-plane. This line segment is the slip linear representing the direction of slip on a fault.

Step 4: If the sense of movement on the fault is known, an arrow can be added to the line segment indicating the relative movement of the hanging-wall block. In this example, the hanging-wall block moved toward the southwest, so the slip linear points toward the southwest.

Step 5: To interpret an arrow on a slip linear, you must keep in mind the orientation of the fault plane (indicated by the pole position) as you look at the line segment or arrow. From just the slip linear in Figure 12-14b, we know that fault plane dips moderately to the east and that the hanging-wall block moved up and to the southwest. Therefore, we immediately know that it is an oblique-slip fault on which there has been a component of right-lateral shear and a component of reverse shear.

If a large number of slip linears from faults in a region are plotted on a single equal-area plot, the diagram graphically indicates the kinematics of movement on the array. For example, Figure 12-14c shows a slip linear plot for faults in a portion of southern Pyrenees of Spain (from Anastasio, 1987). From this diagram we see that many faults dip to the south (cluster A); on these faults, the movement of the hanging-wall block is directly up-dip, so the faults are reverse faults. In addition, many faults (e.g., cluster B) dip steeply to the northwest. These faults display slip lineations that point to the southwest, and, therefore the faults are strike-slip faults. In general, radial slip linears indicate dip-slip faults and arrows parallel to the primitive are strike slip.

Another plot that provides kinematic data on faults can be constructed by first plotting the pole to the M-plane and then by indicating the sense of movement as a rotation around the pole. The points on such a plot are called kinematic rotation axes. For a description of the method, see Wojtal (1986).

In addition to attitude data on fault arrays, a description of an array should also provide information on the size of faults, spacing of faults, the magnitude of displacement on individual faults, and the character of the fault surface (see Chapter 11). You should also describe the relationship of the fault array to other structures in the region, such as folds and larger faults.

Determination of Principal Stress Directions from Fault Arrays

The pattern of slip on a simple conjugate array of mesoscopic faults is directly related to the state of stress at the time of faulting. The basis of this interpretation is

called the *Anderson theory of faulting* (after Anderson, 1942), which assumes that σ_2 lies in the plane of the fault, and that σ_1 is oriented at 30° - 45° to the fault, and σ_3 is orientated at 45° - 60° to the fault; the configuration of the three principal stresses with respect to the surface of the earth determines whether the fault initiates as normal, thrust, or strike-slip (Fig. 12-15).

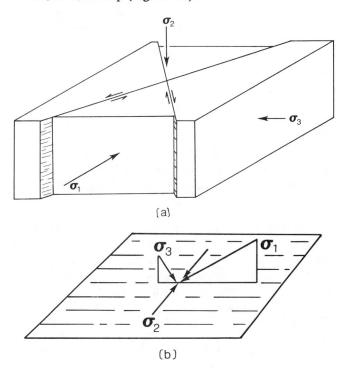

Figure 12-15. Ideal orientations of fault planes with respect to principal stresses. (a) Block diagram showing the orientation of principal stresses with respect to two conjugate strike-slip faults; (b) diagram showing principal stresses with respect to slip lineations on a single fault plane.

Problem 12-3

Two fault sets are observed in a region. Set A is oriented N88°W,40°NE and has slip lineations oriented 26°,N56°E. Set B is oriented N44°W,82°SW and has slip lineations oriented 61°,S30°E. What were the orientations of the principal stresses which produced these fault sets?

Method 12-3

If the faults define a simple conjugate array, then σ_1 bisects the acute angle of intersection, σ_2 is parallel to the intersection of the two sets, and σ_3 bisects the obtuse angle of intersection.

Step 1: Plot the faults as great circles on an equal-area net (Fig. 12-16). Plot the slip lineations as lines (L_a and L_b) that lie in the fault planes (Chapter 5).

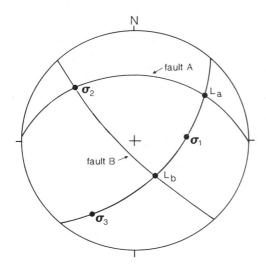

Figure 12-16. Equal-area plot showing estimation of principal stresses from data on two faults of a conjugate system. L_a and L_b are slip-lineation attitudes.

Step 2: Construct a great circle perpendicular to the intersection of the two fault sets (Figure 12-16). This great circle is oriented N42°E,62°SE. The slip lineations should lie on or near to this great circle. If they do not, then the fault geometry is not truly conjugate.

Step 3: By counting along the great circle drawn in Step 2, find the acute bisectrix of the two fault sets. This line, which is 53°,N86°E, gives the orientation of σ_1.

Step 4: Along the same great circle, find the obtuse bisectrix of the two fault sets. This line, which is 23°,S29°W, is the orientation of σ_3.

Step 5: The line of intersection between the two faults sets gives σ_2.

The geometry described in Problem 12-2 is that of a conjugate fault array. However, it is necessary to examine additional criteria before treating the calculation of stress orientation with confidence. Specifically, it is important to check that the sense of slip on the fault sets is appropriate; in general, slip on fault set A should be right-lateral with a component of normal motion, and slip on fault set B should be left-lateral with a component of reverse motion. With such a sense of slip, the acute wedges between the faults move toward each other.

If we assume that the σ_1 direction is at a specified angle from the fault plane, we can estimate the principal stress orientations from a single set of lineations.

Problem 12-4

A single set of mesoscopic faults occur in an area. The faults are oriented N30°W,80°NE. Slip lineations on the fault are oriented 19°,N26°W. The fault has a right-lateral sense of shear. Estimate the orientations of

the principal stresses which gave rise to this fault set. To help visualize the problem, refer to Figure 12-15b.

Method 12-4

Step 1: Plot the orientation of the fault set as a great circle and the slip lineations as a point (L) lying on that great circle (Fig. 12-17). We assume that σ_1 is oriented at 30° to the fault plane and lies in the M-plane (the plane that is perpendicular to the fault plane and contains the slip lineations). We assume that σ_2 lies in the fault plane 90° from the slip lineations.

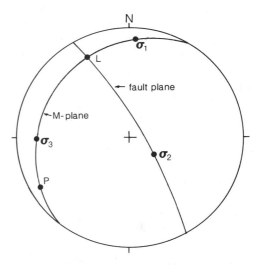

Figure 12-17. Equal-area plot showing estimation of principal stresses from a single set of slip lineations.

Step 2: Align the great circle representing the fault plane with a great circle on the equal-area net. Count along this great circle exactly 90° from the orientations of the slip lineations; this point represents theoretical orientation of σ_2 (68°,S54°E).

Step 3: Draw a great circle, which has the σ_2 orientation as its pole. All points along this great circle are 90° from the σ_2 orientation, and thus σ_1 and σ_3 must lie along the great circle. Note that this great circle contains the pole to the fault plane and the slip lineation, and thus represents an M-plane. We could have constructed the M-plane without first locating σ_2.

Step 4: We assume that the angle that σ_1 makes with the faults is 30°. Count along the M-plane great circle 30° from the slip orientation to find the σ_1 orientation and an additional 60° to find the σ_3 orientation. The direction in which you count depends on the sense of motion on the fault. In this problem, the fault is right-lateral, so σ_1 is 12°,N04°E and σ_3 is 19°,S90°W.

A number of techniques exist for determining the orientations of principal stress directions and or deformation directions (i.e., directions of shortening and elongation) associated with faults of a complex array. For discussion of these techniques, refer to papers by Compton (1966), Arthaud (1969), Arthaud and Choukroune (1972), Angelier and Melcher (1977), Reches (1978; 1983), Angelier (1979), Etchecopar et al. (1981), and Aleksandrowski (1985).

Some of the techniques are based on plots of M-planes and slip linears. These techniques require orientation and slip data on a large number of faults. The essence of the M-plane method is as follows; you should study the details of the method and its limitations in the primary references before using it for a real field study.

Problem 12-5 (Simplified M-plane stress calculation)

Given the attitude of 100 fault planes in a complex array and the attitude of the slip lineations on the fault planes, estimate the attitudes of principal stresses defining the state of stress at the time of faulting.

Method 12-5

Step 1: Plot the great circle traces representing the M-planes of the faults on an overlay placed over an equal-area net (Fig. 12-18a). If the faulting in the region is part of a complex array, the M-plane great-circle traces will intersect at a limited number of common locations (called "common intersection points") on the equal-area plot; the number of intersections depends on the variability of fracture orientations and on the nature of strain in the region. Two of the common intersection points of the M-planes may represent σ_1 and σ_3 (Fig. 12-18b). σ_2 should occupy a region of the net in which there is a low density of M-plane intersections.

Step 2: On the separate overlay, draw the slip linears for the fault array (Fig. 12-18c). In an ideal case, the slip linears point away from the σ_1 direction and toward the σ_3 direction, and therefore allow you to isolate which common intersection points are representative of principal stress directions.

Step 3: Choose the appropriate common intersection points from your M-plane plot. The σ_2 orientation is defined by the pole to the plane containing σ_1 and σ_3.

We caution that the above method should not be applied blindly. Aleksandrowski (1985) discusses limitations of the method and shows how to increase the reliability of the results.

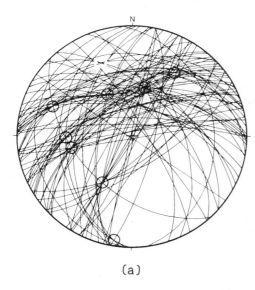

(a)

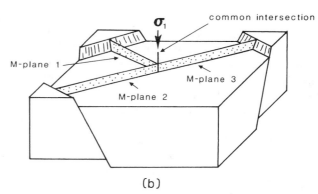

(b)

Figure 12-18. M-plane method of calculating principal stresses from a complex fault array. (a) M-plane great-circle traces for members of a complex array. Circles show the common intersection points (from Aleksandrowski, 1985); (b) block diagram showing how the common intersection of three M-planes may be related to a principal stress; (c) slip linear plot for the faults of plot 'a'. Note that the slip linears point toward σ_3 and and away from σ_1 (from Aleksandrowski, 1985).

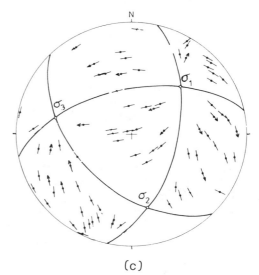

(c)

12-6   LINEAMENT-ARRAY ANALYSIS

Nature of Lineaments

Lineaments are straight or nearly straight topographic alignments that are visible on remote-sensing images (e.g., air photographs and satellite images) or on topographic maps. A set of lineaments that occur in a given area can be called a *lineament array*.

The recognition of a lineament is merely a judgment of the viewer; any two observers may or may not agree on the presence of a specific lineament or may draw a given lineament with a slightly different azimuth and/or length. Some lineaments are, in fact, optical illusions whose presence or absence may reflect only illumination direction (Wise, 1982). It is often difficult to verify the presence of a lineament or to assess its geological significance from field work. Wise et al. (1985) have noted, however, that although experienced observers might disagree about a single lineament, they will usually find the same overall lineament trends on a single image, and these trends probably have geological significance. Because the observation and interpretation of lineaments is highly subjective, the topic of lineament analysis is one that is approached with justifiable skepticism by most geologists.

The description of a lineament pattern in an area is of little value if the geologic significance of the lineament is unknown. Commonly, lineaments represent the trends of fault zones or the trends of major systematic joints. Joint systems stand out as lineaments particularly if they control the drainage in a region. Lineaments may also represent the structural grain of bedrock (orientation of deformation fabric), the distribution of bedrock or cover units, the position of linear intrusives or igneous fissures, nonstructurally controlled topography, or human-made constructions. Assessment of the geologic significance of a lineament must be determined through field analysis.

Practical Suggestions for Lineament Analysis

In order to make the best of a lineament study, the technique for drawing lineaments should be standardized for a given study, so that results may be repeatable. Following are a number of suggestions to achieve this goal:

1. In studying lineaments exposed in a region, use images of the same type that have been processed in the same way. Do not mix study of satellite images with study of low-altitude photographs.

2. Obtain images that are illuminated obliquely, for it is the shadowing effect of topography that permits lineaments to be observed; the sun should not be directly overhead. Maximum enhancement of a linear feature occurs when the feature is illuminated at an angle of 15° from parallel (measured in a horizontal plane) (Wise et al., 1985).

3. To get consistent results it is best to use images of adjacent areas that were made at the same time of day. On a given image, however, lineaments oriented 15° from the sun azimuth or radar look direction tend to stand out. Therefore, to avoid biasing results, it is best to repeat measurements on two or more images of the same area taken at different times.

4. When drawing lines, use an overlay. Try not to think of the geology of the area covered by the image. Prejudice is subtle.

5. Do not work on an image for too long a period of time. The longer you stare at an image the more lineaments you will see. The first 10 to 45 minutes of observation will allow you to find the most obvious lineaments that are probably the most geologically significant.

6. When studying an image, rotate it frequently so that you do not always look at the image in the same direction. We tend to see lineaments parallel to the direction we are looking.

7. Draw only what you see. Do not extend a lineament to intersect another unless you actually observe it to do so.

8. Do not mix observers. One observer should make the measurements on all the images of the area. It is of value, however, to have more than one observer repeat measurements of the same images. The results of two or more observers can be overlain on top of one another, and a synoptic overlay showing only the lineaments indicated by all observers can then be produced. The synoptic overlay may be less noisy than the individual overlays.

Once an overlay showing the traces of lineaments has been produced, it can be analyzed with a rose diagram, histogram, or running average. Equal-area nets, of course, cannot be used, as the data are only two-dimensional. Remember that the high degree of subjectivity involved in analyzing lineaments means that the results of a lineament study should be viewed with skepticism until the lineaments can be related to real geology.

12-7   ACKNOWLEDGEMENTS

We thank Russell Wheeler, Steven Wojtal, and Deepak Srivastava for helpful comments on the manuscript. Discussions with Terry Engelder improved our treatment of joint interpretation. Scott Wilkerson provided data for the second exercise.

EXERCISES

1. As an initial exercise, your instructor will set up a field-based study of jointing. Unless the area that you live in is completely covered by glacial drift or vegetation, there should be an outcrop nearby that contains joints. Prepare a field form or use the one provided. If you have time, use several methods of sampling joints, and compare the results. If your time is limited, use only the traverse method. Begin your sampling traverse at some recognizable location, and record data on joint orientations, spacings, and style elements. Make preliminary data plots (rose diagram, histogram, or equal-area plot) at the outcrop. Construct a synoptic diagram to indicate the major joint sets in the outcrop. While still at the outcrop, try to determine the relative timing of formation of different joint sets.

2. The data in Table 12-M1 are strikes of vertical joints measured along a traverse.

Table 12-M1
Joint Orientations

N03°W	N04°W	N69°W	N03°W
N06°W	N03°W	N03°W	N03°E
N53°E	N36°W	N09°W	N03°W
N03°W	N27°W	N03°E	N09°W
N69°W	N21°E	N18°W	N72°W
N18°W	N21°E	N69°W	N63°W
N07°W	N18°E	N04°W	N72°W
N63°W	N09°W	N27°W	N04°E

(a) Make a histogram of these orientations using a 1° class interval. Plot a second histogram with a 5° class interval. Compare the two plots and describe the orientations of prominent joint sets along the traverse. Make a rose diagram of the data with a 5° class interval. Which of the plots do you find to be most easy to interpret?

(b) Apply a traverse correction for these data using as the strike of a class interval, the mid-point of that interval. Portray the corrected data as a histogram with a 5° class interval. Describe the differences between the corrected and uncorrected plots.

(c) Smooth the data in the interval between a strike of N10°W and a strike of N04°E. Use the original 1° class interval plot for this calculation. Use a 10° running average with a 2° step. Describe the differences between the smoothed and raw plots.

3. Slip lineations on fault set A of a conjugate pair are 18°,N70°E and slip lineations on fault set B are oriented 30°,S30°E. What were the orientations of the principal stresses responsible for faulting?

4. Mesoscopic faults in an area have the following dominant orientations: Set A is N17°W,70°SW, and Set B is N21°E,52°SE.

(a) For these two sets to be considered conjugate, what should be the orientations of the slip lineations on them?

(b) What would be the senses of motion on the two sets if they were conjugate?

(c) What would be the orientations of the principal stresses that gave rise to a conjugate set with the specified geometry? Speculate about the tectonic setting that could have generated such stress orientations.

5. The image in Figure 12-M1 is a synthetic aperture radar image of part of the island of Luzon in the Phillipines. This type of image enhances topography by using the "shadowing" effect of radar. Several different types of linear features can be seen on this image. First, a major linear boundary cuts diagonally across the image and separates the northwest corner of the image from the rest of the image. This boundary, which trends east-northeast, may be a fault. Within the rough topography, short linear valleys and ridges can also be seen. These lineaments are typical of lineaments in heavily vegetated regions and are commonly interpreted as reflecting etching of pervasive systematic joints by erosion (Wise et al., 1985).

Figure 12-M1. Synthetic-aperture radar image of part of Luzon, the Philippines. Flight lines ran north-south, and radar illumination direction is east-west. Scale bar is 5 km. (Courtesy Aero Service.)

(a) Draw all the linear features that you can see on the image. Try to differentiate between those lineaments which may be major faults, and those which are related to the etching of erosion (*Hint:* Look for offsets of topographic features).

(b) Plot a histogram of the erosional-etching lineaments. Note that this image was produced using an east-west illumination direction. Do the prominent lineation directions show a relationship to the illumination direction?

(c) Compare your tracing and your diagrams from parts (a) and (b) with those of other students. Do individual students have distinctly different styles of drawing lineaments? Do all students show the same lineaments? Are there some dominant directions that all students show? Based on your answers, do you think that lineament analysis can produce geologically significant results?

6. Figure 12-M2 is a LANDSAT image of a portion of the Makran Range in southwestern Pakistan. Comparison of the image with a geologic map of the area shows that the major east-northeast-trending lineaments in the image correspond to outcrop belts of strata that have been involved in formation of a fold-thrust belt. Place an overlay on the image and spend about 10 minutes tracing the lineaments that cut across the ENE-trending outcrop belts. Interpret your results (*Hint:* In the southern quarter of the image, there are some well defined plunging folds. Consider the geometry of the lineament array with respect to these folds.)

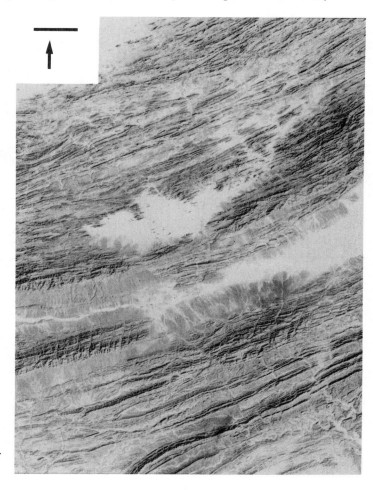

Figure 12-M2. LANDSAT image of a portion of the Makran Range in southwestern Pakistan. Scale bar is approximately 5 km.

CHAPTER

13

OBJECTIVE METHODS
FOR CONTRUCTING PROFILES
AND BLOCK DIAGRAMS
OF FOLDS

Steven Wojtal

13-1 INTRODUCTION

A slice or section through any three-dimensional object provides a useful visual image of its interior. Geologists often construct sections of the earth to illustrate its internal structure. A section that is oriented perpendicular to the surface of the earth is called a *cross section* (Appendix 1). Block diagrams (Appendix 1) combine data from maps and cross sections to provide a perspective image of a three-dimensional block of the earth.

Cross sections and block diagrams are influential tools that you can use to convey your ideas about the geologic structure of an area to other geologists. They are also often used *as data* when analyzing the tectonic history or resource potential of an area, so it is important that they be as accurate and truthful as possible. Constructing cross sections and block diagrams tests your understanding of the geometry of deformed rocks. The central problem that you encounter when constructing a cross section or block diagram is, How can surface data be extrapolated to depth? Extrapolation, as you will see, depends in part on objective geometric techniques of projecting structures and in part on subjective interpretation.

In this chapter we examine some of the objective geometric techniques (Busk method, kink method, dip-isogon method, and down-structure projection) used to project surface data on fold geometry to depth. We will also see how to incorporate drill-hole and seismic data in such sections. Finally, we will describe how to represent accurately the three-dimensional configuration of rock

structures in block diagrams that have geologic maps on their upper surfaces and geologic cross sections on their sides. Additional aspects of cross-section construction are introduced in Chapter 14. It is important to emphasize at the outset that the *reproducibility* of cross sections and block diagrams drawn using the techniques described in this chapter must not be confused with the *truthfulness* of these representations. Natural geologic structures rarely conform to ideal geometries; thus, real geology may deviate markedly from sections drawn using geometric models.

13-2 FOLD STYLES AND SECTION LINES

Cylindrical and Cylindroidal Folds

Objective techniques for projecting fold geometry to depth can be applied only to folds whose shapes have a certain degree of regularity. *Cylindrical* or *cylindroidal* folds are two types of folds whose shapes are sufficiently regular that data on fold shapes at the surface can be used to characterize fold shapes at depth. The techniques are not feasible in regions where fold shapes are very irregular (folded layers thicken or thin dramatically, and fold trains are disharmonic); in such regions, knowledge of fold geometry at the surface will not help us predict fold geometry at depth.

As described in Chapter 8, the axis of a cylindrical fold is a straight line that, when moved parallel to itself, can "trace out" the folded surface (Fig. 13-1). Because of this

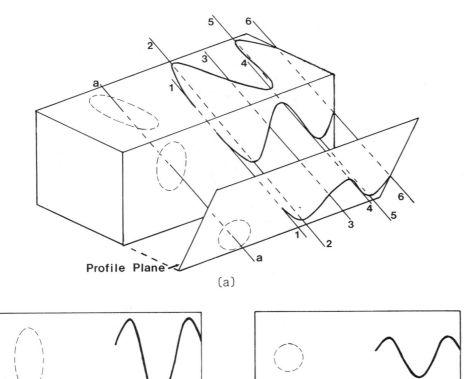

Figure 13-1. Map, cross section, and profile views. (a) If the axis of a circular cylinder (line a) plunges, the map pattern of the cylinder on a horizontal surface is an ellipse, and the intersection of the cylinder with a vertical cross-section plane is an ellipse. Likewise, the shape of a plunging fold is distorted in the map and cross-section planes; (b) distorted shapes in the map plane; (c) true shapes in the profile plane.

property, the shape of a cylindrical fold can be projected orthographically along the fold axis onto a plane that is normal to the axis. Few real folds are cylindrical, but many real folds are *cylindroidal*. In a cylindroidal fold, segments of the hinge line are nearly straight lines, but no single straight line can trace out the entire fold. In practice, we can construct representations of cylindroidal folds by assuming that they are composed of several cylindrical segments, where the axis of each segment is not exactly parallel to the axes of the adjacent segments (Ramsay and Huber, 1987; Langenberg and others, in press).

Choosing the Line of Section

You will recall from earlier chapters that the inclination of a dipping layer in a vertical section equals the layer's true dip only if the section is oriented perpendicular to the layer's strike. Likewise, the truest representation of any cylindrical fold is a section taken perpendicular to the fold's axis (Fig. 13-1; see also Suppe, 1985). A section that is oriented perpendicular to fold axes in a region is called a *profile section* or, more commonly, a *profile*, and a section that is oblique to the axis of a structure is called an *oblique section*. The traces of the folded layers exhibit their maximum curvature in a profile plane. The profile plane of a nonplunging cylindrical fold is vertical and strikes

perpendicular to the strike of the folded layers. The profile plane of a plunging cylindrical fold must be an inclined plane. Since the fold is cylindrical, profiles drawn at all points along the fold axis must be identical. If a fold is cylindroidal, folded layers exhibit different shapes in different sections along the length of the fold. Sections that show the maximum curvature of the folded layers at different points along the fold axis have different strikes and dips. There is, therefore, no such thing as a single profile of a cylindroidal fold. We can draw a section that approximates a fold profile for an individual segment of a cylindroidal fold by positioning the section normal to the local fold hinge.

In a given map area it is best to draw sections of folds so that the surface trace of the section, called the *line of section*, crosses regions where surface geology is well constrained and/or there are seismic or drill-hole data available. The line of section should intersect several attitude measurements; the spacing between attitude measurements along the line of section must be less than the wavelength of the folds.

Parallel and Nonparallel Folds

In some geologic settings layered sequences of rock are folded in such a manner that (1) individual layers are not appreciably thickened or thinned during folding, and the thickness of an individual layer measured perpendicular to its local dip is nearly the same at all points around the fold; and (2) successive layers in the fold are conformable or harmonic. Folds that fit these criteria are called *parallel folds*. Some parallel folds have smoothly curved broad hinge zones (Fig. 13-2a), whereas others have narrow angular hinge zones that separate domains in which layers have nearly constant dips (Fig. 13-2b). The Busk method of section construction (Busk, 1929) is appropriate for smoothly curved parallel folds, whereas the kink method (e.g., Suppe, 1985) is appropriate for parallel folds with angular hinge zones and straight limbs. These methods produce reliable cross sections only if the assumption of parallel folding is valid; they cannot be used for extrapolation to depth in regions of nonparallel or disharmonic folding. If we cannot assume that folds are parallel, we use the dip-isogon method or one of several orthographic projection techniques to construct sections. Keep in mind that there are geologic settings where none of these techniques yields a truthful representation of subsurface structure.

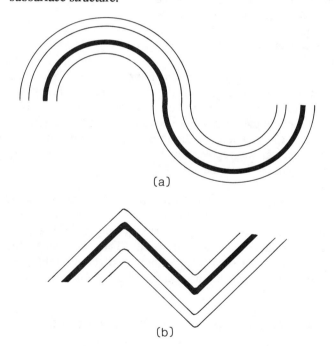

Figure 13-2. Styles of parallel folds referred to in the text. (a) Concentric parallel folds; (b) angular parallel folds.

13-3   BUSK METHOD OF CONSTRUCTING SECTIONS OF NONPLUNGING FOLDS

The Busk method is the most popular method for constructing sections of parallel folds with smooth, rounded hinges. An alternative method that uses "evolutes"

and "involutes" (types of curves; see Bronshstein and Semendyayev, 1973) was proposed by Mertie (1947; see also Roberts, 1982), but this method is considerably more difficult and usually does not yield cross sections that are sufficiently more accurate than those constructed with the Busk method to be worth the trouble.

The Busk method permits us to reconstruct the traces of layers in a section plane from surface or subsurface measurements of the attitudes of the folded layers. The geometric basis of this method is the assumption that folded layers are everywhere tangent to circular arcs. In practice this assumption means that (1) the trace of each folded layer in a profile plane can be divided into a number of segments each of which is either a portion of a circular arc or a straight line. Along each circular-arc segment, dip values change smoothly and continuously. Adjacent circular-arc segments are connected either by inflection points or by straight-line segments; and (2) Folding is harmonic and the traces of adjacent layers in profile are concentric arcs of different radii (Fig. 13-2a).

Busk Method for Two Points

Problem 13-1 (Busk method using data at two points)

Stations A and B lie 140 m apart along a N45°W-trending section line across a horizontal parallel fold; the elevation of A is 5 m higher than the elevation of B. The attitude of bedding at A is N45°E,10°SE, and the attitude at B is N45°E,35°SE. Use the Busk method to reconstruct the segments of folded layers that pass through A and B.

Method 13-1

Step 1: Draw a profile plane to scale. In this problem the plane is vertical and is perpendicular to the axis of the fold. If stations fall directly on the section line, as in this problem, just plot them at their appropriate relative elevations and lateral spacing. (If a station does not fall directly on the section line, plot its projection on the section line by drawing a projection line parallel to the fold axis from the station to the section line. The intersection of the projection line with the section line gives the station's relative position along the section line; plot a projected station at the same elevation as the original station.)

Step 2: In this problem the section line is perpendicular to the strike of the the beds, so the true dips of the beds can be shown in the profile plane. If the section line is not perpendicular to strike, we can still carry out a Busk construction, but the dips indicated in the profile plane must be apparent dips. Once the appropriate apparent dip value in the profile plane has been determined, indicate the dip values on the profile plane by short line segments drawn in ink (e.g., line segments 1 and 2 in Figure 13-3a).

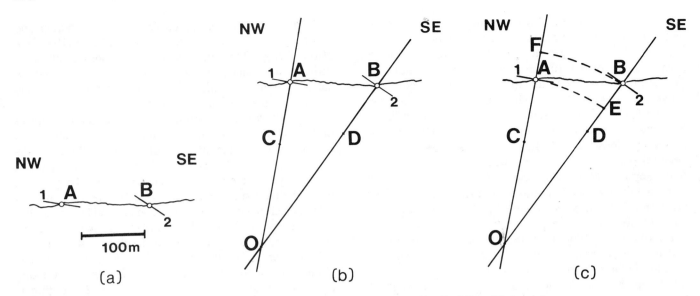

Figure 13-3. Illustrations for Busk construction Problem 13-1. (a) The location of stations A and B plotted along a line of section. Line segments 1 and 2 give the dip values in the plane of section at A and B, respectively; (b) cross section completed in step 3; (c) completed Busk reconstruction of layers passing through A and B. The dip value increases from A to B, so the circular arcs are antiformal.

Step 3: We assume that the traces of beds passing through A and B are segments of concentric circular arcs. We must now find the common center of these circular arcs. To do this, recall that the radius of a circle is perpendicular to the trace of the circle at all points along the circle. Draw line AC perpendicular to line segment 1 at A and BD perpendicular to line segment 2 at B. Extend lines AC and BD to intersect at point O (Fig. 13-3b). Point O is the center of the concentric circular arcs.

Step 4: Next, obtain a compass and place its anchor needle on point O. With the compass's pencil, draw one circular arc that passes through point A and intersects radius OB at point E, and a second circular arc that passes through point B and intersects the extension of radius OA at point F. The two circular-arc segments, AE and BF, are the profile traces of bedding that we desire (Fig. 13-3c). Note that arc segments AE and BF are concentric, but have different radii, and that points A and B do *not* lie on the same arc segment. It is important that the arc segments not be extended beyond their intersections with rays OA and OB, for the traces of layers outside the rays are fixed by dip readings at other station locations.

In Problem 13-1 note that if, for example, point A was the outcrop of a stratigraphic contact, arc segment AE would be the Busk reconstruction of that stratigraphic contact. Note also that the layer thickness is uniform along the fold segment between stations A and B (i.e., AF = BE). The relative dip values at stations A and B

determine whether the center of curvature for circular-arc segments passing through the two points lies below the ground surface (as in Figure 13-3c, where the dip at B is greater than the dip at A) or above the ground surface (as in Figure 13-4a, where the dip at B is less than the dip at A). The circular arcs representing the folded layers drawn using these centers will be either antiformal (Fig. 13-3c) or synformal (Fig. 13-4a).

In Problem 13-1 the dips at our two stations were different. If the dip readings at two adjacent stations are identical, the normals to the line segments representing the dipping layers do not intersect. The "arc" segments connecting these two dip stations will be straight lines (i.e., they will have an infinite radius of curvature; Fig. 13-4b).

Busk Method for Three or More Points

Problem 13-2 (Busk method using data at three points)

Stations A, B, and C lie on a N45°W-trending section line across a horizontal parallel fold. The horizontal distance between A and B is 190 m, and the horizontal distance between B and C is 200 m. The elevation at A is 150 m, at B is 160 m, and at C is 170 m. The attitude of bedding at A is N45°E,10°SE, the attitude at B is N45°E,35°SE, and the attitude at C is N45°E,70°SE. Use the Busk method to construct the traces of the folded beds passing through A, B, and C.

Figure 13-4. Busk construction, continued. (a) If dip values decrease from **A** (where the attitude is N45°E,40°SE) to **B**, (where the attitude is N45°E, 10°SE), the center of curvature for the arc segments passing through A and B lies above the points, and the folded surfaces are synformal; (b) if strike and dip readings at two adjacent stations along a line of section are the same (N45°E, 30°SE at both A and B), the limb segments between the two points are straight-line segments.

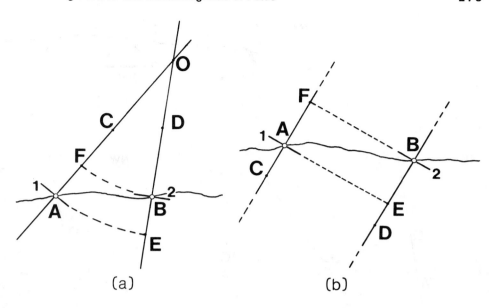

(a)   (b)

Method 13-2

Step 1: Draw your section line, locate the positions of the stations along the section line, and plot line segments representing the dips at these stations (Fig. 13-5a). In this problem the dips indicated on the line of section are true dips.

Step 2: Draw the normals to line segment 1 at A and line segment 2 at B, lines AD and BE respectively (Fig. 13-5b). Extend these lines to intersect at point O.

Step 3: Place the anchor needle of your compass at point O and draw circular arcs AF and GB. These arcs represent the segments of the folded layers between stations A and B.

Step 4: Draw the normal to line segment 3 at C, line CH. Extend line CH to intersect the radius OB at O' (Fig. 13-5c). The point O' is the center of concentric circular-arc segments that are tangent to line segment 2 at B and line segment 3 at C. The centers of the two sets of circular-arc segments must lie on a single straight line normal to line segment 2 at B (e.g., line OB in Fig. 13-5c).

Step 5: Move the anchor needle of the compass to point O', and draw arc segments BJ and CI. To finish the drawing, return the anchor needle back to point O and draw arc segment IL (Fig. 13-5c).

Once again, the relative spacing of stations A, B, and C and the relative magnitudes of the dips at these stations determine the locations of the centers of concentric circular arcs passing through three stations along a single line of section (see Fig. 13-6).

So far, we have not worked with problems where a single marker bed is exposed at several localities along a line of section. If we have an insufficient number of surface or subsurface dips along the line of section, the

Busk method may predict surface exposures of a horizon that do not correspond with known surface exposures. If a particular marker bed appears at the ground surface two or more times long the line of section, it is possible to test the consistency of the Busk construction of the fold with the surface data. The following method is taken from Billings (1972).

Problem 13-3 (Busk method applied to a marker horizon)

We know the attitude of bedding at four points along a N45°W-trending line of section across a horizontal parallel fold: N45°E,15°NW at point A, N45°E,40°SE at point B, N45°E,60°NW at point C, and N45°E,15°SE at point D. A distinctive conformable stratigraphic contact crops out at points A and D along a section line. Use the Busk method to reconstruct the folded layers between points A and D. Be sure that your solution conforms with known positions of the contact.

Method 13-3

As we will see, if we simply apply the steps of Method 13-2 to the data given here, we obtain a cross section that cannot be correct (Fig. 13-7a). The arc segment that passes through point A does not connect with the arc segment passing through point D. Next, we illustrate how such a mistake can be corrected.

Step 1: Draw the line of section, locate the measurement stations, and plot the appropriate line segments 1, 2, 3, and 4 representing the dip values at each station.

Step 2: Draw the normal to line segment 1 at A and the normal to line segment 2 at B, and extend them to intersect at O. Place the anchor needle of a compass at O, and draw circular-arc segment AE. Point E does not coincide with point B, but based on the field data that we

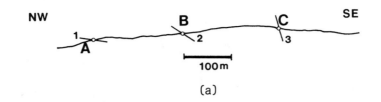

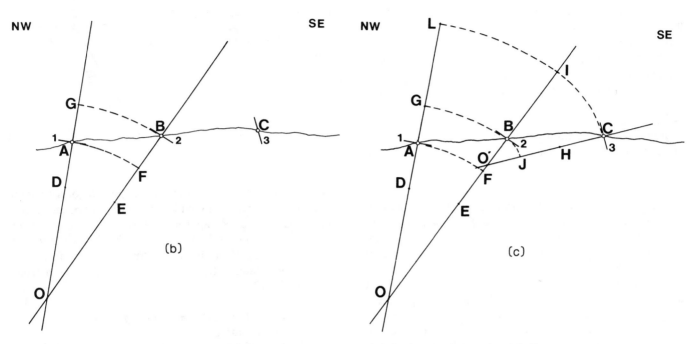

Figure 13-5. Busk construction for Problem 13-2. (a) Stations A, B, and C, plotted on a cross section plane; (b) partial construction after completing step 3 of Method 13-2; (c) completed cross section.

have, we do not expect the layer passing through A to crop out at B.

Step 3: Draw the normal to line segment 3 at C, and extend the normal to line segment 2 at B (line OB) to where they intersect (point P). Use a compass to draw circular-arc segment EF. Once again, F and C do not coincide, but this is consistent with our field data.

Step 4: Draw the normal to line segment 4 at D, and extend it to intersect the normal to line segment 3 at point Q. Place the anchor needle of your compass at Q, and draw circular-arc segment FG (Fig. 13-7a). G does not coincide with D. Field data indicate, however, that the same contact crops out at both points A and D. Therefore, barring faults, point G *should* coincide with point D. Obviously, our section at this stage is not correct, and we must modify it.

We alter the construction to conform with surface data by interpolating a dip value between the two stations that are most widely spaced (B and C) and by replacing the

single arc segment (EF) between those two stations with two arc segments.

Step 5: Place the anchor needle of your compass at Q, and draw arc DW (Fig. 13-7b).

Step 6: Draw line WX perpendicular to QP, and draw line EY perpendicular to OP. Lines WX and EY intersect at point R.

Step 7: Draw a straight line between points E and W. Drop a perpendicular from the line EW to point R, and extend it to intersect the ground surface at U. The inclination of the line EW is, in effect, an interpolated dip value at U for the layer passing through A and D.

Step 8: Extend line RU upward to intersect the extension of OB. The intersection of these two lines is point S. Line SR, which is perpendicular to the interpolated dip at U, intersects line PQ at point T.

Step 9: With the compass anchor at S, draw circular-arc segment EH. With the compass anchor at T, draw circular-arc segment HW. Curve AEHWD (Fig.

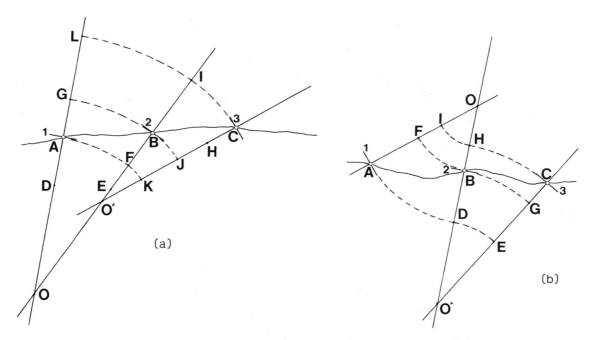

Figure 13-6. Illustration showing that the relative positions of the centers of concentric arc segments passing through three stations in a Busk construction change as distances between stations and dip values at the stations change. (a) Dip values in the plane of section increase from A (10°) to B (35°) to C (60°). Comparing this figure with Fig. 13-5c, we note that the center O' may fall between points F and B or between points O and F, depending upon the dip values at stations and the spacing between stations; (b) construction for the case where the dip at the intermediate station (station B) is less than the dip values at the two ends. Dip values are 60° at A, 10° at B, and 40° at C. Curvature changes from synformal to antiformal. Curvature will change from antiformal to synformal if the dip at a station is greater than dips at the two adjacent stations.

13-7b) is the trace of a surface that fits the known dip data and passes through both points A and D.

Problems with the Busk Method

Figure 13-8 shows a cross section drawn by applying the Busk method to seven dip readings along a line of section. Remember that, by definition, the hinge of a fold is the line along which the curvature of the fold is a maximum. If the trace of a layer is drawn as a circular arc (i.e., OP in Fig. 13-8), the curvature of the layer is constant, so technically, we cannot define a unique hinge. In Busk constructions, we arbitrarily place the hinge at the midpoint of an arc. ZZ' is the trace of one antiform's hinge surface in Figure 13-8.

Look again at Figure 13-8. At points U and X along line ZZ', concave-up arc segments of adjacent synforms intersect (there is no intervening antiformal arc). The trace of the the folded layer has infinite curvature at such points. Points of infinite curvature, called *singularities*, often appear in Busk constructions of folds whose wavelengths are short relative to the thickness of the layered sequence. Singularities, such as those that occur at points U and X, are a consequence of the assumptions that folded layers are concentric circular arcs. Singularities are rarely observed in outcrop, so we must question whether Busk-constructed folds actually represent reality. Badgley (1959) suggested replacing singularities in Busk reconstructions with curved line segments, but, as Ragan (1985) noted, this alters the appearance of the reconstruction without necessarily making it more truthful. If several singularities appear in a Busk construction of an area but none are observed in surface exposures, we probably should find an alternative method to reconstruct the subsurface geology of the area.

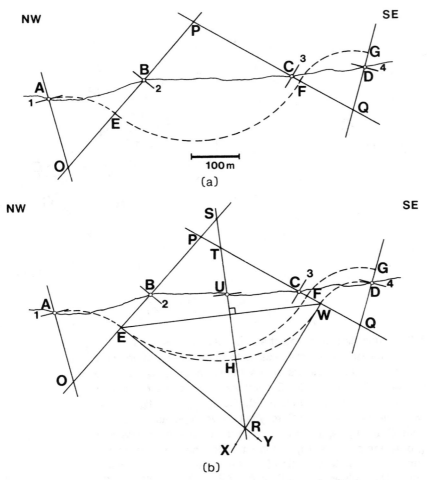

Figure 13-7. Steps in the Busk construction of Problem 13-3. (a) Incorrect cross section drawn using dip values at four stations; (b) interpolating a dip reading at an intermediate location in the Busk construction (after Billings, 1972).

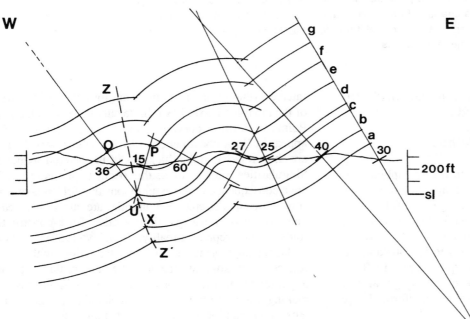

Figure 13-8. Completed Busk reconstruction of folded Devonian strata near Kingston, New York. (See Marshak and others, 1986.) sl = sea level; a through g are different stratigraphic horizons. Line ZZ' is a fold hinge. Note the infinite curvature (singularities) at points U and X along the hinge.

13-4   KINK-STYLE CONSTRUCTION OF NONPLUNGING FOLDS

Kink Geometries

In recent years geologists have recognized that many folds, particularly those in fold-thrust belts, have straight limbs and angular hinges (cf. Faill, 1969, 1973; Laubscher, 1977; Thompson, 1981). Angular folds produce *domainal dip patterns* on maps. A *dip domain* on such a map is an area in which strata have nearly constant dips. Adjacent dip domains are separated by narrow belts in which dips change abruptly (Fig. 13-9). The formation of these angular folds is often accommodated by interlayer slip, and it often occurs without appreciable thickening or thinning of strata. We model these angular folds as kink folds and use a method that relies on the geometric properties of kink folds to draw cross sections of regions exhibiting domainal dip patterns.

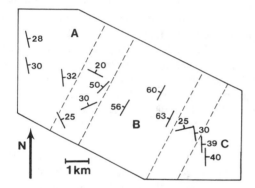

Figure 13-9. Dip domains on a map of angular folds. The map in this figure shows attitude measurements on bedding in a small area. The map can be divided into three distinct dip domains (A, B, and C) where layer dips are fairly constant. Belts between domains are fold-hinge zones (i.e., domain boundaries).

In an ideal kink fold, layering abruptly changes its attitude across an imaginary planar boundary called a *kink plane* (Fig. 13-10), which is the fold's axial surface. Layers in adjacent limbs of a kink fold meet along a *kink axis*; the change in attitude between adjacent limbs can be described by a rotation around this axis. If layers do not thicken or thin during folding, the kink plane bisects the angle between adjacent limbs. If the layer thickness does change, the kink plane does not bisect the angle between adjacent limbs.

A kink band is composed of two parallel kink planes whose kink axes are parallel but have the opposite sense of rotation (Fig. 13-10). Intersecting kink bands produce folds with straight limbs and angular hinges (Fig. 13-11a).

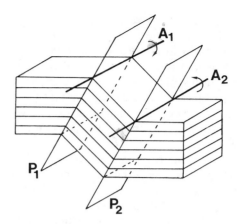

Figure 13-10. Idealized kink band, showing folded layers, two kink planes (P_1 and P_2) and kink axes A_1 and A_2. (From Faill, 1969.)

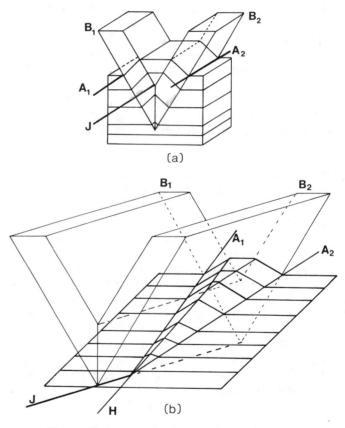

Figure 13-11. Geometry of folds formed by intersecting kink bands. (From Faill, 1969.) (a) A cylindrical angular fold is generated when two kink bands, B_1 and B_2, with kink junction axes A_1 and A_2 intersect along a kink junction axis J that lies in bedding; (b) if the kink junction axis J is oblique to bedding, intersecting kink bands form noncylindrical folds.

The line at which the two kink bands join is called the *kink junction axis* (line J in Figure 13-11a). If the kink junction axis lies in the plane of layering (Fig. 13-11a), the resulting fold is cylindrical, but if the kink junction axis is not in the plane of layering, noncylindrical folds result (Fig. 13-11b). In a section through either type of angular fold, the traces of contacts are straight, parallel lines whose inclinations change abruptly at fold hinges (the kink planes).

We can use the Busk method to construct sections of kink-style folds, but by doing so, we disregard the unique properties of kink folds. The kink method (Faill, 1969, 1973; Suppe, 1985) takes advantage of the unique properties of kink-style folds. In the kink method we draw the traces of layers as straight-line segments between adjacent fold hinges. As we will see, we locate and use known stratigraphic contacts to draw our cross section at the outset, thereby avoiding difficulties like those encountered in Problem 13-3. Because the kink method allows us to construct cross sections of angular folds rapidly and reliably, it has become popular in recent years. The boundary-ray method of section construction (Badgley, 1959, after Coates, 1945, and Gill, 1953) was also developed to accommodate angular folds, but because it requires data other than strike and dip readings and because it is quite complex, it is not widely used. The kink method is the method of choice for drawing sections of angular folds.

Kink Method Applied to Folds with Constant Layer Thickness

Problem 13-4

The stratigraphic contact between a limestone and a shale crops out at points X and Y along a N75°W-trending line of section across an angular parallel fold (Fig. 13-12a). The contact between the shale and a sandstone crops out at points W and Z, and the top of the sandstone crops out at point V. Line segments 1, 2, and 3 at points A, B, and C, respectively, give the dip values of different domains. Points V and W fall in dip domain 1 (N15°E,50°W), point X falls in dip domain 2 (N15°E, 10°W), and points Y and Z fall in dip domain 3 (N15°E,25°E). Use the kink method to draw a profile of the fold.

Method 13-4

Step 1: First, we locate the fold hinges. We assume that the limestone/shale contact that passes through point X is a straight-line segment. This line must have an inclination in the plane of the section that corresponds to the dip value of this domain, so we draw this segment of the contact parallel to line segment 2 at B. Likewise, draw the segment of the limestone/shale contact that passes

through point Y parallel to line segment 3 at C. The two segments of the limestone/shale contact intersect at point L (Fig. 13-12b). Point L is the intersection between the trace of a kink hinge plane and the limestone/shale contact.

Step 2: To determine the trace of the hinge plane that passes through point L, bisect angle XLY. The hinge-plane trace is line ab (Fig. 13-12b).

Step 3: Draw the segment of the shale/sandstone contact trace passing through Z as a straight line parallel to YL. This line intersects hinge ab at point M. To continue the trace of this contact beyond the hinge, draw a line from M that is parallel to XL.

Step 4: Next, we position the second hinge-plane trace. The segment of the shale/sandstone contact passing through W must have an inclination in the plane of section that corresponds to the dip value for its domain. Draw the segment of this shale/sandstone contact passing through W as a straight-line segment parallel to line segment 1 at A. This portion of the shale/sandstone contact intersects the segment of the contact passing through M at the point N. Bisect ∠ WNM to determine the orientation of the second hinge-plane trace, which is line cd (Fig. 13-12c).

Step 5: Now we can complete the outer portion of the fold. The limestone/shale contact intersects line cd at point O. We can extend this contact beyond hinge cd as line segment OP, where angle POd = ∠ XOd. Draw the upper contact of the sandstone (QV parallel to PO, etc.).

Step 6: When any two kink-fold hinge traces intersect, they are supplanted by a single hinge trace that bisects the angle between the remaining opposing limbs (provided the layers' thicknesses remain constant). Hinge traces ab and cd intersect in the subsurface at point e. To position the traces of layers in the core of the fold, draw eS parallel to OP and eT parallel to LY. Draw fold hinge ef so that it bisects the angle between eS and eT.

Kink Method Applied to Folds with Changing Layer Thickness

In many cases surface data indicate that corresponding layers on opposing limbs of folds have different thicknesses. In these cases the fold hinges cannot bisect the interlimb angle, so the *axial angle* between each limb and the fold hinge is different for opposing limbs. Consider a kink fold where layer thickness changes abruptly from T on one limb (Fig. 13-13) to T' on the other limb. The two axial angles are γ and γ', where

$$\gamma = \text{MBD} \neq \gamma' = \text{MBE}.$$

A comparison of triangles MBD and EMB indicates that

$$T/\sin \gamma = \text{BM} = T'/\sin \gamma' \qquad \text{(Eq. 13-1)}$$

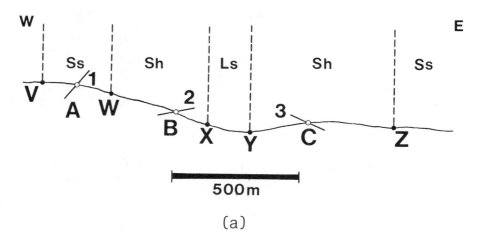

(a)

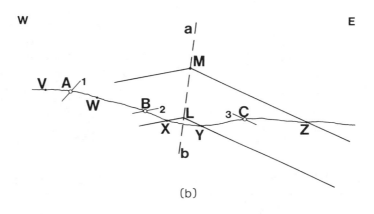

(b)

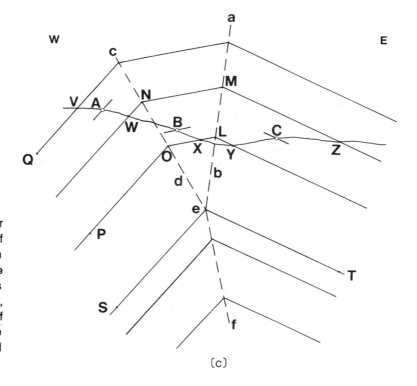

Figure 13-12. Illustrations for kink-method construction of Problem 13-4. (a) Raw dip data on a cross-section plane. Line segments 1, 2, and 3 give the dips at points A, B, and C, respectively, and represent the attitudes of different domains; (b) fold profile after completing step 3 in Method 13-4; (c) completed fold profile.

(c)

or,

$$T'/T = \sin \gamma'/\sin \gamma \qquad \text{(Eq. 13-2).}$$

If we can locate a fold hinge and we know the dips of the opposing fold limbs, we can measure the axial angles γ and γ' and can use Equation 13-2 to find the relative orthogonal thicknesses of layers on opposing limbs. If, alternatively, we know T and T' from field data, we can use Equation 13-2 to calculate the axial angles and orient the fold hinge in our reconstruction.

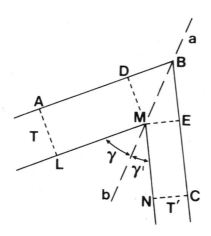

Figure 13-13. An angular fold with different layer thicknesses (T and T') on opposing limbs and unequal axial angles ($\angle$ DBM $\neq$ $\angle$ MBE).

Problem 13-5

Field mapping has established that a single stratum in a region of angular parallel folds crops out at points A and C along a line of section (Fig. 13-14a). A fold hinge is exposed at point h between the two limbs. Line segments 1, 2, and 3 give domainal dip values at points A, B, and C, respectively. Dipmeter readings in the borehole at D are also plotted on the figure. Note that two distinct dip domains are defined by the measurements in the drill hole. Use the kink method to draw a profile of the fold.

Method 13-5

Step 1: First, draw the trace of the fold hinge passing through point h. Locate point i in the drill hole halfway between two nonparallel dip measurements in the bore hole, and draw line ih. Extend this line into the sky.

Step 2: Draw a straight-line segment parallel to line segment 3 through point C, and extend it to intersect the fold-hinge trace at point G. Draw a straight line parallel to line segment 2 through point B, and extend it to intersect the fold-hinge trace at E.

Step 3: Draw line GJ parallel to BE and line EF parallel to CG. Note that the axial angle JGE is not equivalent to the axial angle CGE and that layer thickness JB is not equivalent to layer thickness CF. By inspecting Figure 13-14b, we see that the eastern limb was thinned to 83% of the thickness of the flat-lying limb during folding.

Step 4: To complete the outer portion of the fold profile, we must position the fold hinge between dip domains 1 and 2 in the western portion of the profile. Changes in layer thickness in the eastern portion of the profile make us wary about applying the parallel folding assumption elsewhere in the profile. We need field data, however, on the relative thicknesses of layers in the limbs of this open fold to draw this western fold hinge. Lacking such data, we *assume* here that layering neither thickened nor thinned as this open fold formed. This is reasonable because changes in layer thicknesses in angular folds are often restricted to folds with relatively tight interlimb angles. Thus, draw a straight line parallel to line segment 1 through A, and extend it to intersect the continuation of GJ at K. Bisect angle AKJ to find the trace of the second fold hinge (line lm). Extend BE to intersect lm at N, then draw NO parallel to AK. The shallower levels of the fold can be traced out by drawing lines parallel to established contacts.

Step 5: A problem arises below the depth at which fold-hinge traces hi and lm intersect (point S). Because of the thinning of the eastern limb, we cannot simply bisect the angle between opposite limbs to determine the hinge trace below point S. In order to complete the fold profile, it is necessary to know how much layer thinning occurred at depth. For the sake of argument, assume that the ratio of thicknesses in opposing limbs in the subsurface is equal to that observed in the outer portion of the fold. With this assumption, we can position the fold hinge graphically by drawing lithologic contacts at depth parallel to established contacts while maintaining a fixed ratio between the orthogonal thicknesses of the eastern and western limbs (Fig. 13-14c).

Alternatively, we could substitute this ratio of orthogonal thicknesses into Equation 13-2. We know that the two axial angles together must equal angle RST, measured (with a protractor) to be 94°. We have, then, two equations (Equation 13.2 and $\gamma + \gamma' = 94°$) with two unknowns and can solve for the values of the two axial angles in this portion of the fold. We begin by substituting $94° - \gamma$ for γ' into Equation 13.2. Using the trigonometric identity for the sine of the difference between two angles, we can rewrite this equation in terms of $\sin \gamma$ and $\cos \gamma$. Combining terms and rearranging, we have $\gamma = \arctan [\sin 94°/(0.84 + \cos 94°)]$, or $\gamma = 52.3°$ and $\gamma' = 41.7°$, the same values we determined graphically.

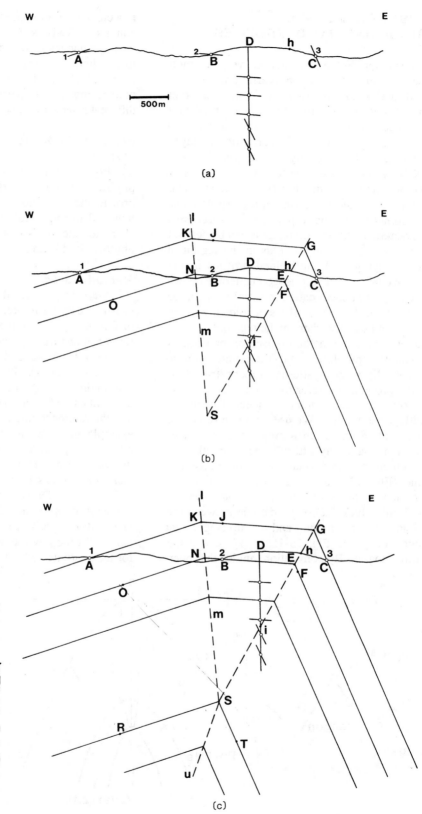

Figure 13-14. Illustrations for kink-method construction of a fold with nonconstant layer thickness, described in Problem 13-5. (a) Raw data plotted on a cross-section plane. A drill hole at point D provides subsurface data; (b) partially completed fold profile after step 4 in Method 13-5; (c) completed fold profile.

13-5 DIP-ISOGON METHOD OF CONSTRUCTING FOLD PROFILES

If we have information on the characteristic way in which layer thicknesses vary in a region (variations that are commonly a function of lithology), we can construct reasonable profiles of the folds in the region, even when it is not possible, because of the layer-thickness variation, to use Methods 13-3, 13-4, or 13-5. Variations in layer thickness are readily described by *dip-isogon patterns* (Ramsay, 1967), which indicate the relative curvatures of the outer and inner arcs of folded layers (see Chapter 11). Remember that a *dip isogon* is a line in the profile plane that connects points with equal dip values on successive contacts. Characteristic dip-isogon patterns can be derived by examining well-exposed minor folds in the region of interest or by studying well-constrained profiles in the region. Once the characteristic dip-isogon pattern is known for a sequence of beds at one location in a region, it can be used as a guide in constructing profiles of folds involving similar sequences of beds elsewhere in the region.

The following problem/method (based on procedures outlined by Ramsay and Huber, 1987) illustrates how to use characteristic dip-isogon patterns to reconstruct fold profiles. In the problem we refer to two angles: (1) δ is the angle between a specified reference line and the tangent to any folded layer at a point. We use the same reference line to measure all δ values in a single profile. The reference line may have any attitude, but normally we choose a horizontal line or a normal to the fold hinge as a reference line (Fig. 13-15a). If a fold-hinge plane is vertical, the normal to the hinge is horizontal, and δ values equal the local dip values. (2) ø is the angle between the normal to a folded layer at a point and the dip isogon that passes through that point (Fig. 13-15a). We call ø the *deflection* of the isogon. By convention, angles that open

in a clockwise sense are negative, and angles that open in a counterclockwise sense are positive. The deflection angle usually changes as we trace an isogon from one rock type to another; ø is usually greater in less-competent layers (Fig. 13-15b). In plots of ø against δ, the curves connecting (ø, γ) pairs for points on the surfaces of different layers are generally different.

Problem 13-6a (Characterizing dip-isogon patterns)

Figure 13-15b is a profile of folded quartzites and phyllites seen in a vertical roadcut. Assume that this profile indicates how quartzites and phyllites typically behaved during folding in this area, and determine the characteristic dip-isogon patterns for folded quartzites and phyllites in this area.

Method 13-6a

Step 1: Draw the hinge-surface trace ab on the profile. Draw a reference line for measuring δ values at different points on the layers. In Figure 13-15b the reference line RL is perpendicular to the fold hinge ab.

Step 2: At several points around the fold (at regular intervals, such as 5° or 10° increments, along each lithologic contact), draw a short line segment tangent to the contact. Measure the angle between this line segment and the reference line (Fig. 13-15a). In Figure 13-15b, for example, the outer arc of the phyllite bed has a dip value of δ = +60° at point X, the outer arc of the quartzite bed has a dip value of δ = +60° at point Y, and the inner arc of the quartzite bed has a dip value of δ = +60° at point Z.

Step 3: Draw straight-line segments across layers connecting equal dip (δ) values. These segments are the layers' dip isogons. In Figure 13-15b XY, is the +60° dip isogon for the phyllite bed, and YZ is the +60° dip isogon for the quartzite bed. Draw normals to layering where each

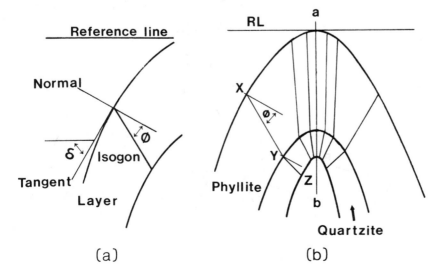

(a)

(b)

Figure 13-15. Illustrations of dip-isogon method of fold construction described in Problem 13-6. (a) Convention for measuring dip (δ) values with respect to a given reference line and dip-isogon deflection ø; (b) profile of a well-exposed minor fold in the western portion of the Great Smoky Mountains, Tennessee Appalachians, involving a quartzite layer and a phyllite layer. The fold hinge is ab; several isogons are shown. XYZ is the +60° isogon in this fold; (c) plot of dip-isogon

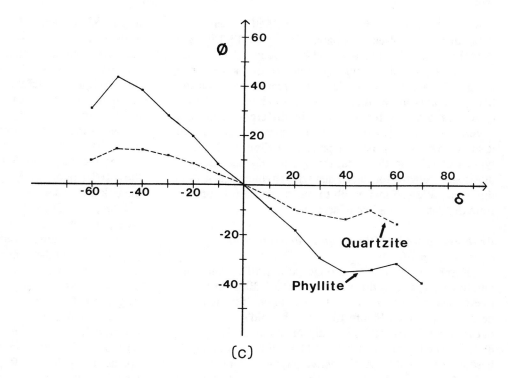

(c)

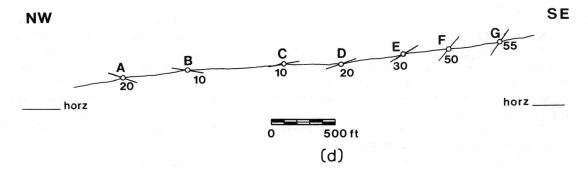

(d)

deflection (ø) versus dip values (δ) for the quartzite layer (dashed line) and the phyllite layer (solid line) in (b); (d) topographic profile across Bates Mountain (from King, 1964), which is near the location of the fold shown in (b), with dip values at several locations. Quartzites crop out between D and E, a distinctive phylllite layer crops out at G, and the fold hinge (with strike and dip N60°E,80SE) crops out at C; (e) profile after step 5 in Method 13-6b.

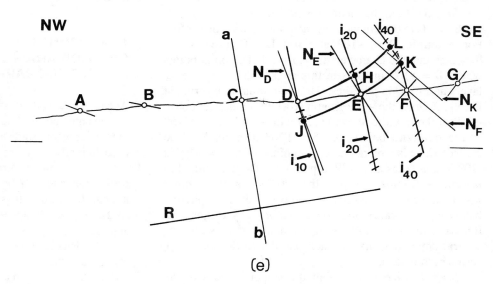

(e)

isogon intersects a lithologic contact, and measure the deflection angle of each isogon. In Figure 13-15b, ø = -32° for isogon XY, and ø = -15° for isogon YZ.

Step 4: Tabulate values of δ and ø for each layer in the profile, and plot these data on a graph with δ on the horizontal axis and ø on the vertical axis. Note that the the ø-versus-δ curve for phyllite is different from the ø-versus-δ curve for quartzite in Figure 13-15c. If we assume that the dip-isogon pattern in Figure 13-15b is typical of folds in this area, this graph *characterizes* the dip-isogon patterns for folded quartzite/phyllite sequences in this area. We use this graph to reconstruct a fold profile from dip data along a line of section.

Problem 13-6b (Dip-isogon construction of a fold profile)

Figure 13-15d is a topographic profile showing outcrop patterns and dip values along a N30°W-trending line of section across a folded quartzite/phyllite sequence near the location of Figure 13-15b. The fold's hinge crops out at point C; the strike and dip of the hinge is N60°E, 80°SE. The top of a quartzite bed crops out at point D; its base is exposed at E. A distinctive phyllite layer crops out at G. Assume that the fold does not plunge, and construct a profile of this fold.

Method 13-6b

Step 1: First, draw a reference line on the section with an orientation comparable to that in our characteristic profile (i.e., normal to the fold hinge). Draw the trace of the hinge surface, ab, on the profile, and draw reference line R perpendicular to ab.

Step 2: Measure the angle δ between the dip mark at each station along the line of section and the reference line R, and tabulate the measurements. For example, δ = +10° at point D, δ = +20° at point E, and δ = +40° at point F.

Step 3: Find the characteristic isogon deflection that corresponds to the dip value at each point along the profile. For example, δ = +10° at point D. Figure 13-15c indicates that the characteristic isogon deflection at points where δ = +10° in folds in this area is -5° in quartzite layers and -9.5° in phyllites. Draw a normal to layering at D (N$_D$). Draw the +10° isogon (i$_{10}$) with a deflection of -9.5° in the phyllite above D and with a deflection of -5° in the quartzite below D. δ = +20° at point E. The characteristic isogon deflection at points where δ = +20° is -10° in quartzites and -18° in phyllites. Draw a normal to layering at E (N$_E$), and draw the +20° isogon (i$_{20}$) with ø = -10° in the quartzite above E and ø = -18° in the phyllite below E. Dip values are constant along any isogon. As Ramsay and Huber (1987) suggest, indicate this by drawing several small tick marks, each parallel to the local dip, across each isogon.

Step 4: We assume that the isogon pattern in the fold in our section in Figure 13-15d is comparable to that in the characteristic profile. The phyllite/quartzite contact that crops out at D must have a δ value of +20° when it crosses isogon i$_{20}$ above E. Using a french curve or a flexible ruler, extend the phyllite/quartzite contact to point H on the +20° isogon, making sure that this contact parallels the tick marks across the isogon at H. Similarly, the quartzite/phyllite contact exposed at E must have a δ value of +10° when it crosses isogon i$_{10}$ below D. Extend the quartzite/phyllite contact from E to point J on the +10° isogon, making sure that this contact parallels the tick marks across the isogon at J.

Step 5: Draw a normal to layering through point F (N$_F$), and then draw the +40° isogon through this point (i$_{40}$) with the appropriate deflection angle for phyllites (-33°). Extend the quartzite/phyllite contact to point K on the +40° isogon, making sure that it is parallel to the tick marks across the isogon. The deflection of the +40° isogon must change from a value appropriate for phyllite to one appropriate for quartzite above K. Draw a normal to layering at K (N$_K$), and extend the +40° isogon above K with ø = -13°. We can then extend the phyllite/quartzite contact from H to L. Repeat steps 4 and 5 until you have completed the profile of the fold.

Fold profiles constructed by the dip-isogon method will show changes in layer thickness similar to those seen in the characteristic profile. Because the dip-isogon method uses the shapes of folds seen in well-constrained profiles as models for other profiles instead of assuming that layer thickness does not change around folds or changes abruptly across fold hinges, this method is conceptually more attractive than the Busk and kink methods. As Ramsay and Huber (1987) note, however, fold profiles drawn by the dip-isogon method become less reliable as we extend our profiles farther from data constraints.

13-6 CONSTRUCTING PROFILES OF NONPARALLEL FOLDS BY ORTHOGRAPHIC PROJECTION

Accurate and well-constrained profiles of nonplunging folds can be constructed directly from observational data obtained in regions where folds are cylindrical and topographic relief is sufficiently high that large portions of folds are exposed.

The technique introduced next involves orthographic projection and can be used for either parallel folds or nonparallel folds. It is particularly useful for nonparallel folds, for which the Busk and kink methods cannot be applied.

To visualize how a cross section of a nonplunging fold can be constructed, consider the patterns defined by the intersections between a cylindrically folded surface and the ground surface. A cylindrically folded surface intersects a horizontal ground surface in a series of straight lines that parallel the fold axis (ab, ef, and ij are parallel to cd and gh in Fig. 13-16a). This map pattern, a series of straight lines, tells us little about the shape of the fold in profile. If the folded surface intersects a vertical quarry wall,

however, the trace of the fold on the vertical surface gives the true shape of the folded layer (Fig. 13-16b). If the ground surface is irregular, the folded surface intersects the ground along an irregular trace (Fig. 13-16c). The shape of each segment of this irregular trace is controlled by the strike and dip of one portion of the folded layer. We can use this map trace to construct a profile of that folded contact. To define a folded layer, we need to know the map trace of both the upper and lower boundary of the layer.

Next, we show how to construct a profile of a folded contact from its trace on an irregular topographic surface. This problem is the inverse of the problem of calculating outcrop traces that was described in Chapter 2. Note that this method can also be used for parallel nonplunging folds that are exposed in regions of high relief.

Problem 13-7

Figure 13-17a is a map of a cross-bedded quartzite (stippled) that is overlain by marble (M) and is underlain by slate (S). The attitude measurements on the map indicate that this meta-sedimentary sequence has been folded, and a stereogram of the poles to bedding (not shown) indicates that folding is cylindrical around a horizontal axis. The arrow FA above the map gives the bearing of the fold axis. Draw a profile of the folded layers.

Method 13-7

Step 1: Draw a folding line (F1) perpendicular to the fold axis at the edge of the map, and swing up the cross-sectional view into the plane of the map projection. Draw a suite of lines parallel to the folding line on the rotated cross section. These lines are spaced to represent the difference in elevation between contour lines. The vertical scale on the cross section must be the same as the map scale.

Step 2: Locate the points on the map where the top and bottom contacts of the quartzite layer cross contour lines. From each point, draw a straight line parallel to the fold axis (and, therefore, perpendicular to the folding line) to the corresponding contour line in the rotated cross-sectional plane. For example, the quartzite/marble contact at A_1 on the map projection outcrops at 100 m; we draw a straight line from A_1 across to the 100-m grid line on the rotated cross section. This point plots as point A' on the grid. Notice that point A_2 on the map also plots as A' on the grid. Points B_1 and B_2 on the map plot as point B' on the rotated cross section.

Step 3: Repeat the procedure of step 2 for a sufficient number of points to define the profile trace of the lower contact. (For example, points C and D on the map plot as C' and D' on the rotated cross section). Then, repeat the procedure for points on the upper contact.

Step 4: Connect the points on the rotated cross

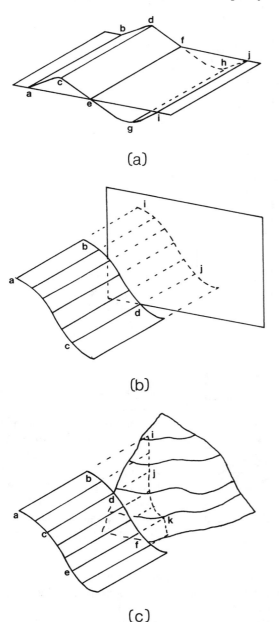

(a)

(b)

(c)

Figure 13-16.   Traces of folds. (a) The traces of nonplunging cylindrical folds on a horizontal surface are straight lines; (b) trace of a cylindrically folded surface abdc on a vertical face is curve ij; (c) trace of cylindrically folded surface abfe on an irregular topographic surface is curve ijk.

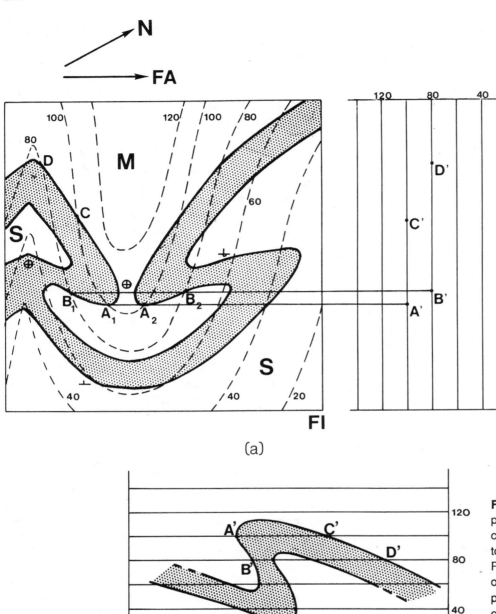

(a)

(b)

Figure 13-17.   Illustration of profile construction of a horizontal cylindrical fold in a region of topographic relief, as described in Problem 13-7. (a) Map showing the outcrop belt of a unit (stipple pattern). The grid at the right is the cross-section plane rotated into the map plane around folding line F1; (b) completed profile of the fold.

section to trace out the upper and lower folded surfaces (Fig. 13-17b). The resulting profile automatically shows the variation in the thickness of the quartzite layer around the fold.

13-7 CONSTRUCTING PROFILES OF PLUNGING FOLDS

We stated earlier that the truest image of the shape of a cylindrical fold is a *profile* of the fold, drawn on a plane normal to the fold axis. The profile plane for a plunging fold is necessarily inclined. Sections other than profiles (e.g., vertical cross sections, oblique sections, or maps) yield fold forms with distorted limb thicknesses, distorted interlimb angles, and incorrect hinge positions (Fig. 13-18; see also Roberts, 1982; Ramsay and Huber, 1987). Fortunately, the very fact that the fold plunges makes it possible for us to see large portions of the fold on a map. Map data can, therefore, allow us to *construct* a profile of a plunging fold. Likewise, subsurface data in a drill hole that is not perpendicular to the fold axis can also be used to construct fold profiles, if it is available. Next, we illustrate how to construct profiles of plunging folds from

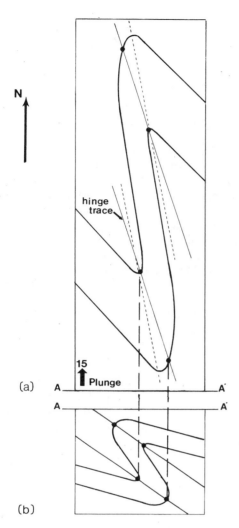

Figure 13-18. Illustration of the distortion of a plunging fold that occurs on a map. (See Schryver, 1966.) (a) Map of a plunging asymmetric fold. The lines connecting zones of maximum curvature in the map traces (thin dashed lines) do not give the true positions of the hinge traces (thin solid lines); (b) profile of the fold.

map and subsurface data. The specific technique that must be used depends on whether the fold is exposed in a region of high topographic relief or in a region of low topographic relief.

Constructing Profiles of Plunging Folds from Maps of Regions with Low Topographic Relief

In regions of low topographic relief (i.e., the heights of hills in the map area are significantly less than the amplitude of the folds in the area), the map plane is essentially an oblique section through the fold. To

visualize this principle, refer back to Figure 13-1. Note how the intersection of the plunging circular cylinder with both the map plane and the vertical cross-sectional plane is an ellipse; only on a profile plane oriented perpendicular to the cylinder do you see a circular section. There are two ways to use the map pattern to construct a profile.

(1) Down-Structure Viewing and Freehand Sketching: The first method is generally referred to as *down-structure viewing*. To obtain a down-structure view of the plunging circular cylinder in Figure 13-1, simply orient your line of sight so that it parallels the axis of the structure. When viewed from this angle, the ellipse will appear to be a circle. In the field, to obtain a down-structure view of a fold, you should place yourself so that you are looking down (or up) the hinge of a fold (Fig. 13-19a). Sometimes, it is necessary to get into an awkward position in order to properly view a structure (Fig. 13-19b)! When positioned properly, you will see the profile form of the fold. It takes practice to do down-plunge viewing easily. It may help to relax your eyes or close one eye and trick yourself into ignoring your natural depth perception, so that the oblique section of the fold on the outcrop surface appears foreshortened onto a single plane that is oriented perpendicular to the hinge. Sketch the shape of the fold that you see freehand.

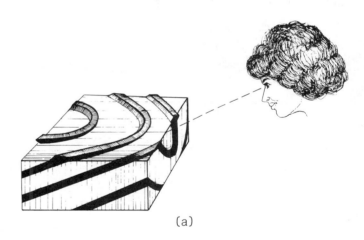

(a)

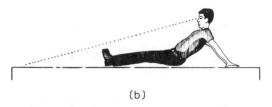

(b)

Figure 13-19. Illustration of down-structure viewing. (a) Observer positioned so that her line of sight is parallel is coincident with the hinge of the fold; (b) observer positioned to view a shallowly plunging fold that intersects the ground surface.

We can also use the technique of down-structure viewing to view maps (Mackin, 1950). The only difference between down-structure viewing of a map and down-structure viewing of an outcrop is that in order to orient yourself properly to view the map, you must read the attitude symbols on the map so that you know the plunge direction of the folds. Place the map on a table, and position your line of sight to view it down structure. As before, you can sketch the profile of the fold freehand as you view the map from this angle. Typically, the true amplitude of a fold (the amplitude in profile) will be much less than the apparent amplitude that is indicated in the map plane, and apparent thickening in the hinge may vanish (Figs. 13-1b and c, 13-18). Also, the true symmetry or asymmetry of a fold will be obvious in the profile view.

(2) Grid Method of Profile Construction: The *grid method* (Roberts, 1982; Ragan, 1985) is a graphical technique that allows us to construct *accurate* profiles of plunging folds in regions of low relief from a map of the fold. To understand the basis of this method, again consider the plunging circular cylinder of Figure 13-1. We saw that the map-plane image of the plunging circular cylinder is an ellipse. The line WX across the ellipse (Fig. 13-20), which is perpendicular to the bearing of the cylinder axis, is the same as the diameter (D) of the

circular profile of the cylinder, but the line YZ across the ellipse, which has the same bearing as the cylinder axis, is greater than the diameter (D) of the circular profile. If we are given the length of YZ (called D'), we can calculate D from the equation

$$D = D'(\sin \mu) \qquad \text{(Eq. 13-3)},$$

where μ is the plunge of the axis of the cylinder.

Now consider a map of a plunging fold. Distances between points on the fold surface measured along lines that are perpendicular to the bearing of the hinge are undistorted, whereas distances measured along lines that are parallel to the bearing of the hinge will be greater than they would be in profile. The distance in profile between any two points along a line on the map that is parallel to the bearing of the fold axis can be found by applying Equation 13-3. In other words,

$$\begin{array}{l} \text{Profile distance between} \\ \text{two points on the map} \end{array} = \begin{array}{l} \text{Distance observed} \\ \text{on the map} \end{array} (\sin \mu)$$

$$\text{(Eq. 13-4).}$$

Keeping this equation in mind, we can transfer contact positions from a square map grid onto a profile grid in which one direction is foreshortened according to Equation 13-3, as demonstrated in the following problem (see also Roberts, 1982; Ragan 1985).

Problem 13-8

Figure 13-21a is a map of a plunging fold in an area where the topographic relief is small relative to the amplitude of the folds. Stereographic projections of poles to bedding from this map area lie along a single great circle, indicating that the folds are cylindrical. The fold axis (the normal to the great circle on the π-diagram) is oriented 30°,040°. Draw an accurate profile of this fold by using the grid method.

Method 13-8

Step 1: On a transparent overlay large enough to cover the map area, draw a square grid composed of mutually orthogonal suites of lines. The spacing between lines in a suite is an arbitrary distance S (S should be chosen so that a reasonable number of grid lines are drawn; i.e., it should be possible to locate points on the folded contacts accurately with respect to the grid). Use a thicker pen to draw the lines at the left edge of the grid and at the bottom of the grid; these two lines are reference lines.

Step 2: On a separate piece of drafting paper, construct a rectangular grid (here called the *foreshortened grid*) composed of two mutually orthogonal suites of lines. The lines in one suite should be spaced at a distance S apart, and the lines of the second suite should be spaced at a

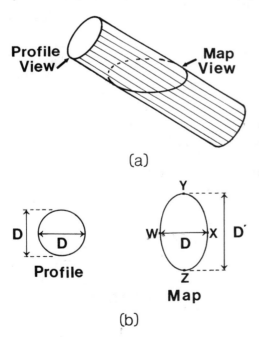

(a)

(b)

Figure 13-20. Profile versus map views of a plunging circular cylinder. (a) A plunging circular cylinder produces an elliptical trace on a horizontal map; (b) a comparison of the map and profile sections of the circular cylinder.

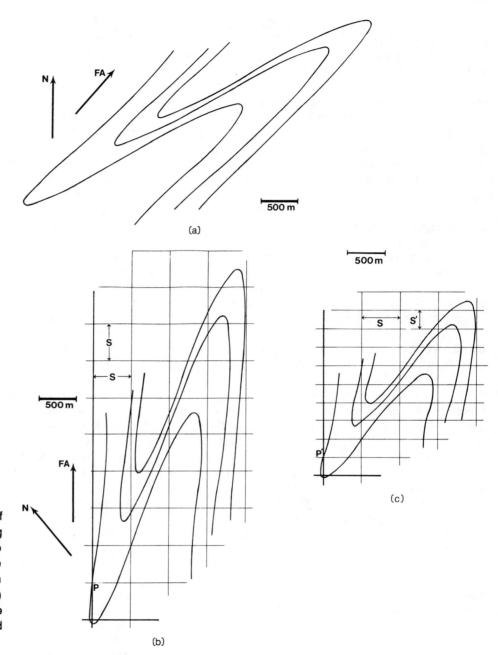

Figure 13-21. Illustration of the grid method for constructing profiles of plunging folds. (a) Map of a fold that plunges 30° in the direction given by FA; (b) map with square grid superimposed on it; (c) the profile plane illustrating the foreshortened grid. See Method 13-8 for more explanation.

distance S' apart, where S' = S(sin 30°) = 0.5S. There should be the same number of lines in the rectangular grid as there are in the square grid. Use a thicker pen to draw the reference lines.

Step 3: Secure the map to your drafting table with the up-plunge portions of the fold nearest to you. Place the square grid over the map with one set of lines parallel to the trend of the fold axis, and secure the overlay in place (Fig. 13-21b). Secure the foreshortened grid to the table next to the map and overlay; the foreshortened grid should be positioned so the suite of lines that are S' apart are oriented perpendicular to the fold-axis bearing (Fig.

13-21c). The lines spaced S' apart are horizontal lines in the profile plane. The lines spaced at a distance S apart are parallel to the bearing of the fold axis; they are parallel to the dip direction of the inclined profile plane--they do not represent vertical lines.

Step 4: Now we use the square grid to locate points on the map traces of the folded layers, and the rectangular grid to position the corresponding points in a profile plane. For example, locate point P in Figure 13-21b. It coincides with the intersection of two grid lines; one line is two lines up from the bottom reference line and the other line is the left reference line. We plot the image of P on the profile plane by finding the corresponding location on the rectangular grid (Fig. 13-21c). Point P' on Figure 13-21c is also located at the intersection of the left reference line with a line two lines up from the bottom reference line.

Step 5: Locate other points on the map of the fold trace with respect to the square grid, and plot the corresponding image points on the profile plane (i.e., the rectangular grid). Using the positions of the points plotted on the rectangular grid as a guide, trace out the image of the folded surface on the rectangular grid. This image (Fig. 13-21c) is the properly foreshortened image of the fold.

An alternative approach is to slide the foreshortened grid over the square grid and, with one of the "horizontal" lines (those spaced distance S' apart) over the corresponding line in the square grid, make tick marks where each contact crosses the line. Repeat this for each "horizontal" line in the profile plane, and use the tick marks to trace out the image of the folded contacts.

Constructing Fold Profiles from Maps of Regions with High Relief

When topographic relief is high (the heights of hills approach or exceed the amplitudes of folds in the area), our map is no longer a single oblique section through a plunging fold. Different portions of the map may be different oblique sections through the fold, but the composite map image is not simply related to the fold shape in profile. If, however, we know the orientation of the fold axis, we can pass a straight line parallel to the fold axis through each point along the map trace of a folded layer and extend these lines to pierce a profile plane. The piercing points collectively define the layer's trace on the profile plane. Next, we introduce a graphical method for finding the piercing points on a profile plane and constructing a profile of the fold. This method works both for parallel and nonparallel cylindrical folds.

Problem 13-9

Figure 13-22 is a map of a folded marble (M), sandstone (Ss), and shale (Sh). An equal-area plot of poles-to-bedding readings from this region (not shown) indicates that the attitude of the fold axis is 30°,045°. Draw an accurate profile of these folded layers.

Method 13-9

Step 1: Align the map so that the plunging fold axis points away from you. Place a sheet of tracing paper over the map.

Step 2: On the right side of the overlay, draw line AB parallel to the bearing of the fold axis. Next, draw lines AD and BC normal to AB, with AD across the up-plunge edge of the map, and BC across the down-plunge edge of the map. Rectangle ABCD outlines that part of the map that we will project onto the profile plane.

Step 3: Let line AB be a folding line (F1), and swing down a vertical cross-sectional plane around F1 (note that we are swinging down an imaginary plane that had extended up into the sky). Draw a suite of parallel lines in the rotated cross-section plane that are parallel to AB and whose spacing, at the scale of the map, equals the map's contour interval. The lowest contour (in this case the 100-m line) should be placed closest to AB. Draw line BQ in the rotated cross-section plane so that it makes an angle of μ (that is, the plunge of the fold) with AB. Line BQ represents the fold axis; note that it plunges to the NE.

Step 4: Find the perpendicular to line BQ that passes through A, and extend it to intersect BQ at point J. We now have right triangle AJB inscribed in our vertical cross-section plane. Line AB is horizontal, and line JB is parallel to the fold axis. Line AJ, which is perpendicular to JB, is the trace of a profile plane on the vertical section. You may wish to erase the contour lines in the cross-section plane that are outside this right triangle.

Step 5: Draw a second rectangle EFGH at the up-plunge end of the map, with EF (farther from you) and GH (closer to you) parallel to DA and equal in length to DA. FG and EH are colinear with CD and AB, respectively. If rectangle EFGH were positioned so that GH coincided with DA and EH coincided with JA, then we could consider EFGH to be a "frame" in the profile plane through which we could view the plunging folds. Because EFGH represents an inclined plane, lines parallel to EF are horizontal, but those parallel to FG are inclined.

Step 6: To draw the fold profile in the "frame" of EFGH we first find a point on the map plane where the trace of the geologic contact intersects a contour line. We choose point L, which lies on the 200-m contour. Through point L we draw two lines; one parallel to AB and one parallel to DA. Extend the line parallel to DA across triangle ABJ to point M, which lies on the 200-m contour line of the rotated cross-section plane. Extend the line parallel to AB across rectangle EFGH; this line intersects EF at N and intersects GH at O.

Step 7: Return to the triangle ABJ. Draw line MP parallel to BJ. Measure the length of segment JP along line AJ. Point L', which is the projection of L in the profile plane (EFGH), lies along line NO; the length of segment NL' equals the length of segment JP.

Step 8: Repeat the procedure for many other points. For example, S' is found by drawing lines SV and ST. Line SV intersects rectangle EFGH at U and V. Line ST ends where it crosses the contour line in triangle AJB in the vertical section plane whose elevation equals the elevation at S. Draw line TW parallel to BJ. Plot point S' along line UV so that the length of segment US' equals the

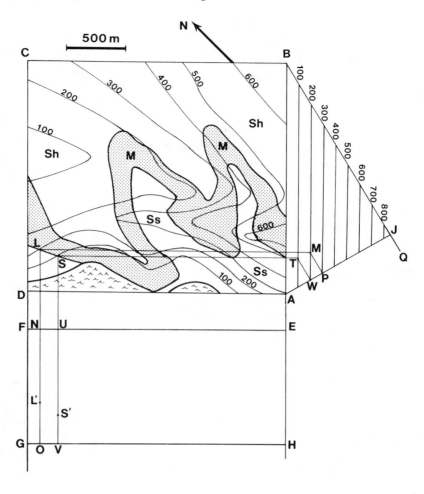

Figure 13-22. Illustration of method for constructing profiles of plunging folds that crop out in areas of high relief, as described in Problem 13-9. Map of folded sandstone (Ss), marble (M and stippled), and shale (Sh) sequence; topographic contours in meters. Triangle ABJ at right is a vertical cross section folded down into plane of the map. Rectangle EFGH is a profile plane positioned in the plane of the map.

length of segment JW along AJ. By connecting the profile images of several points along a particular contact, we can trace out the profile image of that contact (Fig. 13-23).

In the preceding method, lines EF and GH were used as reference lines to determine the positions of points L' and S'. Note that we can draw contour lines on the profile plane, but that the spacing of the contours will not be the same as the spacing on triangle ABJ. Remember that ABJ represents a vertical plane, whereas EFGH represents an

inclined profile plane. Therefore, the spacing of contours on EFGH is

$$\text{spacing} = d/\cos \mu \qquad \text{(Eq. 13-5)},$$

where d is the distance between contours in triangle ABJ, and μ is the plunge of the fold axis.

The graphical procedure just described is very tedious if there are many points to be transferred from the map to the profile plane. Section 13-8 provides an algebraic version

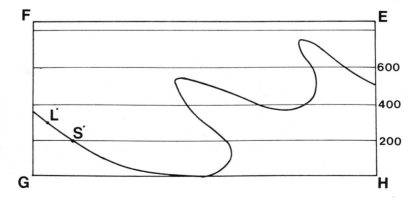

Figure 13-23. Fold profile of the contact between the marble (M) and the sandstone (Ss) shown on the map in Figure 13-22. L' and S' are points determined in Method 13-9.

of the projection procedure, which can be easily converted into a computer algorithm, thereby making it possible for a computer to construct the profile.

13-8 CONSTRUCTING BLOCK DIAGRAMS

A block diagram, with a geologic map on its top face and geologic cross sections along its side faces, is an effective means of portraying geologic structures. In this section we examine how to draw block diagrams with any orientation that correctly portray geologic structures in perspective. We also learn how to plot geologic data on the diagrams. The methods require the use of an *orthographic net* (Fig. 13-24).

The Orthographic Net

An orthographic projection of a sphere can be constructed by simply passing a suite of parallel projection lines through the sphere so that they intersect a projection plane;

the projection plane must be perpendicular to the projection lines (Fig. 13-24a). To help you visualize an orthographic projection, consider that the view of the moon that we have from earth is essentially an orthographic projection of half the moon's surface. Imagine a sphere on which lines of latitude and longitude have been drawn. An orthographic projection of this graticule (see Appendix 1) onto any vertical great circle will appear as a grid of lines within the circle. The lines of latitude (which are small circles) appear as straight lines parallel to the equator, and the lines of longitude (which are great circles) appear as elliptical arcs running from pole to pole. This grid is called an *orthographic net* (Fig. 13-24b) or *orthonet*. We can plot lines and planes or rotate geometric elements on an orthographic net exactly like we have done on a stereonet. As is the case with a stereonet, we portray only lower-hemisphere spherical projections on an orthonet, so the projection lines are vertical, and the projection plane is horizontal.

The properties of an orthographic net allow you to rotate figures to simulate the effect of changing your line

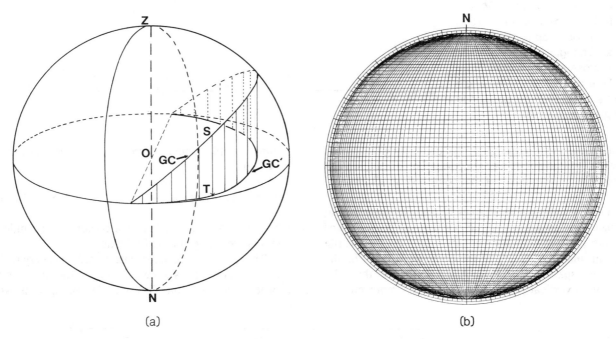

(a)   (b)

Figure 13-24. The orthographic net. (a) Construction of an orthographic net. O is the center of the projection sphere, Z is its zenith, and N is its nadir. GC is the upper-hemisphere spherical projection of a plane passing through the center of the net. We find the orthographic projection of any point on the surface of the projection sphere (such as point S) by drawing a line that is parallel to ZN through the point and finding where it pierces a plane perpendicular to ZN. In this way we can draw the orthographic projection of the great circle (GC'). We show the upper-hemisphere projection only because it is easier to see in this drawing; (b) a completed orthographic net.

of sight; thus, it is very useful for constructing isometric block diagrams (Appendix 1; also see McIntyre and Weiss, 1956).

Constructing a Cube with an Orthographic Net

Let us begin by examining how to render a cube with a horizontal top face that is viewed along different lines of sight. If we view the cube along a normal to its top face, we see the cube as a square (Fig. 13-25a). Other faces, edges, and vertices on the cube are hidden by this face. We can see the cube's other components only by rotating the cube or by changing our viewing axis. Next, we show a method outlined by Lisle (1980) to draw, in proper perspective, a cube of any orientation viewed along any line of sight.

Problem 13-10

Begin with a cube whose top face is horizontal and whose side faces are aligned northeast and northwest. Draw how the cube would look when viewed along a line oriented 30º,005º.

Method 13-10

Step 1: Prepare an overlay for use with the orthographic net as you did for the stereonet (i.e., push a pin through the center of the net and puncture the overlay with the pin; see Chapter 5). On the overlay, trace the primitive of the orthographic net, and draw a tick mark to indicate north. Place the north mark on the overlay over the north mark on the net.

Step 2: First, plot three points, each representing the orientation of one edge of the cube (these are called the principal directions of the cube). One edge is vertical and is represented by point V at the center of the the net (Fig. 13-25b). The northeast-trending horizontal edge plots as point X on the primitive, and the northwest-trending horizontal edge plots as point W on the primitive.   By drawing VX, VW, and line segments parallel to them, we have an image of the cube viewed along a vertical line of sight, with north at the top of the page (Fig. 13-25b).

Step 3: To plot the point representing the line of sight, mark the bearing of the line on the primitive of your overlay. Revolve the overlay so that the bearing (5º east of north) mark lies on the equator (or over the north mark on the grid). Count in from the primitive by 30º to locate point L. Because grid lines are very closely spaced near the primitive, it is useful to check the location of any point by counting out the complement of this angle (60º) from the center of the net.   Point L is the lower-hemisphere orthographic projection of the line of sight (Fig. 13-25b).

Step 4: To obtain a cube that appears to be viewed along our new line of sight (L), we rotate L about a horizontal axis so that it moves to the center of the net. To do this, we let the north-south axis of the net be the rotation axis and revolve the overlay so that point L lies on the equator. Move L along the equator by 60º to the center of the net. This rotation brings the point representing line of sight to the center of the net (i.e., the line of sight becomes vertical). We also rotate V, W, and X through the same angle about the same horizontal axis. To do this, X, V, and W move along small circles by 60º to their rotated positions at X', V', and W' (Fig. 13-25c). The spherical angles between L, V, W, and X are not changed by rotating them, but their new positions on the overlay indicate where the spherical projections of these elements would fall if projected orthographically onto a plane normal to the line of sight (L).

Step 5: Draw lines from the center of the net to each rotated edge of the cube (lines OV', OW', and OX' on Fig. 13-25d). These lines have the appropriate attitudes and relative lengths to be a perspective rendering of the three principal directions of the cube as viewed along of line of sight oriented 30º,005º when you revolve the overlay so that OV' appears vertical (Fig. 13-25d).

Step 6: Draw the cube by drawing line segments parallel to, and equal in length, respectively, to OV', OW', and OX' (Fig. 13-25e).

Constructing Geologic Block Diagrams with an Orthographic Net

Next, we show how to portray geologic features on the face of the cube in such a way that angular relationships are correctly portrayed. First, we consider how to project geology onto the top face of the cube, then we consider how to project geology onto the side faces of the cube.

Problem 13-11

The trace of a contact is shown on a map (Fig. 13-26a). Portray this geology on the top surface of a cube whose principal directions are vertical, north-south, and east-west. The cube is to be viewed along a line oriented 40º,050º.

Method 13-11

Step 1: Draw a square grid on the map. The grid lines should be parallel to the edges of the proposed block (Fig. 13-26a).   A point along the map trace of any geologic contact has unique coordinates with the square-grid reference frame.

Step 2: Construct the block in the proper orientation following Method 13-10. On the top surface of the block draw the map grid in the appropriate orientation. In this example, the north-south grid lines must parallel the

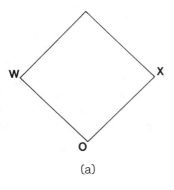

(a)

Figure 13-25. Construction of a cube with an orthographic net, as described in Problem 13-10. (a) The cube viewed normal to one face appears as a square; (b) the orthographic projections of a desired line of sight (L) and of the principal directions of the cube (V, W, and X); (c) configuration of projections after rotations; (d) orientations of rays OV', OW', and OX', which give the orientations and relative lengths of the cube's principal directions viewed along the line of sight L, if we align OV' vertically; (e) an isometric projection of the cube as viewed along the line oriented 30°,005°.

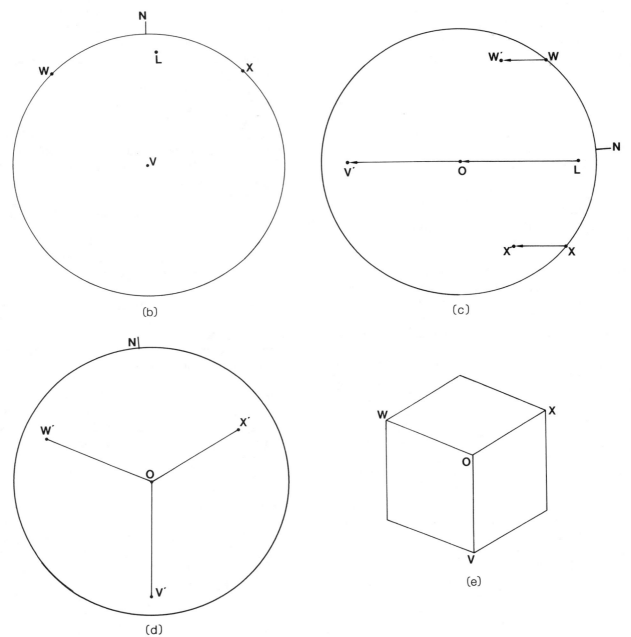

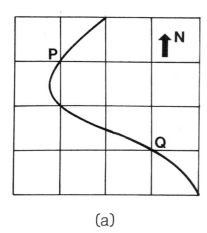

(a)

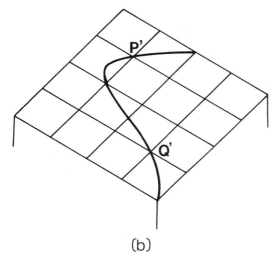

(b)

Figure 13-26. Projection of a map onto the top surface of a block diagram, as described in Problem 13-11. (a) Square grid drawn on a map. Points P and Q fall on a geologic contact; (b) the corresponding grid shown in perspective after rotation on an orthographic net. P' and Q' correspond to P and Q, respectively. The geologic contact is shown in perspective on the top of the block diagram.

north-south edge of the block, and east-west grid lines must parallel the east-west edge of the block. The distance between adjacent east-west grid lines is not the same as the spacing of north-south grid lines; the spacing must be proportional to the lengths of the cube edges. In other words, if each edge of the map is divided into four equal segments by construction of the grid, then each segment of the cube edge must be divided into four equal segments by construction of the grid.

Step 3: To determine the trace of the contact on the

top of the block, simply determine the coordinates of many points on the contact with reference to the map grid. Transfer these coordinates to the grid on the surface of the cube, and retrace the contact (Fig. 13-26b).

Problem 13-12

Draw an isometric block diagram of the geologic map shown in Figure 13-27a. Show all the geologic features in proper perspective on the sides of the block diagram. The block is to be viewed along a line of sight plunging 30° in the direction 315°. The strike and dip of bedding in the southern limb is N85°E,65°N; the strike and dip of bedding in the northern limb is N37°E,35°S. The azimuth and plunge of the fold axis is 23°,078°.

Method 13-12

Step 1: Prepare an overlay and place it over an orthographic net. Plot the points representing the line of sight and the edges of the cube. In this case the edges of the cube are vertical, east-west, and north-south, respectively. The line of sight plots as point L within the primitive, and the edges of the cube plot as points N and W on the primitive.

Step 2: Plot the points representing poles to bedding in the two limbs of the fold and the fold axis. The southern limb of the the fold plots as Sl, the northern limb plots as Nl on the overlay, and the fold axis plots as F.

Step 3: Rotate all structural elements and principal cube directions by an appropriate amount around a horizontal axis such that L becomes vertical. To do this, revolve the overlay so that L lies on the east-west diameter of the orthonet. Move L by 60° along the diameter to the center of the net. All other points move 60° along appropriate small-circle traces. Remember that if a point reaches the primitive during a rotation, it reappears on the diametrically opposite side of the orthonet. Sl moves to Sl', Nl moves to Nl', F moves to F', etc. (Fig. 13-27b).

Step 4: Draw the properly oriented cube. Transfer the geologic contacts from the map to the top of the cube, using the technique described in Method 13-11.

Step 5: We use a method described by Lisle (1980) to draw lines on the sides of the cube indicating the dipping bedding surfaces. Consider an overlay (Fig. 13-27c) that shows only the points representing the edges of the properly oriented cube (W', X', V'), the now-vertical line of sight (O), and the rotated points representing the structural elements (Sl', Nl', F'). Trace the three great circles that represent the three principal planes of the cube. Each of these great circles is determined by aligning two of the principal axes of the cube along a great circle on the orthonet. Next, draw the great circles that represent bedding in fold limbs as viewed along the desired line of sight. Each great circle is normal to the rotated position of

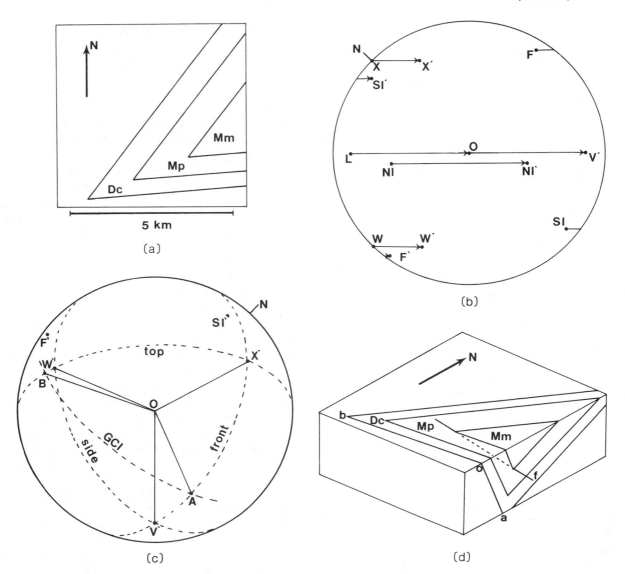

Figure 13-27. Construction of a block diagram of a fold, as described in Problem 13-12. (a) Generalized map of the west end of the Cove syncline, Pennsylvania Appalachians. (After Dyson, 1967.) The dip and dip direction of beds on the northern limb is $35^{\circ}, 127^{\circ}$. The dip and dip direction of beds on the southern limb is $65^{\circ}, 355^{\circ}$. We take the line of intersection of the beds, $23^{\circ}, 078^{\circ}$, to be the fold axis; (b) orthographic projection of elements from the map. V, W, and X are principal directions of the block diagram. F is the fold axis, NI is the pole to bedding of the northern limb, SI is the pole to bedding of the southern limb, and L is the line of sight $(30^{\circ}, 315^{\circ})$. All the primed letters have been rotated through 60° around the axis that brings L to the center of the net; (c) rays OV', OW', and OX' are the edges of the block diagram. Great circles denoting top, front, and side face are shown. GCI is the great circle normal to SI'. GCI intersects the great circle representing the front of the block at A and that representing the top of the block at B. F' is the rotated fold axis; (d) completed block diagram. When OV' is aligned parallel to the vertical edges of the block diagram, oa parallels OA, ob parallels OB, and f parallels OF'.

a pole to bedding in a fold limb. For example, GC1 is the great circle normal to the rotated pole to bedding (Sl') in the southern limb. Great circle GC1 intersects the great circle representing the front of the block diagram at A and the great circle representing the top of the block diagram at B (Fig. 13-27c). Rays from the center of the orthographic net to each intersection (OA and OB on Fig. 13-27c) give the orientations, on the block diagram, of the lines representing the intersection of the two planes. Repeat this procedure for bedding in the northern limb.

Step 6: Now that we know the orientations of the beds in the vertical walls of the block, it is easy to complete the block diagram. For example, locate the intersection of the Mp/Dc contact on the southern limb of the fold with the east edge of the block (we know this location from step 4). The trace of this contact in the east wall of the block in Figure 13-27d is a line that is parallel to line OA in Figure 13-27c. The orientations of other contacts are drawn in the same manner. The fold axis in Figure 13-27d, which appears to pass through the block, is drawn parallel to line OF' in Figure 13-27c.

The block diagrams shown in the preceding examples do not show topography on the top surface. Topography, and other embellishments that make a block diagram more realistic, can be added following techniques described by Lobeck (1958) and Goguel (1962).

13-9 APPENDIX: USE OF A COMPUTER FOR DOWN-PLUNGE PROJECTIONS

In order to use a computer to construct down-plunge projections of geologic structures depicted on a map, we must recast the construction process in an algebraic form (Charlesworth and others, 1976; Kilby and Charlesworth, 1980; Langenberg, 1985; Langenberg and others, in press). We first define Cartesian axes with the x-axis parallel to a horizontal north-south line on the map, the y-axis parallel to a horizontal east-west line on the map, and the z-axis vertical (i.e., perpendicular to the map surface). Place the origin of this coordinate frame at one corner of the up-plunge end of the structure and orient the coordinate frame to be right-handed. Any point on the topographic surface or in the subsurface has unique coordinates (x,y,z) relative to these axes. All coordinate values must be measured in the same units (e.g., feet, meters, or miles). It is usually easiest to use the map scale to convert all horizontal distance measurements to those used to measure elevation.

To project a point onto a profile plane, we perform a coordinate transformation to new Cartesian axes x', y', and z'. x' is a horizontal line whose bearing is perpendicular to the bearing of the fold axis. y' has the same bearing and plunge as the fold axis. z' parallels the true dip direction of the profile plane. If the bearing of the fold axis is θ, and the plunge of the fold axis is μ, the point with coordinates (x, y, z) relative to the east-north-vertical coordinate frame has coordinates (x', y', z') relative to the new coordinate axes. The values of x', y', and z' are given by

$$x' = x[\cos \mu \cos(\theta - 90°)] - y[\cos \mu \sin(\theta - 90°)] - z \sin \mu$$
$$\text{(Eq. 13-A1)}$$

$$y' = x[\sin(\theta - 90°)] + y[\cos(\theta - 90°)] \qquad \text{(Eq. 13-A2)}$$

$$z' = x[\sin \mu \cos(\theta - 90°)] - y[\sin \mu \sin(\theta - 90°)] + z \cos \mu$$
$$\text{(Eq. 13-A3)}$$

or, in matrix form,

$$\begin{bmatrix} x' \\ y' \\ z' \end{bmatrix} = \begin{bmatrix} [\cos \mu \cos(\theta - 90°)] & [\cos \mu \sin(\theta - 90°)] & \sin \mu \\ [\sin(\theta - 90°)] & [\cos(\theta - 90°)] & 0 \\ [\sin \mu \cos(\theta - 90°)] & [\sin \mu \sin(\theta - 90°)] & \cos \mu \end{bmatrix} \begin{bmatrix} x \\ y \\ z \end{bmatrix}$$
$$\text{(Eq. 13-A4).}$$

Since the y'-coordinate axis is parallel to the fold axis, the x'-z' plane is a profile plane. We can project points onto a profile plane by collapsing all points onto a single plane that parallels the x'-z' plane. To do this, we simply ignore the y'-coordinate values and plot all points on a two-dimensional Cartesian coordinate frame with abscissa x' and ordinate z' using only their x'- and z'-coordinate values. The plot of points on the x'-z' frame is the profile of the structure.

Computer construction of profiles involves simply (1) digitizing points along the trace of a structure on a map (i.e., determining their x,y,z coordinates), (2) calculating the coordinates in x', y', z' space, by using the preceding equations, and (3) having the computer plot the transformed x'- and z'-coordinates on an x'-z' profile plane. The algorithms are relatively simple, so the procedure can be accomplished with a desktop computer.

EXERCISES

1 . Choose an appropriate line of section across the map in Figure 13-M1.

2. Point A is 150 m northwest of point B along a N45°W-trending section line; the elevation at B is 15 m higher than the elevation at A. The strike and dip of

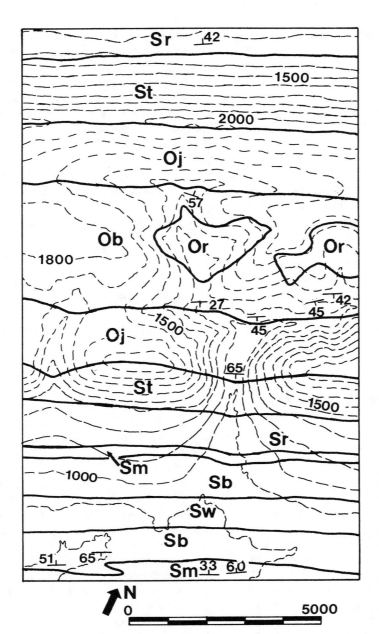

Figure 13-M1. Map of a portion of the Mifflintown Quadrangle, Pennsylvania Appalachians, USA. (From Conlin and Hoskins, 1962.) Or = Ordovician Reedsville Formation; Ob = Ordovician Bald Eagle Formation; Oj = Ordovician Juniata Formation; St = Silurian Tuscarora Formation; Sr = Silurian Rose Hill Formation. Topographic contours in feet.

bedding is N45°E,27°NW at A and N45°E,52°SE at B. Use the Busk method to reconstruct the folded layers passing through points A and B.

3. Points A, B, and C fall along a N88°E-trending section line. The distance from A to B is 550 m, and the distance from B to C is 200 m; all three points have the same elevation. The strike and dip of bedding is N02°W,22°E at A, N02°W,45°E at B, and N2°W, 54°W at C. Use the Busk method to reconstruct the folded layers passing through A, B, and C.

4. Use the Busk method to draw a cross section using the data given in Figure 13-M2.

5. Use the Busk method to draw a cross section from the map in Figure 13-M1.

6. Points A, B, and C fall along a N50°W-trending section line across a region of angular folds. The distance from A to B is 220 m, and the distance from B to C is 370 m; all three points have the same elevation. A falls in a dip domain where the strike and dip of bedding is N40°E, 27°N; B falls in a domain where the strike and dip of bedding is N40°E, 76°S; and C falls in a domain where the strike and dip of bedding is N40°E, 22°N. Use the kink method to reconstruct the folded layers passing through A, B, and C.

7. Use the kink method to draw a cross section using the data in Figure 13-M2. *Hint:* Draw line segments indicating the domainal dips through each point where a kink fold hinge intersects the ground surface (the "h's" on the profile), and bisect that angle to orient the kink fold hinge.

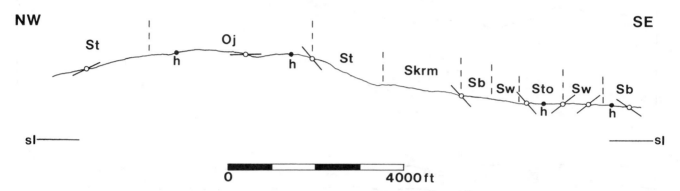

Figure 13-M2. Dip readings along a line of section across folds in the Millerstown Quadrangle, Pennsylvania Appalachians, USA. (After Faill and Wells,1974.) Dashed vertical lines separate outcrop belts of different stratigraphic units. Each filled circle (h) marks the intersection between a kink plane and the ground surface; use them when completing exercise 7. sl = sea level; Oj = Ordovician Juniata Formation. St = Silurian Tuscarora Formation; Srkm = Silurian Rose Hill, Keefer, and Mifflintown Formations; Sb = Silurian Bloomsburg Formations; Sw = Silurian Wills Creek Formation; Sto = Silurian Tonoloway Formation. The Ordovician Juniata Formation is about 1500 ft thick and is underlain by the 750-ft-thick Bald Eagle Formation and the 1500-ft-thick Reedsville Formation.

8. Use the kink method to draw a cross section of the region illustrated in Figure 13-M3.

9. Use the dip-isogon method to complete the fold profile begun in Problem 13-6.

10. Figure 13-M4 is a map of nonplunging, cylindrical folds. Construct a profile of these folds using the orthographic projection method outlined in Problem 13-7.

11. Use the grid method to draw a profile of the fold in Figure 13-M5.

12. Assume that the folds shown in Figure 13-M6 are cylindrical and plunge 11° toward 015°. Use the method outlined in Problem 13-9 to determine the structure in this region.

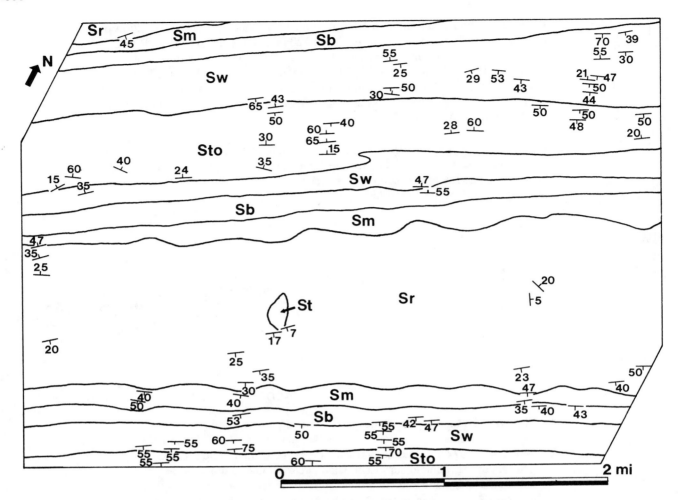

Figure 13-M3. Map of a portion of the Mifflintown Quadrangle, Pennsylvania Appalachians, USA. (From Conlin and Hoskins, 1962.) St = Silurian Tuscarora Formation; Sr = Silurian Rose Hill Formation; Sm = Mifflintown Formation; Sb = Bloomsburg Formation; Sw = Wills Creek Formation; Sto = Tonoloway Formation.

13. Use the orthographic net to draw, in proper perspective, how a cube with a horizontal top and northeast/northwest directed edges would look when viewed along a line of sight 25°,315°.

14. Draw a block diagram viewed along a line of sight 340°,30° of the fold illustrated in Figure 13-M7.

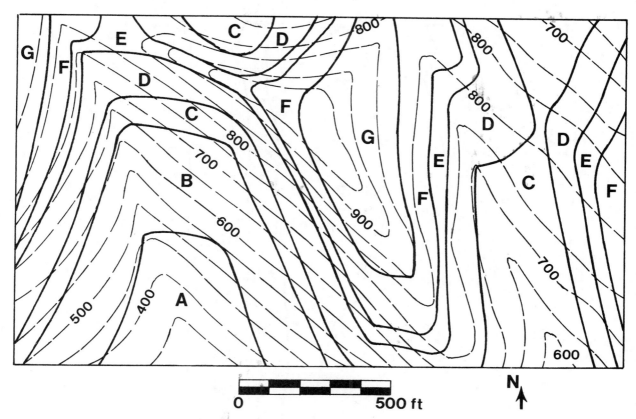

Figure 13-M4. Map of hypothetical nonplunging, cylindrical folds with north-south axes. Topographic contours in feet.

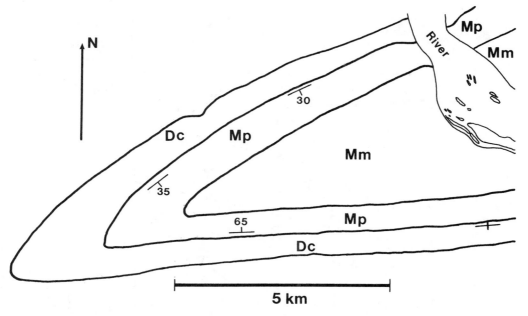

Figure 13-M5. Simplified geologic map of the Cove syncline, New Bloomfield Quadrangle, Pennsylvania Appalachians, USA. (From Dyson, 1967.) Dc = Devonian Catskill Formation; Mp = Mississippian Pocono Formation; Mm = Mississippian Mauch Chunk Formation. Bearing and plunge of fold axis are 085° and 15°.

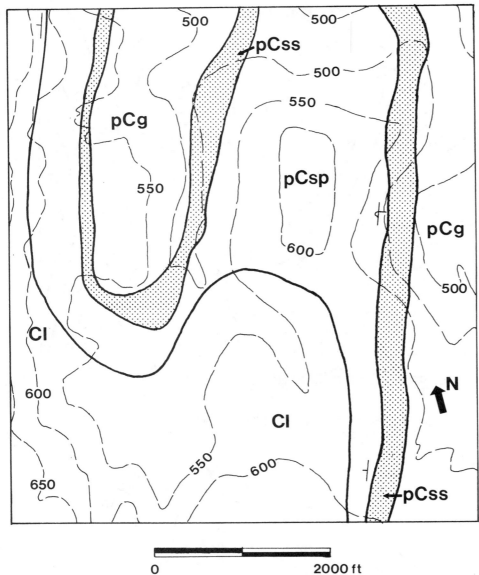

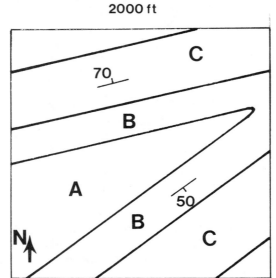

Figure 13-M6. Geologic map of a portion of Blue Ridge, Loudoun County, Virginia, USA. (Adapted from Nickelsen, 1956.) pCg = granitic gneiss; pCss = Precambrian Swift Run Formation sandstone; pCsp = Precambrian Swift Run Formation phyllite; Cl = Cambrian Loudoun Formation. Topographic contours in feet.

Figure 13-M7. Map of a hypothetical fold. Strike and dip of bedding is N77°E,70°NW in the northern limb and N53°E, 50°S in the southern limb. The fold axis plunges 19° toward 071°.

14

INTRODUCTION TO CROSS-SECTION BALANCING

Stephen Marshak
Nicholas Woodward

14-1   INTRODUCTION

Cross sections are very important tools for communicating information about geologic structures, so the interpretation depicted on a cross section must be as close to the truth as possible. The procedure of *cross-section balancing* has become popular in recent years as a means of helping to analyze and improve cross sections. Cross-section balancing permits geologists to test the validity of the structural geometry portrayed on a cross section (see Dahlstrom, 1969, and Elliott, 1983). It requires thoughtful analysis of fault shapes, bed lengths, and cross-sectional areas. One of the key steps involved in the procedure is the *restoration* of the beds depicted on the cross section to the relative positions that they had prior to deformation.

The purpose of this chapter is to introduce elementary aspects of cross-section balancing. We provide simple examples of how a cross section can be constructed, restored, checked, and improved. Study of cross-section balancing is a subject that is rapidly advancing, and so the techniques are constantly being improved. Cross-section balancing is sometimes an excruciating iterative process that relies heavily on intuition and on a broad knowledge of structural geology; it is not suited to a "cookbook" approach. Nevertheless, we hope that by following the "steps" of our simple examples, you will grasp the fundamental goals of cross-section balancing and will be forced to think hard about every line that you draw on a cross section.

The procedure of cross-section balancing has proven to be most valuable in the study of deformed belts in which deformation is largely confined to layers of rock that lie above a subhorizontal *detachment fault* or *decollement* (Rich, 1934; Rodgers, 1949; 1963). In some literature such belts are called *thin-skinned* deformed belts (e.g., Gwinn, 1964; Harris and Milici, 1977). The thin-skinned concept emphasizes that the rock below the detachment need not display the folding and faulting found in rocks above the detachment (Fig. 14-1). The term thin-skinned was used to describe regions where deformation was confined to a stratified sequence (cover) above crystalline basement. If basement was involved in the deformation, the belt was called "thick-skinned." It is now known that at some localities detachments lie in crystalline rock below the basement/cover nonconformity, so the term thin-skinned is not used as frequently these days. To avoid confusion, it is best to state simply whether or not basement rocks are involved in the thrusting.

Deformation involving detachments occurs both in *fold-thrust belts* (e.g., Boyer and Elliott, 1982), in which shortening of the crust is accommodated by the formation of thrust faults and associated folds (e.g., Hossack, 1979), and in *extensional* or *rift terranes*, in which crustal thinning is accommodated by the formation of normal faults and associated folds (Fig. 14-2; also see Gibbs, 1983, 1984; Wernicke and Burchfiel, 1982). This chapter focuses on cross sections of fold-thrust belts, for most cross-section balancing studies to date have been applied to these belts (e.g., Elliott and Johnson, 1980).

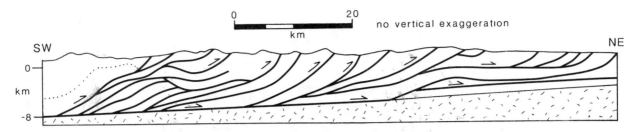

Figure 14-1. Cross section of a segment of the foothills belt of the Canadian Rockies that emphasizes the concept of thin-skinned tectonics. The heavy lines are faults (arrows indicate relative movement; transport is toward the foreland). Faults do not penetrate below a detachment fault that lies in the plane of bedding near the base of the sedimentary sequence. (After Price, 1981.)

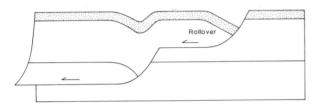

Figure 14-2. Cross section illustrating geometry of folds above a two-step normal fault. Note the anticline and syncline in the hanging-wall block.

14-2   TERMINOLOGY OF FOLD-THRUST BELTS

Structures of Fold-Thrust Belts

Fold-thrust belts develop either at convergent plate boundaries or as a result of continental collision, and they result in shortening or contraction of the crust. Next, we review the structural elements characteristic of fold-thrust belts.

(a) Internal and External Zones: Broadly speaking, orogenic belts in which fold-thrust belts occur can be divided into *internal zones* and *external zones.* The internal zone is the portion of the belt in which plastic deformation dominates, penetrative strains develop, and metamorphism occurs. The internal zone is sometimes referred to as the *hinterland.* The external zone borders the undeformed continental interior and is characterized by less plastic deformation, nonmetamorphic conditions, and nonpenetrative strains. The *foreland* of an orogenic belt refers, in a strict sense, to the undeformed region in front of the thrust belt. Sometimes this term is used with reference to the region of diminishing shortening comprising the most external portion of the fold-thrust belt.

(b) Detachments and Thrust Sheets: A *detachment* or *decollement* is a subhorizontal or shallowly dipping fault along which a sheet of rock has moved relative to the underlying substrate. In a stratified sequence, detachments commonly lie in the plane of bedding. Several detachments may occur in a vertical sequence; in such a case the *basal detachment* is the lowest one. The basal detachment can be a regional fault that separates the entire package of rock undergoing deformation and movement from the unaffected rock below.

In external zones the basal detachment commonly forms at or near the contact between sedimentary units and crystalline basement (Dahlstrom, 1970; Fig. 14-1). In internal zones the basal detachment commonly lies within crystalline basement rocks. Therefore, the hanging wall of the fault in internal zones contains basement (e.g., Harris, 1979; Cook et al., 1979; Stanley and Ratcliffe, 1985). A *detachment horizon* or a *glide horizon* is a stratigraphic interval in which detachments are commonly found. In many cases detachment horizons are composed of relatively weak rock, such as evaporite, but it is dangerous to assume that simply because a unit is weak that it must be a detachment horizon or that all detachments lie in weak horizons. At some localities detachments run through stiff units.

In fold-thrust belts the package of rock above a fault is called a *thrust sheet.* Thrust sheets are named for the fault that underlies them; for example, the body of rock lying above the McEachran thrust is called the "McEachran thrust sheet" (Fig. 14-3a). The *leading edge* and *trailing edge* of a thrust sheet are defined with respect to the *transport direction* of the thrust sheet. The transport direction is a vector in the map plane that defines the direction that the thrust sheet has moved at a given locality. For example, if a thrust sheet has moved to the east, the eastern edge is the leading edge, and the western edge is the trailing edge. A *forethrust* is a thrust fault on which displacement is in the same general direction as regional transport direction, and a *backthrust* is a thrust fault on which displacement is opposite to regional transport direction. Similarly, the *forelimb* of an anticline is the limb closer to the leading edge of a thrust sheet, and the *backlimb* is the limb closer to the trailing edge of the thrust sheet (Fig. 14-3).

(c) Ramp-Flat Geometry: To simplify thrust geometries, it is common to portray thrust faults as smooth planes in cross section. Many thrust faults actually have a step-like profile (Fig. 14-3a) called *ramp-flat geometry*. A thrust fault that cuts up to the syn-deformational erosion surface is called an *emergent thrust*, whereas a thrust fault that dies out in the subsurface is called a *blind thrust* (Fig. 3b). The termination of a blind thrust at which displacement has decreased to zero is called a *tip line* (Boyer and Elliott, 1982). An exposed thrust fault may be an emergent thrust or simply a blind thrust which has been exhumed by erosion.

Over part of its trace in profile a thrust fault lies in or nearly in the plane of bedding and is parallel to bedding; these segments of the fault are called *flats*. At other localities the fault cuts more steeply across bedding; these segments of faults are called *ramps*. In general, flats are much longer than ramps. The intersection between a contact (e.g., a bedding plane) and the fault is called a *cutoff* or *cutoff line;* hanging-wall bedding is truncated by the fault at hanging-wall cutoffs, and footwall bedding-plane contacts are truncated at *footwall cutoffs* (Fig. 14-3a). The acute angle between the bed and the fault at a cutoff is called a *cutoff angle*.

In practice it is necessary to distinguish between different types of ramps and flats: Locations where the fault

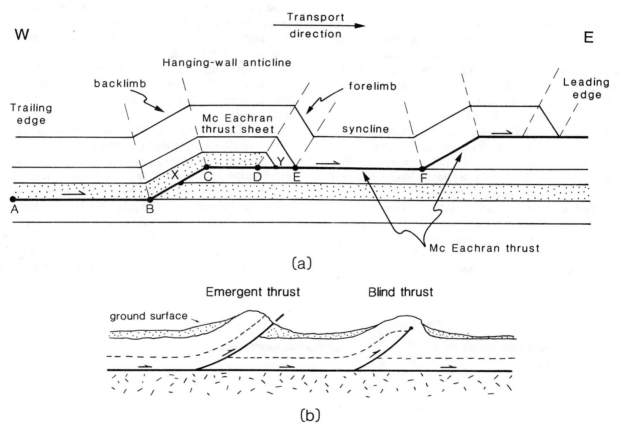

Figure 14-3. (a) Cross section illustrating the step-like geometry of a thrust fault. The definitions of a thrust sheet, leading edge, trailing edge, backlimb, and forelimb are indicated. Footwall cutoff X was originally adjacent to hanging-wall cutoff Y. Fault segment AB juxtaposes a footwall flat and a hanging-wall flat. Fault segment BC juxtaposes a footwall ramp and a hanging-wall flat. Fault segment CD juxtaposes a hanging-wall flat and a footwall flat. Fault segment DE juxtaposes a hanging-wall ramp and a footwall flat. Fault segment EF juxtaposes a hanging-wall flat against a footwall flat. (b) Cross section illustrating the difference between an emergent thrust and a blind thrust (courtesy of D. Anastasio).

is parallel to bedding of the hanging wall are called *hanging-wall flats*, locations where the fault is parallel to bedding of the footwall are called *footwall flats*, locations where the fault cuts across bedding of the footwall are called *footwall ramps,* and locations where the fault cuts across bedding of the hanging wall are called *hanging-wall ramps* (Boyer and Elliott, 1982; Woodward et al., 1985). After thrusting, a single segment of fault can juxtapose either a hanging-wall ramp or a hanging-wall flat against either a footwall ramp or a flat (Fig. 14-3a). For example, a single segment of a fault can be a ramp with respect to the hanging wall and a flat with respect to the footwall. We can describe a locality where such a configuration occurs by saying that, "In this outcrop there is a hanging-wall ramp *on* a footwall flat." A footwall flat is a detachment.

In general, the strike of a ramp is perpendicular to the transport direction of the overlying thrust sheet. If the orientation of a ramp with respect to regional transport direction is such that the strike of the ramp is highly oblique or even perpendicular to the transport direction (i.e., the fault cuts up-section along strike), the ramp is called an *oblique ramp* or a *lateral ramp*.

(d) Fault-Related Folds: Three major classes of folds are associated with the development of ramp-flat fault geometries. (1) The first class includes *fault-bend folds* (Suppe, 1983). A *fault bend* is a change in dip of a fault surface. Fault-bend folds develop in the hanging wall because the hanging-wall block must bend to accommodate changes in the shape of the fault (Fig. 14-3a). A *hanging-wall anticline*, or *ramp anticline,* typically occurs above a ramp. Broad open synclines, whose dimensions are controlled by the distance between ramps, lie between

adjacent ramp anticlines (*Note:* If ramps are very close, adjacent ramp anticlines may merge). Fault-bend folds also form in extensional terranes (Fig. 14-2), in which case they are sometimes called *rollover folds* (see Gibbs, 1984; Hamblin, 1965). (2) The second class of folds includes *fault-propagation folds* (Suppe and Medwedeff, 1984; Suppe, 1985). These folds are the result of flexural bending of a layered sequence of rock in advance of the actual rupture and development of the fault plane (Fig. 14-4a). (3) The third class of folds includes *detachment folds* (Dahlstrom, 1970; Jamison, 1987). These folds develop in response to shortening above a detachment and are not associated directly with ramps (Fig. 14-4b).

The geometry of folds exposed at the surface can be used to predict fault geometry at depth. For example, in most (but not all) localities, regions at the ground surface in which beds dip away from the transport direction (e.g., transport was to the west and the beds dip east) at the surface (region D; Fig. 14-5) reflect areas where the upper block moved upward over a ramp in the footwall. Forward bed dips (i.e., dips toward the direction of transport) at the surface (region B; Fig. 14-5) occur above hanging-wall ramp areas.

(e) Kink vs. Concentric Fold Styles: In many localities folds that develop in fold-thrust belts are not smooth concentric curves in profile but rather are subdivided into several *dip domains* (Usdansky and Groshong, 1984; see also Faill, 1969) in which the beds have a uniform dip (see Chapter 13). The dip domains join with one another at an angular hinge (A, B, C, D, and E are each distinct dip domains in Figure 14-5). In such regions the fault-related folds are said to have a *kink style*. Kink styles are associated with areas where the faults

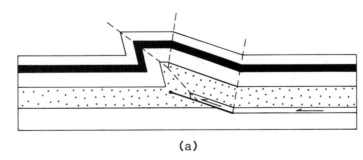

(a)

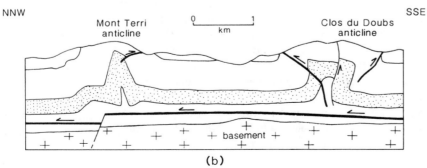

(b)

Figure 14-4. Additional illustrations of fault-related folds. (a) Fault-propagation fold (adapted from Suppe, 1985); (b) detachment folds above a subhorizontal detachment in the Jura Mountains (adapted from Laubscher, 1962).

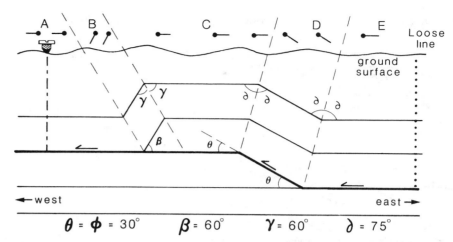

Figure 14-5. A single step in a thrust fault and the associated anticline in the hanging-wall thrust sheet illustrating the terminology used for describing cutoff angles and interlimb angles.

themselves are planar between bends. There are localities, however, where kink styles do not occur, and *concentric fold styles* are more appropriate (see Chapter 13).

It is easier to draw a cross section if the folds are kink style, because limbs of the folds can be drawn as straight-line segments, and if there is no thinning or thickening of beds, the distances between contacts can easily be kept constant. In addition, it is easier to measure bed lengths and bed areas and to determine cutoff angles on kink-style cross sections. For this reason most of the cross sections drawn in this chapter are drawn with kink-style geometries.

(f) Fault-Bend Fold Angles: Suppe (1983) showed that the relationships between cutoff angles and fault-bend angles are not arbitrary. Specifically, for situations in which two flats connected by a ramp are parallel to one another (as in Fig. 14-5), bed length and layer thickness are conserved during faulting, and there is no slip between beds above flats, the relationship between fault-bend angles and cutoff angles can be described by a simple equation:

$$\phi = \theta = \tan^{-1}\{\sin 2\gamma/(2\cos^2\gamma + 1)\} \qquad \text{(Eq. 14-1),}$$

where θ is the angle between the *lower flat* (the flat at the base of the ramp) and the ramp, ϕ is the angle between the *upper flat* (the flat at the top of the ramp) and the ramp, and 2γ is the interlimb angle of the kinks above the hanging-wall cutoff (Fig. 14-5). The angle between a ramp and a flat (θ) is generally less than 30° to 40°, and for every θ there are two possible values for γ (called first mode and second mode). Usually, the shallower limb dips (first-mode value for γ) is observed. Additional trigonometric derivation (see Suppe, 1983) allows you to calculate the hanging-wall cutoff angle (ß), and the interlimb angles (2∂) characterising the backlimb folds. It turns out, as an example, that if $\theta = 30°$, then ß = 60°; these angles used in Figure 14-5.

The equation describing the angular relationship

among angles in fault-bend folds formed over faults in which the upper and lower flats are not parallel is a bit more complex (see Suppe, 1983, 1985). For simplicity most of the cross sections described in this chapter contain parallel upper and lower flats and obey Equation 14-1. On kink-style cross sections, fault-bend angles and cutoff angles can be easily displayed.

(g) Thrust Systems: A *thrust system* is an array of kinematically related faults that developed in sequence during a single regional deformation and are associated with deformation above a basal detachment. There are two basic types of thrust systems (see Boyer and Elliott, 1982, for more detail): (1) An *imbricate fan* is a thrust system in which faults cut up-section from a basal detachment but do not rejoin at a higher stratigraphic level (Fig. 14-6a). (2) A *duplex* is a thrust system in which faults cut up-section from a basal detachment and merge at a higher stratigraphic level to form another continuous detachment (Fig. 14-6b). In a duplex the lower detachment is called the *floor thrust*, and the upper detachment is called the *roof thrust*. The faults that cut up from the floor to the roof thrust surround bodies of rock. These bodies, which are bounded on all sides by faults, are called *horses*. Duplexes occur in a range of scales. Commonly, small duplexes form at the base of larger thrust sheets; in such a position, minor slip on each small fault in the duplex contributes to the overall movement of the thrust sheet (Fig. 14-6c).

Duplex geometry in fold-thrust belts results in substantial structural thickening in the vertical direction and substantial shortening of the section in the horizontal direction. The geometry of duplexes can be quite variable, depending on the relative displacements on the faults within the duplex. Figure 14-7a shows an early stage in the evolution of a duplex in which the relative displacements on successive ramps are such that the roof thrust is a smooth surface and is parallel to the floor thrust. The final duplex composed of three horses is provided as Figure 14-7b. Figure 14-7c shows a duplex configuration in which the successive horses are stacked on top of one

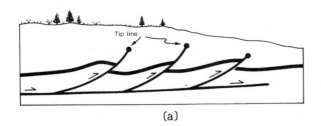

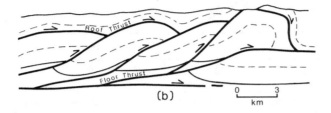

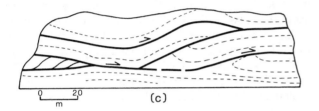

Figure 14-6. Illustrations of thrust systems. (a) Imbricate fan; (b) duplex structure (adapted from Perry, 1978); (c) small duplex at the base of a larger one (adapted from Marshak, 1986).

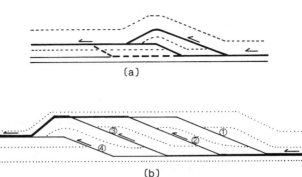

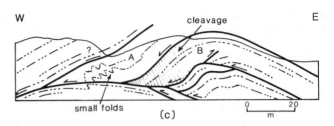

Figure 14-7. Types of duplexes. (a) Early stage in the evolution of a duplex in which the roof thrust ends up being parallel to the floor thrust. The dashed line is an incipient thrust (adapted from Boyer, 1978); (b) later stage in the development of a flat-roofed duplex. Each fault block confined between an upper and lower thrust is called a horse. The thickness of the lines represents the proportion of slip that is transferred along any individual segment of fault. In this figure the total slip on the upper detachment to the east of the duplex equals the slip on the lower detachment to the west of the duplex, but slip on individual ramps in the duplex is only a fraction of the total (adapted from Boyer, 1978); (c) an antiformal stack of horses (adapted from Marshak, 1986).

another; such a configuration is called an *antiformal stack.* The duplex of Figure 14-6b is one in which the roof thrust is not a smooth surface that parallels the floor thrust; such a structure is sometimes referred to as a *lumpy-roofed duplex.*

(h) Internal Strain in Thrust Sheets: In some localities a significant portion of the total strain in fold-thrust belts is accommodated by formation of structures within thrust sheets (e.g., Reks and Gray, 1983). Several different types of structures can be considered in this category: (1) Backthrusts commonly develop above hanging-wall anticlines (e.g., the east-verging thrust faults in Figure 14-7c, which occur in a west-verging fold-thrust belt). (2) *Out-of-the-syncline thrusts* (Fig. 14-8) are faults that die out toward the hinge of a syncline, thereby allowing rock to squeeze out of the core area of the syncline when a *room problem* develops (i.e., there is insufficient space for rock). Out-of-the-syncline faults can be either backthrusts or forethrusts. These faults are sometimes called *accommodation structures.* (3) *Minor faults* are faults on which displacement is about an order of magnitude less than the displacement of principal faults in the thrust system (Price, 1967; Wojtal, 1986). (4) *Minor folds* are folds with amplitudes that are significantly

smaller than the height of the thrust sheet (e.g., the folds in thrust sheet A in Figure 14-7c).

In fold-thrust belts significant strain may also occur by development of cleavage and/or by plastic shape change of grains in the rock. The cleavage formed in fold-thrust belts is typically spaced cleavage or slaty cleavage, the formation of which involves pressure-solution deformation that may result in volume-loss strain (e.g., Marshak and Engelder, 1985). Cleavage is usually not uniformly distributed in fold-thrust belts (e.g., the cleavage in Figure 14-7c is concentrated at the leading edge of thrust sheet B).

Reference Lines

Three lines are commonly used to provide a reference frame for describing relative movement and shortening in fold-thrust belts.

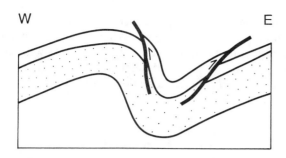

W   E

Figure 14-8.   Out-of-the-syncline thrust faults. Both a backthrust and a forethrust are shown.

(a) Regional Dip and Regional Level: Prior to displacement on faults and development of folds, strata in the external zone of a developing orogenic belt are subhorizontal or only shallowly dipping. The dip displayed by a package of strata prior to folding and faulting is called the *regional dip.* Generally, regional dips do not exceed a few degrees, and the original depositional thickness of strata commonly increases in the direction of regional dip. Folding and faulting not only result in deviation from regional dip (e.g., if the regional dip is 1.5°, and the limb of the fold has a dip of 35°, there is clearly a deviation from regional dip) but also cause beds to be lifted above a reference plane; this reference plane (a line in cross section), which marks the elevation of a bedding surface before deformation, can be called the *regional level.* We will see that one of the steps involved in cross-section balancing requires removing the effects of deformation so that beds "return to regional dip and regional level." In this chapter, we use the term *structural relief* to refer to the difference in elevation between a surface (e.g., a bedding plane) at the top of a structure and the same surface at regional level.

(b) Loose Lines: It is important during the restoration of a cross section to keep track of how much slip between points in originally adjacent beds has developed in different parts of the cross section. This is done by inserting a *loose line,* which is merely a reference line drawn at an angle to bedding *either* in the deformed or undeformed cross sections (Fig. 14-5); a loose line does not extend below the basal detachment and is usually placed near the trailing edge of the cross section. A loose line can be an arbitrary line drawn perpendicular to bedding, it can be the cross-sectional trace of a fault, it can be a vertical line drawn at the trailing edge of a cross section, or it can follow a known vertical drill hole.

A loose line can be considered to be a chain of marker points in the layers of a sequence. If the loose line is marked on a cross section of deformed rocks, then we can observe how it is distorted or how it changes orientation during restoration of the section. Inversely, if the loose line is inserted on the restored section, we can observe how it distorts during our hypothesized development of the

deformation. The distortion or tilting of the loose lines during these operations gives an indication of the angular shear of beds past one another in different parts of a structure. This shear is called *interbed slip* and is a consequence of both shear on detachments and shear related to flexural folding.

(c) Pin Lines: A *pin line* is another type of reference line in a cross section. When we measure bed lengths, we need to decide on some locality in the cross section at which to begin our measurement. A pin line is merely the reference line at which we begin measurement of bed lengths (Fig. 14-5). If our section extends across the boundary between the deformed belt and an adjacent undeformed foreland, we choose a *regional pin line* to lie in the undeformed foreland; a regional pin line can extend all the way down to basement and can penetrate the cross section either from the top or from the bottom.

Many cross sections, however, do not include undeformed rock. In such cases an arbitrary *local pin line* must be chosen. It is best to draw the local pin line perpendicular to bedding in the least deformed part of a thrust sheet (where there has been no interbed slip), where, in addition, there is the most complete stratigraphic section. Therefore, local pin lines are commonly drawn along a long flat, in a broad syncline, or at a fold hinge; a local pin line cannot extend below the detachment at the base of the thrust sheet containing the pin line. Pin lines should be placed along a fold hinge only if rock-fabric data (e.g., the occurrence of an axial-planar cleavage) indicates that the hinge has not migrated (moved along the fault plane).

14-3   CONCEPT OF A BALANCED CROSS SECTION

Types of Cross Sections

A new terminology has recently developed for discussing cross sections; next, we introduce the names that are used to refer to different types of cross sections (see also Elliott memorial volume of the Journal of Structural Geology, 1983, v. 5, n. 2).

(a) Deformed-State Cross Section: A cross section that represents the geometry of structures as they appear today, after deformation, is called a *deformed-state cross section.*

(b) Admissible Cross Section: A deformed-state cross section that depicts an interpretation in which structures look like those that can be directly observed in mountain sides and road cuts (i.e., the cross section depicts realistic-looking fold and fault geometries) is called an *admissible deformed-state cross section.* Whether or not a specific structure should be deemed admissible depends not

only on whether creation of the structure is physically possible but also on whether such a structure is likely to occur in a specific type of deformed belt. Dahlstrom (1969, p. 743) pointed out that, "in a specific geological environment, there is only a limited suite of structures which can exist . . . ." For example, you should be suspicious of a cross section that portrays a large recumbent isoclinal fold with an extremely thickened hinge at the foreland edge of a fold-thrust belt; such a structure would more likely be found in the hinterland.

A restored cross section that depicts admissible structures is called an *admissible restored cross section*. By saying that a restored cross section is admissible, we mean that the fault trajectories (the traces of faults on the cross section) shown on the cross section are possible. Specifically, the angles between ramps and flats should not be in excess of about 35°.

(c) Restored Cross Section: A cross section that has been "pulled apart," such that the fault displacement has been removed and folds have been straightened out, is called a *restored cross section*. The relative positions of rocks shown on a restored cross section should be the relative positions that the rocks had prior to deformation. Traces indicating the position and attitude of surfaces that later became faults are also typically shown on a restored cross section.

(d) Viable Cross Section: If a deformed-state cross section can be restored to an unstrained state such that the predeformation geometry of faults is admissible, bed lengths are conserved and/or bed area is conserved, and bed lengths are consistent, then the deformed-state cross section is said to be viable. An understanding of what is meant by bed "conservation" and "consistent" in this context is critical, so the next part of this chapter is devoted to explaining the application of these terms. Suppe (1985) uses the term *retrodeformable* with much the same meaning as viable.

(e) Balanced Cross Section: A *balanced cross section* is a deformed-state cross section that is both admissible and viable. In other words, a balanced cross section portrays an admissible suite of structures and can be restored such that the restored cross section depicts consistency of bed length, conservation of bed length and/or area, and admissible premovement fault geometries. Additional constraints, described later in the chapter, must also be met in order for a cross section to be balanced.

The difference between a balanced cross section and an "unbalanced" admissible cross section is that the balanced version has been restored and tested for viability. Thus, when someone says that they have "balanced" a cross section, they mean that they have gone to the effort of restoring the section and have tested the restored version.

It is important to keep in mind that by balancing a section, you are *not* checking to see if it is correct; a balanced cross section is still merely an interpretation that is quite possibly incorrect. When new data become available, it is likely that the section will need modification. A balanced section, however, is at least possibly correct, whereas a section that does not balance is probably wrong.

Considerations Involved in Testing the Viability of a Restored Cross Section

In this portion of the chapter we clarify the concepts and requirements for determining whether or not a cross section is viable.

(a) Conservation of Area: If the deformation of a bed or thrust sheet involves only folding and faulting the volume of a bed or thrust sheet will *not* change during deformation. If, in addition, the deformation yields plane strain, then the cross-sectional area of the thrust sheet or bed does not change during deformation. Figure 14-9a shows an undeformed layer of rock containing a ramp on which there has not yet been displacement. When the thrust sheet moves west over the ramp (Fig. 14-9b), its cross-sectional area does not change and is still equal to that shown in Figure 14-9a. In Figure 14-9c the shape of the ramp anticline is different; the thrust sheet illustrated in this figure has a different area than does the original undeformed thrust sheet, and thus this figure illustrates a case where area has not been conserved.

If the area of the thrust sheet in Figure 14-9a equals the area of the sheet in Figure 14-9b, then the *excess area* (area above regional level), which is labeled A_x, must equal A_i. A_i does not equal A_x in Figure 14-9c. Figure 14-9d shows the same fold form as does Figure 14-9c, but the trailing-edge of the thrust sheet has been tilted sufficiently to make A_i equal to A_x and thereby result in conservation of area.

Area conservation cannot be assumed for deformation that involves development of volume-loss strain, such as commonly accompanies the development of spaced cleavage involving pressure solution. As noted earlier, significant pressure-solution cleavage occurs locally in fold-thrust belts, so the measured cross-sectional area of a deformed-state thrust sheet is locally less than the original area. Volume-loss strain is usually not a problem in the external portions of fold-thrust belts.

(b) Area Balance: If area conservation is assumed, the area of a bed or a thrust sheet depicted on the deformed-state cross section must equal the area of the thrust sheet as depicted on the restored cross section. In other words, the restored area of a thrust sheet must "balance" or correspond to the deformed-state area of the thrust sheet. In this chapter we call the operation of comparing deformed-state and restored areas *area balancing*. The area of a bed or thrust sheet can be measured by subdividing it into simple geometric forms whose areas can

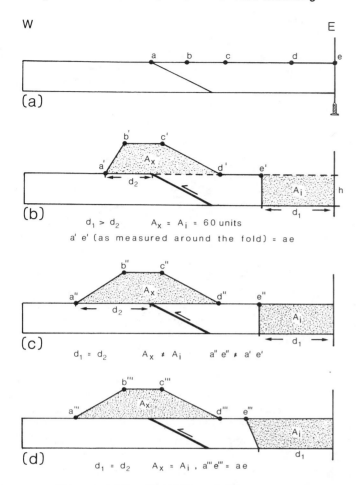

W
E

(a)

(b)

$d_1 > d_2$ $A_x = A_i = 60$ units

a' e' (as measured around the fold) = ae

(c)

$d_1 = d_2$ $A_x \neq A_i$ a"e" ≠ a'e'

(d)

$d_1 = d_2$ $A_x = A_i$, a'''e''' = ae

Figure 14-9. Illustration of the concept of bed-length and area conservation and balancing. (a) Undeformed thrust sheet; (b) thrust sheet after moving over a ramp. Area balance and bed-length balance are evident, and there is not constant slip along the fault; (c) cross section in which area balance and bed-length balance are not evident; (d) cross section in which there is bed-length balance and area balance and constant slip along the fault. Points a, b, c, d, and e are reference points, A_x is the excess area, and A_i is the area between the pin line and the trailing edge of the fault.

be calculated using plane geometry, by using a planimeter, by digitizing, by plotting on graph paper and counting squares, or by cutting out thrust sheet cross sections and weighing them.

(c) Conservation of Bed Length: Conservation of bed length refers to the supposition that the length of a contact in cross section does not change during deformation. Refer again to Figure 14-9a. Reference points along the top surface of the bed are labeled a through e; the pin line is fixed with respect to the footwall, to

provide a reference frame. The length of this contact in Figure 14-9b (line a'e', as measured around the fold) equals the original length of ae. Therefore, bed length was conserved during development of this ramp anticline. The length of contact a"e" in Figure 14-9c is not equal to ae, so bed length was not conserved during development of this structure. The length of contact a'''e''' in Figure 14-9d is equal to ae, so bed length was conserved during development of this structure.

(d) Bed-Length Balance: If bed length is conserved during deformation, the length of a contact (e.g., the top surface of a horse) is the same in both the deformed-state and restored cross sections. Such correspondence is called *bed-length balance.* Whether or not bed lengths in the deformed-state and the restored cross sections *should* balance depends on whether or not the bed changed thickness during deformation and on whether or not the strain was volume constant. We will see that thickening of a layer during deformation means that bed-length balance cannot occur if there is to be conservation of area.

(e) Consistency of Bed Length: The total lengths of each layer in a sequence depicted in a restored cross section should be nearly the same or should vary in a *consistent* manner. In other words, a straight loose line drawn perpendicular to bedding at the trailing edge of a deformed-state cross section should be either straight and perpendicular to bedding in the restored section or smoothly varying in the restored section. Note that if a vertical loose line were drawn at the trailing edge of the thrust sheet in Figure 14-9d, it would be inclined to the west in a restored version of this cross section.

Bed-length consistency can be determined by looking at the restored shape of a loose line. It is important to emphasize that consistency does *not* require that the restored loose line be exactly perpendicular to bedding in the restored and deformed-state cross sections. At present the characteristics of acceptable shapes for restored loose lines are not fully understood; it is fair to say, however, that a sudden discontinuity or zigzag in a restored loose line (Fig. 14-10) should be viewed as indicating that bed lengths are not consistent. Bed-length consistency is also indicated by restored fault geometry, as described next.

(f) Admissible Restored Fault Shapes: In the deformed-state cross section, fault geometries are distorted as a consequence of movement on younger faults. In restored cross sections, faults are depicted with the original shape that they had prior to movement on younger faults. Therefore, faults in the restored section should have reasonable step-like geometries (such as is shown in Fig. 14-3). The abundant literature on fold-thrust belts suggests that when a step-like fault initially forms, (a) ramp angles are generally less than 35° with respect to bedding, (b) faults do not turn back on themselves (i.e., segments of the fault trace do not dip in opposite directions), and (c) faults do not cut down-section.

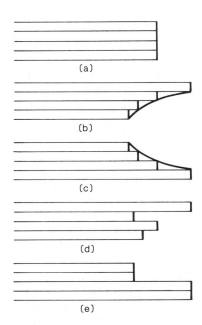

Figure 14-10. Examples of restored loose-line geometry. Each figure shows the trailing edge of a thrust sheet. Transport had been toward the west. (a) Admissible straight loose line; (b) admissible smoothly curving loose line inclined to the west; (c) admissible smoothly curving loose line inclined toward the east; (d) jagged loose line that is not admissible; (e) stepped loose line that is not admissible.

Additional Constraints on Balanced Cross Sections

Next, we provide additional constraints that guide construction of a balanced cross section.

(a) Sequence of Faulting: Many studies of fold-thrust belts have led to the conclusion that not all the faults in such belts form at the same time. In fold-thrust belts the more external faults (farthest toward the foreland) are generally the youngest, whereas the more internal faults are the oldest (e.g., Figs. 14-7a,b and 14-11a). This sequence of faulting is called a *break-forward sequence.* Structurally lower faults (faults at greater depth), therefore, are younger than overlying faults. Because of the spatial arrangement of faults, older faults can be folded by movement on younger faults (e.g., Jones, 1971; Figs. 14-6b and 14-7c). The interior to exterior (hinterland to foreland) sequence of faulting has been compared to the sequence of deformation affecting snow piling up in advance of a snowplow (cf. Davis, Suppe, and Dahlen, 1983); the snow closest to the plow blade deforms first. An arrow pointing in the direction that the rocks move generally points in the direction of the younger faults in a break-forward sequence.

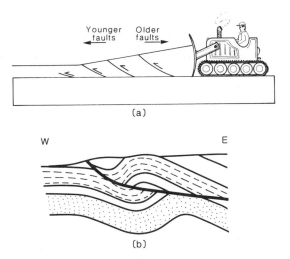

Figure 14-11. Timing of thrusting in fold-thrust belts. (a) Illustration showing the age sequence of faulting in fold-thrust belt using the snow-plow analogy; (b) out-of-sequence fault cutting a preformed fold (adapted from Woodward et al., 1985).

Out-of-sequence faults are faults that form in the more internal parts of the fold-thrust belt at a time when the main thrust system is developing in the more external parts of the fold-thrust belt. Out-of-sequence faults may truncate folds in both the upper and lower fault blocks and may cut down-section locally (Fig. 14-11b), or they may display geometries that are indistinguishable from other faults. Their existence is becoming increasingly well documented (Boyer, 1986). To date, it has been standard practice to assume break-forward sequencing as a working hypothesis and, therefore, to restore fault displacements in order from the foreland toward the hinterland.

(b) Transport Direction: In choosing a traverse direction along which to construct a balanced cross section, it is critical that the section line be within about 5° of the *transport direction* of the thrust. Cross sections in orientations other than the transport direction may be excellent representations of geology, but they are impossible to balance.

In general, the transport direction is perpendicular to the strikes of major thrust faults at a locality (Fig. 14-12). This rule works best where the faults are straight or are only gently curved in plan. Similarly, transport direction is usually perpendicular to fold hinges as defined by an equal-area plot of poles to bedding. Regional analysis of slip lineations on fault surfaces may indicate transport direction, but often such orientations are scattered and can be misleading.

(c) Ramping Directions: In almost all examples ramps that are not cutting across an earlier-formed fold cut up-section in the direction of transport (e.g., Fig. 14-3).

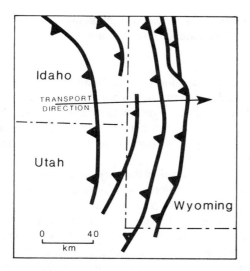

Figure 14-12. Map of a portion of the Wyoming fold-thrust belt showing the transport direction interpreted as the perpendicular to the fault traces and fold hinges. (Geology from Royse et al., 1975.)

depicted in Figure 14-13a, we note that footwall ramp AB coincides with hanging-wall ramp A'B', footwall flat BC coincides with hanging-wall flat B'C', and footwall ramp CD coincides with hanging-wall ramp C'D'. Note that the lengths of the hanging-wall ramps are less than the lengths of the corresponding footwall ramps because the hanging-wall cutoff angle is greater than the footwall cutoff angle. Because all hanging-wall ramps can be matched with footwall ramps, and all hanging-wall flats can be matched with footwall flats, the cross section in Figure 14-13a obeys the template constraint. In contrast, the cross section in Figure 14-13b does *not* obey the template constraint, because it does not depict a hanging-wall flat to correspond to footwall flat BC. The cross section in Figure 14-13b is, therefore, not viable.

As a consequence, ramps place older units over younger units and duplicate section. Faults on which this relationship is observed are sometimes called *older-over-younger faults.*

(d) Plane Strain: In general, it is best to attempt the procedures of cross-section balancing on regions where rock has not moved in or out of the plane of the cross section. This constraint, called the *plane strain constraint,* is generally met for most external fold-thrust belts that are straight and that do not contain a large penetrative strain. The constraint is not met in regions where faults are bent tightly or where there are lateral ramps; such regions are usually associated with complex strain histories.

(e) Template Constraint: Imagine the step-like trace of a fault on which movement has not yet occurred (e.g., Fig. 14-9a). It is not suprising that the hanging-wall block fits against the footwall block with no gaps or overlaps. In other words, before deformation the shape of the base of the hanging wall is identical with the shape of the top of the footwall, and the hanging-wall cutoff of a given contact must be adjacent to the footwall cutoff of the same contact. After deformation, therefore, the shape of the base of the hanging wall that we observe in outcrop must have its counterpart along the top of the footwall somewhere at depth. Likewise, footwall shapes observed in outcrop may be representative of the shape of the hanging wall that has been eroded. This geometric constraint is called the *template constraint.* Whether or not the template constraint is obeyed sometimes becomes apparent only when you try to restore a cross section.

If we apply the template constraint to the thrust fault

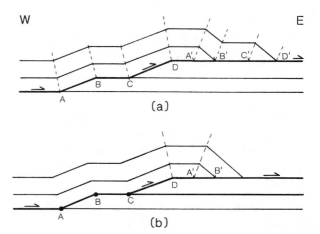

Figure 14-13. Illustration of the template constraint. (a) A cross section drafted so that the hanging-wall ramps and flats can be restored so as to correspond with footwall ramps and flats; (b) a cross section constructed so that the hanging-wall ramps and flats cannot be restored so as to coincide with the footwall ramps and flats.

(f) Conservation of Slip on a Fault: The premise that slip on faults is conserved means that the magnitude of slip on a fault as indicated by offset markers on a cross section can either (a) be constant along the trace of the fault in cross section or (b) vary in an explainable way.

Many "balanced" cross sections are constructed by assuming that slip is constant. If slip is constant, the distance between hanging-wall and footwall cutoffs for different units offset along the same fault is the same. For example, if the top of unit A is offset by 200 m along a fault, then the top of unit C is also displaced by 200 m along the same fault. Obviously, the constant-slip assumption simplifies the process of placing cutoffs on a cross section as you draw it.

Constant slip on a fault, however, is not necessary and in many cases is not even realistic. Good reasons for a change in slip magnitude along the trace include (a) the occurrence of simple shear along bedding planes in the thrust sheet (Fig. 14-14; note that in this example slip on the detachment increases in the direction of transport and that a loose line at the trailing edge of the thrust sheet is inclined in the direction of transport), (b) transfer of slip from another fault into the fault of interest at their mutual intersection (Fig. 14-7b), and (c) the partitioning of strain between fault displacement and other structures such as folds and cleavage.

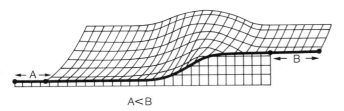

Figure 14-14. Changes in slip magnitude along the trace of the fault resulting from shear along bedding planes in the thrust sheet. Note that displacement on the upper flat is greater than displacement on the lower flat. (Adapted from Elliott, 1976.)

The last statement in the preceding paragraph requires clarification. Look once again at Figure 14-9b. Note that $d_1 > d_2$, where d_1 is the displacement of the trailing edge of the thrust sheet with respect to the pin line, and d_2 is the displacement of the leading edge of the thrust sheet along the upper detachment. The reason for the difference between d_1 and d_2 is that part of the shortening of the thrust sheet was accommodated by formation of the fold. In the case of a fault-propagation fold (Fig. 14-4a), it is particularly clear that displacement progressively decreases until it becomes zero at the tip of the fault.

14-4   DRAWING A DEFORMED-STATE CROSS SECTION

Next we illustrate the construction of a simple deformed-state cross section. This example differs from those described in Chapter 13 in that we demonstrate how to incorporate a fault that is not exposed at the ground surface.

Problem 14-1

Given the strip map of a portion of a fold-thrust belt (Fig. 14-15a), construct an admissible deformed-state cross section that accommodates the map data. Assume that the

ground surface is a horizontal plane. The only unit exposed at the ground surface is unit h.

Method 14-1

Step 1: Examine the strip map (Fig. 14-15a). We know that the map portrays a portion of a fold-thrust belt, so, although there are no faults mapped, we suspect that there may be a detachment at depth and that the fold portrayed may be related to a ramp cutting up-section from the detachment.

Step 2: Determine the direction of transport. We see that there are numerous strike and dip measurements. We take the transport direction to be perpendicular to the regional strike. Therefore, we choose an east-west line (XX') to be our line of section. Usually, we have enough information from regional mapping to know which way the transport vector points. In this example, the vector points west.

Step 3: Draw the topographic profile along the line of section and plot the surface structural data (Fig. 14-15b). Draw the stratigraphic column (presumed to be known from other work) along the leading edge of the cross section to give an indication of the position of contacts in the subsurface. We know the depth of the top of unit g from the well data, and we know that there must be enough room between the ground surface and the basement to fit in the entire stratigraphic section of the region. In fold-thrust belts, faulting thickens the section. Thus, the undeformed thickness of strata in the area gives a minimum depth to basement; basement may be deeper than this minimum but not shallower.

Step 4: There are distinct dip domains, and thus we decide to depict the anticline using the kink style of folding. We sketch in unit g to illustrate the form of the anticline (Fig. 14-15b). Notice that the top of unit g is at the same level on both sides of the anticline, so we assume that the top of unit g represents a regional level and that its dip is regional dip (horizontal). Using the methods described in Chapter 13, we project the axial-plane traces of kinks down to depth. We assume that there is no thinning of strata on the limbs of the folds, so the traces of the axial planes bisect the angle between the limbs of the folds (Fig. 14-15c).

Step 5: To complete the deformed-state cross section, we propose a fault at depth. We begin with the hypothesis that the fold at the ground surface is a ramp anticline. There is no way to know where the fault is at depth based on the information available. Constraints on the level of detachment may be provided by knowledge of regional structural geology. Let's say that we believe that the detachment lies at the base of unit d.

(a) To construct the ramp we work from east to west (in the direction of transport). The easternmost kink hinge SS' must bisect the base of the ramp (point A), and the

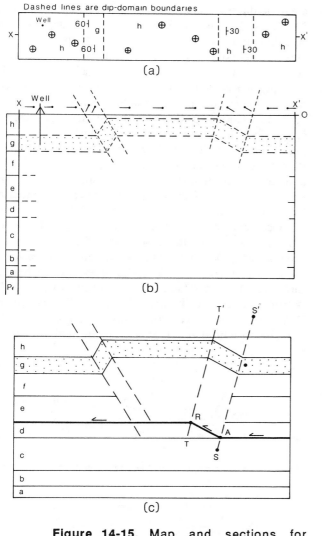

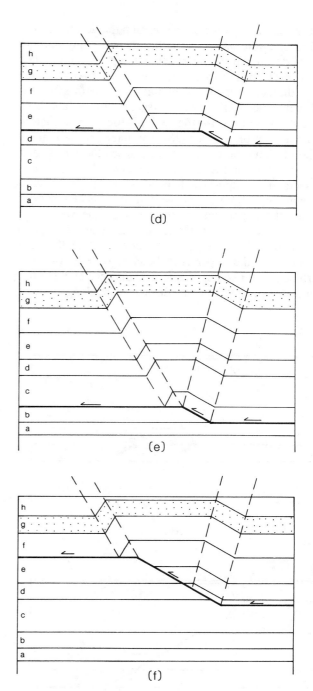

Figure 14-15. Map and sections for problem 14-1. (a) Strip map of a portion of a fold-thrust belt showing the attitudes of bedding and the dip-domain boundaries; (b) cross section showing dips in the plane of the section and the regional stratigraphy to scale; (c) cross section showing hypothetical fault; (d - f) alternative completed cross sections that fit the same original data.

ramp must be parallel to the backlimb dip (Fig. 14-15c). We draw the ramp cutting up-section to where it intersects the second kink hinge TT' (point R). In this interpretation we propose that the fault bends at point R and becomes a footwall flat. Point R happens to fall on the unit d/unit e contact (this coincidence is circumstantial and reflects the geometry of this example).

(b) Now that we have drawn in the fault, we draw in the footwall strata (Fig. 14-15c). We assume that the footwall strata have regional dip.

(c) All that remains now is for us to draw in the core of the anticline. We extend the kink hinges down to where they intersect the fault surface and then draw in the formation contacts so that they have the correct dips in the different dip domains (Fig. 14-15d).

As we noted above, the interpretation shown in Figure 14-15d is *not* a unique interpretation. Alternative cross sections that fit the data in Figure 14-15a are shown in Figures 14-15e and 14-15f.

Considering that there are several possible solutions to the data set, you may ask whether there are any definite constraints on the geometry of a ramp at depth. The answer is, fortunately, yes. Look at Figure 14-16; regardless of the depth of the lower flat, the eastern synclinal axial-plane trace (KK') must intersect the lower fault bend (point A in Figures 14-16a and 14-16b). The eastern anticlinal axial-plane trace (JM) can intersect the ramp at any point along the ramp until it reaches the upper fault bend (point B in Figures 14-16a and 14-16b). If it intersects at the upper fault bend (case 1), the western anticlinal hinge (LM) *and* the western synclinal hinge must intersect the upper flat at point B or to the west (Fig. 14-16a). If JM intersects the ramp at some arbitrary point along the ramp between points A and B (case 2), then LM must intersect the upper fault bend at B (Fig. 14-16b). Clearly, the *ramp height* (the vertical distance between the lower and upper flats) and the displacement along the fault

are quite different depending on which interpretation (Fig. 14-16a vs. Fig. 14-16b) is chosen, even though the lower detachment is at the same depth in both cases. The maximum possible depth for the fault is the depth at which point B becomes point M. If the detachment is at this depth, case 1 becomes identical with case 2, and only one ramp height can occur (shown in Figure 14-15e).

The relationships described in the previous paragraph are a direct consequence of the evolution of a simple fault-bend fold as described by Suppe (1983). Figure 14-17 illustrates three stages in the development of a ramp anticline. Note that the eastern synclinal hinge (ES) is fixed at the base of the ramp, and the eastern anticlinal hinge (EA) migrates up the ramp with increasing displacement. As long as EA intersects the ramp below the upper fault bend, the western anticlinal hinge (WA) is fixed at the upper fault bend. While EA migrates up the ramp, WS moves west along the upper flat. When EA intersects the upper fault bend, it becomes fixed at that position, the distance between WA and WS is locked, and WA and WS move to the west along the upper flat. Notice that the structural relief progressively increases as EA

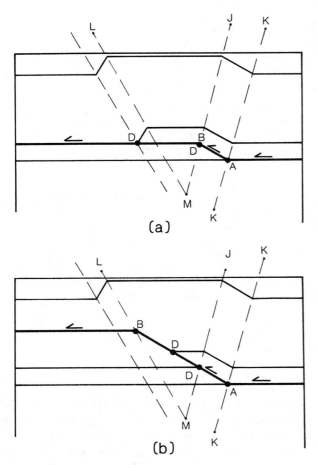

Figure 14-16. Illustration showing the range of possible ramp geometries that could fit the data of Figure 14-15a. (a) Fault bends intersect axial-surface traces KK' and JM; (b) fault bends intersect axial-surface traces KK' and LM.

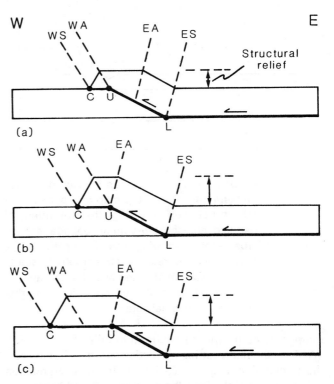

Figure 14-17. Evolution of a fault-bend fold emphasizing the migration and locking of axial trace positions. (a) EA is migrating up the ramp, and WA is fixed at the upper fault bend; (b) EA and WA both intersect the upper fault bend; (c) EA is locked at the upper fault bend, and WA moves along the upper flat.

moves up the ramp. Once EA is at the top of the ramp, the structural relief remains constant even as displacement on the fault increases and the distance between the hanging-wall cutoff (C) and the footwall cutoff (U) increases.

14-5   RESTORING A CROSS SECTION

In this portion of the chapter we describe the procedure of restoring a cross section. Recently, techniques have been developed that allow this procedure to be done interactively with a computer (e.g., Kligfield et al., 1986), but computer procedures cannot be understood unless you are first adept at restoring a section by hand. If bedding thickness does not appear to have been changed during deformation (i.e., bedding thickness is constant around folds), then restoration can be carried out by assuming bed-length balance. If bed-thickness change accompanying deformation is apparent in the deformed-state cross section, then bed-length balance cannot be assumed and you must measure areas of thrust sheets.

When balancing a cross section, it is important that you take into account the sequence of faulting; the youngest fault should be restored first, and the oldest fault should be restored last. This means that you should remove the effect of the displacement on a younger fault before you remove the effect of the displacement on an older fault. Consideration of the sequence of faulting forces you to determine whether you understand the evolution of structures portrayed in a cross section.

Restoration Based on Bed-Length Balance

We start by restoring one of the deformed-state cross sections produced in Method 14-1, then we will deal with more complicated examples. In these problems bedding thickness is constant around folds. Therefore, we can assume that area balance occurs if bed-length balance occurs, and we can restore sections merely by stretching out the contacts to return them to regional level and regional dip. This method is sometimes called the *sinuous-bed method* (Woodward et al., 1985). We do not need to compare areas of beds or thrust sheets. Measurement of bed lengths on the cross-section sketch can be done directly with a ruler or dividers (Fig. 14-18) if the folds are kink style. If the folds are concentric style, you may need to use a planimeter, a waxed thread, or a computer digitizer.

Problem 14-2

Restore the cross section shown in Figure 14-15d. Assume that unit thickness is constant in the section, and assume that the displacement is constant along the fault.

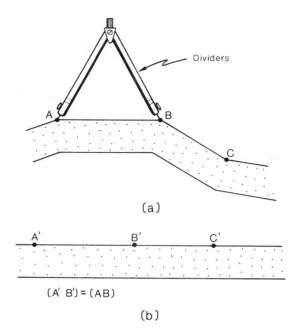

(a)

(A' B') = (AB)

(b)

Figure 14-18. Sketch showing how to use dividers to measure line segments composing a kink-style fold. (a) Measurement of line segment AB on the fold; (b) restored bed. The length of A'B' equals the length of AB.

Method 14-2

Step 1: Draft your cross section carefully; label the points of intersection between the axial-plane traces and the bed traces (Fig. 14-19a). To save space, units a and b are not shown. The distance AB represents the distance between the hanging-wall cutoff and the footwall cutoff on the unit d/unit e contact and thus is the separation on the upper flat.

Step 2: Add reference lines to the cross section (Fig. 14-19a). The cross section portrays only a portion of a deformed belt; the detachment extends all the way to the west edge of the figure, so there is no regional pin line. Draw a local pin line perpendicular to bedding at the leading edge of the cross section. The pin line does not extend below the fault, and because it is only a local pin line, it moves during the restoration. Draw a loose line (dotted line) at the trailing edge of the cross section. The loose line does not extend across the detachment and also moves during restoration.

Step 3: In a space below the deformed-state cross section, draw a set of lines at appropriate spacing to represent the stratigraphic sequence in the undeformed state. We call this set of lines a *stratigraphic frame*. If the stratigraphic thickness does not vary across the length of the cross-section area, the lines are parallel (Fig. 14-19b). The lines should be longer than the deformed-state cross section to accommodate the restored lengths of the contacts.

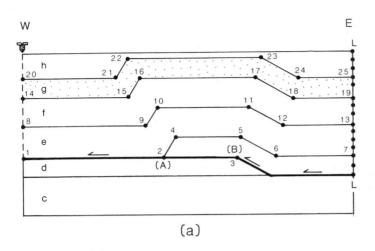

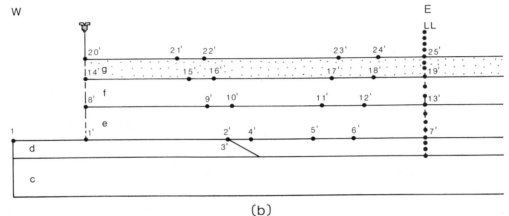

Figure 14-19. Cross sections for problem 14-2. (a) The deformed-state cross section with the reference lines shown; (b) the stratigraphic frame and the completed restored cross section.

Step 4: Note that the detachment is not folded, so the position of the ramp is fixed with respect to the footwall beds. Therefore, we draw the ramp on the stratigraphic frame in the same orientation and position with respect to the footwall beds as it was depicted on the deformed-state cross section.

Step 5: With a pair of dividers, measure the separation on the upper flat (AB). Point A coincides with the hanging-wall cutoff at point 2, and point B coincides with the footwall cutoff at point 3. On the stratigraphic frame measure a distance equal to AB starting at point 1 and extending east along the unit d/unit e contact. Mark point 1'. Point 1' represents the restored position of point 1. On the unit d/unit e contact line of the stratigraphic frame, lay off the distance 1 to 2. Note that the restored position of point 2 (i.e., point 2') lies directly over point 3. These points must coincide because when the displacement on the detachment is removed, the hanging-wall and footwall cutoffs must be juxtaposed. Now, lay off the distances 2 to 4, 4 to 5, 5 to 6, and 6 to 7 along the unit d/unit e contact line to locate the restored positions 2', 4', 5', 6', and 7'. The restoration of the unit d/unit e contact is complete (Fig. 14-19b).

Step 6: Repeat the procedure for the other contacts, and locate the restored positions of points 8 thru 25. Restore from the leading edge to the trailing edge, and start with the lower contacts and move up-section. Draw the restored pin line and loose line. Note that both of these reference lines moved east in this example, because both lay above the detachment. Also note that the horizontal distance between the loose line and the pin line is greater in the restored cross section, because restoration removed the folding that resulted from movement on the detachment.

If the original thickness of units was not constant across the cross section, the stratigraphic frame could not consist of parallel lines. If strata thicken continuously in a given direction, the stratigraphic from should depict a thickening wedge. The exact rate of thickening cannot be specified until the wedge is actually restored. But if you can estimate the approximate amount of shortening depicted on the deformed-state cross section, you can draw an approximate stratigraphic frame by depicting layers that change progressively from the thickness shown at the leading edge of the cross section to that shown at the trailing edge of the cross section (Fig. 14-20).

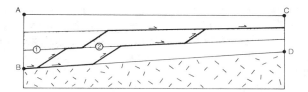

Figure 14-20. Thickening stratigraphic wedge. Layers thicken from east to west, so AB > CD. Restored faults are shown. Note that the thickness of units in thrust sheet 1 is greater than the thickness in thrust sheet 2.

Problem 14-3

Restore the deformed-state cross section shown in Figure 14-21a.

Method 14-3

Step 1: Scan the cross section. It shows three thrust faults that emerge from a basal detachment. Two of the faults bound a horse, and two of the faults are truncated at the ground surface by erosion. The strata below fault A appear to be in place. Thus, a pin line that penetrates these strata at the west edge of the cross section can be considered to be a regional pin line that is fixed during the restoration. Draw the regional pin line and a loose line (Fig. 14-21a).

Step 2: Bedding thickness is constant in the deformed-state cross section, so below the cross section we create a stratigraphic frame by drawing a series of appropriately spaced parallel lines (Fig. 14-21b).

Step 3: Because bed lengths are constant, we can restore by measuring bed lengths (assuming that there has not been significant volume-loss strain). Begin measuring at the regional pin line and work to the east. Start with the beds at the base of the section. Fault A, the youngest fault, plots on Figure 14-21b in the same position and orientation as it did in the deformed-state cross section. The restored positions of points 1 through 5 along the top of the shaded bed are shown. Point 5 shows the restored position of the footwall cutoff of the top contact of the shaded bed.

Step 4: A problem arises when we try to restore contacts in sheet A that lie above the shaded bed; because these contacts are truncated by erosion, we do not know the entire length of these contacts in thrust sheet A. In order to continue our restoration, therefore, we must introduce a local pin line in sheet A. This pin line is drawn perpendicular to bedding at the point where the stratigraphic sequence is most complete and is presumably little deformed (Fig. 14-21a).

Step 5: Now we can continue the section restoration. On the partially restored section, draw the local pin line perpendicular to bedding (Fig. 14-21b). The restored position of the local pin line can be determined from where it cuts the shaded bed. The depiction of the restored pin line as a straight line reflects our assumption that the beds have not slipped past one another in the vicinity of the pin line. Measure bed lengths between the pin line and fault B in the deformed-state cross section, and locate the restored

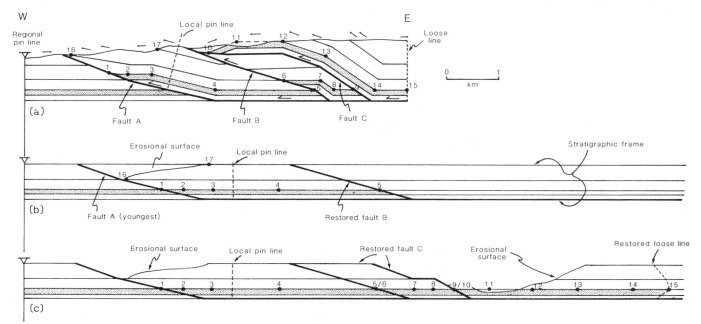

Figure 14-21. Cross sections for problem 14-3. (a) Deformed-state cross section; (b) stratigraphic template for the restoration, and partial restoration; (c) completed restoration.

position of fault B. Measure bed lengths to the west of the local pin line to determine the position of the trace of the erosional truncation surface on the restored section (Fig. 14-21b).

Step 6: Repeat the procedure to restore the remaining thrust sheets, and find the restored position of the loose line (Fig. 14-21c). We discuss problems with this restored cross section later in the chapter.

Restoration Based on Area Balance

In general, if a bed is significantly thickened during deformation, we cannot simply stretch out the bed to restore the cross section; rather, we must restore the cross section such that the areas of restored and deformed-state beds are the same.

Problem 14-4

Consider the rather unusual-looking ramp anticline shown in Figure 14-22a. It is obvious that there has been extreme thickening in the hinge. The area of unit b in the thrust sheet shown in this figure is 322 units^2. Determine the restored length of unit b in the thrust sheet.

Method 14-4

Step 1: First we must create a stratigraphic frame. The trailing edge of the thrust sheet contains unthickened unit b that is 4 units thick. Therefore, we take the undeformed thickness of unit b to be 4 units and create a stratigraphic frame on which unit b has this thickness (Fig. 14-22b).

Step 2: Examine the cross section (Fig. 14-22a) and prove that restoration based on bed-length balance will *not* work. Contact MN on the deformed-state cross section is 50 units long. If we unfold contact MN and place it on the stratigraphic frame, we should get a restored bed that is 50 units long and 4 units thick, with an area of only 200 units^2. The folded bed, however, has an area of 322 units^2, so this restored bed has an area that is 38% too small. Thus, reconstruction by such a procedure would yield a restored cross section that does not meet the constraint of area conservation.

Step 3: To restore unit b of the thrust sheet we measure its area (322 units^2) and determine a bed length on the stratigraphic frame that is long enough to yield this area. Assuming that the original bed thickness is 4 units (as it is under point N), the restored bed should actually be 80.5 units long (Fig. 14-22c). The area of the shaded layer on this figure is also 322 units^2.

14-6 EVALUATING AND IMPROVING A CROSS SECTION

Checking the Balance of a Cross Section

In this portion of the chapter we provide examples of how to check whether a cross section balances and if it does not, how to alter it to achieve balance.

Problem 14-5

Determine if the cross section used in problem 14-3 (Fig. 14-21a) is balanced.

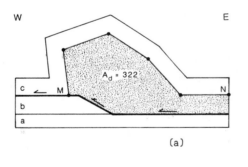

(a)

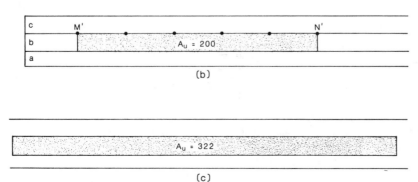

(b)

(c)

Figure 14-22. Cross sections for problem 14-4. (a) Deformed state cross section; (b) restored bed created by simply stretching out the top contact of the shaded layer. The area of this bed is too small; (c) restored bed created by assuming conservation of area.

Method 14-5

Step 1: First, we look at the deformed-state cross section (Fig. 14-21a) and check to see that the structures shown are admissible. Since there are no impossible fault shapes, we accept this cross section as admissible. Now we look to see if the restored cross section is viable.

Step 2: Determine if bed lengths are consistent. To do this we look at the shape of the deformed-state loose line (Fig. 14-21c). It curves back on itself (reverses dip). This shape is worrisome and suggests that the deformed-state cross section may not be viable.

Step 3: Check the shape of the restored fault traces (Fig. 14-21c). All of them have admissible step-like patterns and admissible ramp angles.

Step 4: We conclude, based on the observation that the restored loose line may have an unacceptable shape, that there is a problem with the viability of the restored cross section, and we do not yet accept the original deformed-state cross section as balanced.

Problem 14-6

Determine if the deformed-state cross section provided in Figure 14-23a is balanced.

Method 14-6

Step 1: Examine the deformed-state cross section (Fig. 14-23a). The numbered points are reference points on the cross section, and the patterned bed is a marker bed. At first glance it looks admissible.

Step 2: Look to see if the template constraint is obeyed. It is not. We note, for example, that in thrust sheet A, the length of the ramp that begins at point 12 and cuts up-section across the unit is not matched by the length of the corresponding hanging-wall ramp, even after the difference between the hanging-wall and footwall cutoff angles is accommodated. There are several comparable problems in this section.

Step 3: Construct a restored cross section (Fig. 14-23b).

Step 4: Examine the restored cross section (Fig. 14-23b). Note that the restored loose line looks unreasonable and that the restored faults reverse dip. From these geometries we conclude that the deformed-state cross section is not balanced. By going through the exercise of restoring the section, our attention has been drawn to problems in the section that we might otherwise have missed.

Improving a Cross Section

At this point you are probably asking the most important question of all, namely, What can be done if a cross section does not balance? How can a cross section be corrected or improved? The most common errors that experienced geologists make in constructing a cross section include (1) depicting an incorrect depth to detachment, and (2) depicting insufficient shortening at a given level in the cross section. We will show how to deal with these problems in the following examples.

Problem 14-7

In Problem 14-5 we faced a situation where the restored loose line reversed dip at the base of the restored cross section. Change the deformed-state cross section of Figure 14-21a so that this problem is removed.

Method 14-7

Step 1: Think about the significance of the problem. We conclude that the base of the patterned bed is too short. By referring back to the deformed-state cross section, we notice that if we change the dip of the lower part of fault C, we can increase the length of the shaded bed in thrust sheet B.

Step 2: Make the necessary change (Figure 14-24a). Check the bed-length balance of the new deformed-state cross section by restoring it (Fig. 14-24b). The restored loose line does not reverse dip and thus looks reasonable.

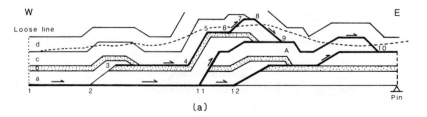

Figure 14-23. Cross sections for problem 14-6. (a) Deformed-state cross section; (b) restored cross section.

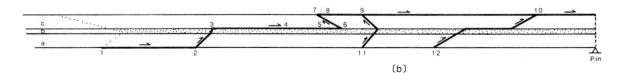

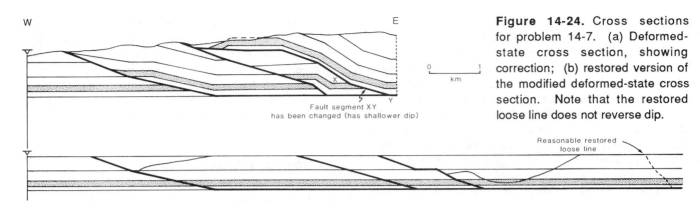

W
E

Fault segment XY
has been changed (has shallower dip)

Reasonable restored
loose line

Figure 14-24. Cross sections for problem 14-7. (a) Deformed-state cross section, showing correction; (b) restored version of the modified deformed-state cross section. Note that the restored loose line does not reverse dip.

Thus, we can say that the restored cross section is viable and thus that the deformed-state cross section is balanced.

The correction described above is not the only possible change that could improve the cross section. For example, rather than increasing the length of beds lower in the section, we could shorten beds higher in the section. A shortening of the beds could be accomplished by changing the hanging-wall cutoff angles.

Problem 14-8

Figure 14-25a shows a deformed-state cross section. This cross section is not admissible. Improve the deformed-state cross section.

Method 14-8

Step 1: Study the cross section (Fig. 14-25a). We immediately see some potential problems. First, the displacement on the thrust *drastically* increases up-dip; the top of unit b is only slightly offset, whereas the top of unit c is offset considerably. Also, the template constraint is not met; the hanging-wall cutoffs are not matched by the footwall cutoffs.

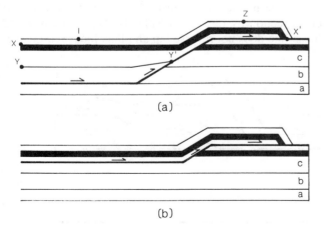

(a)

(b)

Figure 14-25. Cross sections for problem 14-8. (a) Inadmissible cross section; (b) possible cross section.

Step 2: Redraw the section with an appropriate change. We relocate the lower flat to ensure that the template constraint is met (Fig. 14-25b). This cross section is admissible. You may prove to yourself that it is also viable.

Problem 14-9

Figure 14-26a provides dip data along an east-west line of section. Assume that the line of section is parallel to the transport direction. A well drilled along the line of section provides stratigraphic data at depth. Draw a balanced cross section of this area. This example demonstrates that in practice, you should keep in mind the potential viability of the restored cross section while you draw the deformed-state cross section.

Method 14-9

Step 1: Examine the dip data and stratigraphic data provided in Figure 14-26a. Notice that it is possible to recognize dip domains. Also notice that stratigraphic section seems to be repeated at depth. It appears that the well penetrates a location where a hanging-wall flat lies over a footwall flat.

Step 2: We use the stratigraphic data from the well to define unit thickness. With this information plus the dip data we project kink-style folds down to the depth of the base of unit a. Below the well we can draw in the repeated stratigraphy (Fig. 14-26b). Examine the resulting cross section. It appears that the base of unit a is indeed a detachment. The problem remains, How do we fill in the space in the cross section below the detachment?

Step 3: We speculate that along the western edge of the cross section the thrust sheet has been emplaced over flat-lying strata. In Figure 14-26c we draw in the stratigraphic section and extend the contacts to the east. The geometry looks viable below the syncline and below the westernmost ramp. But we still have a bothersome gap to fill between the anticline (indicated by stipple in Fig. 14-26c). Clearly, a solution in which we have only one ramp originating from the basal detachment does not work. We speculate that the space can be filled by inserting a

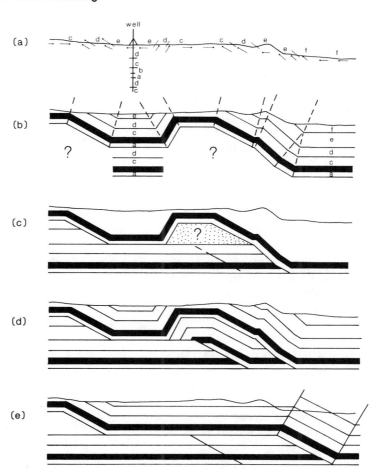

Figure 14-26. Maps and cross sections for problem 14-9. (a) Dip-domain data, positions at which contacts cross the ground surface, and a well showing stratigraphy at depth (a - d are stratigraphic units) provided along a section line; (b) partial construction of a deformed-state section, showing folded layers and a folded thrust. Question marks indicate space that must be filled; (c) partial solution showing how the space at the west edge of the section can be filled with flat-lying strata. Stipple indicates area that still needs to be filled; (d) solution showing a horse filling the space below the anticline; (e) partially restored cross section showing the restored shape of the horse.

horse, so we sketch in a possible ramp that could form the base of the horse (heavy dashed line in Fig. 14-26c). This line is drawn to be parallel to the backlimb of the anticline.

Step 4: We complete the deformed-state cross section by completing the ramp at the base of the horse and by filling in the strata within the horse (Fig. 14-26d). The deformed-state cross section looks admissible.

Step 5: We restore the cross section. Figure 14-26e provides a partial restoration that shows the geometry of the region before movement on fault A occurred. Notice that the anticline disappears, and the trailing edge of the horse becomes a simple ramp; the restored cross section looks viable. Based on this partially restored section, we feel that the deformed-state cross section of Figure 14-26d is probably balanced. We do not have the data to restore the major thrust sheet, because we have no exposures of hanging-wall cutoffs and therefore do not know the displacement on the fault. We do recognize, however, that the displacement is substantial, because section is doubled across the entire length of the cross section.

Notice that in Method 14-8 we used the word *speculate* several times. That is because the process of constructing the deformed-state cross section shown in Figure 14-26d is

an intuitive process guided by our prior knowledge of geometries in fold-thrust belts. Our end product *looks* reasonable, but we have no way of confirming that the cross section is correct unless we can obtain sufficient drill-hole or seismic data. Note how the procedures of cross-section balancing have made us think about every line drawn in the section.

One final comment about correcting cross sections: In some regions a duplex forms at depth and results in significant shortening of the lower part of the stratigraphic sequence. Units above the roof thrust do not appear to show evidence of comparable shortening. For example, the restored length of unit Dc in Figure 14-27a is much less than the restored length of Dm (Fig. 14-27b). This discrepancy reflects the fact that in the deformed-state cross section (Fig. 14-27a) Dm is duplicated by faulting, whereas Dc is only folded. There are real examples of formations where the shortening of the "folded-only" strata above the roof thrust is much less than the shortening of a fault-imbricated sequence below the roof thrust. This discrepancy can be explained in a number of ways, one of which is shown in Figure 14-27c; in this cross section, shortening of unit Dm is shown to involve not only the large folds but also mesoscopic folding and cleavage

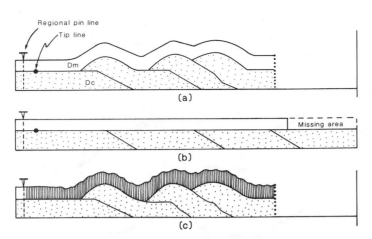

(a)

(b)

(c)

Figure 14-27. Alternative solutions to the problem of discrepancy in the amount of shortening between strata above and below a roof thrust. (a) Deformed-state cross section. Regional pin line is shown to the west of the tip line; (b) restoration, showing the difference in bed length between the supra- and subroof thrust strata; (c) discrepancy explained by the formation of volume-loss cleavage. (Adapted from Geiser, in press.)

formation, which did not appear in cross section 14-26a. This extra shortening makes up the missing area. For discussion of additional explanations, see Geiser (in press).

14-7 DEPTH-TO-DETACHMENT AND REGIONAL SHORTENING CALCULATIONS

The principle of area conservation that we described earlier can be applied to calculate the depth to the detachment beneath a detachment fold. This calculation works only if there are no additional thrusts between the folded surface and the detachment.

Problem 14-10

A shallow cross section of a folded surface is provided in Figure 14-28a. Determine the depth to the detachment below this structure, assuming that it is a detachment fold.

Method 14-10

Step 1: Consider the folded surface between points A and G to be a marker line (Fig. 14-28a). The length of this folded line is L_u, which represents the undeformed length of this bed (the distance between points A and G before deformation). Trace the folded marker line onto a new piece of paper.

Step 2: Connect two points along the line by a straight chord (the dashed line between A and G in Fig. 14-28a). This chord represents regional level and should be parallel to the detachment. The length of the chord is L_d (and represents the distance between A and G in the deformed state). The difference between L_u and L_d represents the amount of shortening of the bed that resulted from formation of the fold. Measure L_u and L_d. $L_u = AB + BC + CD + DE + EF + FG = 39$ units, and $L_d = AG$ measured along the chord = 32 units. Therefore, $L_u - L_d = s = 7$ units.

Step 3: The area between the trace of the folded surface and the dashed line between points A and G is called

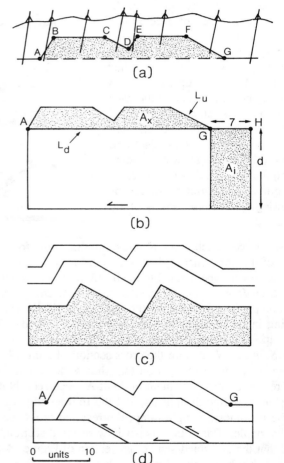

(a)

(b)

(c)

(d)

Figure 14-28. Cross sections for problem 14-10 (depth-to-detachment calculation). (a) A folded marker horizon. Excess area is shaded; (b) construction for depth-to-detachment calculation; (c) cross section showing properly located detachment; (d) alternative solution showing two imbricate horses. Note that in this solution the depth to the detachment is much less.

the *excess area* (A_x). We measure the area by counting squares on a grid placed under the figure and determine that $A_x = 101$ units^2.

Step 4: Extend chord AG to point H; the position of point H is determined so that the length of straight line AH is equal to L_u. If there is conservation of area, the area (A_i) between the line segment GH and the detachment at depth d must also equal the excess area, A_x. This relationship indicates that if the detachment remains at a given depth, shortening of the bed resulted in movement of a point from G to H and creation of the excess area A_x above chord AG. Therefore,

$$s = L_u - L_d$$

$$A_x = A_i = (s)(d)$$

$$d = A_x/s \qquad \text{(Eq. 14-2).}$$

Step 5: Using this equation, we can calculate the value of d (14.4 units), which is the depth to the detachment. Note that all we needed to measure was the lengths L_u and L_d and the area A_x under the fold. All these values can be obtained directly from the cross section.

Step 6: Redraft the cross section showing the detachment at depth (Fig. 14-28c).

Note that in the solution shown in Figure 14-27c we filled the space between the folded marker bed and the detachment with folded rock. The cores of the anticline could contain ductile evaporite or shale that squeezed upward during shortening or a layered sequence that is folded disharmonically with respect to the overlying strata. We cannot know for sure without additional data.

If, alternatively, imbrication on thrust faults had occurred between the detachment and the folded marker bed, our calculation of depth to detachment would be incorrect. In Figure 14-28d we show the same folded marker layer, but now we show it to be above the roof thrust of a duplex. Note that with this geometry the depth to detachment is much less. In other words, a depth-to-detachment calculation works only if there are no faults between the folded surface and the detachment. Furthermore, note that in the duplex interpretation of Figure 14-28d the bed above the roof thrust slipped to the left out of the cross section.

In cases where it is not possible to calculate the depth to detachment, this depth may be predicted from knowledge of regional stratigraphy. Also, the age of the oldest rocks brought up as a hanging-wall flat is a clue to the identity of the unit in which the detachment lies.

If faults are present beneath a fold, the excess area above regional level can still indicate the amount of shortening resulting from movement of the thrust sheet

(Laubscher, 1962; Gwinn, 1970). Consider Figure 14-9b as an example. If the fault geometry *and* depth to detachment are known, and area has been conserved, then the excess area (A_x), which can be measured on the cross section, determines the value of A_i, and the amount of shortening (d_1) is merely A_i/h, where h is the depth to the deeper detachment.

14-8   APPLICATIONS OF BALANCED CROSS SECTIONS

As we have emphasized already, the principal use of balancing procedures is in providing constraints on cross sections that must be constructed from incomplete data. Because balanced cross sections are more likely to be correct, they are obviously useful in guiding resource exploration. The application of cross-section balancing techniques to petroleum exploration is evident in many papers, such as Bally, Gordy, and Stewart (1966), Dahlstrom (1969, 1970), Royse, Warner, and Reese (1975), and Lamerson (1982).

Balanced cross sections are also useful in regional tectonic analysis, because they provide more reliable constraints on the production of palinspastic maps. A *palinspastic map* is a map that shows the distribution of stratigraphic units prior to their deformation. Such maps are constructed by removing the effects of fault displacement and folding (see Kay, 1945; Dennison and Woodward, 1963; Dennison, 1968). A palinspastic map gives a much clearer image of the spatial distribution of stratigraphic units and/or of early structures that formed prior to the restoration of the structures. For example, a palinspastic map of the Valley and Ridge fold-thrust belt of the Appalachians gives a much clearer image of the dimensions of the Paleozoic sedimentary basins that existed along the eastern margin of North America prior to the occurrence of convergent tectonics (e.g., collisional events) in the region. A palinspastic map of the Cenozoic Basin and Range province in the North American Cordillera can be used to show the spatial relationships of major Mesozoic and Paleozic deformed belts and depositional belts of the Cordillera. The restored cross section that is constructed during cross-section restoration provides data on the positions of units and markers before deformation and thus allows construction of a palinspastic map.

14-9   ACKNOWLEDGMENTS

The reviews of David Anastasio, Sandy Figures, Scott Wilkerson, and Steven Wojtal greatly improved this chapter. We appreciate additional comments from Tim Byrne, Peter Geiser, and Steven Usdansky.

EXERCISES

1 . Complete the deformed-state cross section of the Bristol Peak region shown in Figure 14-M1. Assume that bed thickness is constant across the map area, that the Precambrian/Cambrian contact is not folded, and that the ground-surface trace gives the shape of a simple ramp anticline.

2 . Figure 14-M2 shows a simple cross section of a ramp anticline above the Mashpee thrust.

(a) Label the hanging-wall and footwall ramps and flats on this cross section. Does this cross section obey the template constraint?
(b) If not, construct a cross section that displays the same stratigraphy and the same fault geometry and obeys the template constraint.

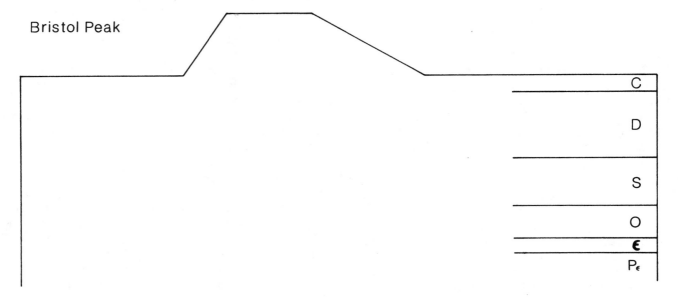

Figure 14-M1.   Cross-sectional sketch of the Bristol Peak area for exercise 14-1. (Adapted from Woodward et al., 1985.)

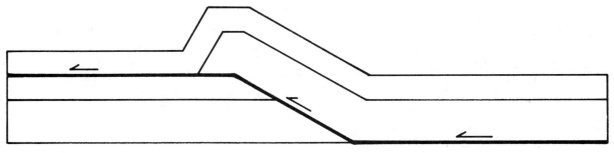

Figure 14-M2.   Cross-sectional sketch of the Mashpee thrust for exercise 14-2.

3. The geologist working on a cross section of the Alice Hills region has been called off on another project. You are left to develop a cross-sectional interpretation of the region. Your predecessor has left several useful hints to make the task easier; you have access to a partially completed cross section (Figure 14-M3). Note that the region displays kink-style folds and that the dip-domain boundaries are already labeled. The dip angles in the nonhorizontal domains are indicated. Note that there are well data at three localities. If you measure closely, you will see that the Silurian layer in hole 3 is *almost* twice as thick as in the other two holes. The upper interval of Silurian strata is not as thick as the lower interval. Finally, studies elsewhere in the region suggest that beds have uniform thickness in the region and that detachment horizons occur along formation contacts.

(a) Complete the cross section by showing the structure between the ground surface and the Precambrian/Cambrian contact. You must choose detachment levels, position ramps, and indicate the amount of slip on each ramp.
(b) Restore your completed section. Does it balance?

4. Figure 14-M4 shows a cross section of the Appalachian Mountains fold-thrust belt extending across Virginia and West Virginia from the Blue Ridge out onto the Appalachian Plateau. The original version of this cross section is by T. H. Wilson and appears in Woodward (1985). The real geology has been modified a little in drawing Figure 14-M4 in order to make this exercise easier to work.

(a) Describe the type(s) of thrust system(s) that are displayed on this cross section.
(b) In comparison with the structural geometries shown in the chapter, are the structures displayed on the cross section admissible?
(c) What is the direction of transport? Where do the principal detachments occur?
(d) Number the thrust sheets that contain the Cambro-Ordovician carbonate strata. Assuming a break-forward sequence, in what order were the thrust sheets emplaced?
(e) Assume that the leading edge of the cross section is a regional pin line and stays fixed during deformation. (This assumption is flawed, because the cross section indicates that the upper flat extends at least to the west edge of the cross section). Restore the cross section. If you wish, you may assume constant bed thickness to simplify the restoration, even though it is clear that the Cambro-Ordovician carbonate sequence thickens to the southeast. You may restore assuming bed-length balance.
(f) Is the deformed-state cross section balanced?
(g) Using your restoration, compare the shortening of the strata above the roof thrust with the shortening of the strata below the roof thrust. Discuss the discrepancy. Remember that the pin line you used is probably not a regional pin line.
(h) Calculate the percent shortening of the Cambro-Ordovician sequence by comparing your restored and deformed-state cross sections. If you wish, repeat the calculation using the excess-area method.

5. Figure 14-M5 provides a cross section of the Appalachian fold-thrust belt in the vicinity of the Powell Valley anticline, Tennessee.

(a) Briefly describe the thrust system depicted in this cross section.
(b) Restore this cross section and check the balance. (The geology has been modified slightly to facilitate this exercise). Be sure to choose a regional pin line.
(c) Calculate the shortening of the patterned layer in the Knox Group.

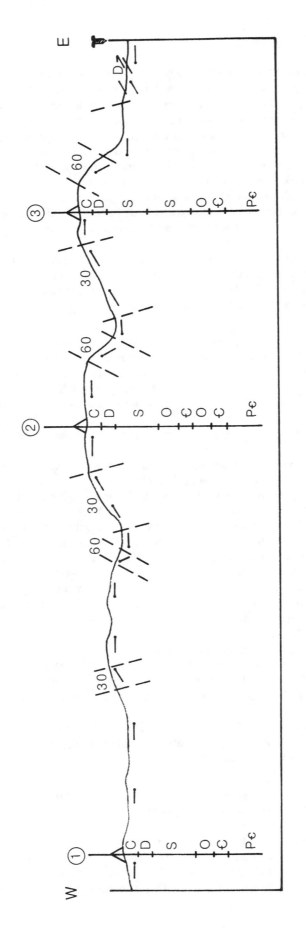

Figure 14-M3. Structural data along a line of section in the Alice Hills region for exercise 14-3.

Alice Hills

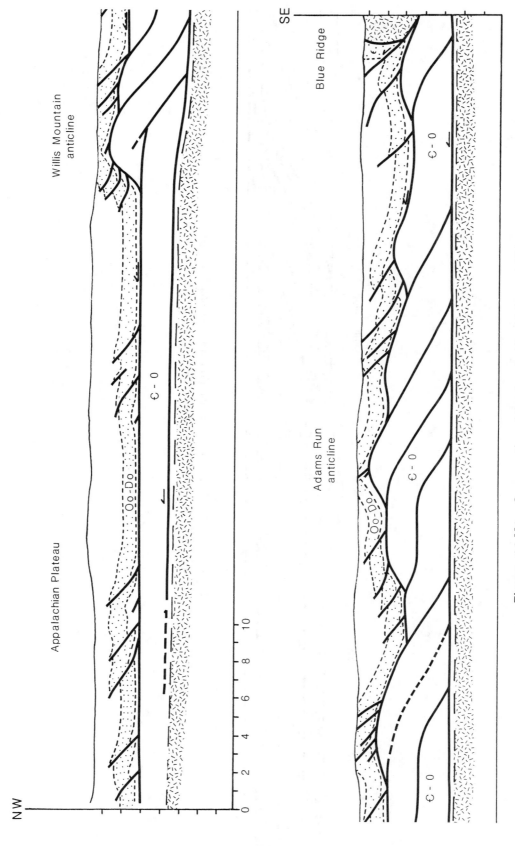

Figure 14-M4. Cross section of the Appalachian Valley and Ridge Province for exercise 14-4. (adapted from Woodward, 1985.)

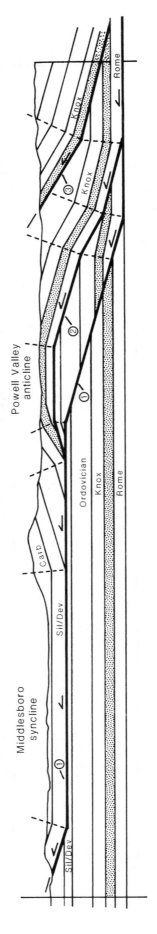

Figure 14-M5. Cross section of the Powell Valley Anticline region for exercise 14-5. (adapted from Suppe, 1985.)

6. Construct two restored versions of Figure 14-M6 (Kingston Ridge region), one using bed-length balance and the other using area balance. Compare the two and explain which you think is the more appropriate technique for restoring this particular cross section.

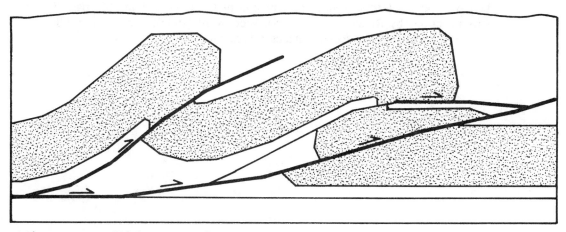

Kingston Ridge region

Figure 14-M6. Cross section of the Kingston Ridge region for exercise 14-6.

7. Restore the cross section of Figure 14-M7 (Ben Jesse region). Note that the stratigraphy in the section increases in thickness toward the trailing edge of the thrust system. You will need to insert local pin lines in order to complete the restoration. Be sure to show the position of the erosion surface on the restored version.

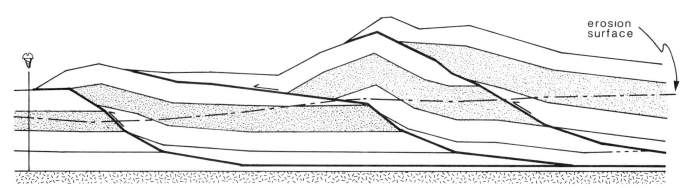

Ben Jesse region

Figure 14-M7. Cross section of the Ben Jesse region for exercise 14-7 (adapted from Woodward et al., 1985).

8. Calculate the depth to detachment beneath the top of the stippled layer in Figure 14-4b using the method described in Section 14-9. Is the detachment level shown in Figure 14-4b correct?

9. Once the exercises in this chapter have been completed, we recommend that students be challenged to produce balanced cross sections of real map areas. Due to

cost, we could not provide real maps with this book. Maps that provide appropriate data for cross-section balancing exercises include U.S.G.S. quadrangle maps I-686 (Afton 30' quadrangle) and I-1129 (Cokeville 30' quadrangle), which are maps of the Idaho-Wyoming fold-thrust belt by W.W. Rubey, and the 1:125,000-scale maps of the Tennessee portion of the Valley and Ridge fold-thrust belt by Rodgers (1953; Tennessee Division of Geology, volume 58, part II). Cross sections by Roeder et al. (1978), Woodward (1985), and Suppe (1980) provide a variety of cross sections at a useful scale for study.

15

ANALYSIS
OF TWO-DIMENSIONAL
FINITE STRAIN

Carol Simpson

15-1   INTRODUCTION

Strain is either a recoverable (elastic) or a permanent
(plastic) change in the shape and/or size of a body of rock
caused by stress changes within the earth. The state of
strain that we observe in the rock is called the *finite* or
total strain; it has been built up from a succession of very
small strain *increments* over a period of time. Strain
measurements compare the geometry of the deformed rock
before and after deformation; measurements are therefore of
changes in length (area, volume) and angles. We will see
later that certain naturally occurring objects, such as
pebbles and fossils, enable us to determine finite strain.

 The purpose of this chapter is to introduce various
techniques and procedures that have been developed to help
visualize strain and to permit direct measurement of strain
in rock. We are concerned primarily with permanent
strains that develop as a consequence of plastic
deformation. This chapter is limited to strain analysis in
two dimensions because two-dimensional strain is the most
common strain measured in the field and is easier to
analyze than three-dimensional strain. Methods for the
calculation of three-dimensional strain are covered in more
advanced courses (Ramsay and Huber, 1983).

 For the purpose of brevity, I have used the same
examples in more than one exercise. This approach allows
the student to compare relative ease and accuracy of each
method. However, I strongly recommend that wherever
possible the student be given different real examples with
which to work. If none are readily available, a casual

search through the literature should produce any number of
excellent photographs of deformed objects. Ramsay and
Huber (1983), Weiss (1972), and Borradaile et al. (1982)
are recommended as good source materials to start with.

15-2   DISPLACEMENT-VECTOR PATTERNS

In the general case, when a rock deforms each particle
within it changes its position with respect to each other
particle. Thus, a reference frame (see Chapter 1) is needed
to describe the movement of these points. In two
dimensions the point (x, y) in Figure 15-1a is *displaced*
with respect to Cartesian axes to its new position (x', y') as
the square is deformed into a parallelogram. The straight
line connecting point (x', y') with its starting point at (x,
y) is that point's *displacement vector*. Note that the actual
displacement path followed by any given point does not
necessarily coincide with its displacement vector. If the
displacement vectors for many such points in the rock can
be drawn, then the result is a *displacement vector field*
(Fig. 15-1b).

 It is possible for a body of rock to be displaced
without undergoing any internal strain (Fig. 15-2a,b), but
it is not possible for the material to be strained without the
displacement of its constituent particles (Fig. 15-2c). In
Figure 15-2a the square that represents a body of rock has
undergone a *rigid body translation* to its new position; this
can occur across a brittle fault, for example. In Figure
15-2b the square has undergone a *rigid body rotation.*

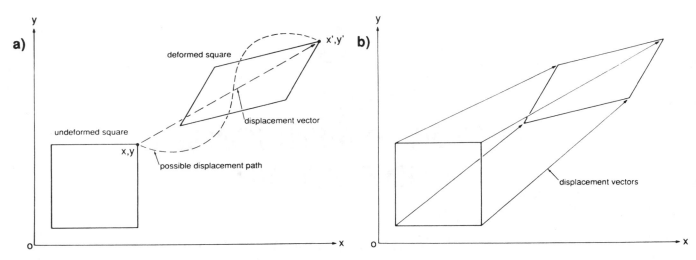

Figure 15-1. Displacements within the xy-coordinate reference frame. (a) A displacement vector is not necessarily the same as a displacement path; (b) a displacement vector field. (After Ramsay and Huber, 1983.)

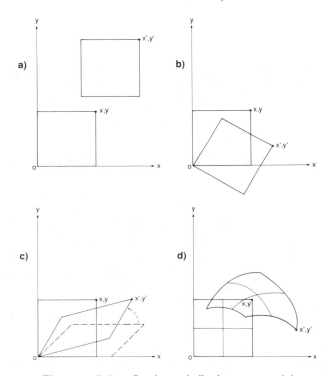

Figure 15-2. Strain and displacement. (a) A rigid body translation without internal strain; (b) a rigid body rotation without internal strain; (c) homogeneous strain; (d) heterogeneous strain. (After Ramsay and Huber, 1983.)

Such rotations are sometimes indicated by paleomagnetic data. Note that in Figure 15-2c the finite strain state represented by the solid parallelogram could be achieved by a single deformation, or indirectly by first forming the dotted parallelogram and following this with a rigid body rotation. In practice it is seldom possible to distinguish among several such *deformation paths*.

In order to analyze finite strain we must first find the scale at which the rock deformed *homogeneously*. In homogeneous strain, lines that were initially straight and parallel remain so, although the angle between two nonparallel lines will generally change (Fig. 15-2c). In real rocks the strain is more likely to be heterogeneous; originally straight lines become curved, and parallel lines become nonparallel (Fig. 15-2d). However, if we consider a small enough portion of the heterogeneously deformed rock, an infinitesimally small point in fact, then the deformation can be said to approximate homogeneous strain. Clearly, in naturally deformed rocks it will be necessary to examine the finite strain state at many different points in order to build a picture of the overall heterogeneous strain.

Exercise 15-1

The deformed grid in Figure 15-3b has the same area as the undeformed grid in Figure 15-3a. It is an example of a general homogeneous strain. Determine the displacement-vector field as follows:

Step 1: Trace the nodal points of the grid in Figure 15-3b onto a sheet of tracing paper. Now position the paper over Figure 15-3a such that the (0, 0) points coincide and the reference axes remain parallel.

Step 2: On the tracing paper draw the displacement vectors for each nodal point. Repeat this exercise on a fresh piece of paper, changing the points of coincidence to (2, 2) and (2', 2'). Compare the two displacement vector fields that you have created. Note that neither field is

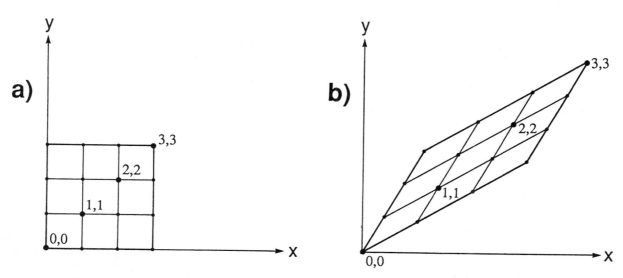

Figure 15-3. A marker grid (a) in its undeformed state and (b) after homogeneous deformation. See Exercise 15-1.

simple, even though the strain pattern in the grid appears simple.

Step 3: On a fresh piece of paper create for yourself a constant-displacement vector field (i.e., all vectors the same length and orientation), using the nodal points in Figure 15-3a as starting positions. Which of the following best describes the resultant grid you produce: rigid body rotation; rigid body translation; homogeneous strain; or heterogeneous strain?

15-3   STRAIN MEASUREMENT

Length Changes

The change in length of a line after strain is known as *longitudinal* strain. It is measured as the *extension*, ε, which is defined as:

$$\varepsilon = (l_f - l_0) / l_0 = \Delta l / l_0 \qquad \text{(Eq. 15-1)},$$

where l_f is the final length and l_0 the original length of the line, or as the *stretch* (S), where

$$S = l_f / l_0 = (1 + \varepsilon) \qquad \text{(Eq. 15-2)}.$$

The square of the stretch is known as the *quadratic stretch (quadratic elongation)*, λ. Written as an equation,

$$\lambda = (1 + \varepsilon)^2 \qquad \text{(Eq. 15-3)}.$$

The value of ε is positive where the line has lengthened and is negative where the line has shortened. For a line that has not changed length, S is equal to 1.0 ($l_f = l_0$). S

always has a positive value even where the line has shortened, since shortening a line of original length 1.0 by an amount $\varepsilon = -1.0$ would cause the line to vanish!

Angular Changes

In general, two lines originally at right angles change their angular relationship after strain by an amount equal to the *angular shear*, ψ. The sign convention for description of angular shear is as follows: where ψ is measured clockwise, it is positive, and where ψ is measured counterclockwise, it is negative (Fig. 15-4a). The tangent of the angular shear is called the *shear strain*, γ:

$$\gamma = \tan \psi \qquad \text{(Eq. 15-4)}.$$

This relationship is readily seen in Figure 15-4b, where the undeformed line OA has length unity.

Shape Changes and the Strain Ellipse

It is common practice to represent a finite strain as a change in the shape of a hypothetical circle of unit radius (or a sphere of unit radius in three dimensions) or of a square of unit side length (a unit cube in three dimensions). If a rock has deformed homogeneously, the original sphere will be changed into an ellipsoid, the *strain ellipsoid*. The principal axes of this ellipsoid are called the *principal strains*. In two dimensions, if we start with an undeformed circle of unit radius (for the *unit circle*, $x^2 + y^2 = 1.0$) and subject it to an increment of homogeneous strain, the result is an ellipse, the *strain ellipse*. Figure 15-5a shows a strain ellipse whose long axis is parallel to the x-axis of

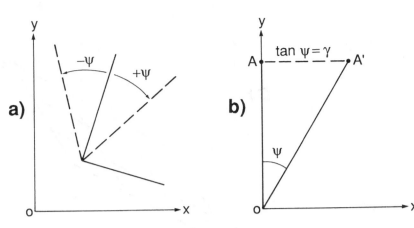

Figure 15-4. Changes in angles. (a) Sign convention for angular shear, ψ; (b) definition of shear strain, γ.

the reference frame. The general equation for an ellipse in this orientation is

$$(x^2/\lambda_1) + (y^2/\lambda_2) = 1.0 \qquad \text{(Eq. 15-5),}$$

where $\sqrt{\lambda_1}$ is the major semiaxis (also equal to $(1 + \varepsilon_1)$) and $\sqrt{\lambda_2}$ is the minor semiaxis (also equal to $(1 + \varepsilon_2)$). The *ellipticity,* R, of the ellipse is defined as the ratio of the two semiaxes:

$$R = (1 + \varepsilon_1)/(1 + \varepsilon_2) \qquad \text{(Eq. 15-6).}$$

This quantity is often referred to as the *strain ratio,* R_s. The two semiaxes of the strain ellipse, $\sqrt{\lambda_1}$ and $\sqrt{\lambda_2}$, are called the *principal strains* (or *principal stretches*), and their orientations are *principal directions of strain.*

Principal directions of strain are, by definition, 90° apart, and we can see from Figure 15-5a that they also form a right angle BOA in the undeformed unit circle. We will examine this important property of the principal directions of strain more thoroughly in the next exercise. Points C' and D' are the intersection of the unit circle with the strain ellipse; therefore, lines OC' and OD' are equal in length to the unit circle radius. They have not changed their length in this increment of strain and, therefore, are known as *the lines of no longitudinal strain.*

15-4   TYPES OF HOMOGENEOUS STRAIN

Pure Shear

The strain shown in Figure 15-5a is a special type in that there was no rotation of the principal directions of strain with respect to the axes of the reference frame during the development of the strain. This is known as *irrotational deformation.* If, in addition, there was no change in area during development of the strain, then the strain is called *pure shear.*

If a second increment of strain is added to the strain

ellipse in Figure 15-5a, such that the principal directions of the new incremental strain ellipse are the same as the principal directions of the first-formed ellipse (Fig. 15-5b), the deformation is said to *accumulate coaxially.* With the second increment, OC" and OD" become the lines of no longitudinal strain (Fig. 15-5b). We could draw other lines within the unit circle and see that they too will rotate with each new increment of strain. Thus, when we speak of coaxial or irrotational deformation, we mean only that the principal directions of strain do not rotate--any other line in a different orientation is forced to rotate as the strain progresses.

General Strain

The preceding discussion relates only to the case of pure shear. In more general deformation the principal directions of strain do not remain coincident with the axes of the reference frame but rotate (always preserving their orthogonality) such that the major principal strain axis makes an angle θ' with the x-axis. In the undeformed unit circle this line makes the angle θ with the x-axis. The relationship between θ and θ' is shown in Figure 15-6. The difference between the initial angle θ and the final angle θ' gives the *rigid rotation* component, ω of the deformation. We can now see that in order to fully describe the state of strain in a rock we need more than one number. Strain is really a *tensor* quantity, as described thoroughly by Means (1976), Ramsay and Huber (1983), and references therein. For the purposes of this chapter we will restrict ourselves to solving fairly simple problems in two dimensions, but it should be kept in mind that more complicated strain states exist in nature.

Simple Shear

We will now consider a second very special kind of deformation known as *simple shear.* In Figure 15-7a an undeformed unit circle (radius = 1.0) has superimposed upon it a unit square (side length = 1.0). A dashed line is

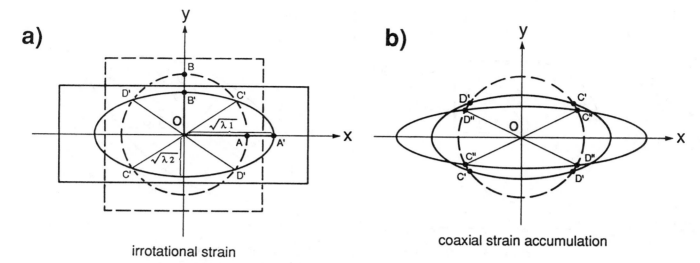

Figure 15-5. Pure shear. (a) Irrotational strain--principal directions of strain do not rotate with respect to axes of reference frame. Unit circle deformed into ellipse, and square deformed into rectangle. Long axis of ellipse is $\sqrt{\lambda_1}$, and short axis is $\sqrt{\lambda_2}$. Lines OC' and OD' are lines of no longitudinal strain. Points A and B move to new positions A' and B'; (b) coaxial strain accumulation. Strain ellipse in (a) has second increment added. Principal directions of new incremental strain ellipse are the same as for first formed ellipse. Points C' and D' move to new positions C" and D".

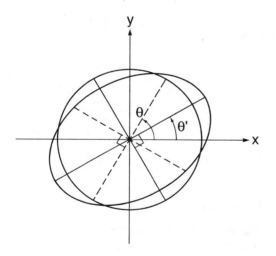

Figure 15-6. Relationship between θ and θ' for general deformation in which principal directions of strain rotate with respect to reference frame.

drawn across the diagonal of the square for reference. In Figure 15-7b the unit circle and square have been subjected to a *right-lateral (dextral)* shear parallel to the x-direction. The long axis of the strain ellipse makes an angle θ' to the x-axis. Note that the long axis does *not* pass from corner to corner of the parallelogram. Two equations can be solved to find γ from Figure 15-7, namely,

$$\gamma = \tan \psi \qquad \text{(Eq. 15-7)}$$

$$\gamma = 2/\tan 2\theta' \qquad \text{(Eq. 15-8)}.$$

Unlike progressive pure shear, progressive simple shear accumulates *noncoaxially*, which means that principal directions rotate with respect to the reference frame, and each accumulation of strain has different principal directions. The vertical side of the square in Figure 15-7 has lengthened and rotated through angle ψ. The horizontal side of the square has neither lengthened nor rotated; it is in a special orientation parallel to the shear direction x and is one of the two lines of no longitudinal strain for simple shear (Fig. 15-8, lines OA' and OB'). If we extend line OB' to intersect P-P' at R' and then "unstrain" point R' to its original position R, we will find that the angle ROP = POR'. In other words, the line of no longitudinal strain, OB', has its origin along the direction OR, and the two lines OR and OB' are symmetrically disposed about the y-axis of the reference frame. At this point you may wish to repeat Exercise 15-1 to demonstrate the pattern of displacement vectors characteristic of simple shear.

The simplest method for demonstrating the basic principles of simple shear is to use a wooden box designed to hold a deck of large cards, such as that illustrated by Ramsay and Huber (1983, p. 2). Right-triangular wooden

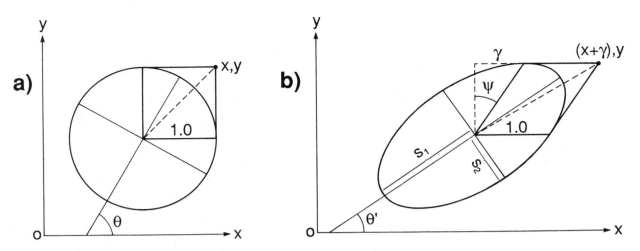

Figure 15-7. Simple shear. (a) Undeformed unit circle and unit square; (b) situation after an amount of right-lateral simple shear. Point x, y has moved to new position $(x + \gamma)$, y.

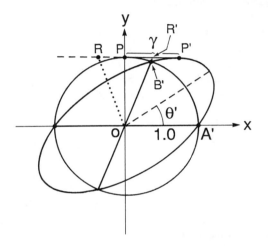

Figure 15-8. The two lines (OA' and OB') of no longitudinal strain in simple shear. Line OB' has its origin along the direction OR; line OA' has not changed.

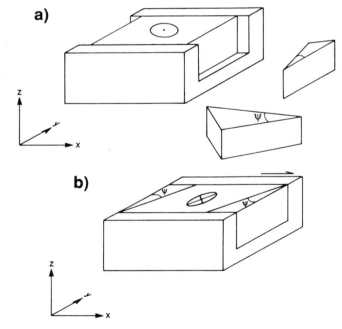

Figure 15-9. Simple-shear box for use with card deck. (After Ramsay and Huber, 1983.) (a) Undeformed card deck with unit circle drawn on it. Each card lies in xz-plane of reference frame; (b) after an increment of simple shear has been imposed by means of wooden templates.

templates are used to deform the card deck as shown in Figure 15-9. Ramsay and Huber (1983) recommend a series of ψ angles for the templates that require very high precision workmanship. It is quite satisfactory to cut ψ angles as close as possible to 10°, 20°, 30°, 40°, 50°, and 60° and to record the exact angle of the finished template before using them. The exercises that follow use one of these card-deck shear boxes. Computer programs exist (e.g., McEachran and Marshak, 1985) that will simulate these exercises.

Exercise 15-2

Step 1: Pack a simple card-deck shear box with cards so that they will slide easily past one another parallel to the x-axis (shear direction, Fig. 15-9a). Draw a circle

onto the top surface (the xy-plane) of the cards and mark the circle center. Measure the radius, r, of the circle.

Step 2: Choose a pair of templates with a low value of ψ and use them to deform the circle into an ellipse. Draw in the principal strains (the major and minor axes of the strain ellipse) using a colored pencil. Note the difficulty of exactly positioning the long axis of a fat ellipse. Precise measurement of low strains is never an easy task!

Step 3: Measure the lengths of the major and minor semiaxes of the ellipse (a' and b') and the angle θ' that a' makes with the x-direction. Tabulate your data as shown in Figure 15-10 and calculate the following:

(a) The stretches, S_1 and S_2, of the major and minor semiaxes of the strain ellipse.

(b) The ellipticity, R_S, of the strain ellipse.

(c) The shear strain, γ_i, *imposed,* from $\gamma = \tan \psi$.

(d) The shear strain, γ_c, *calculated,* from $\gamma = 2/(\tan 2\theta')$.

In the ideal case the two values for γ should be identical, and the product of S_1 and S_2 should equal 1.0, since there has been no area change.

Step 4: Now restore the cards to their starting position. The two colored lines are still mutually perpendicular but have a different orientation in the xy-frame. Measure the angle θ that line a (the unstrained line a') makes with the x-direction. Calculate the following:

(a) The rotational component of the strain, ω.

(b) Tan ω.

(c) $\gamma/2$.

Step 5: Choose a second set of templates with a higher value of ψ and redeform the circle, keeping the shear sense the same (i.e., if you used right-lateral shear before, then you must use right-lateral shear in this and all subsequent strain increments). Mark the new principal directions of strain in a different color. Make the same measurements as previously, but *before* you "undeform" the cards look carefully at the pair of colored lines that were the axes of the first strain ellipse. Not only are they no longer in the correct orientation for the new strain ellipse, they are also no longer at right angles to each other. After

restoring the cards to their starting position, check that the two new colored lines are mutually perpendicular.

Step 6: Repeat the experiment twice more, using progressively higher values of ψ. The principal directions of strain for each increment of strain should be mutually perpendicular before and after deformation, but they will *not* form a right angle at any other value of strain. Thus, the lines you draw as principal directions of strain for an angular shear of, say, $\psi = 20°$ will not form a right angle at any other value of angular shear.

Step 7: Restore the cards to their undeformed state and note the sequence of the unstrained ellipse axes; with each higher value of ψ the angle θ has increased. Subtract 45° from each value of θ and then subtract each value of θ' from 45°. The two columns of numbers should be identical, illustrating that the rotational component, ω, of the strain is symmetrically disposed about a line drawn at 45° to the x-axis. What would be the orientation of θ' if the increment of angular shear were less than $\psi = 1°$?

The following important properties of simple shear can be seen by inspection of your completed table:

1. Doubling the amount of shear strain, γ, gives more than twice the strain ratio, R_s.

2. Although the long axis of the strain ellipse approaches the x-direction with increasing strain, it can never reach it (unless $\gamma = $ infinity).

3. The orientation of the long axis of the ellipse is not parallel to the sheared-over edge of the card deck (i.e., $\psi \neq 90° - \theta'$), but there is a relationship between them such that

$$\tan \omega = \gamma/2 \qquad \text{(Eq. 15-9)}$$
or
$$\tan (\theta - \theta') = (\tan \psi)/2 \qquad \text{(Eq. 15-10).}$$

unit circle radius r=..cm	γ_i	θ'	γ_c	major semi-axis (a') cm	minor semi-axis (b') cm	S_1=a'/r	S_2=b'/r
$\psi_A = 10°$							
$\psi_B = 20°$							
.							
.							
.							

$S_1 \times S_2$	R_s=S_1/S_2	θ	ω=($\theta - \theta'$)	$\tan \omega$	$\gamma/2$	$\theta - 45°$	$45° - \theta'$
.							
.							

Figure 15-10. Table for use with Exercise 15-2.

4. Although the principal directions of strain rotate with respect to the external xy-frame, the cards themselves do not rotate.

Exercise 15-3

Step 1: Using a fresh set of cards, draw a new circle onto the xy-plane and measure its radius. Draw a bilaterally symmetrical fossil, such as a brachiopod, onto the card-deck surface in any orientation (it is easier if you use a stencil or cutout shape for this). Construct the line of bilateral symmetry for the fossil (e.g., Fig. 15-11a).

Step 2: Select one set of templates with which to deform the card deck, and deform the "fossil" and circle. Measure the following:

(a) θ', the angle between the ellipse long axis and the shear direction.

(b) ψ_{FOSSIL} (measure the deflection of the median line from 90°).

Step 3: Now calculate the following:

(a) γ for the whole card deck, from $\gamma = 2/(\tan 2\theta')$.

(b) γ_F *for the fossil,* from $\gamma_F = \tan \psi_{FOSSIL}$.

There are only two orientations of the fossil (and two mirror images of them) in which the two values of γ and γ_F will be identical. What are they?

The final geometric feature we need to examine with the card-deck model is the change in orientation and length of a line drawn in any orientation not parallel to a principal direction of strain. In nature such a line might be a plant stem in coal or an aplite dike in a granite.

In Figure 15-12a the edge of the card deck lies along

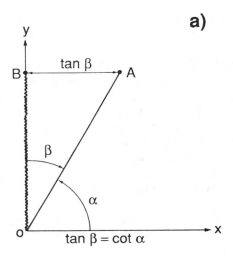

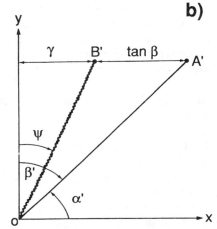

Figure 15-12. Change in length and orientation of line OA on top surface of simple shear box. (a) Starting position - edge of card deck lies along OB, marker line OA makes angle β from y-axis; (b) after an increment of simple shear, B moves to B', A moves to A', and β becomes β'.

line OB and a line of length OA makes an initial angle β from the y-axis, or α from the x-axis. After an increment of simple shear (B moves to B'), the line OA has a new length OA' and a new orientation β' (or α' from the x-axis). The line has clearly lengthened, and the relationship between the shear strain and the new orientation of the line is given by:

$$\tan \beta' = \tan \beta + \gamma \qquad \text{(Eq. 15-11)}$$

or

$$\cot \alpha' = \cot \alpha + \gamma \qquad \text{(Eq. 15-12).}$$

Exercise 15-4

Step 1: Using a fresh deck of cards, draw a marker line OA such that the angle α when measured

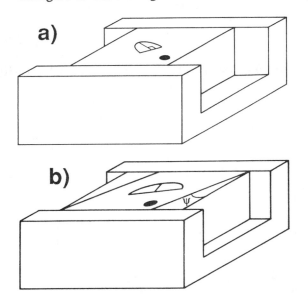

Figure 15-11. Simple shear box with bilaterally symmetrical fossil (a) in undeformed state and (b) after an increment of simple shear.

counterclockwise from the x-axis is greater than 90° (this is equivalent to a negative value of β measured from the y-axis). Measure length OA.

Step 2: Subject the card deck to a right-lateral simple shear until line OA' is exactly perpendicular to the x-axis (α' = 90°). Remeasure the line OA', which should now be shorter than OA. Continue to shear the cards in a right-lateral sense to some new orientation of the line at α". The line will lengthen again.

This simple experiment demonstrates two important aspects of a generally oriented line in simple shear: (1) A line may shorten and then lengthen, but it may never lengthen and then shorten in progressive simple shear, and (2) there will always be two lines that will have the same length after deformation as they had before. One of these lines of no longitudinal strain remains parallel to the x-axis or shear direction at all times. The other line starts at some angle -β to the y-axis and ends at a new angle +β (Figs. 15-8, 15-13).

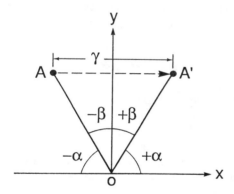

Figure 15-13. Line of no longitudinal strain OA' at angle +β from y-axis was originally at angle -β from y -axis.

Heterogeneous Simple Shear (Shear Zones)

In the foregoing series of exercises the shear strain applied was homogeneous (each card moved in exactly the same way as each other card) and *plane strain* (no change in the third dimension). There was also no change in area of the material. In many naturally deformed rocks that have undergone strain approximating simple shear, the deformation is heterogeneous and is confined to relatively narrow, planar, and parallel-sided *shear zones* (Fig. 15-14). These may be *ductile* in nature (Fig. 15-14a), *brittle* (Fig. 15-14b), or somewhere in between. In ductile shear zones the shear zone boundary (SZB) marks the boundary between undeformed wall rock and the zone of heterogeneous shear strain. In a deformed granite, for example, a new foliation is developed within the shear zone such that biotite and

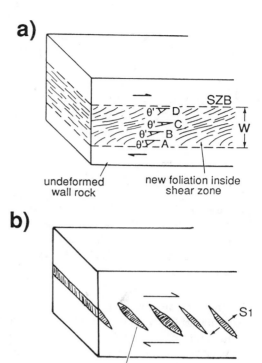

Figure 15-14. Three-dimensional views of ideal shear zones in which deformation is heterogeneous simple shear. (a) Ductile shear zone; (b) "brittle" shear zone.

quartz grain aggregates change their shape and orientation to lie along the principal direction of elongation. By measuring the angles θ'_A, θ'_B, etc., between the new foliation and the shear zone boundary, we can compute γ at each position across a shear zone of width W (Fig. 15-14a). It is then a simple matter to integrate these values of γ to find the total displacement, D, of one side of the shear zone with respect to the other side, by using the equation

$$D = \int_0^W \gamma \ dW \qquad \text{(Eq. 15-13),}$$

or by plotting a graph of γ versus W, as in Figure 15-15, and finding the area under the curve.

Exercise 15-5   (Strain in a shear zone)

In Figure 15-16 a small ductile shear zone crosses a granodiorite. Originally equidimensional aggregates of quartz grains (pale gray), feldspar (white), and biotite (black) record the shear strain by becoming elongate, platy aggregates inside the shear zone. Assume that the trace of the new foliation defined by the platy aggregates is parallel to the long axis of the strain ellipse.

Step 1: Calculate the displacement D across this zone. Note that you must first define the shear zone

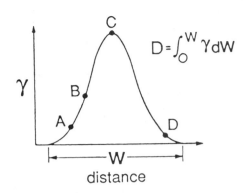

$$D = \int_0^W \gamma \, dW$$

Figure 15-15. Shear strain (γ) versus distance (W) between shear zone boundaries for an ideal ductile shear zone. Points A, B, C, and D refer to positions on Figure 15-14a where θ' was measured.

boundaries (not as simple as it first seems!) and then take as many readings of θ' across the zone from one side to the other as you can in order to create a smooth γ versus W curve. At very high strains (low values of θ') very slight changes in θ' can make enormous changes in γ and D estimates. For example, if θ' in the zone center is 3.5°, this gives γ = 16.7; if θ' is 2.5°, γ = 22.9; and if θ' is 1.5°, γ is 40.0! You will probably find that your γ versus W curve is not as smooth as Figure 15-15, but the more positions at which you measure θ', the smoother it will be.

Shear zones such as the one in Figure 15-14b are often referred to as *en-echelon "tension" gash arrays.* Strictly speaking, each calcite- or quartz-filled vein is really an *extension* gash that has its long axis parallel to the direction of incremental shortening and that opens in the direction of elongation in the rock (Fig. 15-14b). Although these arrays can seldom give us complete information on the finite strain state, they are useful for the determination of sense-of-shear, for incremental strain histories, and for volume change estimates (Ramsay and Huber, 1983, pp. 48-51).

Under certain conditions of deformation a rock may change in volume during the development of a heterogeneous strain. Vein arrays or dike swarms usually indicate a volume increase, and volume decrease is manifest by *pressure solution* structures such as *stylolites* (surfaces that have a characteristic toothlike morphology; Ramsay and Huber, 1983, Figure 3-23; Davis 1984, Figure 10-20). We can simulate the effect of volume reduction in simple shear by using the card-deck model and removing every second (or third, or fourth) card after the unit circle has been drawn. Similarly, an area (or volume) increase can be simulated by the addition of one new card every second (or third, etc.) card within the circled area.

15-5   STRAIN MARKERS

In the foregoing exercises we have concentrated on the special case of simple shear for the sake of simplicity. Of course, all rocks do not deform in this way; in fact, "simple shear zones" such as in Figure 15-16 are rather rare. In the following sections we present exercises that demonstrate how to calculate two-dimensional strain where

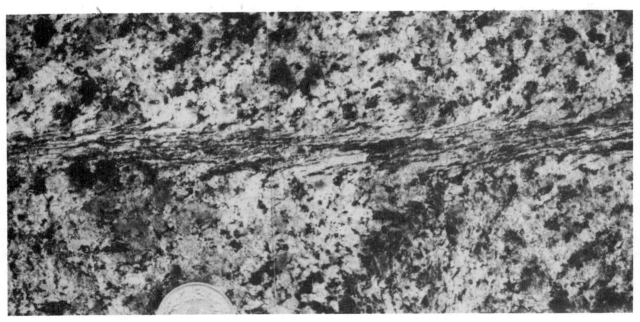

Figure 15-16.   Photograph of ductile shear zone in granodiorite from Penninic Alps, Switzerland. Coin is 2.4 cm in diameter.

it is not known beforehand whether the strain formed by simple shear or pure shear. Such measurements rely on recognition of strain markers.

It is rare to find objects whose original lengths are known for certain, so we usually compute the ratios and orientations of the principal stretches. To do this it is necessary to find objects that have the following properties:

1. *Original spherical or equidimensional form,* such as some ooids, reduction spots and concretions; some foraminifera and radiolaria; amygdules and mineral aggregates in hornfels and certain igneous rocks.

2. *Original circular cross section on a known surface,* such as worm burrows (in particular *Skolithos*), crinoid ossicles, or fossil raindrops on a bedding surface.

3. *Original nonspherical shapes (approximating ellipsoids),* such as most pebbles in conglomerates and several of the examples cited under (1).

4. *Original linear form,* such as belemnites, rutile needles in quartz grains, tourmaline crystals, plant stems, and the trace of pegmatite or other veins and dikes on an outcrop surface.

5. *Bilaterally symmetrical fossils,* such as brachiopods, trilobites, echinoderms, plants.

6. *Known initial angles between two or more lines or planes,* such as cross bedding in sediments, worm burrows, graptolites, polygonal mud cracks, polygonal cooling joints, symmetry planes in plant or animal fossils, other objects of known initial geometry, such as gastropods and pillow basalts.

There are many more potentially useful strain markers than those listed here, and the more complex the initial geometry, the more complex is the mathematical treatment required to find their strain state. We do not have the scope in this chapter to examine each category listed but will begin with the very simplest case where lines of known original length, such as belemnites, lie on a deformed bedding surface. We will generally work with the minimum number of deformed objects that are necessary for analysis in order to keep the exercises short, but it should be kept in mind that strain analysis is like orientation analysis (Chapter 8) in that the more data gathered, the more reliable will be the solution obtained.

15-6 USE OF ORIGINALLY LINEAR STRAIN MARKERS

Figure 15-17 shows a belemnite that has fractured during deformation of the limestone in which it lies. The segments of broken belemnite are of different lengths along the maximum principal stretch (S_1) direction, and fibrous crystals of calcite fill the intervening spaces. The fibers are elongate in the direction of S_1. Often we do not see the entire fossil preserved and so cannot be sure of the original

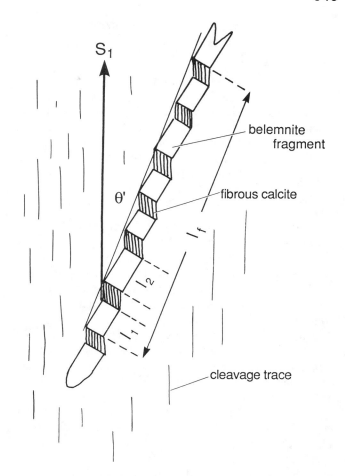

Figure 15-17. A stretched belemnite. (Adapted from Ramsay and Huber, 1983.)

length, but the stretch can still be determined by measuring the lengths of an equal number of segments and fibrous regions for each fossil. In fact, it is better to ignore the ends of the deformed fossil altogether, as the strain is unlikely to be homogeneously distributed at these positions. Also, the fiber orientations are not always parallel to S_1, so whenever possible, an independent estimate of the S_1 direction, such as the trace of the cleavage planes on the outcrop surface, should be sought.

Exercise 15-6

Calculate the longitudinal strain of the belemnite in Figure 15-17 in the following way:

Step 1: Measure the lengths (l_1, l_2, etc.) of six fragments in a direction parallel to the long dimension of each fragment (this corresponds to the assumed original orientation of the fossil).

Step 2: Sum the values for l_1, l_2, etc., to give l_0, the original length.

Step 3: Measure the final length, l_f, along the direction that corresponds to the new overall orientation of the fossil. This direction lies at angle θ' to the S_1 direction.

Step 4: Measure θ'.
Step 5: Calculate ε, S, and λ for the belemnite.

If there is only one such fossil on the outcrop, then no further information can be obtained. However, it is fortunate for us (and correspondingly unfortunate for the belemnites) that these organisms tended to die in groups and lie in a random orientation on the bedding surface. Such an assemblage after deformation is shown schematically in Figure 15-18.

Exercise 15-7
Step 1: Calculate S_A, S_B, etc., for each of fossils A to H in Figure 15-18.
Step 2: Determine the S_1 direction by inspection of the fossils and use this to find the angles $θ'_A$, $θ'_B$, etc. Angles measured in a clockwise direction from S_1 can be counted as positive, and those measured in a counterclockwise direction as negative.
Step 3: On a sheet of graph paper draw two lines to represent principal stretches S_1 and S_2 (see Fig. 15-19a). Represent each value of S_A, S_B, etc., obtained by measuring the appropriate angle $θ'_A$, $θ'_B$ . . . from the S_1 axis and plotting points S_A, S_B, . . . at their correct distances from the origin. For example, fossil A lies at angle $θ'_A = 31°$ to S_1 and is unstretched. Point A on Figure 15-19a therefore lies at a distance $S_A = 1.0$ from the origin and at angle $θ'_A = 31°$ from S_1. Fossil B lies at angle $θ'_B = -20°$ to S_1 and has a stretch value of $S_B = 1.3$.

On the graph, point B lies at a distance $S_B = 1.3$ from the origin and at angle $θ'_B = -20°$ from S_1.
Step 4: Find the best fit ellipse to all the data by inspection (see Figure 15-19b) and compute the strain ratio, $R_S = S_1/S_2$. To facilitate your construction note that the strain ellipse, including the lines of no longitudinal strain, is mirror symmetric about both the S_1 and S_2 axes.

Belemnites (or tourmaline crystals, or rutile needles) do not necessarily have the same strength as their matrix. They tend to be rigidly rotated toward the finite stretch direction and only fracture and extend after they enter the extensional field of strain (i.e., on the S_1 side of the lines of no longitudinal strain), and then only after a certain amount of strain has built up in the matrix; so the results obtained are minimum estimates of the total strain in the rock. In addition, tourmalines do not deform easily in the compressional field of strain (on the S_2 side of the no longitudinal strain lines), and although belemnites may undergo pressure solution in this field, such behavior is uncommon. Information about a large part of the strain ellipse is therefore unobtainable. Finally, Figure 15-20 shows one of the most common problems with linear markers. Several large tourmaline crystals have stretched in a recrystallized quartz vein matrix. There are no fibers between these fragments of tourmaline, and there is no obvious mineral elongation lineation on the rock slab surface, so the S_1 direction cannot be independently determined. In this case a more sophisticated analytical technique is required.

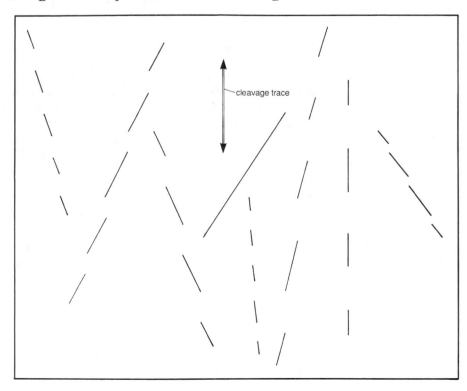

cleavage trace

Figure 15-18. Schematic representation of a group of stretched belemnites for use in Exercise 15-7.

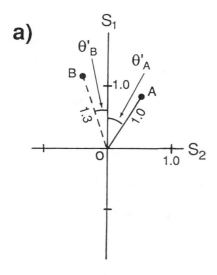

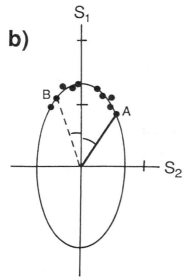

Figure 15-19. Method of plotting stretch values measured from Figure 15-18. (a) Plot of S_A and S_B within S_1 and S_2 reference frame; (b) plotted points should fall on an ellipse.

15-7   USE OF BILATERALLY SYMMETRICAL FOSSILS

Many fossils have bilateral symmetry, a feature that is utilized in several methods to compute the various strain parameters. We will concentrate here on brachiopods, but the methods can be applied to all other suitable fossils. The specimens to be used must be carefully chosen--some species of brachiopods are quite fat originally and have strongly curved median lines, and these could cause problems in strain calculations after they are deformed. The best specimens to use are thin ones that are approximately planar on the bedding surface.

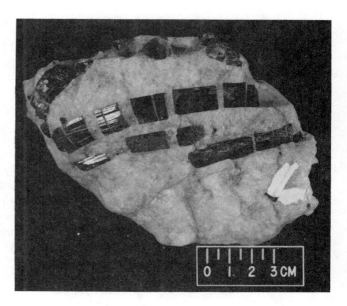

Figure 15-20. Stretched tourmaline crystals in recrystallized vein quartz. Principal stretch directions cannot be independently determined.

Unfortunately, brachiopods die at different growth stages, so all the fossils are not the same size. This means we cannot simply use changes in length of hinge line or median line, or even the ratio of hinge to median line lengths. The only reliable parameter we can use is the angle between hinge and median line; in the undeformed fossil this is always 90°.

The choice of method depends upon the material available. Where there are many specimens on the bedding surface it may be possible to use the classical method that combines the ideas of S. Haughton (1856) and H. Breddin (1956) (see Ramsay and Huber, 1983, p. 143).

The Haughton-Breddin Method

In the Haughton-Breddin method it is necessary that at least a few of the fossils *in the deformed state* have mutually perpendicular hinge and median lines. This means that these special fossils have their hinge and median lines aligned parallel to the principal directions of stretch, and they should occur in two forms--narrow and broad (Fig. 15-21a, b). The narrow form has its hinge line parallel to S_1, and the broad form has its median line parallel to S_1 (Fig. 15-21b).

Exercise 15-8
Figure 15-22 shows several deformed brachiopods on a bedding surface.
    Step 1: Find the specimens that have mutually perpendicular hinge and median lines and determine the directions of S_1 and S_2.

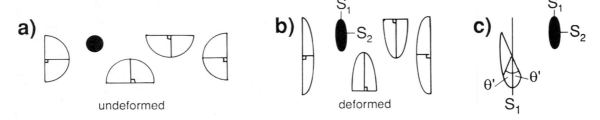

Figure 15-21. The principle behind the Haughton-Breddin method. (a) Four undeformed and mutually perpendicular brachiopods and a unit circle on a bedding plane; (b) situation after a deformation in which principal stretch directions were parallel to hinge and median lines of fossils; (c) a specimen in which hinge and median lines were originally at 45° to S_1 deforms into a symmetrical form (S_1 bisects hinge and median line in deformed state).

Figure 15-22. A suite of deformed brachiopods on a bedding surface. See Exercise 15-8.

Step 2: Examine the remaining specimens and find those whose hinge and median lines are bisected by the S_1 direction (see Fig. 15-21c). Since in these specimens the hinge and median lines were originally at 45° to the S_1 direction (i.e., $\theta = 45°$), use the formula

$$\tan \theta = R_S \tan \theta' \qquad \text{(Eq.15-14)}$$

to find the strain ratio, R_S, for this outcrop.

Step 3: If you have access to ellipse templates, draw a correctly proportioned and oriented strain ellipse onto Figure 15-22. Now tilt the paper away from you while looking down the S_1 axis until the strain ellipse appears to be a circle - the fossils should now appear undeformed as well. We will see later that this trick of "undeforming" the strain ellipse has some very useful consequences!

The Haughton-Breddin method is extremely easy to use, but it does require the preservation of many deformed

fossils and yet uses only a select few of them. It is not generally possible to locate the special narrow and broad forms necessary for this technique to work.

Wellman's Method

If there are a number of fossils *(at least six)* then a good method to use is that of H. W. Wellman (1962). The basis of this method is illustrated in Figure 15-23. In the undeformed state (Fig. 15-23a) a line a_1 drawn parallel to the hinge line of fossil A, and a line a_2 drawn parallel to the median line of A, are projected from the diameter of a circle and form the right angle α at the circle's perimeter. Similarly for fossil B, the hinge line b_1 and median line b_2 are projected to give the right angle β at point B on the circle's perimeter. Other undeformed fossils in different orientations, if plotted in the same way, would produce a series of points that define the circle. In the deformed state the circle becomes an ellipse, and the line OP becomes the

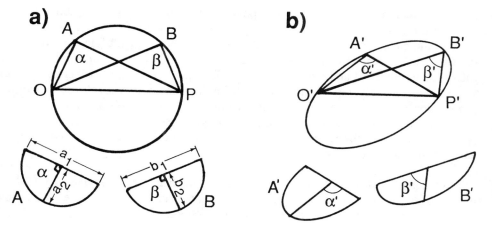

Figure 15-23. The principle behind Wellman's method (After Wellman, 1962.) (a) Two fossils and their projection in undeformed state; (b) after deformation.

ellipse diameter, O'P'. The right angles are deformed to new angles α' and β', and the lines connecting A' and B' to O'P' represent the new orientations of the hinge and median lines of the two fossils.

Exercise 15-9

We will use Wellman's method to analyze the suite of deformed fossils in Figure 15-22 and will compare the answers with the one we obtained using the Haughton-Breddin method.

Step 1: Number each fossil on Figure 15-22 and draw a reference line across the figure in any arbitrary direction (see Fig. 15-24a).

Step 2: On a separate sheet of tracing paper draw another reference line and mark two reference points anywhere along it so that they are several centimeters apart (Fig. 15-24b).

Step 3: Overlay the tracing sheet onto Figure 15-22 so that (a) the two reference lines are parallel and (b) one of the reference points is centered on the hinge/median line intersection of fossil number 1. In pencil, lightly draw the hinge and median lines onto the tracing paper. The line lengths you draw are not important, but they should be extended several centimeters in both directions (Fig. 15-24c).

Step 4: Move the tracing paper until the second

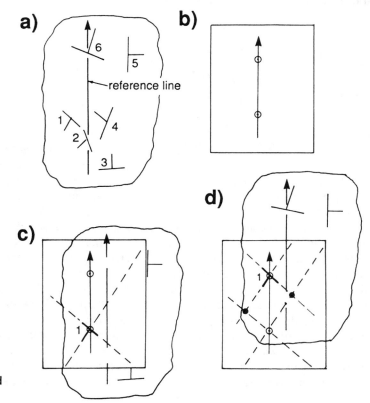

Figure 15-24. Wellman's method - procedure.

reference point is centered on the same point of the same fossil, making sure the reference lines are parallel. Again draw in the median and hinge lines so that they intersect the first pair of lines at two points (Fig. 15-24d). Label each point as number 1 and erase your construction lines.

Step 5: Repeat steps 3 and 4 for each fossil on the page. You should now have 12 points, some of which may coincide. Draw the best fit ellipse through these points and measure the strain ratio, R_S. Compare your answer with that from Exercise 8.

Wellman's method becomes less certain as the number of fossils drops, and it should not be used for fewer than six objects. Its main advantage over the Haughton-Breddin method is that it is much easier to define the ellipse accurately because *all* fossils are used, not just a few in favored orientations.

Ramsay's Method

In some localities there may be only three or four fossils and no reliable indication of the principal direction of stretch. Here we can use Ramsay's (1967) method, which employs the *Mohr circle for strain.* It is possible to use this method with only two fossils, but extreme care is needed with the sign conventions. For safety, it is better to use a minimum of three specimens.

The Mohr circle for strain is illustrated in Figure 15-25a. Full explanations and derivations are available in

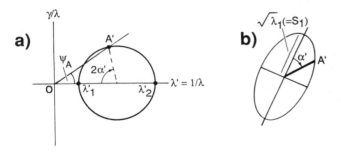

Figure 15-25.   (a) The Mohr circle for strain;
(b) the corresponding strain ellipse.

most structural geology textbooks; we recommend Ramsay and Huber (1983, pp. 93-96) and Davis (1984, pp. 121-24). The axes of the Mohr construction that we will use are γ/λ and $\lambda' = 1/\lambda$, which enables us to deal with angular shear as well as longitudinal strain. Recall that $\lambda_1 = (S_1)^2$, and $\lambda_2 = (S_2)^2$. The center of the circle is at distance $(\lambda_1' + \lambda_2')/2$ from the origin, and its diameter is $(\lambda_2' - \lambda_1')$. Thus, the circle represents an infinite number of points, such as A' on Figure 15-25a, each of which represents the strain of a line A' in two-dimensional space. Figure 15-25b shows the line A' in its more familiar setting

within a strain ellipse. Note that the angle α' between A' and the long axis of the ellipse becomes *twice* that angle ($2\alpha'$) in the Mohr diagram.

In order to use this Mohr circle for strain analysis, we must adhere strictly to sign conventions for angles. Here we will use the convention that the angle α', between the ellipse long axis and the deformed line, if measured *clockwise* on the outcrop becomes angle $2\alpha'$ measured *clockwise positive* from the λ_1' side of the Mohr circle (Fig. 15-25).

Exercise 15-10
Figure 15-26 shows three deformed brachiopods on a bedding surface. We do not know the S_1 orientation. The hinge lines of each fossil have been extended so that the angles between hinge lines can be measured.

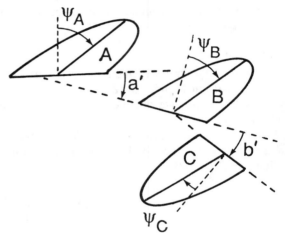

Figure 15-26. Deformed brachiopods on a bedding surface for use in Exercise 15-10. All angular shears are clockwise in this example.

Step 1: Measure ψ_A, ψ_B, and ψ_C for the fossils.
Step 2: On a sheet of graph paper construct the axes γ/λ and $\lambda' = (1/\lambda)$. Choose any linear scale for your axes. Draw three lines of any length through the origin at angles ψ_A, ψ_B, and ψ_C (see Fig. 15-27a).
Step 3: Measure angles α' and β' between the hinge lines of A and B, and B and C, respectively.
Step 4:   On a separate sheet of tracing paper draw a circle of any diameter and subtend angles $2\alpha'$ and $2\beta'$ from its center to intersect the circumference at points A', B', and C' (see Fig. 15-27b).
Step 5: Superimpose the tracing paper onto the graph paper and move the circle so that the three lines ψ_A, ψ_B, and ψ_C intersect the points A', B', and C', and the center of the circle is on the λ'-axis. Make sure that the intersection points are in the same order as they were on the outcrop (i.e., point C is clockwise from B, which is

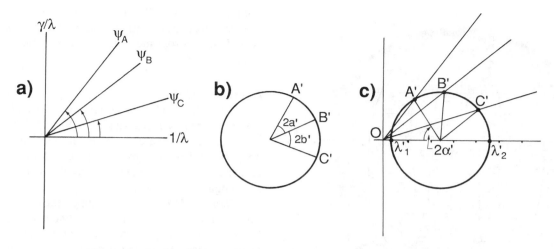

Figure 15-27. Construction of Mohr circle for strain for Figure 15-26.

clockwise from A) otherwise you will obtain incorrect answers because you will have effectively interchanged two of the angular shear readings.

Step 6: Read off the values of λ_1' and λ_2'. Note that the scale is not important because you will compute the ellipticity or strain ratio, R_S. Convert λ_1' and λ_2' to S_1 and S_2 values and find R_S.

Step 7: To find the orientation of the strain ellipse on the outcrop, measure angle $2\alpha'$ on the Mohr circle between the λ_1' point and point A'. The strain ellipse has its long axis at α' degrees to the hinge line of fossil A. Remember that if A' in the Mohr diagram is clockwise positive from the λ_1' point, then on the outcrop the hinge line of fossil A must be clockwise from the long axis of the strain ellipse.

Ramsay's method is rapid and requires few fossils. However, very careful observance of sign conventions is necessary. The method becomes tricky to use with fewer than three fossils.

De Paor's Method

The final method for analysis of deformed bilaterally symmetrical fossils is for the case where only one or two fossils occur on the outcrop. This method was devised by De Paor (1986) and uses a special kind of stereographic projection known as an *orthographic net* or *orthonet* (Fig. 15-28). For this work it is best to use a large version of the orthonet, such as the one provided at the end of this book. The orthographic projection is described in De Paor (1986). For our purposes it is important to note that (a) the small circles on the grid are a series of straight lines, and (b) the great circles on the grid are semiellipses. Recall that in Exercise 15-8 we found that the strain ellipse and deformed fossils could be made to appear undeformed by

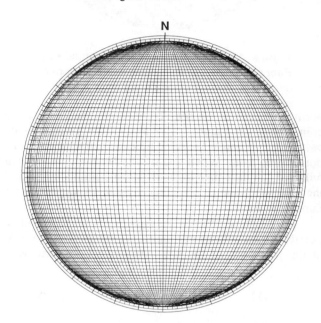

Figure 15-28. The orthonet (orthographic net). (After Ragan, 1985.)

tilting the paper and viewing the ellipse down its long axis. De Paor's method uses the orthonet to simulate the operation of tilting the paper. The angle δ through which the paper is tilted is related to the strain ratio by

$$R_S = 1/\cos\delta \qquad \text{(Eq. 15-15).}$$

Direct estimation of δ is likely to be highly inaccurate, but we can use this basic principle in the form of the orthonet. Figure 15-29 illustrates this principle further. You can view the 25-cent coin as deformed, with an axial ratio of $R_S = 1.55$, or you can view it as undeformed but tilted through angle $\delta = 50^\circ$ about the

Figure 15-29. The principle behind De Paor's method. Coin can be made to appear circular by tilting the page away from you.

north-south axis of the net. Try it for yourself with a real coin. Now tilt Figure 15-29 away from you about its east-west axis and at the same time look straight down the north-south axis. When you tilt the net through 50° you will find that the coin appears to become circular again and the stretched facsimile of George Washington appears more familiar in form.

The elliptical great circles on the orthonet represent all possible strain ellipses starting from the undeformed circle at the periphery of the net. Any two lines that were

initially $\alpha°$ apart will still be $\alpha°$ *in pitch* apart along *one* of these great circles after deformation. All we need to do is find the appropriate great circle and measure its axial ratio to find R_S.

Exercise 15-11

One deformed brachiopod has been found on an outcrop that also contains a cleavage trace (Fig. 15-30a).

Step 1: Measure the angular relationship between hinge (h) and median (m) line and transfer this information to a sheet of tracing paper, making the two lines, h and m, any length (Fig. 15-30b).

Step 2: Measure the cleavage trace orientation with respect to either hinge or median line and draw a labeled line on the tracing paper to represent it. Make sure that the cleavage trace line goes through the hinge/median line intersection point (Fig. 15-30b).

Step 3: Overlay the tracing paper onto the orthonet so that the cleavage trace is aligned with the north-south axis of the net. Do not rotate the tracing paper overlay at any subsequent stage.

Step 4: Inspect the great circles until the one is found where the median and hinge lines are 90° in pitch apart (Fig. 15-30c). Draw in this great circle on both sides of the axis to form an ellipse.

Step 5: Measure the strain ratio, R_S, directly from the ellipse.

Step 6: Repeat the method to find the strain ratio for the fossil in Figure 15-31.

Exercise 15-12

The outcrop surface depicted in Figure 15-32 contains two brachiopods but no cleavage trace.

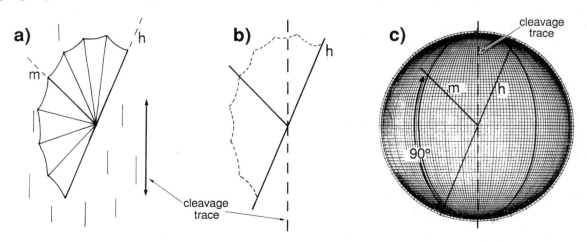

Figure 15-30. De Paor's method - procedure. (a) Deformed brachiopod on bedding surface with obvious cleavage trace; m = median line and h = hinge line of fossil; (b) angular relationships of (a) transferred onto tracing overlay; (c) tracing overlay on orthonet with cleavage trace aligned north-south.

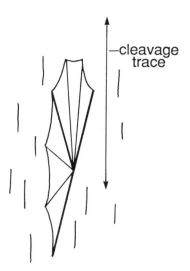

Figure 15-31. A deformed brachiopod on bedding surface for use in Exercise 15-11.

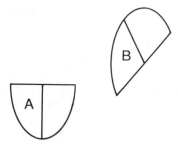

Figure 15-32. Two deformed brachiopods on bedding surface with no visible cleavage trace. See Exercise 15-12.

Step 1: Transfer the hinge and median line orientations for fossil A onto a piece of tracing paper as in step 1 of Exercise 15-11. Use a colored pencil and label these lines h_A and m_A.

Step 2: Follow the same procedure for fossil B, using the same piece of tracing paper but a different colored pencil, and making sure that the two hinge/median line intersection points for the two fossils are superimposed.

Step 3: Put the tracing paper onto the orthonet so that the pin goes through the two hinge/median line intersections.

Step 4: Rotate the tracing paper overlay until it is positioned so that *one* great circle *simultaneously* gives a 90° pitch angle between h_A and m_A, and between h_B and m_B. Note that the same great circle must be used for each side of the net axis.

Step 5: Draw in this great circle on both sides of the axis and measure R_s.

DePaor's method is extremely simple and works well with only one specimen, provided that there is an independent estimate of the principal stretch direction, such as the cleavage trace. If there is no independent means to determine the principal stretch direction, then two fossils are required, and if neither of these has maintained bilateral symmetry after deformation, a simple adaptation of the method in Exercise 15-12 will provide a unique solution (see De Paor, 1986, pp. 90-91). The method also works well for other geometrically regular fossils such as graptolites (if the initial angle between two lines was 60°, then find the unique great circle that contains 60° of pitch between the two lines in their deformed state) and for deformed sedimentary structures such as cross bedding and mudcracks, as long as the original angle between two lines in the structure is known.

15-8 USE OF ORIGINALLY ELLIPSOIDAL MARKERS

In the final series of exercises we will discuss methods to calculate strain by observing the deformation of particles that were initially ellipsoidal. Of course, if we knew for certain that particular objects were originally spherical, then we would need only to find the principal stretch directions and measure the axial ratios directly. But most objects were not originally completely spherical, and the superimposition of a strain ratio onto an initially elliptical shape will produce different final shapes depending on the relative orientations of the ellipses (Fig. 15-33). Other factors that influence the final shape of the objects are their composition with respect to that of their matrix, the mechanism of deformation (crystal plasticity, pressure

Figure 15-33. A strain ellipse 's' superimposed on an initial elliptical shape 'i' produces different final shapes 'f' depending on relative orientation of s and i. (a) s and i are orthogonal; (b) s and i are parallel; (c) s and i are oblique.

solution, cataclasis, etc.), and the degree of interaction among the objects.

In the examples that follow we will assume that the objects have a uniform composition and are not significantly different in strength from their matrix (e.g., quartz pebbles in a quartz-sand matrix or ooids in a limestone). We will not attempt to analyze objects that have a significant strength contrast with their matrix, such as granite pebbles in a shaly matrix, because in such rocks the strain becomes distributed differently in the two lithologies.

First, consider what we would see on a two-dimensional plane cut through a group of three-dimensional ellipsoids. There will be elliptical shapes with two variables to consider: R_f, the final shape ratio, and ϕ, the final orientation of the long axis of each ellipse. Note that the observed area variations do not necessarily indicate original variations in the volume of the objects. A two-dimensional cut goes through the centers of some objects but may just skim the edges of others. We must also consider the initial distribution of the objects in space - Was ϕ random in the undeformed state? Or was there some preferred orientation of the ellipsoid axes, such as is often found in pebble layers in river beds or on beaches?

We can also think of the objects as a collection of central points, ignoring the shape of the objects for the moment. The strain of the whole rock may be represented by the spatial distribution of the centers. In a two-dimensional section the majority of pebbles or ooids are a certain minimum distance apart. Rarely, very small pebbles or ooids may occur side by side to give a very small separation of their centers. This kind of distribution of points that tend to be a minimum distance apart is called an *anticlustered* distribution. If the ellipsoids have a *random orientation* in three dimensions, then the minimum distance tends to be *uniform*, i.e., the same in all directions. If the sediment is imbricated, then the minimum distance may be longer in one direction than another.

There are various methods for the analysis of deformed ellipsoids. We will use the three simplest and yet most practical of them: the Fry (1979) method, which uses particle center distributions; the R_f/ϕ method (e.g., Lisle, 1985) and De Paor's (1987) adaptation of it, which use axial ratios and their orientations; and the Robin (1977) method, which uses irregularly shaped objects such as pillow forms in lava flows or enclaves in granitic rocks.

The Fry Method

The Fry method is based on the assumption that an initially uniform anticlustered distribution of points will change after deformation into a nonuniform distribution. Distances between points become increased in the

extensional field and decreased in the contractional field of strain (Fig. 15-34). Maximum distances between points occur parallel to the principal stretch direction, S_1; minimum distances occur parallel to S_2. So if we can measure the distances between the centers of objects in the deformed state, we can use these data to calculate the finite strain.

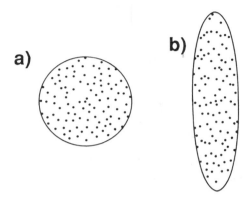

Figure 15-34. The principle behind Fry's method. (a) Initial uniform anticlustered distribution of points inside a unit circle; (b) after strain the distribution becomes non-uniform.

Exercise 15-13

Figure 15-35 is a thin-section photomicrograph of deformed ooids from South Mountain, Maryland. Note that few of the particles have truly elliptical shapes, so measuring axial ratios directly would give an incorrect result.

Step 1: Take a tracing paper sheet and mark the center point of each ooid on it. Number the points as you go, and draw a reference line somewhere to the side of the points. In some of the examples, calcite crystals obscure the ooid center, so you will have to estimate the central position.

Step 2: Mark a central reference cross on a separate sheet of tracing paper. Overlay this sheet onto the first so that the cross coincides with point 1. Trace the reference line through onto the top sheet. Trace the position of all other points (2, 3, 4, . . .) onto the overlay.

Step 3: Move the top sheet so that the cross is on point 2 and the reference lines remain parallel. Trace the position of all other points (1, 3, 4, 5, . . .).

Step 4: Repeat the procedure for all the points on the lower sheet. You will end up with an empty space around the reference cross and a concentration of points just outside this space. It is the shape of the space that is of interest to us; a circular space means a uniform distribution of points (i.e., no strain), and an elliptical shape is a direct representation of the strain ellipse. After moving the overlay four or five times you will see at each station that those points at a considerable distance from the cross do

Figure 15-35. Thin section photograph of deformed ooids, South Mountain, Maryland. Scale bar 0.5 mm. From the archives of E. Cloos, Department of Earth and Planetary Sciences, Johns Hopkins University, with permission.

more, to produce a reasonably ellipse-shaped space. Ellipticity estimates can be quite inaccurate, especially at low point concentrations or where there is a significant size distribution of particles. It is quite common to find "stray" points inside the elliptical space, and one must decide whether to draw the ellipse inside all points, to include the first points at the edge of the space, or to draw the ellipse just outside the space. Finally, the method cannot be used where it is suspected that the ellipsoidal particles had an initial preferred orientation of their axes before deformation.

The R_f/ϕ Method

A principal aim of strain analysis is to find the initial shape of deformed objects such as fossils and pebbles. In a deformed conglomerate, initial pebble shapes, if known, could give important clues to the sedimentary and tectonic environments at the time of their deposition. The R_f/ϕ method of strain analysis assumes that the ellipsoidal object is deformed together with its matrix. On a two-dimensional cut through the rock, the initial *shape factor*, R_i, is changed to the finally observed axial ratio, R_f. The orientation of the ellipse long axis is changed from θ in the undeformed state to ϕ after deformation (Fig. 15-36). The equations that govern this transformation are presented in Lisle (1985) and derived by Ramsay (1967, pp. 205-9). The R_f/ϕ method allows calculation of the ellipticity of the strain ellipse (R_s) by measuring the

not play a role in defining the space and so need not be plotted.

Step 5: After all the points have been covered, remove the upper tracing sheet and find the best fit ellipse to the space around the reference cross.

Step 6: Measure the strain ratio, R_s, and note its orientation with respect to Figure 15-35.

The Fry method is extremely simple and relatively rapid. It can be carried out on rocks that have considerable pressure solution along particle-particle contacts. However, it requires at least 25 points, and preferably many

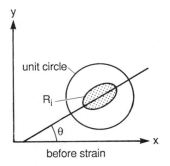

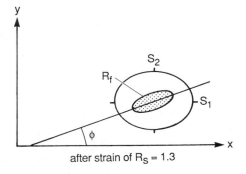

Figure 15-36. Deformation of elliptical shape, R_i, by strain of $R_s = 1.3$ to produce a new ellipse, R_f. (Modified from Lisle, 1985.)

observed ellipticities (R_f) and orientations (ϕ) of a number of objects. Note that in the undeformed state, $R_s = 1.0$.

The final axial ratio, R_f, and orientation, ϕ, of any object will depend on the relative orientations of the initial ellipse, R_i, and the strain ellipse, R_s. Refer back to Figure 15-33 to verify this for yourself. If we start with a group of ellipses such as in Figure 15-37a, having identical R_i values but variable orientations, θ, and deform them by an amount R_s, the final R_f and ϕ values will all vary as in

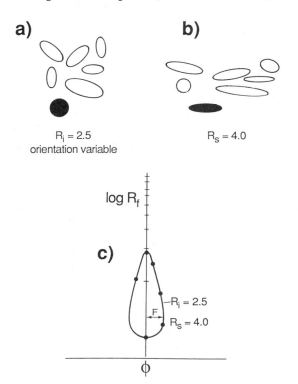

Figure 15-37. Elliptical markers with a constant R_i but variable orientation and their representation after strain. (Modified from Lisle, 1985.) (a) Undeformed state, $R_i = 2.5$; (b) after a strain of $R_s = 4.0$; (c) R_f and ϕ values for ellipses in (b) plotted on log R_f/linear ϕ graph paper. F = fluctuation.

Figure 15-37b. If we now measure R_f and ϕ for each ellipse and plot the values on a log R_f / linear ϕ graph such as in Figure 15-37c, the points should form a characteristically onion-shaped plot (in practice, the angle ϕ is measured with respect to an arbitrary reference line; in Figure 15-37c the reference line was drawn parallel to the principal stretch). The long axes of the deformed ellipses *fluctuate* by an amount F on either side of the principal stretch direction. F is least at the greatest R_s values (i.e., where the "onion" is thinnest). Maximum R_f values occur where R_i and R_s ellipse long axes coincide ($R_{f\,max} = R_s R_i$), and minimum R_f values occur where R_s and R_i

ellipses are mutually perpendicular ($R_{f\,min} = R_s/R_i$ *or* R_i/R_s, whichever is the greater). In these two extreme cases the ϕ value will be zero; hence, the onion-shaped curve is ideally biaxially symmetrical about the $\phi = 0°$ axis (Fig. 15-37c).

If we take a second series of ellipses with R_i values identical to one another, but different from those in Figure 15-37a, and subject them to the same R_s, we will find that their R_f and ϕ values also plot as an onion-shaped distribution. We can build up a series of "onion-ring" R_i curves for any given value of R_s. Figure 15-38 shows one typical series of curves for $R_s = 4.0$ and $R_i = 1.25$, 1.5, 1.75, 2.0, 7.5, 3.0, 4.0, and 6.0.

Suppose that the series of ellipses are all in the same orientation initially but that they have different initial shape factors (Fig. 15-39a). Superposition of a strain, R_s (Fig. 15-39b), will produce R_f/ϕ values that fall on a differently shaped curve (Fig. 15-39c) known as a *theta-curve* (Lisle, 1977). A series of theta-curves for values $\theta = 0°$ to $\theta = 90°$ (Fig. 15-40) can be drawn for every value of R_s. The curve $\theta = 45°$ is also known as the 50%-of-data curve (Fig. 15-40). Together with the $\phi = 0°$,

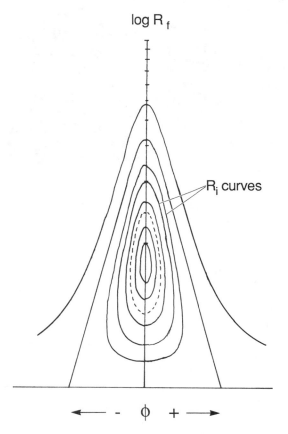

Figure 15-38. R_f/ϕ curves for a strain ratio $R_s = 4.0$. Innermost curve is for $R_i = 1.25$, dashed curve is $R_i = 1.75$, outermost is $R_i = 6.0$. (Adapted from Lisle, 1985.)

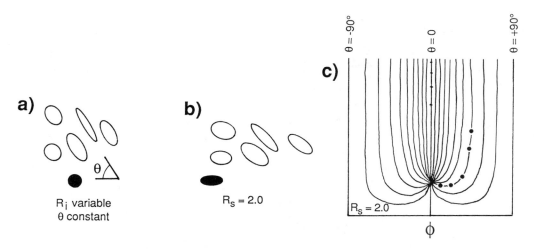

Figure 15-39. Elliptical markers with a constant initial orientation but variable R_i and their representation after strain. (Modified from Lisle,1985.) (a) Undeformed state; (b) after a strain of $R_s = 2.0$; (c) R_f and ϕ values for ellipses in (b). When plotted on log R_f / linear ϕ graph paper, the points fall on a θ curve.

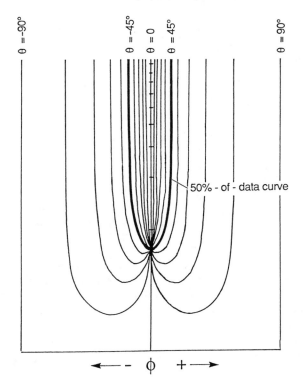

Figure 15-40. A typical set of θ curves for R_f/ϕ analysis. Strain ratio $R_s = 4.0$. (Modified from Lisle,1985.)

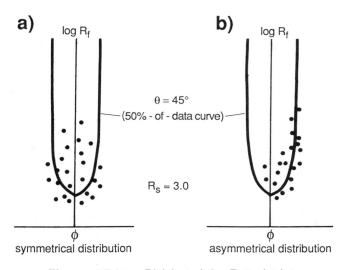

Figure 15-41. Division of the R_f/ϕ plot into four quadrants by use of the 50%-of-data curve ($\theta = 45^0$). (a) A symmetrical distribution - 25% of points fall into each quadrant; (b) an asymmetrical distribution. (Modified from Lisle, 1985.)

$\theta = 0°$ line it divides the plot into four quadrants (Fig. 15-41), each of which should contain 25% of the data points *provided* that the original ellipsoids were randomly oriented. Thus, a *symmetrical* distribution of points about the $\phi = 0°$ line (Fig. 15-41a) is usually taken to indicate

the absence of an original sedimentary fabric, whereas an *asymmetrical* distribution (Fig. 15-41b) is thought to result from an initial fabric such as imbricated pebbles in a stream bed. For a further discussion of interpretation of R_f/ϕ patterns, see Lisle (1985).

Exercise 15-14

A single set of R_f/ϕ curves for the value $R_s = 4.0$ is shown in Figure 15-42. We need similar curves for all

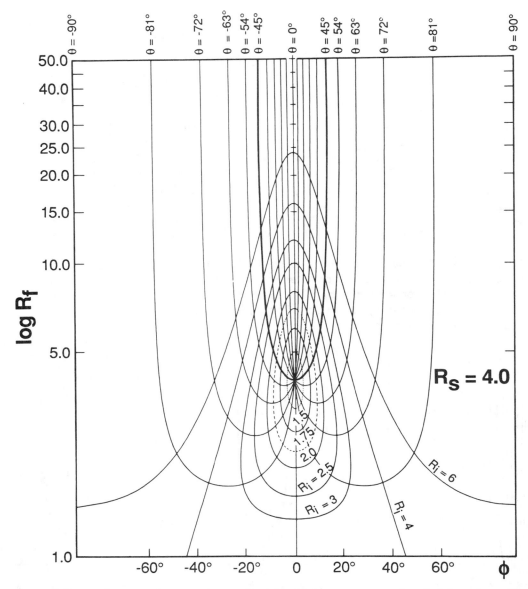

Figure 15-42. R_f/ϕ and θ curves for $R_s = 4.0$. (Modified from Lisle, 1985.)

values of R_s, and a complete set is provided by Lisle (1985). The method for their use is as follows:

Step 1: Use a sheet of tracing paper over Figure 15-35 and draw a reference line in any orientation. You will find it helpful to draw the line close to the apparent maximum principal stretch axis, but the orientation is not critical.

Step 2: Measure the longest and shortest axis of each ooid and the orientation, ϕ, of each long axis with respect to the reference line (ϕ measured clockwise is positive, counterclockwise is negative). Tabulate your data and calculate R_f for each ooid.

Step 3: Plot the R_f and ϕ values for each ooid onto transparent log/linear graph paper (or use an overlay) with axes as shown in Figure 15-41. For use with Lisle's (1985) curves, you will need a ϕ scale of $10^{\circ} = 1$ cm and an R_f logarithmic axis with a 12.5-cm cycle.

Step 4: Fitting the R_f/ϕ curves to the data. First,

we need to find the orientation of the long axis of the strain ellipse. If your reference line was exactly parallel to S_1, then the data should be symmetrical about the $\phi = 0^{\circ}$ line. If your reference line lay at some angle to S_1, then find the axis of symmetry of the plot and this will give ϕ_s, the orientation of S_1 with respect to your reference line. Now center your plot so that its symmetry axis ϕ_s lies along the $\phi = 0^{\circ}$ line.

You must now go through the different R_f/ϕ curves of Lisle (1985) for the various R_s values and find the set that divides the data equally. In other words, 25% of the data should fall in each of the four quadrants formed by the $\theta = 0^{\circ}$ and $\theta = 45^{\circ}$ curves. Ideally, an equal number of points should fall between each pair of adjacent θ curves. This is a time-consuming exercise and one that cannot be carried out if the data are not symmetrically distributed about the ϕ_s line. In this case more sophisticated techniques are required (see Dunnet and Siddans, 1971, and Lisle, 1985).

Step 5: When the appropriate R_f/ϕ curves have been found, simply record their R_s value.

Step 6: The initial shape factor, R_i, for each pebble can now be read directly from the graph by noting the position of each point plotted relative to the R_i contours.

This now traditional approach to strain analysis can be extremely accurate but it is not rapid, nor is it easy to use if the sediment had an initial fabric, as many sediments do! However, the detailed information obtained is well worth the extra effort involved. One drawback is that a suite of different graphs must be examined before the R_s value can be found, and even then it is seldom easy to select among two or three that could equally well fit the data.

De Paor's Adaptation of the R_f/ϕ Method

A simpler, more rapid technique for dealing with the R_f/ϕ data has been developed by De Paor (1988). The complete set of θ curves for all values of R_s have been combined onto one diagram known as a *hyperbolic net* (Fig. 15-43). There are two halves to the net. One hemisphere is labeled R and the other $\mathcal{E}$, the natural strain, where $\mathcal{E} = 0.5 \ln(R)$. We need only concern ourselves here with the R side of the net. The R-axis of the net is divided on a logarithmic scale and represents any axial ratio (R_s, R_i, or R_f). ϕ is measured evenly around the periphery of the net (Fig. 15-44). Points are plotted in this "R_f/ϕ space" just as they would be on a more familiar stereonet: $R_f = 1.0$ occurs at

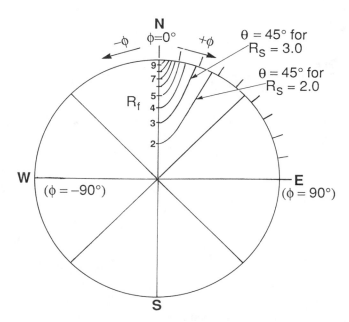

Figure 15-44. Axes and scales of the R side of the hyperbolic net shown in Figure 15-43. (After De Paor, 1988.)

the net center; high values of R_f are nearest the periphery. The acute hyperbolas represent Lisle's $\theta = 45°$ (50%-of-data) curves for every value of R_s from 1.0 to 10.0.

Exercise 15-15

We will use the R_f/ϕ data that you collected in Exercise 15-14.

Step 1: Place a tracing sheet over the hyperbolic net and mark the north axis, N. Plot the R_f and ϕ values for each ooid as follows: (a) Mark the ϕ value on the periphery of the net; (b) rotate the tracing paper until the desired ϕ value is at the north axis of the underlying net; (c) count out the R_f value along the north-south axis starting from the center pin which is at $R_f = 1.0$ (note that any R_f values greater than 10.0 would plot between the two perimeter circles) . Repeat this procedure for each point.

Step 2: After all points are plotted, rotate the tracing paper until the N-S axis becomes the axis of symmetry for the data set. Draw this line through the data set and label it ϕ_S (Fig. 15-45a).

Step 3: Examine the $\theta = 45°$ curves until you find one that divides the population of points in half (Fig. 15-45b). The intersection of this 50%-of-data curve with the north-south axis gives the value R_s for the outcrop. Mark this point and read R_s (*Note:* the scale for R_s is the same as that for R_f).

Step 4: Rotate the tracing paper back to its starting position and read the angle ϕ_S around the periphery. You now have the strain ratio, R_S, and its orientation with respect to your reference line.

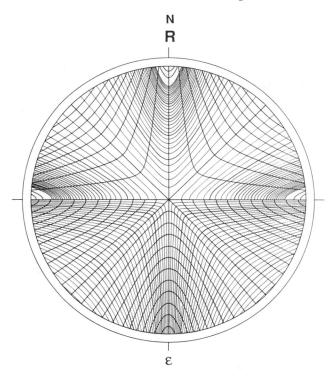

Figure 15-43. The hyperbolic net. (After De Paor, 1988.)

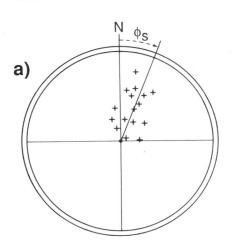

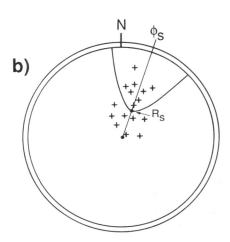

Figure 15-45. The hyperbolic net - procedure.

Step 5: To test for symmetry in the undeformed rock, count the number of data in each quadrant defined by the ϕ_S and $\theta = 45°$ curves (Fig. 15-45b). If there is not a more-or-less equal number of data in each segment, then the original ellipsoids had a preferred orientation. Such a result would invalidate the estimation of ϕ_S in step 2, and a further sequence of steps would then be required in order to accurately determine R_S (see De Paor, 1988).

The main advantage of the hyperbolic net over the traditional R_f/ϕ and theta-curves is its convenience and ease of application, especially in the field. Determination of R_i values for each object is also possible (see De Paor, 1988). A minimum of 16 data are necessary for a valid analysis, in comparison to the traditional methods for R_f/ϕ, which require on the order of 50 data or more.

The final method of strain analysis in two dimensions that we will cover is Robin's (1977) method, which is ideally suited to irregularly shaped strain markers.

Robin's Method

For homogeneous strain the center of gravity, or centroid, of any unstrained object remains its centroid in the deformed state, however irregular the initial shape of the object (Fig. 15-46). We can therefore use the change in length of lines drawn through the centroid and parallel to the principal stretch axes (Fig. 15-47) to calculate the axial ratio of the strain ellipse.

Exercise 15-16

For this final exercise we will again use the deformed ooids in Figure 15-35.

Step 1: Estimate the centroid for each ooid. This can be done fairly accurately by finding the midpoint between the two farthest-apart points on the circumference of the ooid. In the case of a field example with pebbles

several centimeters across, you could carefully trace the pebble shape onto paper, cut out the shape equally carefully, and find the centroid by balancing the paper shape on the eraser tip of a pencil (provided there is no wind that day!), or by suspending the paper by a thin string, marking a vertical line, resuspending from another position, and finding where the two vertical lines cross.

Step 2: Draw two lines, x', and y', parallel to the S_1 and S_2 axes and through the centroid, for each ooid (Fig. 15-47b).

Step 3: Measure x' and y' and calculate the ratio x'/y' for each ooid.

Step 4: Find the strain ratio, R_S, from the geometric mean of the x'/y' ratios, using:

$$R_S = \sqrt[n]{((x\,1'/y\,1')(x\,2'/y\,2')\ldots(x\,n'/y\,n'))} \quad \text{(Eq. 15-16),}$$

where n is the number of ooids measured. (*Note:* If your calculator is not very sophisticated, you will have to find the geometric mean by using logarithms.)

Robin's method is a simple and rapid technique that can be used in the lab or field, on outcrop or thin-section scale, and on many different kinds of strain markers. No assumptions about initial shape are needed, but estimation

Figure 15-46. Principle behind Robin's method. (a) A spherical object; (b) an irregularly shaped object such as a pillow of lava. Centroid of undeformed object, on left-hand side of each figure, remains the object's centroid after deformation.

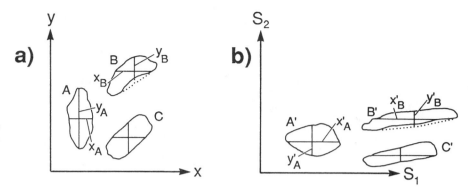

Figure 15-47. Robin's method - procedure. (a) Irregularly shaped objects in undeformed state; (b) after deformation.

of the centroid is not always easy, especially for highly irregular shapes. The assumption of a random initial orientation of the shapes is probably not valid in many cases, but this can be tested by standard statistical techniques (see Robin, 1977).

15-9   CONCLUDING   REMARKS

The exercises in this chapter have covered the most commonly used of the many methods of two-dimensional finite strain analysis. Some have a more complex theoretical background than others, but all are really very simple to carry out. A geologist should, therefore, be able to quantify the strain in any rock, provided, of course, that it contains some sort of strain marker - and we can usually find something, somewhere if we look hard enough and use a little imagination! Converting the two-dimensional data into three dimensions is less simple, but the tensor operations required are straightforward with the help of a computer. The simplest way to convert two-dimensional strain data into three dimensions is to measure the strain on two or more principal planes of strain, i.e., the S_1/S_2, S_2/S_3, and S_1/S_3 planes of the strain ellipsoid. In some field locations it may be possible to select the planes on which two-dimensional strain is measured so that strain in the principal planes is measured directly. Equations to convert such data into three-dimensional form can be found in Ramsay (1967), Owens (1984), and references therein. We strongly recommend that the interested student read Ramsay and Huber's (1983) excellent book for a more in-depth treatment of modern strain analysis techniques.

What do we do with the numbers once we have them? The first step is to convey information about the strain in an area to other geologists. Two-dimensional strain data are readily represented by drawing a correctly proportioned and oriented strain ellipse at a point on a map or cross section where the strain measurement was made. A regional map or cross section on which a number of such strain ellipses are drawn shows graphically the variation in strain with location. If all three principal strains are

known, the strain ellipsoid can be drawn, using the techniques of perspective drawing, or the strain can be represented as a point on a *Flinn diagram* or a related strain plot (see Ramsay and Huber, 1983; Davis, 1984; and Suppe, 1985 for examples).

One of the main applications for strain analysis is in producing balanced cross sections across orogenic belts (Chapter 14). Clearly, in a deformed continental shelf sequence, the strain will tend to vary in intensity from the relatively strong limestones and sandstones to the rather weak shale units. So if we wish to create a palinspastic reconstruction of a fold-and-thrust belt, we will have to "unstrain" each unit separately.

Another major application of strain analysis is in "unstraining" deformed objects such as fossils, so that we can identify the species (e.g., Bambach, 1973), or pillow lavas, so that we can identify the original "way-up" of the rock (e.g., Borradaile and Poulsen, 1981). There are many such applications in the geologic literature.

Last, but by no means least, we wish to understand how the tectonic forces on our planet operate. To do this we must also understand exactly how each small piece of rock became deformed. One of the keys to such understanding lies in unraveling the sometimes rather complex strain history recorded in the rock.

15-10   ACKNOWLEDGMENTS

Many of the exercises and much of the theory presented in this chapter are adapted from the publications of J. G. Ramsay, whose influence, although indirect, has been significant in the production of this work. The advice of D. G. De Paor on all aspects of the work has been of equal significance. W. D. Means read the preliminary version of this chapter and pointed out several potential pitfalls, some of which my students had already discovered the hard way! Without the help of M. L. Wayne and T. E. Wilson, drafting the figures would have caused a considerable strain. To all these people I extend my thanks.

16

INTERPRETATION OF POLY-DEFORMED TERRANES

Sharon Mosher
Mark Helper

16-1 INTRODUCTION

A *poly-deformed terrane* is a region where rocks have undergone more than one phase or episode of deformation. Such terranes are common, for example, along ancient and modern-day convergent margins, in Precambrian shield regions, or in low- to high-grade metamorphic terranes. In these areas superimposed ductile deformations are often accompanied by metamorphism. The coupling of metamorphism and deformation can produce distinctive rock fabrics that are powerful tools for unraveling complex geologic histories. Fold geometries, fold interference patterns, and fold styles are also useful for such analysis.

The objectives of studying a poly-deformed area are (1) to isolate the individual phases of deformation and metamorphism, (2) to determine the temporal and spatial relationships between the phases of deformation (i.e., to decide whether they comprise one or more deformational events), and (3) to determine the kinematic significance of deformation phases. The purpose of this chapter is to discuss techniques for analyzing and interpreting poly-deformed terranes.

16-2 NOMENCLATURE

A *generation of structures* refers to the suite of structures (lineations, foliations, cleavages, folds, faults) that form during the same time interval in response to the same stresses. A *phase of deformation* is the time interval during which a single generation of structures is produced.

During a phase of *progressive deformation* several generations of structures can be produced. A *deformational event* consists of one or more phases of deformation that are temporally and genetically related. The structural generations that define a deformational event form sequentially over a specific interval of time. An *orogeny* comprises one or more deformational events associated with a major period of tectonism or mountain-building. Thus, a poly-deformed terrane can result from (a) a phase of progressive deformation; (b) a multiphase deformational event; (c) two or more deformational events that may, but need not, be time separated; or (d) two or more orogenies.

A shorthand notation is commonly used to classify structures by type and by generation. A capital letter denotes the type of structure, and a numerical subscript indicates the generation. For *linear* features, F_1, F_2, . . ., F_n denote fold axes, and L_1, L_2, . . ., L_n denote other lineations (e.g., mineral or intersection lineations, crenulation axes, rods, mullions, elongated detrital clasts or grains). For planar *surfaces*, S_0 denotes bedding, and S_1, S_2, . . ., S_n denote cleavages, foliations, or axial planar surfaces. If a lineation lies on an S-surface, two subscripts are used. The first subscript gives the generation of the lineation, the second the generation of the surface. L_{31}, for example, denotes a third-generation lineation on an S_1 surface. The term *structural element* is commonly used for a structural feature that can be labeled using the preceding notation. Finally, D_1, D_2, . . ., D_n indicate *deformational events*, which, as noted above, may include one or more generations of structures.

16-3 MAPPING AND DATA ANALYSIS

Structural analysis of poly-deformed terranes begins with mapping and collection of structural data. Mapping of poly-deformed terranes differs from mapping of less deformed or undeformed areas in three important respects. First, it is best to map poly-deformed terranes at a large scale (greater than 1:10,000; for example, 1:2000) so that complex relationships and data for small structures can be shown in adequate detail. Second, it is best to make outcrop maps rather than contact maps of poly-deformed terranes (see Appendix 3) so that all structural interpretations shown on the map can be objectively evaluated. Third, it is better to map lithologic units or distinctive lithologies (marker beds) rather than stratigraphic units because they show the structure best and because unique stratigraphic units cannot, in some cases, be defined. When you begin mapping, the relative ages of structural elements will be unknown, so the earliest generation observed should be labeled with letter subscripts (S_A, F_A, etc.). Once the relative ages have been established, these subscripts can be changed to numbers (S_1, F_1, etc.).

Treatment of structural data in poly-deformed terranes differs from that in simpler regions in the following respects:

1. In addition to measuring the orientations of structural elements in the field, you should collect oriented specimens (Appendix 3) from which oriented thin sections can be made. Microscopic examination of the thin sections can help to substantiate your interpretation of the relative ages of different structures, for magnification will allow you to see crosscutting relations more clearly.

2. Equal-area or equal-angle plots (hereafter referred to in this chapter informally as *stereoplots*) must be used to illustrate structural relationships, to check hypotheses, and to find relationships that show up only regionally. For example, a stereoplot allows you to test graphically whether a given fold axis lies in a given cleavage plane and thus that the cleavage and the fold are probably of the same generation. A stereoplot may also emphasize geometric relationships that are subtle or obscure in the field. Data from across a region, for example, may show that cleavages are folded by large, regional folds that have no small-scale counterparts, and a generation of lineations that locally maintains a common orientation may regionally define a great or small circle, indicating that they are folded or were formed on folded surfaces.

3. The familiar angular relationships of cylindrical folds with planar axial planes (specifically, that poles to folded surfaces lie on a great circle 90° from the fold axis, and axial-planar cleavages contain the fold axis; see Chapter 8) usually do not hold on a regional scale in poly-deformed

terranes. Superposition of two or more generations of folds can cause one or more fold generations to be noncylindrical and can result in folded axial planes. It is necessary, therefore, to divide poly-deformed terranes into *domains* in which folds are roughly cylindrical in order to display the characteristic angular relationships on stereoplots. Domain boundaries are usually related to late-phase folds or to faults. Figure 16-1 (see also Fig. 16-23b) shows a map that has been divided into domains and provides corresponding equal-area plots for three of the domains. The plots can be combined in a synoptic diagram that shows the overall geometry of the structure in the area.

One of the premises in studying any deformed region is that the *orientation* and *style* of small, outcrop-scale structures (hereafter referred to as *minor structures*) mimic

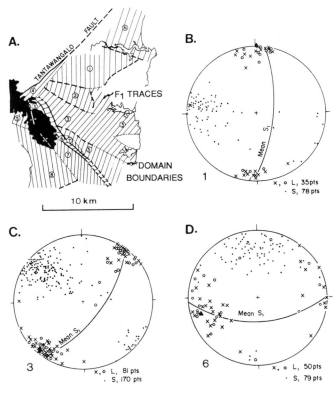

Figure 16-1. An example of the use of structural domains from the Lachlan Fold Belt, Australia. (Adapted from Powell et al., 1985.) (a) Map of area divided into nine domains. Late-stage kink bands have folded the earlier structures. The axial traces of the kink bands are the domain boundaries. F_1 fold axes and S_1 change orientation across boundaries; (b), (c), (d) lower-hemisphere, equal-area nets for domains I, 3, and 6 on the map. Orientations of an early cleavage (S_1) and cleavage-bedding intersections (L_1) are shown.

the orientation and style of regional-scale structures. This premise is a particularly useful guide in studying poly-deformed terranes, because by characterizing the orientation and style of an individual generation of minor structures in outcrop, you have a clue to the orientations and styles of regional structures of the same generation.

As a final note we emphasize that when mapping and collecting structural data in poly-deformed terranes it is imperative to determine as much as possible about the relative ages, orientations, and styles of different generations of structures *while in the field*. There is a tendency for beginning students to collect orientation data on structures without regard to the relative age or significance of the structures, with the mistaken belief that later analysis with stereoplots will somehow distinguish among structural generations. This practice inevitably leads to confusion.   Successful field work of any type, particularly in poly-deformed terranes, involves the continual testing and refinement of multiple working hypotheses (see Chapter 9) during the field-work stage of research rather than at some later date.

16-4   SUPERIMPOSED MINOR FOLDS

Interference Patterns

A *refolded fold* is one whose fold axis or axial plane is itself folded by a younger fold. Superimposed folds that have axes parallel to one another are called *coaxial*; all others are *noncoaxial*. Refolding of folds produces characteristic *interference patterns*. The type of interference pattern provides information on the orientations and angular relationships of the two fold generations. Three "end-member" types of interference patterns can be recognized on the basis of the orientation of fold axes and axial planes of the superimposed folds (Fig. 16-2; Ramsay, 1967). The three types can be remembered by the visual pattern they create when viewed in a particular orientation, namely, eyes or domes and basins (*type 1*), mushrooms (*type 2*), and zigzags (*type 3*). Superimposed folds produced during refolding are coaxial for type 3 patterns and noncoaxial for types 1 and 2.

Although most fold interference patterns have

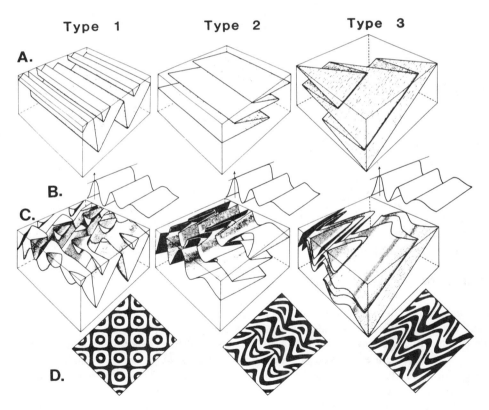

Figure 16-2. Three end-member types of interference patterns produced by superimposing two generations of folds. (Adapted from Ramsay, 1962; 1967.) Rows A and B show the geometry of first and second folds, respectively; row C the resulting interference pattern; row D an idealized plan (map) view of each pattern, similar to what would be seen looking down on the top of each box in row C.

geometries intermediate to those of these end members, it is helpful to understand the angular relationships of the three end members. Examples of the large number of possible interference patterns can be found in Ramsay (1967) and in Thiessen and Means (1980).

One of the best ways to visualize these angular relationships is by folding paper (Fig. 16-3). It is important to realize that the visual appearance of an interference pattern changes when viewed in different orientations. Thus, the actual pattern observed depends on the orientation of the outcrop face with respect to the folds. Try to visualize or draw the patterns that would result from erosion that exposed or cut the refolded folds in Figure 16-2 along different planes.

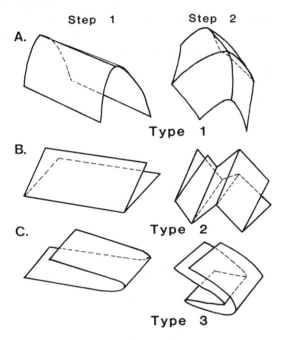

Figure 16-3. Folding paper to produce the three end-member interference patterns.

Stereoplot Patterns for Refolded Folds

The angular relationships of the three end-member interference patterns are well displayed on stereoplots. These plots can help determine the type of refolding in regions where interference patterns are not well exposed. The distinctive stereoplot patterns of end-member interference patterns are described next with the aid of paper-model examples (Fig. 16-3). The sheet of paper represents a folded layer (S_0). As you read the following descriptions, fold a piece of paper as directed, then study the corresponding stereoplot of structural elements (Fig. 16-4).

Type 1 (Eyes or domes and basins): Create an upright anticline (vertical axial plane, S_1). Bend this fold

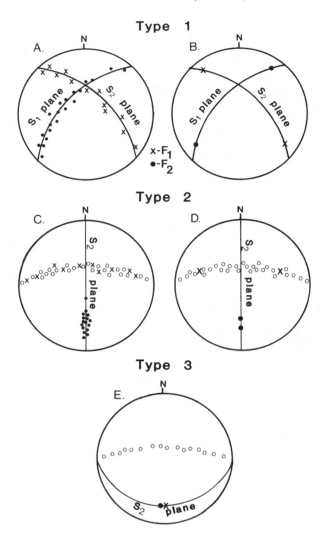

Figure 16-4. Idealized stereoplots for the three end-member interference patterns and the effects of curved (a, c) and planar (b, d) limbs on the distribution of F_1 and F_2 fold axes. Open circles are poles to S_1; X's and dots are F_1 and F_2 axes, respectively.

around an imaginary second upright anticline whose hinge is perpendicular to the hinge of the first fold. The resulting form resembles a dome (Fig. 16-3a). In this dome the axial planes of both fold generations are not folded (i.e., planar) and are nearly vertical, and F_1 and F_2 are curved lines, each lying in its respective axial plane (S_1 and S_2). The stereoplot in Figure 16-4a shows the orientations of the domes' axial planes and fold axes. (To match the figure, hold the fold so that the S_1/S_2 intersection plunges northward and S_1 strikes northeast.) The amount of clustering of fold axes depends on the shape of the folds; if both folds have curved limbs, fold axes are evenly distributed along the S_1 and S_2 great circles (Fig. 16-4a). If the limbs are planar, fold axes cluster (Fig. 16-4b).

Type 2 (Mushrooms):   Create a recumbent fold (horizontal axial plane, S_1). Bend this fold into an upright anticline-syncline pair whose hinges are perpendicular to the hinge of the first fold (Fig. 16-3b). Hold so F_2 plunges slightly southward. The F_2's lie in the axial plane of the F_2 folds (S_2) but are concentrated along only part of the great circle (see representative stereoplot, Fig. 16-4c). F_1's lie on a great circle perpendicular to the average F_2 axis and the S_2 plane; this great circle also contains the S_1 poles. The amount of spread of fold axes on the net depends on the shape of the other generation fold limbs. If F_2's have planar limbs, then there will be two clusters of F_1's (Fig. 16-4d), otherwise F_1's spread along the great circle (Fig. 16-4c). If the limbs of the F_1 fold are planar, then the F_2's will describe two point concentrations (Fig. 16-4d).

Type 3 (Zigzags):   Create a recumbent fold with a piece of paper (horizontal axial plane, S_1). Bend this fold to form a second recumbent fold with an axis parallel to the first fold axis (Fig. 16-3c). Hold so F_1 and F_2 plunge slightly southward. On a stereoplot (Fig. 16-4e) F_1's and F_2's fall in a single point concentration and lie in the S_2 plane. The poles to S_1 should lie on a great circle that is perpendicular to the F_1 and F_2 axes.

Note that if the refolded paper in the preceding examples is rotated about any axis, the orientations of all structural elements will change (except those parallel to the rotation axis). The appearance of the stereoplot pattern must also change, as must the appearance of the interference pattern on a map. The angular relationships among structural elements, however, remain constant. To make sure that you understand the basic types of interference patterns, work through the following exercises.

Problem 16-1

(*a*)   Create a type 2 (mushroom) fold out of paper, as just described. Make the fold limbs of both fold generations planar and the F_1 fold tight. Make the F_2 folds open (Fig. 16-3b).

(*b*)   Position the refolded fold so that the F_2's trend north and the F_2's on the upper F_1 limb have no plunge. Sketch a stereoplot of F_1 and F_2 axes and poles to S_1 and S_2 planes. Label them F_{1a}, F_{2a}, S_{1a}, and S_{2a}.

(*c*)   Rotate the upper F_2 so it plunges 45°S and trends north. Plot the same structural elements as in (a) on the same overlay. Label the new orientations of the structural elements F_{1b}, F_{2b}, S_{1b}, and S_{2b}.

(*d*)   Rotate the upper F_2 to vertical, and plot the same structural elements on the stereoplot again, labeling as before.

(*e*)   Is it ever possible to have F_2's plunging both north and south?

(*f*)   Which, if any, structural element(s) did not change orientation?

(*g*)   If F_2 was isoclinal, could you mistake these for coaxial folds? Why or why not?

(*h*)   Why do you think this pattern is referred to as a mushroom fold?

Problem 16-2

(*a*)   Create a type 1 paper fold. Make the limbs of the paper dome curved and folds of both generations open, as in Figure 16-3a.

(*b*)   Hold the fold so that the intersection of the F_1 and F_2 axial planes is oriented 30°N and S_1 strikes northeast. Sketch a stereoplot of F_1 and F_2 axes, S_1 and S_2 axial planes, and S_1 and S_2 poles.

(*c*)   Rotate the S_1 and S_2 axial planes toward each other until they are 60° apart. Plot the same structural elements as in (a) on a different sketch.

(*d*)   Repeat as in (b) but with the axial planes about 30° apart.

(*e*)   As the angle between S_1 and S_2 decreases from perpendicular, what structural element(s) plot differently and in what way?

Real field data, when plotted on stereoplots, are usually not as easily interpreted as the preceding examples. There are several reasons, including the following:

1. Lithology, previous orientations of S-surfaces, and faults can influence the orientation of folds in ways unrelated to the refolding geometries themselves.

2. The size (amplitude and wavelength) of a fold generation has an effect on the type and amount of data collected and, therefore, on the stereoplot pattern. If a fold is observed only on an outcrop or smaller scale, you may not measure enough data to completely define the geometry of the fold on a stereoplot. For example, both folds in Figure 16-5 have the same orientations, but in case (a), F_1's are larger than F_2's, whereas in case (b) the opposite is true. If the minor effects of the smaller generation of folds are ignored when taking measurements, the stereoplot patterns will be quite different, even though the orientations of the F_1's and F_2's in both (a) and (b) are the same.

3. Data can be incorrectly measured or recorded, or outcrops can be out of place.

Use of Fold Asymmetry

If interference patterns are recognized in outcrop, similar patterns may be present regionally (see Figs. 16-19a and 16-23a). One way of locating regional-scale fold axial traces from outcrop-scale interference patterns is to use parasitic fold asymmetries (S- and Z-shapes). A *parasitic fold* is one that forms on the limbs of a larger fold; parasitic folds can occur at all scales. When viewed

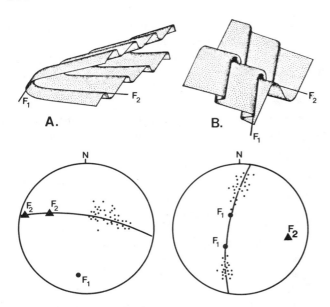

Figure 16-5. Effect of fold size on data collection and the resultant stereoplot patterns. Folds in (a) and (b) have the same orientations, but different relative sizes. In each case, one fold generation has a minor effect on bedding and the other generation fold axes. Unless care is taken in the field, the minor effect of a fold generation will not be measured, resulting in different stereoplot patterns, as in (a) and (b). Unlabeled points on stereoplots are poles to bedding. (Fold sketches from Turner and Weiss, 1963.)

down-plunge, a parasitic fold is shaped like an S, Z, or M, depending on its location on the major structure (Figs. 16-6a and b). The axial trace of the regional fold separates a domain of minor S-folds from a domain of minor Z-folds.

Characteristic patterns result from interference of two generations of asymmetric folds (Fig. 16-6c). For example, if first-generation folds (those deforming only bedding) are Z-shaped, and second-generation folds (those deforming bedding and the axial planes to first folds) are S-shaped, then the resulting interference pattern looks like an S-shaped fold superimposed on a Z-shaped fold (called S on Z fold; Fig. 16-6c).

To use parasitic fold asymmetries, define areas or domains on the map which contain S on Z, Z on S, S on S, or Z on Z folds. The domain boundaries are the axial traces of the two generations of larger folds (Fig. 16-6c). Each type of regional fold interference pattern (type 1, 2, or 3) has a distinctive pattern of parasitic-fold domains (Ramsay, 1967, Fig. 10-19).

Use of Vergence Boundaries

When two or more generations of noncoaxial folds are mapped, fold vergence is often more useful than fold asymmetry in determining the location of regional fold axial traces (Bell, 1981). *Fold vergence* is the horizontal direction, within the plane of the fold profile, in which the upper, long limb of an asymmetric fold appears to have moved to cause the rotation of the short limb (Roberts, 1974; Bell, 1981; Fig. 16-7a). Fold vergence is defined with respect to a geographic reference frame (north, south, east, and west). The vergence of an asymmetric minor fold does not, therefore, depend on the plunge of the fold; folds of the same generation that verge in the same direction are classified together even if they plunge in opposite directions. In Figure 16-7b, for example, both minor folds verge north, but when viewed down-plunge, one is an S-fold and the other is a Z-fold, although both are on the same limb of the earlier fold.

A *vergence boundary* separates parasitic folds of different vergence and represents the axial trace of a larger fold. Vergence boundaries can be extended to areas of multiple folding in a manner completely analogous to fold asymmetry (S on Z, etc.) domain boundaries. It should be

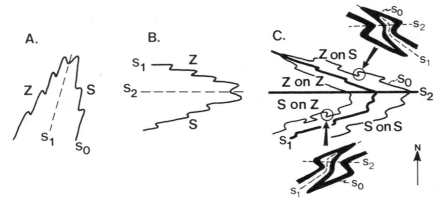

Figure 16-6. An example of domainal mapping on the basis of asymmetric minor fold interference patterns. Interference of large-scale first folds (a) and second folds (b) produces four domains (c) that are separated by the axial traces of the first and second folds, labeled S_1 and S_2. Each domain is characterized by the type of small-scale interference pattern present (i.e., Z on Z, Z on S, S on S, S on Z). The circled areas are enlarged to show examples of Z-on-S and S-on-Z folds.

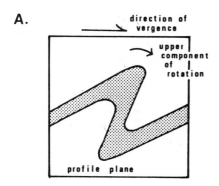

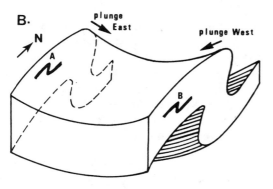

Figure 16-7. Fold vergence. (From Bell, 1981.) (a) Sketch illustrating the definition of fold vergence; (b) relationship between vergence and Z and S folds. Viewed down plunge, fold A is an S and fold B is a Z, but both verge N.

noted that the technique of using either fold asymmetry or vergence to locate regional fold axial traces is useful only where patterns of minor folds are well exposed over a large area and are easy to recognize and interpret.

Problem 16-3

What is the vergence direction of each fold generation for the S-on-Z-fold in Figure 16-6c? The Z-on-S-fold?

Solution 16-3

For the Z-on-S-fold, the Z-fold is east-verging and the S-fold is west-verging. For the S-on-Z-fold, the S-fold is west-verging, and the Z-fold is east-verging.

16-5   UNRAVELING MULTIPLE FABRICS

Use of Foliations

Foliations or cleavages are useful in understanding polyphase deformation in two ways: (1) In areas where depositional features are still preserved, cleavage/bedding relationships coupled with the direction of sedimentary younging can be used to unravel complex fold geometries; and, (2) Multiple foliations provide cross-cutting relationships that can be used to derive the relative age of different generations of structures.

In rocks that have been folded only once, the relationship of cleavage and bedding can indicate whether a unit is right-side up or overturned where both cleavage and bedding dip in the same direction (Fig. 16-8a). On the right-side-up limb of a fold, cleavage is steeper than bedding, whereas on the overturned limb of a fold, bedding is steeper than cleavage. Sedimentary features (e.g., graded bedding) will show the same *younging direction* (i.e., direction of stratigraphically younger rocks) as the cleavage/bedding relationships. In rocks that have been folded more than once, sedimentary younging and cleavage/bedding may disagree. If the generation of the cleavage can be determined (e.g., by cross-cutting relationships), then it is possible to determine which part of the larger refolded structure the outcrop represents, as shown in Figure 16-8b.

Problem 16-4

At an outcrop, bedding is overturned, cleavage is steeper than bedding, and bedding and cleavage dip in the same direction. From what location on Figure 16-8b is the outcrop if (a) the cleavage is S_1; (b) the cleavage is S_2?

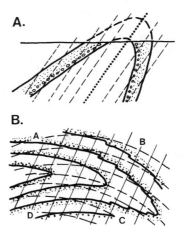

Figure 16-8. The relationship of cleavage, bedding, and younging (stippling shows direction of sedimentary grading). (a) Single-fold generation; when cleavage and bedding dip in the same direction, cleavage dips steeper than bedding on right-way-up limb, shallower than bedding on overturned limb; (b) refolded recumbent fold showing two cleavages. (Adapted from Wilson, 1982.) A, B, C, and D are locations referred to in Problem 16-4.

Solution 16-4

(a) The outcrop is at location C; (b) the outcrop is at location D. Note that at location C, S_2 is also steeper than bedding but dips in the opposite direction.

The fold *facing direction* is the younging direction measured normal to the fold axis on an axial-planar surface (Fig. 16-9; Shackleton, 1957). It is equivalent to the younging direction in the hinge of a fold. In rocks that have been folded only once, the fold facing direction will always be the same, regardless of where it is measured. In rocks that have been folded more than once, fold facing directions may change across an area or even on a single axial-planar surface (Fig. 16-9). The change in fold facing can be used to locate the axial traces of folds that predate those for which facing has been determined. The axial

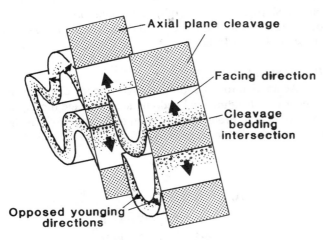

Figure 16-9. Refolded recumbent fold showing fold facing and younging relationships. On opposite limbs of the early fold, the fold facing direction changes. The axial trace of the recumbent fold is located where the facing direction changes. Stippling represent graded bedding. (Adapted from Borradaile, 1976.)

trace of the earlier fold is located where the facing direction changes (Fig. 16-9). Note that in all cases the younging directions on opposite limbs of a fold will point in opposite directions.

Problem 16-5

The incomplete cross section shown in Figure 16-10 is of folded quartzites and schists at Bonnet Shores, Rhode Island. F_2 folds that can be traced in outcrop are shown; the folded surface is bedding. The facing directions on S_2 surfaces are shown where measurable. Late-stage, open, monoclinal folds and their axial planes (dashed lines) are also shown. The depths of F_2 fold hinges were determined using down-plunge projection (see Chapter 13). Finish the cross section, using the facing information and the style of folding shown. Connect all layers with solid lines below the surface and dashed lines above. Is this a unique solution? Do you have enough information to say in which direction F_1 folds verge? Do you have enough information to say in which direction F_2 folds verge? If so, what is the vergence direction?

Where multiple foliations are present, their relative ages can be determined by studying cross-cutting relationships. Cross-cutting relationships are sometimes easy to determine, as when an early cleavage is crenulated to form another. At other times relationships must be determined using thin sections, or by correlation with fold generations. In many terranes, especially metamorphic terranes, relict and incipient cleavages that are rarely observed in the field can be identified in thin section. For example, Figure 16-11 shows a rock that displays two macroscopic foliations: a metamorphic foliation (S_1) cut by a differentiated crenulation cleavage (S_2). Microscopic examination, however, indicates that the micaceous minerals in the sample are aligned in another direction, defining an S_3 fabric. S_1 and S_2 are both defined by concentrations of micaceous minerals, and S_3 is defined by the alignment of the micaceous minerals.

If more than one foliation exists over a large area, then more than one phase of folding has, in most cases,

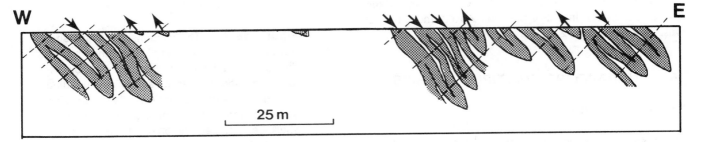

Figure 16-10. An incomplete cross section across Bonnet Shores, Rhode Island. (From Reck, 1985.) Tight folds of layering are F_2; open, monoclinal folds with axial planes dashed are F_4. Arrows point in the direction of facing of the tight folds.

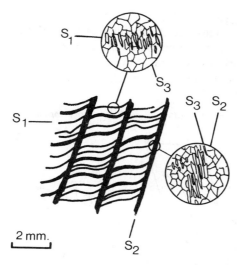

Figure 16-11. Sketch of three cross-cutting cleavages from the Rough Ridge Formation, Llano Uplift, Texas. (Mosher, unpublished data.) S_1 is a metamorphic layering, S_2 is a differentiated crenulation cleavage, and S_3 is an alignment of micas in a third orientation. Expanded views: elongate grains are biotite and muscovite; equant grains are quartz and feldspar.

occurred. Exceptions to this generalization occur in wide shear zones (e.g., S-C fabrics, as described in Chapter 11) and where more than one cleavage has been produced by progressive deformation during a single phase of folding.

Use of Lineations

Many different types of lineations can develop during deformation, including intersection lineations, slip lineations (e.g., slip fibers and grooves), mineral lineations, crenulations (the crests and troughs of a crenulation define a lineation; the crenulation cleavage itself is a planar structure), and elongated clasts or grains. Mineral lineations can be defined by stretched crystals, pressure-shadow overgrowths, alignment of inequant grains, or streaks of microcrystalline mineral clusters. Most lineations form either in the direction of extension during shearing (stretching lineations) or parallel to the long axis (X-direction) of the finite strain ellipsoid during folding. In general, the study of lineations can provide information on the kinematics of deformation, can be used to determine relative ages of superimposed structures, and can help to trace the orientations of fold axes in an area. There are three common applications of lineation analysis:

1. Multiple generations of shear-related lineations may indicate changes in the direction of motion during shearing and can be used to reconstruct complex movement histories within and across shear zones.

2. Intersection lineations that are defined by the intersection of a foliation with a folded surface (e.g., bedding with S_1) give the trend and plunge of fold axes.

3. In pelitic rocks several generations of nonparallel crenulations may develop (Fig. 16-12). Cross-cutting relationships among crenulations may be discernible on the crenulated surface but usually must be determined using thin sections. If the crenulation cleavages associated with the crenulation lineations are axial planar to larger folds, the orientations and crosscutting relationships observed for the crenulations indicate the relative ages and orientations of the larger folds.

Figure 16-12. Cross-cutting crenulations in the Condrey Mountain Schist, northern California. Pencils parallel crenulation axes.

16-6 CORRELATION OF STRUCTURAL GENERATIONS

One of the most difficult aspects of working in poly-deformed terranes is establishing age equivalency among minor structures from outcrop to outcrop or area to area. Assigning a structure to a particular generation is simple when all generations of structures are present, and mutual cross-cutting relationships can be observed. This situation is, unfortunately, the exception rather than the rule. A combination of information on orientation and style of minor structures, coupled with knowledge of the conditions of deformation during which the structures formed, provides the soundest basis for correlation. Structures cannot be correlated on the basis of a single criterion. We discuss next the individual components of a structural correlation.

Correlation Using Orientation

Correlation of structures between outcrops using orientation assumes that the orientation of a generation of structures is constant across the region. If this assumption is accepted, all folds of a single generation, for example, should be coaxial and coplanar, and the axes and axial surfaces of parasitic folds should parallel those of larger-scale folds (Pumpelly's rule). Regional constancy of orientation can occur only when the directions and magnitudes of stresses are uniform across an area. Although uniform stresses rarely prevail along or across an entire orogenic belt, it is often reasonable to assume that stresses were uniform within a discrete portion of the belt. Several other factors that are discussed next can also affect regional constancy of orientation.

Problem 16-6

Cleavage planes in a quartzite unit are oriented N10°E,45°SE; N11°E,40°SE; N05°E,48°SE; N13°E, 43°SE; N04°E,42°SE. Folds of the same generation cannot be found. The cleavage in the quartzite resembles a cleavage (in terms of conditions of formation) found in a nearby schist unit. The cleavage in the schist is axial-planar to mappable folds. The axes of the folds in the schist are oriented S40°E,28°SE, and the axial-planar cleavage N11°E,46°SE. Can the two cleavages be of the same generation?

Method 16-6

Plot the poles to the cleavages and the fold axis on a stereoplot. All cleavage poles plot in the same point concentration. The fold axis lies in the plane of cleavage. On the basis of these relationships, you can tentatively conclude that the cleavages in both units are of the same generation.

Two factors must be considered before the orientation of minor linear structures can be used for correlation. First, only initial orientations can be compared. If, for example, an area contains F_1 through F_3, then the present orientation of F_1 is probably of little value for correlation. The orientation of F_2's might be usable, however, if they all fall on a great or small circle with F_3 as the pole or axis. Second, the orientations of fold axes are influenced by the orientation of the surfaces that they fold. If two limbs of an early fold (e.g., F_1) are in different orientations relative to the stresses that produce the next fold generation (e.g., F_2), then the orientations of the fold axes on the two

opposing limbs will be different (Fig. 16-13). This effect can cause unexpected variations in stereoplot patterns as well as misidentification of the generation of structure. When early folds are tight to isoclinal, the effect of initial surface orientations on later folds is usually negligible.

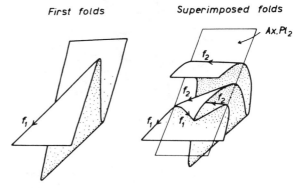

Figure 16-13. Effect of the orientation of first-fold (F_1) limbs on the orientation of second-fold axes (F_2). (From Ramsay, 1967.) (a) First folds with divergent limbs; (b) superimposed second folds. F_1 axis orientations differ on opposite F_1 limbs. AxPl$_2$ is the F_2 axial plane.

When correlating planar elements, it is also important to compare only initial orientations. The orientation of the last-formed S-surface generally shows minor variation in orientation, unless it is a crenulation cleavage or a cleavage that fans around folds. The orientation of a crenulation cleavage depends on the orientation of the surface that is crenulated. The orientation of cleavages in sedimentary rocks can be strongly affected by lithology; cleavages in competent beds tend to fan around folds and therefore have the same strike as the axial plane, but have different dips. If the last-formed S-surface is not a crenulation cleavage and does not fan on a mesoscopic scale, yet its poles scatter on a stereoplot, then there has probably been subsequent deformation.

Other factors that must be considered when using orientation of linear and planar structures for correlation of structural generations between outcrops include (1) the presence of faults that locally reorient folds, (2) the presence of rigid objects (e.g., plutons) that can locally perturb the stress system and thus the orientations of structures, (3) the formation of coeval cross- or conjugate-folds with two orientations of axes and/or axial planes, and (4) the *transposition* (reorientation by deformation) of earlier structures into parallelism with the later structures. These complicating factors can usually be detected, and where appropriate, the affected areas can be treated as separate domains.

Correlation Using Style

The *style* of a fold refers to all the morphological features of a fold, such as the interlimb angle, the fold shape in profile or in three dimensions, and the type of axial-planar foliation. When analyzed with down-plunge projection techniques, fold styles derived from the study of minor structures in different lithologies provide an additional constraint on regional-scale fold geometries at depth. Correlation of structures between outcrops using fold style assumes that folds belonging to a given generation share a characteristic range of styles that reflect the conditions of deformation. For example, all F_1 folds in a region can be isoclinal and have highly thinned limbs and thickened hinges, whereas all F_2 folds can be tight similar folds, and all F_3 folds can be open chevron folds (Fig. 16-14a). Fold style is, however, strongly influenced by lithologic factors (unit composition, layer thickness, and ductility contrast between layers) and variations in strain magnitude.

Several generalizations should be remembered when analyzing fold style. First, well-foliated rocks will tend to form angular, concentric folds regardless of generation or conditions of deformation. In the preceding example, well-foliated schists and phyllites will be likely to form chevron folds during both F_2 and F_3 (Fig. 16-14a and b), whereas massive quartzites will probably not form chevron folds during F_3 (Fig. 16-14c). Second, interlimb angle depends on both lithology and the degree of shortening; thickly layered and competent units will tend to form open folds regardless of generation. Third, fold tightness can also reflect the number of times the fold has been refolded and the location of the fold relative to later structures; the observed tightness does not necessarily reflect the original tightness. For example, first-generation folds are often isoclinal, second tight, and third open. Thus, in the preceding example, if F_1 folds were originally closed, present F_1 folds on limbs of F_2 folds and away from F3 folds can be closed or tight, whereas all others can be isoclinal. Fourth, fold shapes are affected by the orientation of the surfaces they fold. For example, the shape of an F_2 fold may be different on nonparallel limbs of an F_1 fold (Fig. 16-15).

One of the most useful aspects of style for correlation is the type of axial-planar foliation. For example, the axial-planar foliation of F_1 folds may be a slaty cleavage, whereas the axial-planar foliation of the F_2 folds may be a crenulation cleavage. Caution must be used in correlating folds using their associated foliations, for the formation and

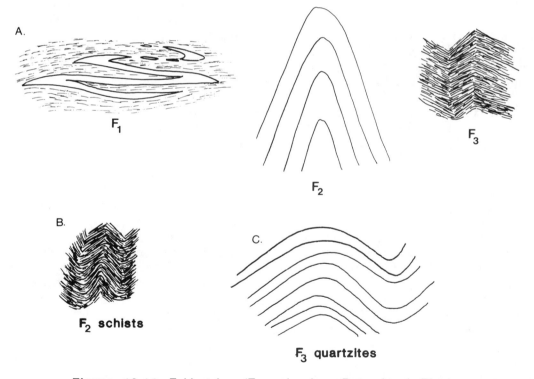

Figure 16-14. Fold styles. (Examples from Beaverhead, Rhode Island; Mosher, unpublished data.) (a) Typical styles for F_1, F_2, and F_3 folds; (b), (c) effect of lithology on fold style; (b) is example of F_2 fold style in schists (compare F_2 in (a)), (c) is example of F_3 fold style in thickly bedded quartzites (compare F_3 in (a)).

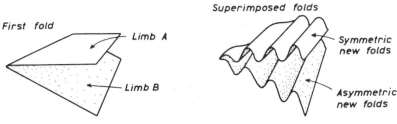

Figure 16-15. Effect of the orientation of first-fold limbs on the shapes of second folds. (From Ramsay, 1967.)

type of a foliation is influenced by lithology and the magnitude of strain. Crenulation cleavages, for example, are common in schists and phyllites and rare in quartzites, because the former contain a preexisting, closely spaced layering.

Problem 16-7

Broken Hill, Australia, is one of the leading lead-zinc-silver districts in the world. Ore minerals are found in poly-deformed Precambrian rocks. Original bedding is completely transposed, and the oldest planar element (labeled S_0) is a metamorphic compositional banding defined by quartzo-feldspathic layers and quartz-mica schist layers. The schistosity in the schist (S_1) is sometimes parallel to S_0, but not always. Figure 16-16 shows carefully drawn fold profiles from the area (Anderson, 1971). The geologist studying these folds divided them into groups on the basis of fold style and the relationship between folds and foliations.

(a) Study the sketches of fold profiles and describe completely the style of one fold from each group. Include the relationship between S_0 and S_1 and any other foliations.

(b) What criteria do you think the geologist used to divide the folds into the three groups? Do you disagree with any assignments? If so, why? In what order did the groups of folds form (i.e., which group is F_1)?

Correlation Based on Conditions of Deformation

Another, often neglected, method of correlating generations of structures is the grouping of structures by the conditions of deformation prevalent during their formation. Clues to deformation condition include (1) the type of rock fabric present and (2) the metamorphic grade indicated by minerals formed along foliation planes or that define lineations. The relative ages of structures can be constrained by recognition of the time(s) of mineral growth with respect to a structural element. Next, we briefly outline approaches used for correlation using conditions of deformation. To fully appreciate the possible complexity of rock fabrics in poly-deformed terranes requires specialized training (see books by Spry, 1969; Nicolas and Poirier, 1976; and Hobbs, Means, and Williams, 1976).

Under the same conditions of deformation, similar lithologies should contain similar fabrics. If, for example, metamorphic conditions during F_1 were upper amphibolite facies and during F_2, lower greenschist facies, then distinctly different fabrics would be associated with each fold generation. S_1 might be defined by aligned muscovite and sillimanite, and S_2 might be a pressure-solution cleavage. In different lithologies, S_1 and S_2 might be different (e.g., S_1 defined by aligned hornblendes and S_2 a crenulation cleavage with no associated new mineral growth in mafic schists), but in each lithology, S_1 and S_2 formed under a different set of metamorphic conditions. Thus, each fabric can be correlated between outcrops despite fabric differences in different lithologies. When conditions during two deformations are similar, such as during a phase of progressive deformation, or when successive metamorphic events cause partial or complete replacement of an earlier fabric by mimetic recrystallization (cf. Williams, 1985), correlation by fabric style will be difficult. In these situations correlation of minor structures on the basis of fabric type requires careful study of thin sections that show cross-cutting relationships and critical mineral assemblages. Deformation conditions can also be assessed by studying the deformation mechanisms (e.g., pressure solution versus dynamic recrystallization) affecting rocks during a given event. This too requires careful thin-section work.

Grouping of structures by their ages relative to periods of metamorphic mineral growth provides another way of correlating minor structures. Structures are classified as pre-, syn-, or post- a given metamorphism on the basis of whether, in thin section, metamorphic minerals are aligned along or cross-cut a foliation. This approach can yield detailed chronologies of conditions during deformation and fabric formation in successive deformational (and metamorphic) events. In some lithologies porphyroblasts may overgrow foliations and trap inclusions that provide a glimpse of an earlier fabric. The matrix to the porphyroblasts (particularly if quartz-rich) records the latest event(s). Many micaceous minerals will often define relict and incipient foliations.

When using deformation conditions to correlate generations of structures across an area, it must be remembered that different conditions can occur during a single phase of deformation, so that this is not a unique tool for correlation. Changes in metamorphic grade across an area, for example, may be accompanied by changes in

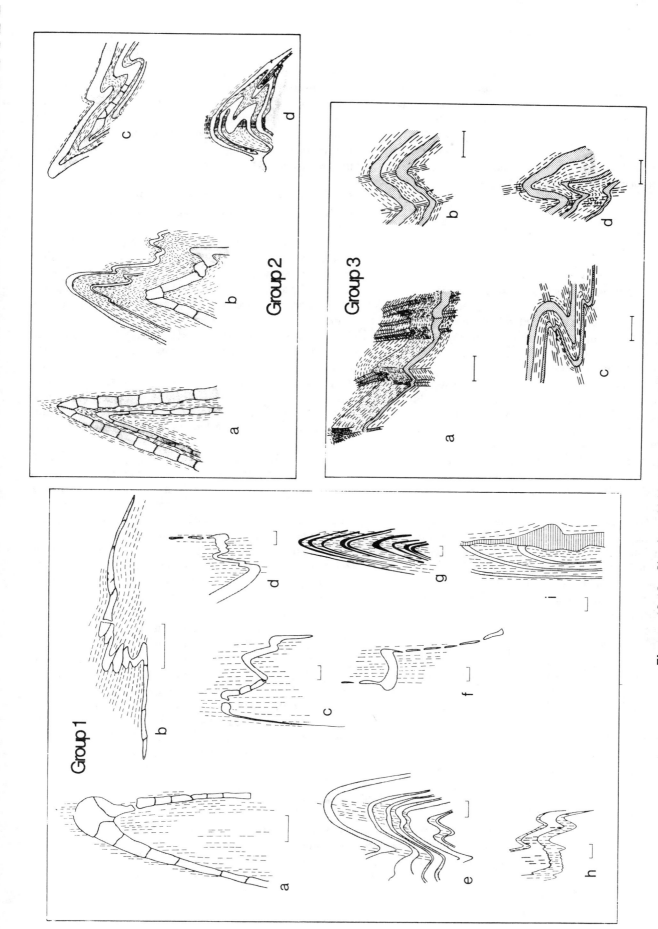

Figure 16-16. Sketches of fold profiles for three groups of folds from Broken Hill, Australia. (From Anderson, 1971.) S_1 is dashed. Bar scale is 0.3 m in length.

the type of rock fabric developed during a *single phase of deformation* if the deformation and metamorphism were contemporaneous. As another example, a fault may juxtapose rocks that were being deformed near the surface with rocks that were deformed at greater depths. Units on opposite sides of the faults may have experienced the same phases of folding, yet the conditions they experienced during deformation could be quite different.

16-7 POSSIBLE ORIGINS OF POLYPHASE STRUCTURAL PATTERNS

Polyphase Patterns Formed During a Single Phase of Deformation

In some regions rocks that display more than one generation of structures are the result of a single phase of nonprogressive deformation. For example, a type 1 interferencelike pattern may merely reflect differential shortening along the length of a fold or extension parallel to the length, which results in doubly plunging folds. A pattern formed from doubly plunging single-generation folds can be distinguished from a pattern formed from two overprinted folds by studying the foliations. The former pattern can have only a single axial-planar foliation, whereas the latter pattern may contain two. The case for two stages of folding will be further supported if it can be shown that the two fold generations formed at different times relative to metamorphism. *Sheath folds*, which are highly noncylindrical tongue-shaped folds, also display eye-type interference patterns but form in shear zones during a single stage of shearing. These folds do not reflect the overprinting of two noncoaxial folds. A sheath-fold pattern can be distinguished from a true interference pattern by studying the structural setting in which it occurs; sheath folds are commonly associated with other structures, such as mylonitic foliation, indicative of their position in a shear zone.

Two cleavages can form during a single phase of folding. Early, layer-parallel compression can produce a cleavage that is reoriented during continued folding. At a later stage in the folding, a second axial-planar cleavage may develop, which will overprint the first cleavage (cf. Boulter, 1979). Flattening of folds during the final stage

of a deformation event may reorient an early cleavage and generate a second cleavage. The absence of two fold orientations in a region of two cleavages is a clue that both cleavages formed during the same phase of folding. In addition, it is likely that there will be only one orientation for intersection lineations and that the cleavage planes themselves will not be folded.

Distinguishing Progressive Deformation from Time-Separated Deformation

One of the most difficult challenges that a geologist faces when working in poly-deformed terranes is determining whether multiple generations of structures resulted from a single phase of progressive deformation or discrete, time-separated phases of deformation. Often this task is impossible.

Progressive deformation is usually the result of a phase of shearing. Such a deformation can produce coaxial or noncoaxial structures depending on the orientation of layering with respect to the shear stresses. In areas involved in thrust-related ductile shearing, coaxial or nearly coaxial superimposed folding predominates. In ductile shear zones related to strike-slip or transpressive motion, noncoaxial folds form.

Progressive deformation can clearly be ruled out if minerals related to different fabrics can be dated, using isotopic techniques, as having formed at distinctly different times. If an area is known to have undergone more than one orogeny (or an orogeny is known from work in other areas to have distinct events), and different structures clearly formed under different conditions and/or at distinct times relative to metamorphic minerals, it may be possible to interpret (but not prove) that the deformations are the result of separate orogenies (or events). It is usually easier to determine whether a deformation is not progressive than whether it is. All the following criteria should be true for a deformation to be progressive: (1) all structural generations are formed under the same metamorphic conditions or show evidence of a gradual change in metamorphic conditions as the deformation progresses; (2) the age of all structures relative to metamorphic minerals will be the same or again show a gradual change; and (3) all generations of structures must be able to be produced by the same or a gradually changing stress field.

EXERCISES

Equal-angle plots are provided for the following exercises. To answer many of the questions, you will need to transfer some information onto an overlay. Trace the primitive great circle and label north on all overlays. For most exercises, you can draw a dashed line around the cluster of points in question or sketch in the average great circle or pole. A small equal-angle net at the same scale as the figures is provided as Figure 16-17.

1. Figure 16-18 is a photograph of a fold interference pattern in an outcrop of the Condrey Mountain Schist, northern California. Lay a piece of tracing paper over the photograph and sketch the white, quartz-rich layers. Draw and label the axial planes to all fold generations. How many generations of folds are there in this profile? What type of interference pattern does this appear to be? What additional information would you need in order to verify this conclusion?

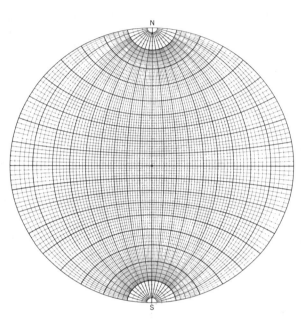

Figure 16-17. Wulff net for use with the exercises.

Figure 16-18. Fold interference pattern in the Condrey Mountain Schist, northern California.

2. The map in Figure 16-19a (Barberton Mountain Land, Eastern Transvaal, South Africa) shows a sequence of Archean slate, quartzite, metaconglomerate, metagreywacke, and metachert. The metagreywacke (unit 3) is the oldest unit in the area; slates (unit 1) and quartzites and metaconglomerates (unit 2) are interlayered. The outcrop pattern shows the interference of two generations of folds.

(a) Draw on the map the axial traces of the two fold generations. Label the earliest generation F_A and the latest generation F_B. Show whether they are anticlines or synclines and indicate the direction of plunge. What type of interference pattern is this? How do you know?

(b) Sketch and label the types of fold interference patterns (e.g., S on Z, and so forth) you would expect to see at map locations A, B, C, and D.

(c) When did the slaty cleavage form relative to F_A? F_B? What is your evidence?

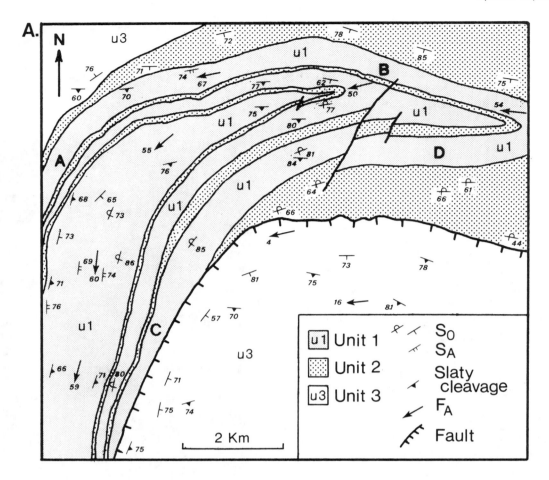

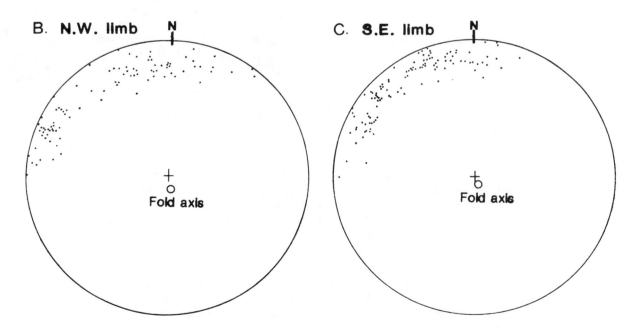

Figure 16-19.

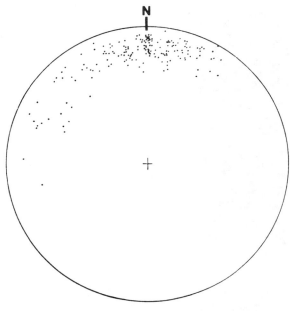

N

D.   **Slaty cleavage**

Figure 16-19. Barberton Mountain Land, Eastern Transvaal, South Africa. (Adapted from Ramsay, 1965.) (a) Geologic and structure map showing interference pattern with localities A, B, C, and D indicated. Units are 1-slate, 2-quartzite and metaconglomerate, and 3-metagraywacke and metachert; (b), (c) lower-hemisphere, equal-angle net for the northwest and southeast F_A fold limbs. Bedding poles define great circles; (d) equal-angle net showing poles to slaty cleavage.

(d) In Figure 16-19b and c, poles to bedding for the northwest and southeast limbs of F_A are plotted on separate stereoplots and define two great circles. The poles to these great circles are F_B axes. What is the angle between these axes? Why are the F_B axes on the two limbs not parallel?

(e) Poles to the slaty cleavage define a great circle (Figure 16-19d). Sketch in the average great circle and find its pole. Has this cleavage been folded by F_B? If so, would you expect the fold axis to coincide with those for the F_A fold limbs? Does it?

3. Figure 16-20 shows photographs of fold styles in Proterozoic units of north central New Mexico. The units in this area are multiply deformed schists and quartzites. The area has been divided into three structural domains. This exercise allows you to follow the actual steps taken when interpreting a poly-deformed region. Each domain will be analyzed separately and then compared to previously analyzed domains. Each new domain may provide information that changes earlier interpretations or clarifies remaining questions. When all domains are mapped, a final interpretation of the entire area is made.

(a) In the first domain, a pronounced metamorphic foliation (S_1) is parallel to bedding (S_0) and is folded by isoclinal, overturned to recumbent, second-generation folds (F_2) with an axial-planar cleavage (S_2) (Fig. 16-20a). Small, open F_3 folds and an axial-planar crenulation cleavage (S_3) clearly cross-cut and deform F_2 and S_2. Equal-angle plots of poles to S_0/S_1, S_2, and S_3, and of F_1, F_2, and F_3 axes are shown in Figure 16-21a. Most F_3 axes on the plot are measurements of crenulation axes on S_0/S_1. Some minor, open, chevron or box folds also fold S_0/S_1, but the relationships between these folds and F_3/S_3 cannot be determined. When plotted on a stereoplot their orientations overlap or are close to that of F_2; for this reason, they have been grouped with F_2 folds.

(i)   What is the orientation of the regional fold axis as defined by the great circle containing the poles to S_0/S_1?

a

b

Figure 16-20. Fold styles of Proterozoic rocks in north central New Mexico. (a) F_2 folds; (b) F_4 folds.

(ii) Poles to S_2 surfaces define a diffuse great circle, but most S_2 poles are concentrated in a point maximum within this great circle. The point maximum represents the predominant (or average) orientation of S_2 in the area. What is the orientation of the average S_2 plane? (Use the second stereoplot. The center of the area with the highest density of points will be most representative of the average orientation). Does this plane contain the minor F_2 fold axes? Does this plane contain the regional fold axes? Where do the minor F_2 folds plot relative to the regional fold axis? What do you think is the generation of the regional fold?

(iii) Now find the best-fit great circle to the S_2 poles. (Use the second stereoplot). What is the orientation of the pole to this great circle? This represents the axis about which the S_2 surfaces are refolded. Does it plot within the cluster of minor F_3 fold axes? Remembering that F_3 fold axes are mostly crenulations of S_0/S_1, how can you explain the divergence of this axis and the measured F_3 fold axes?

(iv) F_3 axes and S_3 poles also show a large amount of scatter. If S_3 surfaces were folded, the fold axis would be the pole to the best-fit great circle. What is the orientation of that pole? Remembering that S_3 is a crenulation cleavage, what are some other possible causes of the scatter?

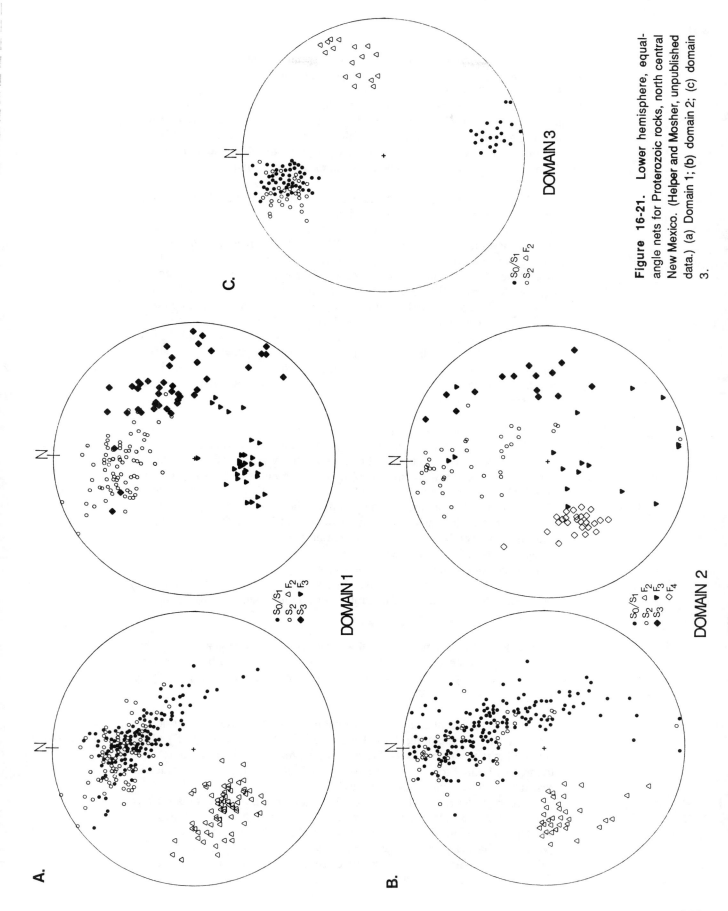

Figure 16-21. Lower hemisphere, equal-angle nets for Proterozoic rocks, north central New Mexico. (Helper and Mosher, unpublished data.) (a) Domain 1; (b) domain 2; (c) domain 3.

DOMAIN 1

DOMAIN 2

DOMAIN 3

• S_0/S_1 △ F_2
○ S_2 ▼ F_3
◆ S_3

• S_0/S_1 ▼ F_3
○ S_2 ◇ F_4
△ F_2

• S_0/S_1 △ F_2
○ S_2

A.

B.

C.

(b) In a nearby domain (domain 2), the same structural elements (Figure 16-21b) show the same styles. In this domain, however, open chevron or box folds of S_0/S_1 locally fold S_3 surfaces, indicating that they are F_4 folds (Fig. 16-20b). (Measurements of F_4 folds are from F_4's that fold S_0/S_1.) These folds overlap in orientation with the isoclinal, overturned to recumbent F_2 folds, but the F_4 folds generally trend S60°W, whereas the F_2 folds generally trend east-west. A cross section of this domain requires two styles of large-scale folds, which mimic F_2 and F_4 minor fold styles, to describe the variations in strike and dip of S_0/S_1.

(i)   What is the orientation of the regional fold axis defined by the great circle containing the poles to S_0/S_1 in this domain?

(ii) Poles to S_2 surfaces also define a very diffuse great circle, within which is an equally ill-defined point maximum. What is the orientation of the average S_2 plane? (Find this plane using the method given above in part (ii) of (a); this plane should contain the minor F_2 fold axes.) Does it contain the regional fold axis? Where do the minor F_2 folds plot relative to the regional fold axis? minor F_3 folds? minor F_4 folds? To what generation(s) of structure(s) do you think the regional folds belong? Why?

(iii) The amount of scatter of S_2 poles on the stereoplot reflects both F_3 and F_4 folding. Find the best-fit great circle to the S_2 poles. What is the orientation of the pole to this great circle? To which minor fold axes, F_3 or F_4, does it plot closest? Does this answer fit the information given about the large-scale folds in this domain?

(iv) What is the orientation of the pole to the great circle containing poles to S_3? Why does this pole diverge somewhat from the measured F_4 fold axes?

(v) Compare the pole to the great circle defined by poles to S_3 in domain 1 with the orientation of the minor F_4 fold axes in domain 2. Now, what do you think is the most likely cause of the scatter in poles to S_3 in domain 1? To what generation of structure do you think the open, chevron to box folds observed in domain 1 belong? If you could go out into the field, what would you check to verify this conclusion, remembering that the relationship to F_3/S_3 in domain 1 cannot be determined?

(vi) After comparing domains 1 and 2, what generation do you think the regional-scale fold in domain 1 is? In domain 2? If they are of the same generation, what common structural feature could cause the difference in orientation?

(c) A third domain is several miles northwest of the other two domains. Here few minor structures are observed, but many regional-scale, east-northeast-plunging folds are defined by the outcrop pattern. Stereoplot data are shown in Figure 16-21c.

(i)   What is the orientation of the regional fold axis in this domain defined by the great circle containing the poles to S_0/S_1? Does the average S_2 plane contain the regional fold axis? Where does the regional fold axis plot relative to measured minor fold axes? Are the regional folds of the same generation of structures as S_2 and measured fold axes?

(ii) Compare the regional fold axis orientation and the minor folds for this domain to minor and regional fold orientations in the other two domains. To what generation of structure do you think the regional folds in domain 3 belong? What is a possible cause for the change in plunge direction?

(iii) Using the information from all three domains, give the sequence of events, including the style and orientations of the structures.

4. Beaverhead and Bonnet Shores are two small areas, approximately 2 km apart, within the Naragansett Basin of Rhode Island. The basin metasediments are polydeformed, and not all structural elements or generations are found at both localities.

(a) In the Beaverhead area, graphitic schists and quartzites show rare, isoclinal F_1 folds of bedding that have an associated well-developed axial planar schistosity (S_1). An S_2 crenulation cleavage of S_1 and small mesoscopic F_2 folds are observed locally. F_3 folds deform S_1 into upright box and chevron folds and reorient the S_2 cleavage. An associated axial-planar crenulation cleavage (S_2) is locally present. Sketches of F_1, F_2, and F_3 fold styles are shown in Figure 16-14.

(i) On a stereoplot (Fig. 16-22a), bedding poles define a weak great circle girdle that is about 90° from F_1, and both bedding and S_1 poles fall on a great circle 90° from F_3. Minor F_2 folds (oriented N10°W,00°) have little effect on the distribution of S_0 and S_1 poles and have been omitted for clarity. What type of interference pattern would cause the angular relationships shown by the stereoplot?

(ii) Assuming that no other deformation occurred after F_3, why do the poles to S_3 plot in three clusters? (*Hint:* Think about the style of F_3 folds.)

(iii) In this area, F_2 and F_3 fold styles are identical in the schists. What two features could you use to distinguish them when mapping?

(b) At Bonnet Shores, there are four distinctly different and cross-cutting cleavages and four generations of nearly coaxial folds in schists and quartzites. The latter preserve younging indicators such as graded- and cross-bedding. F_1, F_2, and F_4 are shown in your completed version of Figure 16-10. F_3 folds (not shown) are small-amplitude, closed folds with axial planes that fan around the F_4 folds. S_1 is defined by aligned biotite, muscovite, and elongate quartz grains, S_2 is defined by

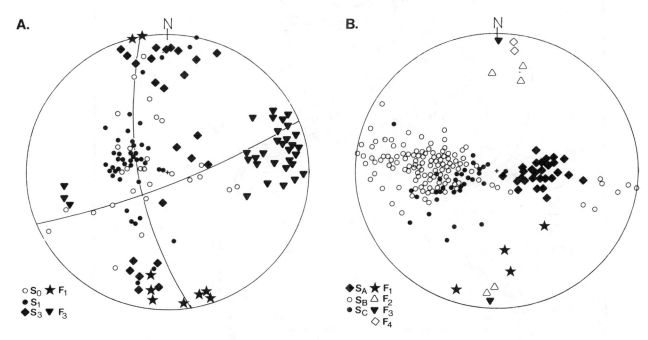

A.

○ S_0 ★ F_1
● S_1
◆ S_3 ▼ F_3

B.

◆ S_A ★ F_1
○ S_B △ F_2
● S_C ▼ F_3
◇ F_4

Figure 16-22. Lower-hemisphere, equal-angle nets of (a) Beaverhead (from Burks, 1981) and (b) Bonnet Shores, Rhode Island. (From Reck, 1985.)

muscovite and biotite, and S_3 is a crenulation cleavage. S_1, S_2, and S_3 were folded by open, monoclinal folds (F_4) with no associated axial-planar cleavage. A crenulation cleavage oriented approximately east-west clearly postdates S_3 and changes orientation across F_4 fold limbs.

(i)   The poles to all cleavages fall on the same great circle (Fig. 16-22b) and cannot be separated. Does this require that the folds be coaxial?

(ii)   The different amounts of scatter of the poles for the three cleavages reflects the relative ages. Indicate the age (i.e., S_1, S_2, and S_3) of S_A, S_B, and S_C (they are not in chronological order) on the stereoplot. Why is this technique for sorting generations usable in this area, even though S_3 is a crenulation cleavage?

(iii) The east-west striking crenulation cleavage that postdates S_3 could be folded by F_4 because it changes orientation across F_4 fold limbs. Could the cleavage have formed after F_4? Why or why not?

(iv)   Using style, orientation, and deformation conditions, which, if any, of these generations could (or are likely to) have formed during a phase of progressive deformation? What are your reasons?

(c)   Construct a table showing style and orientation by generation for both Beaverhead and Bonnet Shores. Use this table to correlate the different generations (e.g., F_*, S_* at Beaverhead correspond to $F_\#$, $S_\#$ on Bonnet Shores).

5.   A portion of the Connecticut Valley Synclinorium near Strafford Village, Vermont, is shown in Figure 16-23a. The inset in the lower right corner is a map

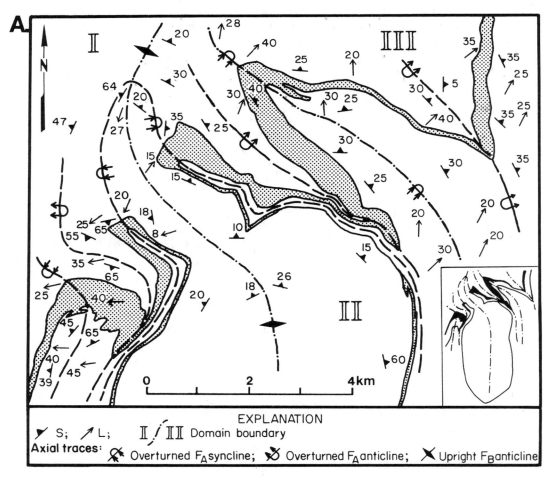

Figure 16-23.

B.

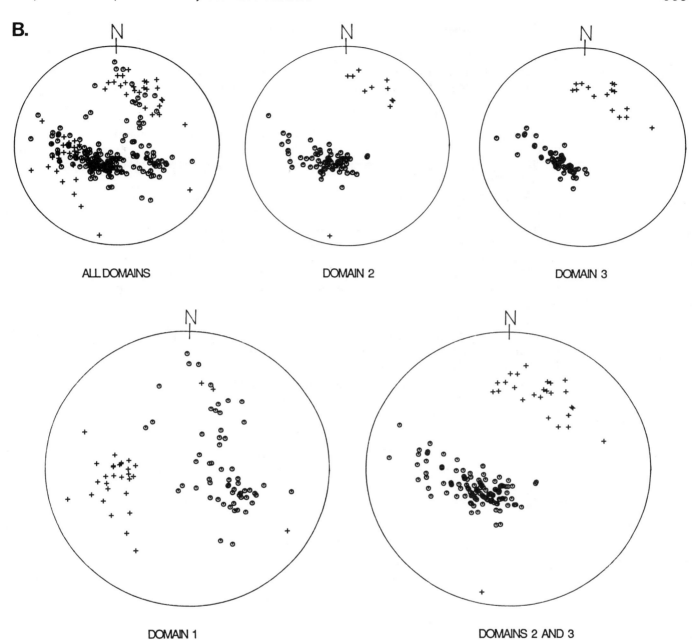

ALL DOMAINS   DOMAIN 2   DOMAIN 3

DOMAIN 1   DOMAINS 2 AND 3

Figure 16-23. Strafford Village area, Vermont. (a) Structure map showing interference pattern; one marker bed shown. (Adapted from White and Jahns, 1950.) (b) **Lower hemisphere,** equal-angle nets showing L (crosses) and poles to S (dots) for entire area and each domain. **Note domains II and III are similar and can be merged to form a single domain.**

of the regional fold pattern for the same units. The units define an interference pattern. Axial traces of the two fold generations are shown (F_A is older than F_B); a single, upright F_B anticline separates a series of overturned F_A anticlines and synclines. Also shown are the orientations of the prominent schistosity (S) and an intersection lineation (L) parallel to the axes of F_A minor folds. The area is divided into three structural domains whose boundaries are parallel to fold axial traces.

(a) In what direction do the axial planes of F_A folds dip in domains II and III? In domain I?

(b) Note the change in trend and plunge of the intersection lineations. In what general direction do the F_A folds plunge in domains II and III? In domain I?

(c) In what direction(s) does the F_B anticline plunge? Look at the outcrop pattern shown on the inset map as well as the structural data.

(d) Fold a piece of paper to produce the interference pattern shown. Use the outcrop pattern, the axial traces, and the structural data. What type of interference pattern do you think F_A and F_B form?

(e) If the schistosity shown on the map is S_1, are the F_A axial traces F_1 folds or F_2 folds? Explain why.

(f) The map pattern and data indicates that another phase of folding (F_C) occurred after formation of the F_A-F_B interference pattern. Give two pieces of evidence.

(g) Figure 16-23b shows stereoplots for each domain. Domains II and III show equivalent patterns and can be merged, whereas domain I shows a distinctively different pattern. What is the orientation of the pole to the great circle containing most of the S-surfaces in domains II and III? In domain I? What is the acute angle between them? Domains II-III and I lie on opposite limbs of a large F_B. If the primary influence on the orientation of S was F_B, then the poles to these two great circles should coincide. They do not, however; thus domain II-III (or domain I) shows the effects of F_A and possibly F_C, and the angle between the poles to the great circles for domains I and II-III is the present interlimb angle for F_B. (*Note:* F_A folds S.) For domain II-III, what is the angular relationship between F_A and S? Is this the correct relationship for only F_A folding? For domain I?

PART

III

APPENDICES

This part includes four appendices. The first provides a concise review of the basic concepts of maps, cross sections, and diagrams and thus can serve as an introduction or a refresher to the chapters of Part I. Appendix 1 also serves as a basic reference to such topics as calculation of map scales and interpretation of map grids. Appendix 2 provides the basic trigonometric functions that are used in Chapters 3 and 4. Appendix 3 provides a summary of suggestions concerning the mapping of geologic structures and can serve as part of an introduction to field trips held in conjunction with a structural geology course. Appendix 4 provides templates for use in plotting geologic data; the pages are perforated so that the templates can be easily removed from the book.

APPENDIX

1

REVIEW OF THE KEY CONCEPTS OF MAPS, CROSS SECTIONS, DIAGRAMS, AND PHOTOS

A1-1 INTRODUCTION

The collection of data is only one step in a geologic study. To complete a project it is necessary to communicate your observations and their interpretation to an audience of interested geologists. For many purposes such communication is done through the use of maps, cross sections, photos, sketches, and block diagrams. In this appendix we review the key components of these tools. This basic information may be familiar to you from a previous course in geology or geography; our purpose here is to refresh your memory so that we can use the information throughout this book.

A1-2 ELEMENTS OF MAPS

Types of Maps

A *map* is a two-dimensional plane on which information about the earth's surface, a portion of the earth's surface, or a portion of the earth's subsurface are displayed (Fig. A1-1). Typically, a map provides a *plan-view*, meaning that the map plane is considered to be horizontal. The discipline that studies ways in which maps can be created and used is called *cartography*.

A variety of types of maps are used in geology, each of which is intended to display a specific class of information. The most commonly used types of maps include the following:

1. *Topographic maps* (landforms)
2. *Surface geology maps* (alluvial deposits and soil)
3. *Bedrock geology maps* (rock units)
4. *Tectonic maps* (interpretation of the distribution of geologic provinces)
5. *Fabric maps* (pattern of foliations and lineations)
6. *Structure contour maps* (shape of a structurally significant surface)
7. *Isopach maps* (variations in unit thickness)

Because a map is a plane, the intersection of a planar geologic feature with a map is a line. Such a line is called a *trace*. For example, fault planes or contacts appear as traces on a geologic map (Fig. A1-1b). The pattern of contacts, fold-hinge traces, and fault traces can convey information on the shape of structures in three dimensions. Such patterns can be emphasized by plotting attitude data (e.g., strike and dip symbols) on the map. A skilled map reader can quickly visualize the shape of geologic structures in a region, the so-called *structural geometry* of a region, by studying a map. There are no set rules as to what information can be included on a given map, but the author of the map should avoid plotting so much data that there will be overlap of symbols.

All maps represent some level of interpretation, so there is no such thing as a "right" map. There are "good" maps, however, which are defined as maps that accurately portray spatial relationships among features in the map area. For example, if a map indicates that Wilkerson Formation occurs at the junction of Highways 4 and 6,

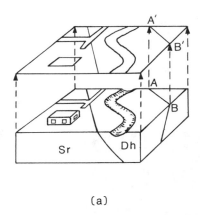

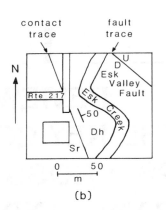

(a)   (b)

Figure A1-1. Map representation of features on the surface of the earth. (a) Projection of features from a block of the earth onto a map plane; (b) a simple geologic map. Note that planar geologic features, such as the fault and the contact, appear as traces on the map.

then an outcrop at this junction should be composed of Wilkerson Formation. If you visit this locality and find outcrops composed of Hodder Limestone, then the map is wrong.

Latitude and Longitude

A line of *longitude* is a line on the surface of the earth that passes through the two geographic poles of the earth and is a *great circle*. A great circle, by definition, represents the intersection between a sphere and a plane that passes through the center of the sphere. On a given sphere the lengths of all great circles are the same (Fig. A1-2a). Lines of *latitude* are parallel to the equator of the earth and are perpendicular to lines of longitude. With the exception of the equator itself, lines of latitude are *small circles*. A small circle represents the intersection of a sphere with a

plane that does not pass through the center of the sphere (Fig. A1-2b). Small circles also represent the intersection of a cone with a sphere (if the tip of the cone is at the center of the sphere), and the intersection of two spheres. The length of each successive line of latitude decreases from equator to pole. Note that it is possible to draw great or small circles that are oblique to the earth's poles, but these circles are not latitude or longitude lines.

The lines of latitude and longitude define a coordinate grid on the earth's surface called a *graticule* (Fig. A1-3). The location of any point on or near the earth's surface can be indicated by specifying its latitude and longitude and its distance above or below a reference plane (usually mean sea level) as measured along a vertical line. Both latitude and longitude are measured in degrees. Longitude is measured as a number between 0° and 180° east or west of Greenwich, England. Lines of longitude are also called *meridians;* the meridian passing through Greenwich is called the *prime meridian.* Latitude is measured as a

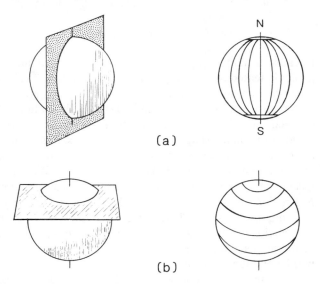

Figure A1-2. Meaning of latitude and longitude. (a) Lines of longitude created by great circles that pass through the poles; (b) lines of latitude created by small circles parallel to the equator.

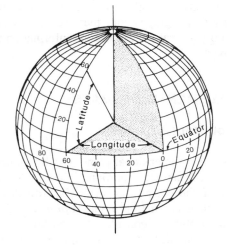

Figure A1-3. The graticule defined by lines of latitude and longitude. (Greenhood, *Mapping,* © 1964, p. 9. Adapted by permission of The University of Chicago Press, Chicago, Illinois.)

number between 0° and 90° north or south of the equator. Lines of latitude are also called *parallels*. For more precise location specification, degrees can be subdivided into minutes (1° = 60'), and minutes can be subdivided into seconds (1' = 60").

Because meridians merge at the poles, the distance between meridians varies as a function of latitude. At the equator, 1° of longitude is about 111 km, at the latitude of New York, it is about 82 km, and at the pole, it is, of course 0 km. The distance between parallels remains almost constant with latitude, and this distance is about 111 km (there is a variation of about 0.5 km due to flattening of the earth at the poles by centrifugal force). In general,

$$1° \text{ longitude} = 1° \text{ latitude} \times \cos \text{ latitude} \qquad (\text{Eq. A1-1}).$$

Time zones are roughly 15° of longitude wide (360°/24 h = 15°), although there are many local irregularities in time-zone boundaries that have been created for political reasons.

Map Projections

A projection is a representation of a three-dimensional surface on a two-dimensional sheet. The shadow that you cast on a wall when you stand in the sun is a familiar example of a projection. The outline of your shadow is the boundary between rays of the sun that intersect your body and rays that do not. Maps are projections; if the map area is small enough, a map projection can be made by passing an array of parallel vertical lines through the features of a map area to where they intersect a sheet of paper (Fig. 3-1). Creation of maps covering larger portions of the earth's entire surface is more difficult because the earth is a sphere. A straight-ray projection of the earth onto a plane, for example, would create a circle. On such a projection, features from opposite hemispheres would be superimposed and there would be great *distortion* (change in shape) along the edge of the projection. A great variety of clever map projections have been invented in order to decrease the

distortion. These projections differ from one another in the type of surface onto which the map is projected and in the configuration of the projection rays. There are three basic types of projection surface (*cylindrical, conical,* and *planar;* Fig. A1-4) and two basic ray configurations (*point source* and *parallel;* Fig. 3-1). The shape and/or area of features on the surface of the globe are distorted to some extent in all these projections, but distortion can be minimized by carefully choosing the projection type that is best suited to the scale and purpose of the map.

One way to produce an undistorted map of the world would be simply to cut up and flatten out the surface of the globe. Such a map would have large discontinuities. Perhaps the most common projection of the entire globe is the *Mercator projection* (Fig. A1-5a), which is a type of cylindrical projection. For a Mercator projection the cylindrical projection plane is parallel to the spin axis of the earth and is tangent to the earth's surface at the equator. The rays used to construct a Mercator projection emanate from a point source at the center of the earth. Adjustments are made to the map to reduce distortions of shapes in high latitudes. Nevertheless, the area of regions in high latitudes is greatly enlarged. For example, on a Mercator projection Greenland appears to be as large as the United States, although in reality it is much smaller. The Mercator projection has the advantage that longitude and latitude lines and any compass direction appear as straight lines, but on a Mercator projection the trace of a great circle on the earth is not a straight line.

Maps that portray the entire United States are commonly drawn using a *Lambert conic projection* (Fig. A1-5b). The projection surface for this projection is a cone whose apex lies in space high above the North Pole. The surface of this cone passes inside the surface of the earth.

Most U.S.G.S. topographic quadrangle maps, which are the standard base map for geologists working in the United States, and Ordnance Survey maps, which are the standard base for geologists working in Great Britain, are constructed on a *Universal Transverse Mercator* (UTM) projection. For a UTM map, the projection cylinder is oriented so that its axis is perpendicular to the spin axis of

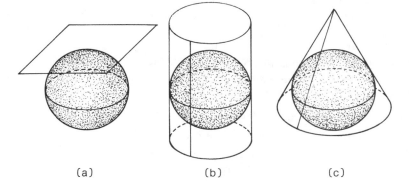

Figure A1-4. The three types of projection surfaces. (a) Planar; (b) cylindrical; (c) conical.

(a)   (b)   (c)

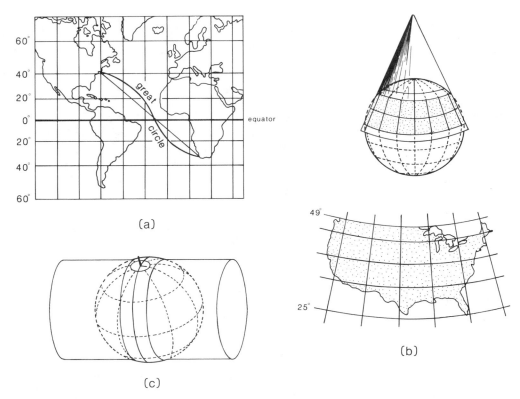

Figure A1-5. Commonly used map projections. (a) Mercator projection, showing the latitude and longitude grid across the Atlantic region. Lines of latitude are labeled. The curved line represents the trace of a great circle; (b) Lambert conic projection. The top part of the figure shows the projection cone intersecting the earth. Shaded interval represents the band in which the projection cone lies beneath the earth's surface. The lower part of the figure shows the grid across the United States; (c) universal transverse Mercator (UTM) projection showing the projection cylinder intersecting the earth. (Greenhood, *Mapping*, © 1964, pp. 130, 139, 134. Adapted by permission of The University of Chicago Press, Chicago, Illinois.)

the earth. The diameter of the UTM projection cylinder is slightly less than the diameter of the globe; therefore, the cylinder cuts the globe and defines a north-south trending ring (Fig. A1-5c). The diameter of the cylinder is chosen so that the ring, which is called a *zone*, is 6° of longitude wide at the equator. Definition of each zone requires a rotation of the projection cylinder around the earth's spin axis. Distortion of areas in the United States on a UTM map is much less than on a standard Mercator projection.

Additional discussion of map projections is available in standard cartography or geography texts (e.g.,Greenhood, 1964; Robinson and Sale, 1969; Raisz, 1962).

Map Scales

In order to represent dimensions and positions of features, a map must include a scale. The *scale* of a map is the ratio between the distance separating two points on the map and the distance separating the same two points in the real world. This ratio is also called the *representative fraction* (RF) or the *scale factor*. For example, if the distance between outcrop A and outcrop B is 10 cm on the map, and the RF of the map is 1:10,000, then the distance between the two outcrops in the real world will be 100,000 cm, or 1 km. Alternatively, if the two outcrops are separated by 1 km in the real world, and the RF of the map is 1:10,000, then the distance between those two outcrops on the map will be 0.00001 km, or 10 cm. If a map is drawn at a scale of 1:1, then the map will be the same size as the area that it is intended to represent.

Problem A1-1

The distance between Illyria and Elysium on a map is 2 in. The RF for the map is 1:24,000. How far apart, in miles, are the two localities?

Method A1-1

The true distance (D) between the two localities is

2 in. X 24,000 = 48,000 in.

48,000 in./12 = 4000 ft

1/5280 = D/4000

D = 0.758 mi.

Clearly, the selection of an appropriate scale is an important step in the production of a map. For example, imagine a small quarry that is 100 m X 100 m in area. On a map whose scale was 1:1, the quarry would be 100 m X 100 m. It would be impossible to find a piece of paper large enough to permit you to draw such a map. Alternatively, if the map scale was 1:1,000,000, the quarry would be invisible and it would be impossible to display the geologic features that occur in the quarry. The geologist studying the quarry could, however, map the quarry at a scale of 1:1000. At this scale, the quarry would occupy an area of 100 cm X 100 cm, a reasonably sized piece of paper.

Problem A1-2

How large a sheet of paper is required to produce a 1:5000 map of a structure whose dimensions are 1 km X 1 km?

Method A1-2

1 km = 1000 m. At a scale of 1:5000, 1000 m will cover a distance of

1000 m/5000 = 0.2 m = 20 cm.

There is frequently confusion about the terms *large scale* and *small scale*. A *large-scale* map is a map the shows a lot of detail in a small area, and a *small-scale* map is one that includes such an extensive area that only general features can be shown on the map sheet. For example, a 1:1000-scale map is a larger-scale map than a 1:1,000,000-scale map. A 1 in. = 1 mi map is a larger-scale map than a 1 in. = 20 mi map. Of course, *large* and *small* are relative terms; a scale that is considered large for one purpose may be considered small for another purpose. Confusion arises when someone uses the term *large-scale study* to refer to the study of a large area, because, as we just noted, a large-scale map generally covers only a small area. To avoid this confusion, we suggest using the terms *broad-scale* or *regional-scale* with reference to studies of a large region, and *local-scale* with reference to studies of a restricted area.

The scale of a map can be indicated in several ways.

First, the representative fraction (e.g., 1:1000) can be written in the corner of the map. Second, a *scale bar* (Fig. A1-6) can be drawn, which translates map distances into real-world distances. Third, the scale can be indicated by latitude and longitude lines or some other survey grid (see below), assuming the map user remembers the translation factor between degrees and kilometers. *Topographic quadrangle maps* for the United States are prepared by the U.S.G.S. and generally come in one of two scales: 7.5' (minute) maps are 7.5' of latitude or longitude on a side, and are usually prepared at a scale of 1:24,000. 15' maps are 15' of latitude or longitude on a side and are usually prepared at a scale of 1:62,500. Recently, U.S.G.S. maps have been issued that use metric measurements and are at a scale of 1:25,000.

Figure A1-6. A scale bar. It is common to subdivide the bar into smaller units to the left of the zero point.

It is possible to translate from one scale system to another just by using simple arithmetic. For example, a scale of 1:63,360 is equivalent to 1 in. = 1 mi, because there are 63,360 in. in a mile. Remember, on small-scale maps, the area shown may be so large that, because of earth curvature and the nature of the map projection, the scale may vary as a function of position on the map.

Problem A1-3

The representative fraction for a map is 1:8,000. At this scale, how long is a scale bar that represents 1 km?

Method A1-3

1 km/8000 = 0.000125 = 0.125 m = 12.5 cm.

Distance Measurements on Maps

Distances on the surface of the earth or elevations above or below the surface of the earth are usually measured by either the English system (inches, feet, miles) or the metric system (centimeters, meters, kilometers). Table A1-1 is a conversion chart between these two. In the English system, a *statute mile* is arbitrarily asigned a value of 5280 feet. The distance around the earth at the equator is 24,902 miles; using Table A1-1, this can be converted into 40,075 km. Navigators often use a different measure called a *nautical mile*, which is 1 second of arc as measured at the equator, and thus is equal to about 6080 ft. A *kilometer* (1000 m) is defined to be 1/10,000 of the distance between the equator and the pole as measured on a meridian.

Table A1-1
Unit Conversions

1 km	=	3281 ft	=	0.6214 mi
1 cm	=	0.0394 in.	=	0.0328 ft
1 m	=	39.37 in.	=	3.281 ft
1 ft	=	30.48 cm	=	0.3048 m
1 in.	=	2.540 cm	=	0.0254 m
1 mi	=	1609 m	=	1.609 km

miles are statute miles

Because metric measures are always divisible by 10, they are much easier to work with. Many maps in the United States, however, are designed to be used with English measurement.

Problem A1-4

What is the width of a single time zone at a latitude 40°? Express your answer in both kilometers and miles.

Method A1-4

At a latitude of 40°, 1° lat = 111 km. Remember, the distance represented by a degree of latitude is constant.

$$1° \text{ long} = 111 \text{ km} \times \cos 40° = 85 \text{ km}$$

$$1 \text{ time zone} = 15 \times 85 = 1275 \text{ km}.$$

To express the answer in miles:

$$1 \text{ mi}/1.61 \text{ km} = X \text{ mi}/85 \text{ km}$$

$$1° \text{ long} = 0.62(85) = 52.8 \text{ mi}$$

$$1 \text{ time zone} = 15 \times 52.8 = 792 \text{ mi}.$$

Geologic Map Symbols

All maps should contain a *legend, explanation,* or *key.* These terms are different names for the table that describes the symbols used on a map. Symbols are used on geologic maps to indicate the attitude of rocks and structures (e.g., bedding, foliation, and joints; Figs. 1-9 and 1-13), the positions of structural features (e.g., faults, fold hinges, and unconformities), and the distribution of rock units.

A sequence of rocks that can be recognized and identified throughout a map area is called a *map unit.* A unit should have a definable top and bottom. The region on the ground in which outcrops of a map unit are exposed is called an *outcrop belt.* A map unit can be thick enough to be represented on a map by a visible band which is labeled, colored, or patterned, and may correspond to a formally defined geologic unit such as a formation, a group of formations, or a tectonic assemblage. On some maps a map unit may merely be a layer of a distinctive lithology that can be traced across the countryside. Such a distinctive traceable layer is called a *marker horizon.* The width a marker horizon may be exaggerated on a map.

On colored geologic maps, different colors may be used to indicate different map units, whereas on black-and-white maps, these distinctions are usually made by using different patterns or shadings. On maps produced in the United States, an abbreviation indicating a unit name is often written directly on the map within the outcrop belt of the unit (e.g., Db for Devonian Becraft Formation). The first letter of the abbreviation indicates the age, and the second letter (and third, if necessary) indicates the formation name (Fig. A1-7). On many European maps the identity of the pattern is specified in the legend by a number, and the unit name and description are given in a figure caption. In general, units are listed in an explanation in order from youngest to oldest, with the youngest unit at the top of the explanation.

Map Reference Frames and Survey Grids

A map is of little use if it cannot be oriented with respect to the real world. The reference frame on a map can be indicated by simply drawing a north arrow on the map. The reference frame is also provided by the locations of landmarks (e.g., roads, towns, mountain peaks, and rivers). *Benchmarks,* which are surveyed location points whose position and elevation are accurately known, can also be shown. On the ground, benchmarks are indicated by a small circular brass plate that is cemented into place, and on a map they are indicated by a small *x* with the abbreviation *BM* and the elevation written next to it (e.g., BM 2135). Most maps also contain an oriented grid, which is a network of lines whose orientation and position have been accurately determined by surveying.

If you have ever used a detailed map, you will notice that it contains a latitude and longitude grid and one or more additional grids. The other grids, which cover only local areas, are called *survey grids* and have been set up by municipalities, states, or countries for the purpose of surveying. A survey grid is simply an array of mutually perpendicular lines whose position has been determined by accurate surveying (Greenhood, 1964). Survey grid positions are usually determined by triangulation and are ultimately pinned to lines of latitude and longitude.

On U.S.G.S. quadrangle maps it is common to find a UTM grid. This is a rectangular grid keyed to the meridian at the center of each 6° zone on a UTM projection. The convention for specifying a location in a UTM grid is as

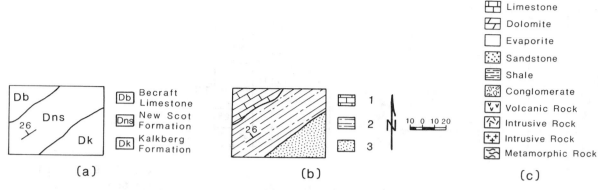

Figure A1-7. Representation of geologic units on a map. (a) American convention; (b) European convention; (c) common lithologic symbols.

follows: The value of the abscissa is called an *easting*, and the value of the ordinate is called a *northing*. Eastings increase to the east and northings to the north. UTM grid lines are specified by a number in meters; the central meridian of each zone is assigned an easting value of 500,000 m, and the equator is assigned a northing value of 0 m. On the east-west map border the UTM reference marks are labeled in the form $^3 10^{000}$ (Fig. A1-8a). This symbol refers to an easting of 310,000 m (i.e., 190,000 m west of the zone's central meridian). On the north-south map border, reference marks in the United States are labeled in the form $^{48} 80^{000}$. This is a northing of 4,880,000 m (i.e., 4880 km north of the equator). The coordinate of the point in the map area that lies at the intersection of these two grid lines can be specified simply by the number 1080. The first two digits indicate the easting, and the second two indicate the northing. The small initial numbers are not needed if you are referring to a known map area. To specify points that are not precisely on the intersection of two grid lines, additional digits can be added. For example, a point P with UTM grid coordinates of 104807 lies to the northeast of point 1080 (Fig. A1-8b).

A second grid found on many maps is defined by the *State Plane-Coordinate System* (SPC), which was set up in the 1930s by the U.S. Coast and Geodetic Survey. Each

state is covered by its own rectangular grid, measured in feet, which is designed to accommodate for the shape of the state, and which is pinned to national survey marks. Many states are divided into zones, each with its own grid. Grid lines in this system are 10,000 ft apart and increase in value to the north and to the east from a surveyed base line.

Land-office grids were created by the federal land office (a federal agency administering public lands) in the 1930s to represent the results of a nationwide land survey. All states, except for 19 eastern and southeastern states, were included in this survey and have public lands. On the basis of this survey, land is divided into blocks called *townships*, which are 6 mi on a side (Fig. A1-9). An east-west row of townships is a *tier*, and a north-south column is a *range*. In order to accommodate the curvature of the earth, the trace of a range line steps over a little where it crosses a tier line, at every fourth range line. Townships are identified by their row and column, with rows increasing in value northward and columns increasing in value eastward. Counting is done with reference to an east-west *baseline* and a north-south *principal meridian*. There are 31 pairs of these reference lines in the United States. A specific township is labeled, for example, T.3N., R.3E., or, in words, Township 3 North (the word tier is not used), and Range 3 East. Townships are further subdivided into

Figure A1-8. A portion of a UTM grid. The eastings and northings to locate point P are indicated. (a) The 6° zones; (b) Subdivisions in a zone.

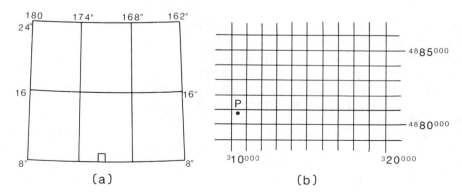

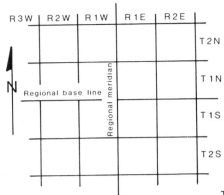

T = township; R = range

Figure A1-9. Township and range designation.

36-mi^2 blocks called *sections*. These are numbered starting from the northeast corner; to count off progressively increasing numbers, first count west along a row, then drop down a row and count east, and so forth.

A1-3 PATTERNS OF SIMPLE GEOLOGIC STRUCTURES ON MAPS

In first-year geology courses most students have the opportunity to see the map patterns of simple structures. Next we review the map patterns of simple structures as they would appear on a surface of no topographic relief, to refresh your memory. As you read the following material, visualize the attitude of the planes and lines that are described so that you can develop a sense for what any structural geometry will look like when projected on a plane.

Homoclinal Strata

On a surface of no relief, the width of an outcrop belt depends on the dip of the bed (Fig. A1-10a). This relationship is described by the equation

outcrop width = true thickness/sin(true dip)   (Eq. A1-1).

From this equation it is clear that a decrease in the dip of a unit leads to an increase in the width of the unit's outcrop belt. The width of a vertically dipping bed is the same as the true thickness, whereas the width of a horizontal bed is "infinite" (i.e., the unit extends in all directions until it pinches out or changes orientation or until topographic relief appears). Note that on a horizontal surface the strike of a bed is parallel to the bearing of the trace of the contact (Fig. A1-10b). On a sloping surface the strike of a bed is parallel to the map trace of the bed only if the strike direction is perpendicular to the slope direction (A1-10c). Otherwise, the strike is oblique to the outcrop trace (A1-10c).

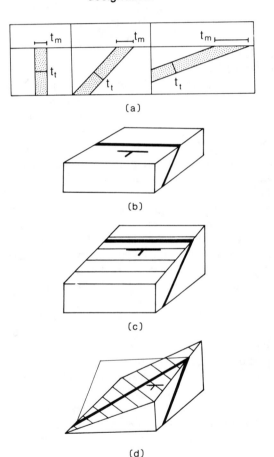

Figure A1-10. Outcrop patterns of contacts and planar layers. (a) Relationship between outcrop width and dip. t_t is the true thickness and t_m is the outcrop thickness on horizontal ground; (b) Diagram showing that the strike is parallel to bed trace on a horizontal plane; (c) Diagram showing that the strike is parallel to bed trace on a slope where the bed trace is parallel to contours; (d) Diagram showing that the strike is oblique to the bed trace on a general slope.

Contacts

A *contact* is the planar boundary between two units. We briefly review the map patterns of four types of contacts. Remember that the relative ages of structures can be determined by *cross-cutting relationships* between contacts. Where contacts intersect, younger contacts always cut across older contacts.

Conformable Contacts: If the contact between two units is parallel to the stratification of both units and does not represent a significant time gap or a surface on which movement has taken place, the contact is said to be *conformable*. Conformable contacts are usually found in stratified sedimentary or volcanic rocks in which the rock layers have been deposited in sequence, with the oldest layer at the base. The symbol for a conformable contact is usually a thin smooth line (Fig. A1-11a).

Fault Contacts: On a geologic map, the presence of a fault is indicated by a heavy line (Fig. A1-11a). If the trace of the fault lies entirely within a single unit, or if the displacement on the fault cannot be recognized in the field (e.g., the fault trace is not well exposed or the offset across the fault is minor) the presence of the fault symbol may be the only indication of the fault. Often, however, the presence of a fault on a map may be indicated by other features, such as the juxtaposition of out-of-sequence map units, by the occurrence of anomalously thin or thick sections of a unit, by the offset of markers (Fig. A1-11b), or by a sudden change in the attitude of units (Fig. A1-11c). Faults are planar structures; thus, a map shows the trace of the intersection between the fault and the ground surface. If the sense of displacement on the fault is known, the symbols shown in Figure A1-12a can be used

to represent the fault trace. The symbols *U* and *D* indicate the relative upthrown and downthrown sides of a fault, straight barbs are used for normal faults (with the barbs on the hanging-wall side), and teeth are used for thrust faults (with the teeth on the hanging-wall side). If information on the attitude and direction of slip on the fault is known, it can be indicated as shown in Figure A1-12b. The strike and dip mark on the fault trace indicates the attitude of the fault plane, and the arrow indicates the plunge and bearing of the slip lineations on the fault plane.

Intrusive Contacts: Intrusive contacts represent the boundary between an igneous body and older rocks or structures. Depending on the type of igneous body, the trace of the contact can be straight (e.g., for sheet intrusions like dikes, sills, and laccoliths) or curved and irregular (e.g., for a granitic pluton or stock). Intrusive contacts can be concordant (i.e., parallel to the layering of the wall rock), as is the case for a sill, or discordant (i.e., nonparallel to the layering of the wall rock), as is the case for a dike or a pluton (Fig. A1-13). Discordant intrusive contacts can be recognized simply from their map pattern (they cut across older contacts), but concordant intrusions may be apparent only from the unit descriptions in the map explanation.

Unconformities: An unconformity is a contact between two units that represents the occurrence of an interval of nondeposition and/or an interval of erosion. Unconformities are delineated either by a thin line on a map that is indistinguishable from the symbol for a concordant stratigraphic contact or by a wavy line. Be sure to check the map explanation for the definition of the symbols that are used; on many Canadian maps the wavy-line symbol is used for faults. Unconformities that

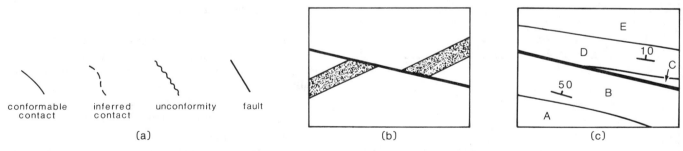

Figure A1-11. (a) Symbols used for contacts; (b) fault trace indicated by offset marker bed (shaded); (c) fault indicated by the presence of an anomalously thin unit (unit C), out-of-sequence units (B adjacent to D), or sudden change in bed attitude.

Figure A1-12. (a) Common map symbols for faults; (b) symbol used to show strike and dip of fault plane and plunge and bearing of slip lineations on the fault plane.

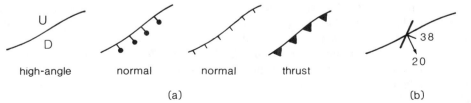

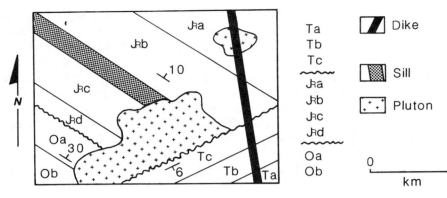

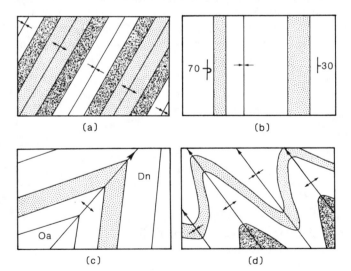

Figure A1-13. Simple geologic map showing the map patterns of various types of contacts. Note that the dike, pluton, and angular unconformity can be recognized because their contacts are discordant with adjacent units, whereas the sill is evident only by reading the legend, and the bed-parallel unconformity is evident only from the map symbol and the absence of section.

separate similarly oriented sections of strata can be recognized only by the age discordance between units above and below the unconformity (contact between Oa and JRd in Fig. A1-13). Angular unconformities cut across the structures of underlying units and thus will stand out on a map (contact between Tc and JRb in Fig. A1-13).

Simple Folds

Opposing limbs of nonplunging upright folds (folds with a horizontal hinge and a vertical axial plane) are mirror images of one another (Fig. A1-14a). Outcrop widths of a given unit on opposite limbs of such a fold are equal if there are no variations in the true thickness of the unit within the map area. Remember that in the case of a syncline, the youngest unit occurs in the core of the fold, and in the case of an anticline, the oldest unit occurs in the

Figure A1-14. Map patterns of simple folds. Note relative ages of units in each sketch. (a) Nonplunging fold train. Blank unit is youngest and dark shaded unit is oldest; (b) nonplunging inclined fold; (c) plunging chevron fold; (d) plunging fold train.

core of the fold. Differentiation between anticlines and synclines is, therefore, possible without strike and dip information, if age relationships are known. The map pattern of some folds shows a variation in the width of outcrop belts on opposing limbs, because of the contrast between the dips of the limbs (Fig. A1-14b).

If a fold is *plunging* (i.e., its hinge is not horizontal), then outcrop belts of a unit on opposing limbs converge with one another, and the map may show a locality where the unit can be traced completely around from one limb to another. If such a *closure* does occur in the map area, the map pattern of the unit will have a *V* or *U* shape (Fig. A1-14d; for horizontal ground), depending on the actual shape of the fold hinge. A *V* shape, for example, indicates that the fold has a chevron profile. The apparent thickness of stratigraphic units in the plane of the map varies from limb to hinge of a plunging fold, because the true dip of the folded layer varies around the fold hinge. The thickness of a unit on opposing limbs is the same if the fold is upright and is different if the fold is inclined. The map pattern of a plunging *fold train*, which is a series of related folds, will appear as a series of zigzags (Fig. A1-14d; for horizontal ground).

Doubly plunging folds are folds whose hinges change in plunge direction along their trace. In a map view a bed can be traced around the hinge of the fold at more than one locality along the hinge (Fig. A1-15a). Remember that if the lateral dimensions of the structure are approximately the same in all directions, then the structure is called a *dome* (if it is concave down) or a *basin* (if it is concave up). The traces of contacts around domes and basins approach a circular form.

Superimposed folding occurs when folded rock is subsequently refolded. A number of map patterns characterize superimposed folds (e.g., Ramsay, 1967; Suppe, 1985), but we present only two simple examples here. The first (Fig. A1-15a) exemplifies dome-and-basin structure, and the second (Fig. A1-15b) exemplifies an anticline refolded around a plunging syncline. Description and interpretation of the patterns of polyphase folds is presented in Chapter 16.

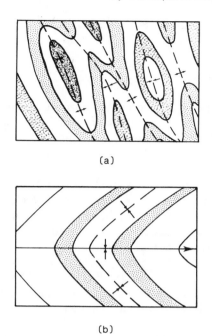

Figure A1-15. Map patterns of simple refolded folds. (a) Geologic map of dome-and-basin structure (adapted from Roberts, 1982); (b) geologic map of a refolded syncline.

A1-4   ELEMENTS OF CROSS SECTIONS AND PROFILES

A *cross section* is a vertical two-dimensional slice through a region. There are several sources of data that can be used in construction of the cross section. Geologic features mapped on the ground can be extrapolated to depth, and control at depth may be provided by drilling data, mine data, seismic-reflection data, or, indirectly, from structure-contour maps.

A line along which a cross section is to be constructed is called a *line of section*. Lines of section are usually labeled X-X', A-A', or something similar. To construct a cross section from a map (Fig. A1-16), place an overlay over the line of section (the line of section forms the edge of the overlay). Transfer topographic data and points where the line of section crosses contacts or layers of known attitude to the overlay. Indicate dip measurements by a *ball and tick mark* inclined by the angle of dip. You may project structure attitudes a short distance along strike onto the line of section. Once you have plotted all the data on the overlay, remove the overlay and transfer the data to a new sheet of paper; imagine that the new sheet of paper represents the plane of the cross section. Construct a topographic profile on this new piece of paper. If the line of section is perpendicular to strike, the dip indicated on the line of section is true dip. Otherwise, the dip indicated will be an apparent dip that you will have to calculate using the methods described in Chapters 3 or 5. You may sketch contacts and other structures on the section, taking care to obey your dip symbols. Based on the data plotted on the line of section, the geology in the plane of section may be extrapolated to depth (Chapters 13 and 14 provide suggestions concerning methods for such extrapolation).

If the vertical scale of a cross section is larger than the horizontal scale, then the section is said to have *vertical exaggeration*. If the vertical scale of a cross section is the same as the horizontal scale, then the section is said to display *no vertical exaggeration*. For example, if the horizontal scale on a line of section is 1:10,000 and the vertical scale is 1:1000, the section displays a vertical exaggeration of 10X. On sections that have vertical exaggeration, dips of units are much steeper than they are in nature. For example, a dip of 45° on a cross section at 1:1 becomes a dip of 84° on a cross section that has a vertical exaggeration of 10X (Fig. A1-17). Stratigraphic thicknesses are also distorted as a consequence of vertical exaggeration. Consider a section with 10X vertical exaggeration. A flat-lying unit is shown to be 10 times its proper thickness. If the same unit had a vertical dip, then its thickness would not be exaggerated, but the bed length would be stretched to 10 times its correct length. Because of the distortion inherent in exaggerated sections, 1:1 sections are generally preferred for structural geology, unless details of stratigraphy must be shown that will not be visible unless the section is exaggerated.

Because of the significance of vertical exaggeration, it is important that both the horizontal and vertical scales be

Figure A1-16. Construction of a cross section. (a) Block diagram showing the relationship between a map and a cross section; (b) key points indicated on a section strip. Note that the section is perpendicular to strike; (c) completed section (no topography is shown).

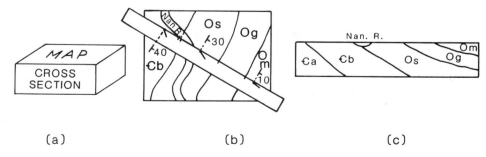

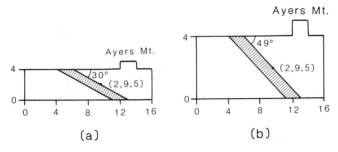

Figure A1-17. Demonstration of vertical exaggeration. (a) Cross section drawn with no vertical exaggeration; (b) same section drawn with 2X vertical exaggeration. Note how the dip of the bed has changed. Coordinates of a point on the bed are given. Also note the change in shape of Ayers Mountain.

indicated on the cross section. It is also important to indicate the orientation of the section with respect to north. This is usually done by putting a compass bearing (or a compass quadrant) at each end of the section and by providing a location map that shows the position of the line of section. If required, there may be bends in a section where the line of section changes direction. These bends are indicated by a vertical line across the section. If appropriate, surface topography and sea level are also indicated on the section.

A profile of a structure is a two-dimensional projection of a structure that is drawn perpendicular to the bearing or strike of the structure. If the structure plunges, then the profile plane is not a vertical plane. Profiles drawn perpendicular to the plunging axis of a fold are called *down-plunge projections* (see Chapter 13).

A1-5   DIAGRAMS, SKETCHES, AND PHOTOS

Block Diagrams

In many cases it is easiest to visualize a geometric problem if it is viewed in three dimensions. It is possible to simulate three dimensions on a two-dimensional sheet of paper by using *block diagrams*. On a block diagram, three faces are visible at one time, with each face containing two coordinate axes. The visual effect of a block diagram is analogous to viewing a photograph of a block. The three coordinate axes, which are mutually perpendicular in the real world, make acute angles with respect to one another in a block diagram (Fig. A1-18a).

There are three basic types of block diagrams. The first is an *isometric diagram* (Fig. A1-18b) and is the easiest to draw. In an isometric diagram, lines that are parallel in the real world are parallel in the diagram, lines that are the same length in the real world are the same length in the diagram, ratios between dimensions in the real world are maintained in the diagram, the sum of the angles defining the corners of the block must add to 360°, and opposing angles on a face of the block must be equal. Isometric diagrams have the advantage that transfer of geologic data onto the block is very easy, but they have the disadvantage that they present a visually distorted image (i.e., they do not look right). More realistic looking block diagrams employ an artistic trick known as perspective. A *perspective diagram* is one in which parallel lines merge toward a vanishing point on the paper, just like parallel lines appear to merge in the distance in the real world. On a *one-point perspective diagram* (Fig. A1-18c), there is one

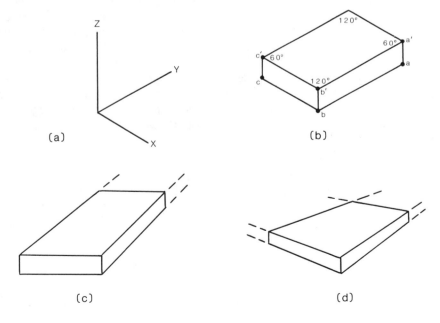

Figure A1-18. Examples of block diagrams. (a) Coordinate axes for a block diagram; (b) isometric diagram; (c) one-point perspective diagram; (d) two-point perspective diagram.

vanishing point, whereas on a *two-point perspective diagram*, there are two vanishing points (Fig. A1-18d).

Sketches and Photos

In many applications a topographic map provides an excellent base map on which geological relations can be represented. The view represented by a map is comparable that of a vertical air photo. In some cases, the vertical air photo itself may prove to be a more useful base for mapping than a topographic map, because surface landmarks of the map area may be recognized more easily (especially if there are shadows), and the outcrops themselves are often visible. It is now possible to obtain inexpensive half-tone vertical air photographs, called *orthophotoquads*, that are at the same scale as 7.5" quadrangle maps. *Stereo pairs* of vertical air photos (photos that overlap in coverage, thereby simulating the parallax created by your eyes that permits your brain to visualize in three dimensions), when viewed with a stereoscope, produce a three-dimensional image of the ground surface in the map area.

A common way to communicate information about

the structure in an outcrop is to present an *outcrop photograph* (Fig. A1-19a). Remember that it is essential that such photographs contain a *scale indicator*. If a scale indicator is not present, a person looking at the photo will have no reference frame with which to judge the size of features in the picture. It will not be clear, for example, if the beds of the outcrop are 1 cm thick or 10 m thick. The best scale indicator is a ruler with clearly defined centimeter or meter increments, but geologists often use familiar objects, such as hammers, knives, coins, pens, people, or notebooks to indicate scale. Outcrop relations over a broader scale can be indicated by a *panorama photograph* (an *oblique photograph* of a fairly large area; oblique means that the line of sight of the camera is not perpendicular to the plane of the outcrop that is photographed). In panoramas (Fig. A1-19b) it is also important that a scale be present (have your field assistant or car present in the field of view). It is also valuable to emphasize contacts or structures in the photo area with lines that are inked onto the print. Sometimes the relations will be more clearly displayed in a *panorama sketch* (a line drawing of an area; Fig. A1-19c). With effort, it may be possible to transfer observations indicated in a panorama sketch or photo onto a map.

Figure A1-19. Examples of geologic photographs. (a) Outcrop photo. Solid line is parallel to bedding; dashed line is parallel to cleavage. Ruler provides scale and is 22 cm long; (b) panorama photo of a roadcut through folded and faulted Devonian strata in New York State; (c) sketch of the photo shown in (b), emphasizing the positions of major faults (solid lines) and the attitude of bedding (dotted lines).

(a)

(b)

(c)

EXERCISES

1. (a) What is the distance in kilometers represented by a degree of longitude at the latitude of Caracas (Venezuela).
    (b) What is the distance in kilometers represented by a degree of longitude at the latitude of Oslo (Norway).

2. (a) On a good atlas map of Italy, determine the distance between Florence and Rome in kilometers, then convert the distances to miles.
    (b) Is the map of Italy at a larger scale or smaller scale than the map of the world that you used?
    (c) What is the representative fraction describing the scale of each map that you have used?

3. (a) Determine the distance in kilometers between New Delhi and Bombay (India). If the representative fraction on a map is 1:80,000, what is the distance between these two cities on the map?
    (b) What is the representative fraction that describes the scale on a map for which 1 cm = 100 km?
    (c) The distance between the islands at the two ends of the Mariana island chain on a given map is 7.5 cm. In reality, these two islands are 750 km apart. What is the representative fraction that describes the scale of the map?
    (d) The distance between Calcutta (India) and Kathmandu (Nepal) is about 600 km. What is the distance between these two cities on a map whose scale is defined by an RF of 1:5000?

4. Figure A1-M1 is a simple geologic map of a locality in central Maine.

    (a) Construct a vertical cross section along line XX'.
    (b) Construct the same cross section with 5X vertical exaggeration. Be sure to adjust the dips of beds and the thickness of units appropriately. Assume that the fold is chevron (i.e., the limbs are straight and the hinge is angular).

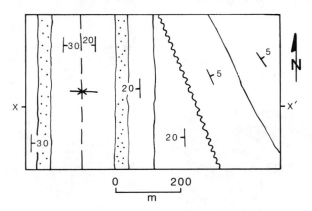

Figure A1-M1. For use in exercise 4.

APPENDIX

2

BASIC TRIGONOMETRY

A2-1 FUNDAMENTAL IDENTITIES

The fundamental identities of trigonometry (sine, cosine, and tangent) refer to the angles and limbs of a right triangle (Fig. A2-1). The identities are presented below for quick reference. If x is the length of the limb adjacent to the angle ø, y is the length of the limb opposite to the angle ø, and h is the length of the hypotenuse, then

$$\sin \phi = y/h \quad \text{(opposite/hypotenuse)} \qquad \text{(Eq. A2-1a)}$$

$$\cos \phi = x/h \quad \text{(adjacent/hypotenuse)} \qquad \text{(Eq. A2-1b)}$$

$$\tan \phi = y/x \quad \text{(opposite/adjacent)} \qquad \text{(Eq. A2-1c)}$$

$$\csc \phi = h/y = 1/\sin \phi \qquad \text{(Eq. A2-1d)}$$

$$\sec \phi = h/x = 1/\cos \phi \qquad \text{(Eq. A2-1e)}$$

$$\cot \phi = x/y = 1/\tan \phi \qquad \text{(Eq. A2-1f)}.$$

If the angle ø is not known, it can be determined from the ratios of the triangle limbs. For example, the angle whose tangent is y/x is the *arc tangent* (abbreviated *arctan*) of y/x or, alternatively, the inverse tangent (abbreviated tan⁻¹) of y/x. In other words,

$$\arctan(y/x) = \tan^{-1}(y/x) = \phi \qquad \text{(Eq. A2-2)}.$$

The terms *arcsin* and *arccos* are used in the same way.

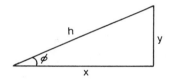

Figure A2-1. Right triangle used for definition of the trigonometric functions.

In addition to the basic identities listed above, it is also useful to keep in mind a number of simple equations that are useful in deriving some of the formulas used in this book. For a complete list of trigonometric equations, check a standard trigonometry textbook or handbook of mathematics.

$$\tan \phi = \sin \phi/\cos \phi \qquad \text{(Eq. A2-3a)}$$

$$\sin^2\phi + \cos^2\phi = 1 \qquad \text{(Eq. A2-3b)}$$

$$\sin(\phi + \beta) = (\sin \phi)(\cos \beta) + (\cos \phi)(\sin \beta) \qquad \text{(Eq. A2-3c)}$$

$$\sin(\phi - \beta) = (\sin \phi)(\cos \beta) - (\cos \phi)(\sin \beta) \qquad \text{(Eq. A2-3d)}$$

$$\cos(\phi + \beta) = (\cos \phi)(\cos \beta) - (\sin \phi)(\sin \beta) \qquad \text{(Eq. A2-3e)}$$

$$\cos(\phi - \beta) = (\cos \phi)(\cos \beta) + (\sin \phi)(\sin \beta) \qquad \text{(Eq. A2-3f)}$$

$$\cos(90^o - \beta) = \sin \beta \qquad \text{(Eq. A2-3g)}$$

$$\sin(90^o - \beta) = \cos \beta \qquad \text{(Eq. A2-3h)}.$$

A2-2 LAW OF SINES
AND LAW OF COSINES

If a triangle does not contain a right angle, trigonometry may still be employed for some problems. Consider a triangle whose three angles are a, b, and c (Fig. A2-2).

Figure A2-2. Triangle used for specification of the Law of Sines and the Law of Cosines.

The side of the triangle opposite a is A units long, the side opposite b is B units long, and the side opposite c is C units long. The following equations apply:

Law of Sines

$$A/2 \sin a = B/2 \sin b = C/2 \sin c \qquad \text{(Eq. A2-4)}.$$

Law of Cosines

$$\cos a = (B^2 + C^2 - A^2)/2BC \qquad \text{(Eq. A2-5a)}$$

$$\cos b = (C^2 + A^2 - B^2)/2CA \qquad \text{(Eq. A2-5b)}$$

$$\cos c = (A^2 + B^2 - C^2)/2AB \qquad \text{(Eq. A2-5c)}.$$

3

SUGGESTIONS
FOR MAPPING
GEOLOGIC STRUCTURES

A3-1   INTRODUCTION

Mapping technique is developed by experience. In that respect, it is somewhat like an art form. Geologists tend to use variations on the basic *theme* of mapping that best suit their interests and goals. In this appendix we list a number of ideas that have helped students faced with a mapping project. Our intent is only to suggest ways to think about mapping, not, of course, to lay out a recipe that must be followed. There is a lot of trial and error in producing a map, and even the most experienced mappers must modify their goals and techniques as they proceed.

A3-2   MATCHING MAPPING TECHNIQUE TO THE PROBLEM

1. The first step of a mapping project is to outline the problem that is to be solved by the mapping. There are a great number of problems that can be solved by mapping, ranging from the general (e.g., What rocks are in this area?) to the specific (e.g., What is the relationship between S_2 cleavage attitude and fold attitude?).

2. Don't worry if, at the beginning of a project, the details of the problem are not clear. As you work, new questions will arise and old ones will become obsolete (see the discussion of *working hypotheses* in Chapter 9).

3. Once you have decided on the nucleus of a problem, make a *reconnaissance* of the potential map area. Try to get a feel for the quality of exposure, the types of rocks

present, the general distribution of lithologies and structures, the ruggedness of the terrane, and the distribution of landowners with dogs. Determine whether your base map adequately represents the topography of the area.

4. From the reconnaissance, estimate the amount of area that you can cover in the time you have available, and estimate the detail necessary to represent the structures.

5. Choose an appropriate scale for your map (see Chapter 9). The scale should be such that the map can show sufficient detail to make the structural relations clear. The area that you plan to cover should be sufficiently large to include representative structural geometries. Don't worry if the boundaries of the map area or the scale must be changed as your work progresses. In some projects different portions of the map area must be mapped at different scales, as a function of the complexity in the area.

6. Based on your reconnaissance, make an initial choice of the type of mapping that is best suited to display the geologic structures that interest you. There are three basic types:

(a) *Outcrop Mapping:* The positions and shapes of discrete outcrops are located on the base map. At each outcrop the unit is identified, and appropriate measurements are made. This method is suitable for regions in which there is little continuity between outcrops. It is often used in metamorphic terranes or in regions with a lot of cover. Such maps are helpful to other geologists who are trying to find outcrops.

(b) *Station Mapping:* Pinpricks or x's are marked on a

map at points where measurements are made, and each point is assigned a station number. The number and the measurements are recorded in a notebook. Some of the measurements are then plotted on the map. This method is suitable for regions where there is a lot more structural detail (e.g., fabric data, mesoscopic structures) at a locality than the geologist can plot on the map.

(c) *Traverse Mapping:* Using this method, a geologist walks out traverses and plots structural measurements directly on the base map as she goes along. The map of the area covered each day is complete at the end of that day. Traverses are commonly planned so that a given contact or structure is traced across the countryside. Such mapping is suited for areas where outcrop is good, the density of data is such that it can be well represented on the map, and structural complexity is such that extrapolation may be dangerous. The positions of contacts are documented by the distribution of attitude measurements on the map.

7. Depending on circumstances, a combination of these techniques may provide the best approach to a specific problem. Don't worry if you start with one method and discover that you have to use another. The main point is that you choose the method that most efficiently lets you document the structural relations in an area.

A3-3   ASPECTS OF MAPPING STRATEGY

8. If you blindly start mapping in an area, you may quickly become frustrated because the outcrops that you stumble across cannot be viewed in any rational context. That is why a reconnaissance is an important first step. Using the information gained from your reconnaissance, start mapping in areas where you can understand the structures and recognize the stratigraphy (or lithologic sequence).

9. If you find yourself in an area where you are lost (geologically), wander around until you find a relation or unit of whose identity you are confident. Then, trace your path back from the "known" into the "unknown." Often such backtracking puts the previously mysterious outcrop into a geologic context, so that it can be understood. Sometimes it helps to approach a difficult outcrop from different directions. Don't worry if the relations at a specific outcrop are unclear; if an outcrop *seems* to be hard to understand, it probably *is* hard to understand. Not all outcrops can be understood. Visit these difficult outcrops again toward the end of your mapping work, and with the

understanding you have developed from the mapping, you may be able finally to figure them out.

10. Early in your work, begin to employ the method of multiple working hypotheses. You should be constantly developing ideas that will be proven or disproven by the next step in your traverse. Plan your traverses around the problems to be solved (not vice versa). By doing this, the process of mapping becomes the process of problem solving and therefore becomes more interesting. Furthermore, your data collection becomes a focused rather than a random process. Needless to say, your hypotheses will evolve as your work progresses. New ones develop and old ones die.

11. Early in your work, characterize the structural assemblage that occurs in your map area (e.g., are you in a fold-thrust belt, a continental platform, a rift terrane, or a polydeformed metamorphic terrane). The reason for doing this is not to bias you observations, but to sensitize you to the geologic problems in the area and help you to formulate working hypotheses based on established concepts. Knowledge of the structural assemblage may help you identify potential problems with your map.

12. Early in your work, sensitize your eyes to recognize variations in lithology. You may find at first that all rocks in your map area look the same, but as you work, you will find that different units become more distinctive. Pay special attention to guide fossils, characteristic lithologies, and unique minerals. These features help you to identify a unit. Don't worry if it takes a long time for you to acclimate to your field area (i.e., to recognize units and structures and formulate working hypotheses). There is always a period of muddling around before you can really make progress.

13. As you test your individual working hypotheses, estimate your level of confidence in each hypothesis. For example, after visiting a specific outcrop (where beds dip east), you may be about 20% sure that you are on the limb of a syncline. You then plan your traverse to test this hypothesis. At the next outcrop, beds dip west, and you become 80% sure that you have crossed a syncline. Without seeing the hinge, however, you cannot be positive, so you predict where the hinge will be and plan your traverse accordingly. If you find the hinge, then you may be 100% positive that the syncline is present.

14. Keep track of which order structure you are working in (see Chapter 11). Are you looking at a first-order regional fold or a local parasitic fold? If you find an isolated outcrop with overturned beds, it does not necessarily mean that the whole ridge you are mapping is overturned.

15. Use your hands (yes, physically use your hands) to help yourself trace out structures along strike. Set your hand to simulate bedding attitude at a starting point, and

then walk down the structure. Your hand reminds you of layer attitude at the at the starting point and thus helps you to detect changes in layer attitude.

16. Use marker horizons to trace out structures. Search for a distinctive layer and walk it out. Be sure to determine whether the layer is parallel to S_1, S_2, or S_3.

17. Visualize structures that you are studying in your mind. If you cannot visualize the structure, you probably do not understand it. Make lots of sketches in your notebook. You do not really understand a structure until you can sketch a cross section of it and can figure out its kinematic evolution. You should not only show the geometry of the rocks in their present configuration but also try to figure out the movement paths that the rocks followed to get into their present configuration. It helps if you have collected data on kinematic indicators.

18. If you have thought about a structure for a long time, have tried diligently to make sketches of it, and still cannot figure out its geometry or evolution, go on to the next outcrop. The structure may be too complex or unusual for even an experienced geologist to figure out at the first try. Remember, you have limited time to complete your map. If you can, return to the mystery outcrop later, and apply the knowledge you gained mapping elsewhere to figuring out the mystery outcrop.

19. Every day, and again upon completion, interpret your map as if it were drawn by someone else. Follow the guidelines described in Chapter 9. This step will help reveal any errors or inconsistencies before you go to the effort of drafting a final copy.

20. It is often preferable to map on a Mylar overlay rather than directly on a paper base map. Mylar is much more durable, doesn't get ruined when wet, and can be erased many times. Be sure to transfer landmarks on your base map onto the overlay so that you can register the overlay with the base map. Label each overlay so that you can rapidly identify the area that it covers. Ink in data that you are confident of so that it cannot be erased.

21. Systematically record fabric data. You never know whether you can get back to an outcrop, so collect this data while you are there for the first time. Be sure you carefully note the locality at which a set of data is collected. Indent, or use some other method of highlighting data in your notebook, so that you can quickly find it later for plotting on a stereonet.

22. Always carry your camera and an extra roll of film. You never know when you will find a photogenic outcrop or an outcrop that displays a critical structural relation. Do not plan on taking all photos in the map area on your last day in the area (it may rain). Be sure each photo has a scale (see Appendix 1). Number and describe each photo in your notebook immediately after shooting it. It may be a while before you have a chance to label the

slides or prints, and by then you may have forgotten why they were taken. Indicate the direction in which the camera was pointing when you took the photo.

23. To the extent possible, document all measurement localities with a symbol on your map. A person who reads your map will judge the certainty of a contact or structure by the density of measurements around it. Of course, don't waste time taking extraneous measurements, but avoid having blank areas in regions where there is outcrop. The blank area on your map might give the impression of lack of outcrop. Even if bedding or foliation maintains the same orientation over a large area, it is important to have measurements to prove that.

24. Do not delay putting measurements on your field map. Do not just record measurements in your notebook and plan to put them on in the evening. By not plotting measurements, you are eliminating the possibility of using your map to guide your traverse or to create or modify working hypotheses. It is important to see how the structural relations develop on your map. Also, by plotting your measurements immediately, you will be able to tell if you made a mistake and therefore will be able to correct it while still in the field. At night, compile your map by transferring the data from the day's overlay to a master clean copy.

25. Any time that you put a structure on your map, be sure that there are enough measurements to document it. Think of the kinds of questions that your audience might have when they study your map.

26. Record measurements in your notebook in addition to putting them on your map to make it easier to plot them on a stereonet later.

27. Make sure that your field notes will be intelligible to other geologists. Use standard terminology for describing structures, be neat, and make sure localities are clearly identified. It should be possible for another geologist to take your notebook and reconstruct your traverses and your thinking. It may help you to make descriptive notes about your traverse in your notebook to remind yourself of where you were. If you have lunch on an outcrop say so, because it may help you remember which outcrop is which. Frequently photocopy your field maps and notebooks so that you will have a spare copy in case the original is lost.

28. If you are collecting samples for a structural study, be sure that they are oriented. It is very frustrating to cut a thin section of an sample and find a beautiful kinematic indicator only to discover that you have no way to define the orientation of the indicator. To take an *oriented sample*, it is best to mark the strike and dip clearly on a structurally significant surface (e.g., a foliation plane) while the the sample is still in the outcrop. Indicate the azimuth of strike with an arrow and mark the azimuth in

degrees at the the tip of the arrow. Indicate stratigraphic up or the up direction on the outcrop with an arrow on the side of the sample. Then remove the sample. Remember that the orientation mark on the sample should be sufficient to allow you to be able to determine the attitude of a thin section that you cut from the sample or to be able to calculate the orientation of a mesoscopic feature that you see in the sample.

29. Ask permission to get on land if you can. If that is not feasible, either stay off the land or stay out of sight.

If you get caught trespassing, act as innocent as possible (presumably you really don't mean any harm anyway).

30. The truth is always right! Be sure to record data even if they seem anomalous and do not fit any of your working hypotheses. If the data are correct, you will eventually be able to explain them, or you will have to admit that they represent an unsolved problem. Never select data so that they fit your favored hypothesis. You may, however, select representative data for plotting on your map if you have only limited space.

4

TEMPLATES
FOR PLOTTING
GEOLOGIC DATA

A4-1 EQUAL-ANGLE (WULFF) NET

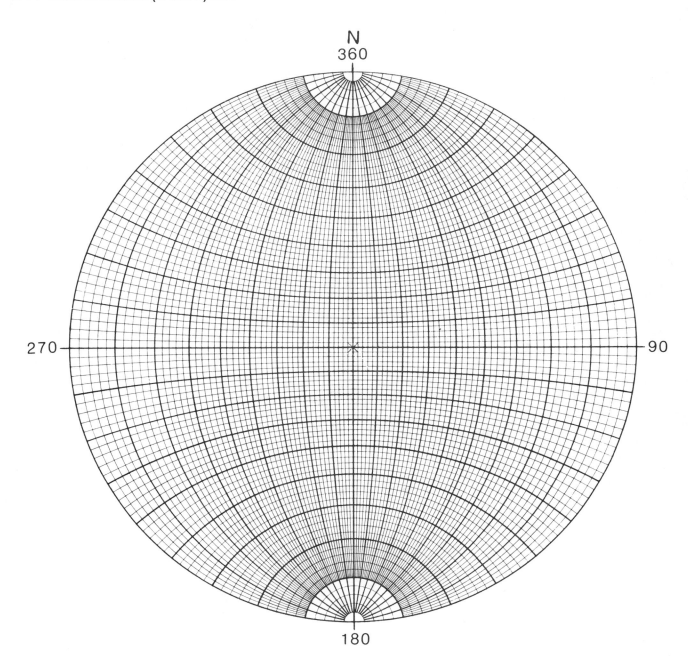

A4-3 ROSE-DIAGRAM GRID

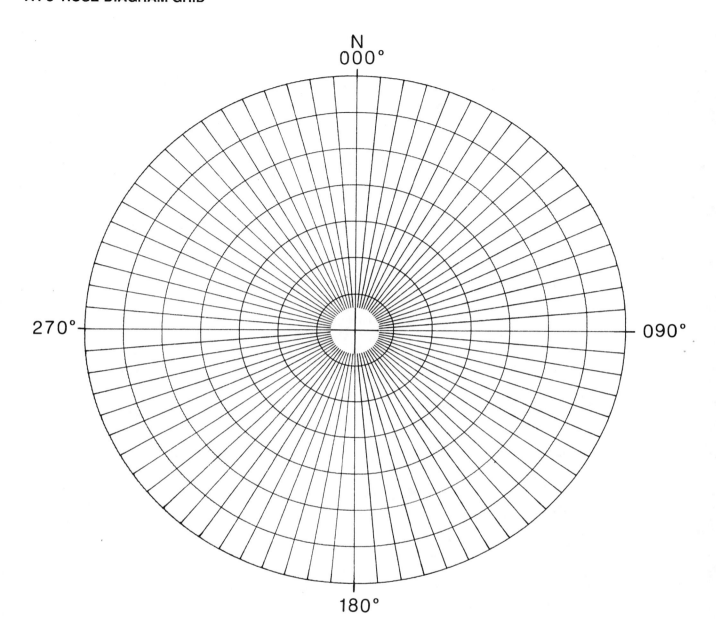

A4-4 LAMBERT POLAR EQUAL-AREA NET

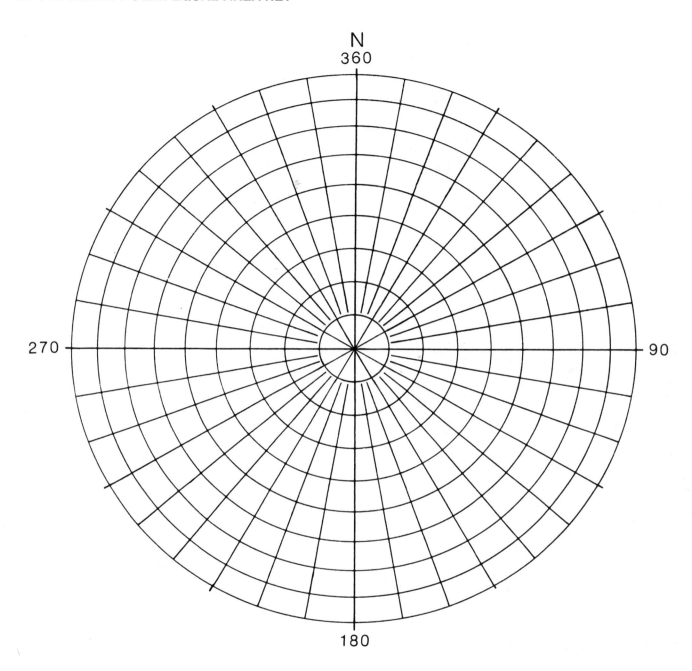

A4-5 KALSBEEK COUNTING GRID

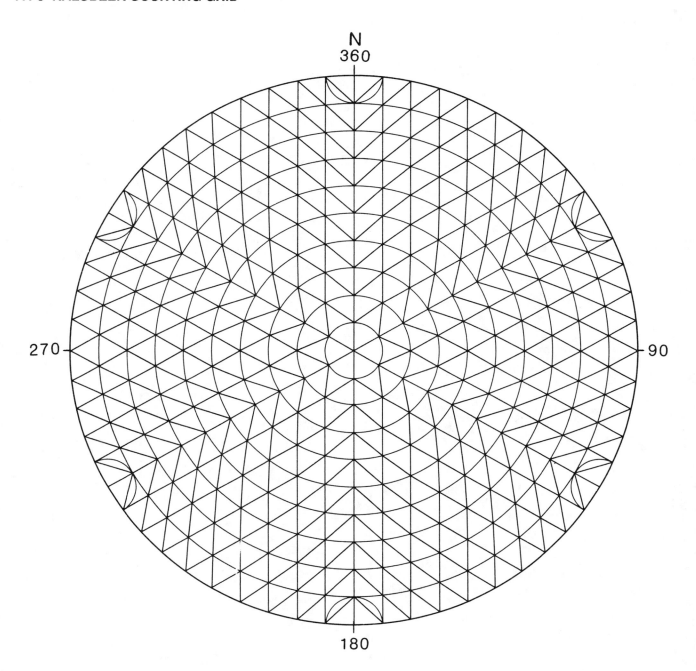

A4-6 SCHMIDT COUNTING GRID

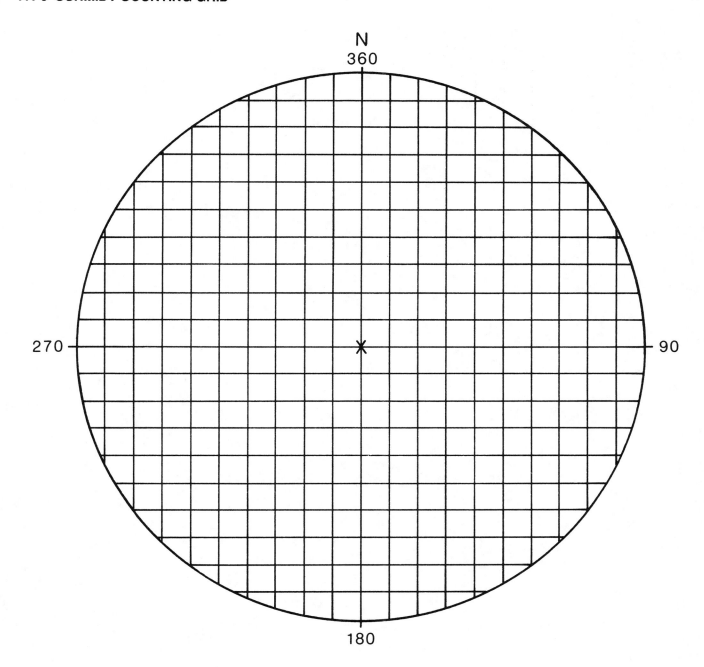

A4-7 SCHMIDT COUNTER

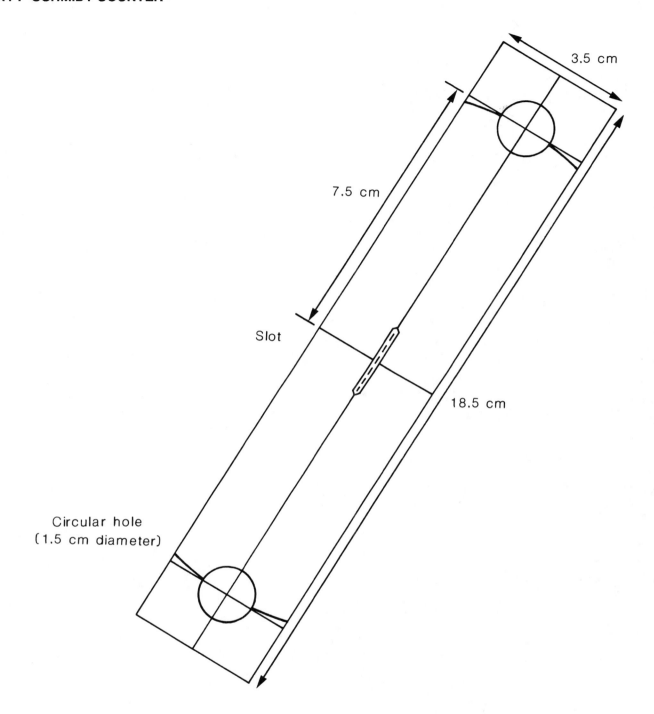

3.5 cm

7.5 cm

Slot

18.5 cm

Circular hole
(1.5 cm diameter)

A4-8 ORTHOGRAPHIC NET

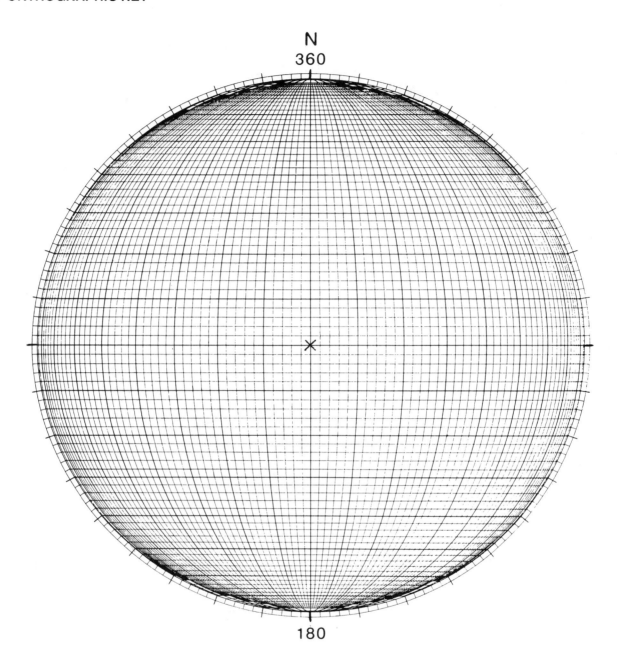

A4-9 HYPERBOLIC NET

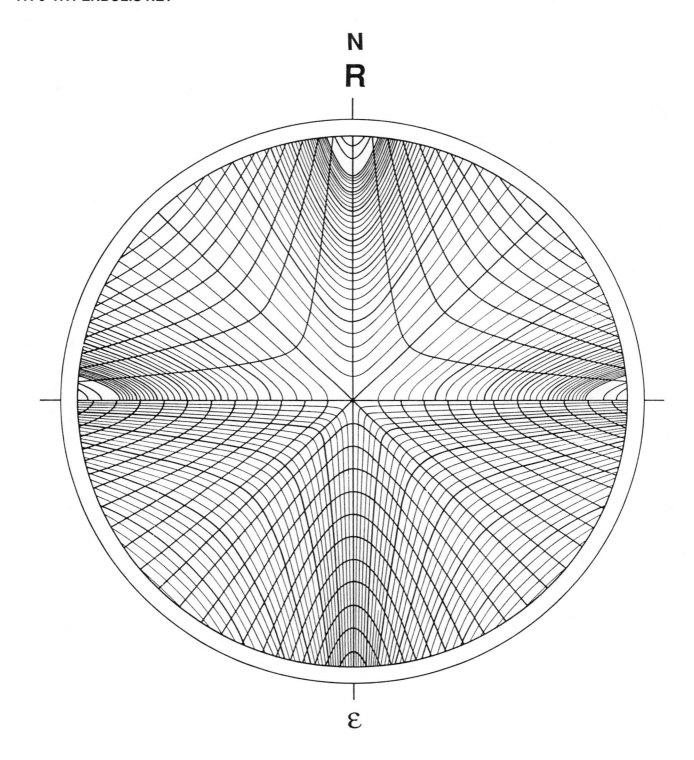

REFERENCES

Aleksandrowski, P., 1985, Graphical determination of principal stress directions for slickenside lineation populations: an attempt to modify Arthaud's method: J. Struc. Geol., v. 7, p. 73-82.

Anastasio, D.J., 1987, Thrusting, halotectonics and sedimentation in the External Sierra, southern Pyrenees, Spain: Ph.D. dissertation, Johns Hopkins University, Baltimore.

Anderson, D.E., 1971, Kink bands and major folds, Broken Hill, Australia: Geol. Soc. Am. Bull., v. 82, p. 1841-1862.

Anderson, E.M., 1942, The Dynamics of Faulting and Dyke Formation with Applications to Britain: Oliver and Boyd, Edinburgh, 191 p.

Anderson, T.B., 1964, Kink bands and related geological structures: Nature, v. 202, p. 272-274.

Anderson, T.B., 1974, The relationship between kink bands and shear fractures in the experimental deformation of slate: J. Geol. Soc. Lond., v. 130, p. 367-382.

Angelier, J., 1975, Sur un apport de l'informatique a l'analyse structurale; exemple de la tectonique cassante: Revue Geogr. Phys. Geol. Dyn., v. 17, p. 137-146.

Angelier, J., 1979, Determination of the mean principal directions of stresses for a given fault population: Tectonophys., v. 56, p. T17-T26.

Angelier, J., and Melcher, P., 1977, Sur une methode graphique de recherche des contraintes principales egalement utilisable en tectonique et en seismologie: la methode des diedres droites: Bull. Soc. Geol. Fr., v. 7, ser. 19, p. 1309-1318.

Arthaud, F., 1969, Methode de determination graphique des directions de reccourcisement, d'allongement et intermediare d'une population de failles: Bull. Soc. Geol. Fr., v. 7, ser. 11, p. 729-737.

Arthaud, F., and Choukroune, P., 1972, Methode d'analyse de la tectonique cassant a l'aide des microstructures dans les zones peu deformees: Example de la plateforme nord-Aquitaine: Inst. Fr. Pet., Rev., v. 27, p. 715-732.

Arthaud, F., and Mattauer, M., 1969, Exemples de stylolites d'origine tectonique dans le Languedoc, leurs relations avec la tectonique cessante: Bull Soc. Geol. France, v. 7, ser. 11, p. 738-744.

Bader, G.E., 1949, Geophysical history of the Anahuac Oil field, Chambers County, Texas: in Nettleton, L.L., (ed.), Geophysical Case Histories, Volume 1 - 1948, Soc. Expl. Geophysicists, p. 66-73.

Badgley, P.C., 1959, Structural Methods for the Exploration Geologist: Harper & Brothers, New York, 280 p.

Baker, W.H., 1960, Elements of Photogrammetry: Ronald Press, New York, 197 p.

Bally, A.W., Gordy, P.L., and Stewart, G.A., 1966, Structure, seismic data and orogenic evolution of southern Canadian Rocky Mountains: Bull. of Can. Petrol. Geology, v. 14, p. 337-381.

Bambach, R.K., 1973, Tectonic deformation of composite-mold fossil Bivalvia (Mollusca): Am. J. of Sci., Cooper Volume, v. 273-A, 409-430.

Bell, A.M., 1981, Vergence: an evaluation: J. Struc. Geol., v. 3, p. 197-202.

Bell, T.H., 1981, Foliation development: The contribution, geometry, and significance of progres- sive bulk inhomogeneous shortening: Tectonophys., v. 75, p. 273-296.

Bell, T.H., and Etheridge, M.A., 1973, Microstructures of mylonites and their descriptive terminology: Lithos, v. 6, p. 337-348.

Bengtson, C.A., 1983, Easy solutions to stereonet rotation problems: Am. Assoc. Petrol. Geologists Bull., v. 67, p. 706-713.

Berry, L.G., and Mason, B., 1959, Mineralogy, Concepts, Descriptions, Determinations: W.H. Freeman and Co., San Francisco, 630 p.

Berthe, D., Choukroune, P., and Jegouzo, P., 1979, Orthogneiss, mylonite and non-coaxial deformation of granites: The example of the South Armorican shear zone: J. Struc. Geol., v. 1, p. 31-42.

Beutner, E.C., and Diegel, R.A., 1985, Determination of fold kinematics from syntectonic fibers in pressure shadows, Martinsburg slate, New Jersey: Am. J. Sci., v. 285, p. 16-50.

Billings, M.P., 1972, Structural Geology, 3rd ed.: Prentice-Hall, Englewood Cliffs, NJ, 606 p.

Birch, T.W., 1964, Maps: Topographical and Statistical, 2nd ed.: Clarendon Press, Oxford, 240 p.

Bishop, M.S., 1960, Subsurface Mapping: John Wiley & Sons, New York, 198 p.

Blenkinsop, T.G., and Rutter, E.H., 1986, Cataclastic deformation of quartzite in the Moine thrust zone: J. Struc. Geol., v. 8, p. 669-681.

Borradaile, G.J., 1976, "Structural facing" (Shackleton's Rule) and the Palaeozoic rocks of the Malaguide Complex near Velez Rubio, SE Spain: Proc. Koninklijke Nederlandse Akad. van Wetenschappen, Amsterdam, series B, v. 79, p. 330-336.

Borradaile, G.J., 1978, Transected folds: A study illustrated with examples from Canada and Scotland: Geol. Soc. Am. Bull., v. 89, p. 481-493.

Borradaile, G.J., 1981, Particulate flow of rock and the formation of cleavage: Tectonophys., v. 72, p. 305-321.

Borradaile, G.J., Bayly, M.B. and Powell, C.McA. (eds.), 1982, Atlas of Deformational and Metamorphic Rock Fabrics: Springer-Verlag, New York, 551 p.

Borradaile, G.J. and Poulsen, K.H., 1981, Tectonic deformation of pillow lava: Tectonophys., v. 79, p. T17-T26.

Bott, M.H.P., 1959, The mechanism of oblique-slip faulting: Geol. Mag., v. 96, p. 109-117.

Bouiller, A.M., and Gueguen, Y., 1975, Origin of some mylonites by superplastic flow: Contr. Min. Petrol., v. 50, p. 93-104.

Boulter, C.A., 1979, On the production of two inclined cleavages during a single folding event, Stirling Range, S.W. Australia: J. Struc. Geol., v. 1, p. 207-219.

Boyer, S.E., 1978, Structure and origin of Grandfather Mountain window, North Carolina: Ph.D. dissertation, Johns Hopkins Univ., Baltimore, Maryland, 306 p.

Boyer, S.E., 1986, Geometric evidence for simultaneous, rather than sequential movement on major thrusts; implications for duplex development: Geol. Soc. Am. Abst. w. Pgms., v. 18, p. 549.

Boyer, S.E., and Elliott, D., 1982, Thrust Systems: Am. Assoc. Petrol. Geologists Bull., v. 66, p. 1196-1230.

Breddin, H., 1956, Die tektonische Deformation der Fossilen im Rheinischen Schiefergebirge: Zeitschrift der deutsche geologische Gesellschaft, v. 106, p. 227-305.

Bronshtein, I.N., and Semendyayev, K.A., 1973, A Guide Book to Mathematics: Harri Deutsch Publishers, Braintree, MA, 783 p.

Bucher, W.H., 1944, The stereographic projection, a handy tool for the practical geologist: J. Geol., v. 52, p. 191-212.

Burks, R.J., 1981, Alleghanian deformation and meta- morphism in southwestern Narragansett Basin, Rhode Island: M.A. thesis, Univ. of Texas, Austin, 93 p.

Burnside, C.D., 1979, Mapping from Aerial Photographs: Granada Publishing, London, 304 p.

Busk, H.G., 1929, Earth Flexures: Cambridge University Press, Cambridge, 106 p.

Byerlee, J., 1978, Friction of rocks: Pure and Appl. Geophys., v. 116, p. 615-626.

Chamberlin, T.C., 1890, The method of multiple working hypotheses: Science (new series), v. 15, p. 92.

Chamberlin, T.C., 1897, The method of multiple working hypotheses: J. Geol., v. 5, p. 837-848.

Charlesworth, H.A.K, Langenberg, C.W., and Ramsden, J., 1976, Determining axes, axial planes and sections of macroscopic folds using computer based methods: Can. J. Earth Sci., v. 13, p. 54-65.

Cloos, E., 1946, Lineation: Geol. Soc. Am. Mem. 18, 122 p.

Cloos, E., 1947, Oolite deformation in the South Mountain Fold, Maryland: Geol. Soc. Am. Bull., v. 58, p. 843-918.

Cloos, E., 1947, Boudinage: Am. Geophys. Union Trans., v. 28, p. 626-632.

Coates, J., 1945, The construction of geological sections: Quart. J. Geol. Mining. Met. Soc. India, v. 17, n. 1.

Cobbold, P.R., and Quinquis, H., 1980, Development of sheath folds in shear regimes: J. Struc. Geol., v. 2, p. 119-126.

Compton, R., 1962, Manual of Field Geology: John Wiley & Sons, New York, 378 p.

Compton, R., 1966, Analysis of Plio-Pleistocene deformation and stresses in northern Santa Lucia Range, California: Geol. Soc. Am. Bull., v. 77, p. 1361-1380.

Conlin, R.R., and Hoskins, D.M., 1962, Geology and mineral resources of the Mifflintown Quadrangle: Penn. Geol. Survey Atlas A126, 46 p.

Cook, L., et al., 1979, Thin-skinned tectonics in the crystalline southern Appalachians: COCORP seismic-reflection profiling of the Blue Ridge and Piedmont: Geology, v. 7, p. 563-567.

Cosgrove, J.W., 1976, The formation of crenulation cleavage: J. Geol. Soc. London, v. 132, p. 155-178.

Crone, D.R., 1968, Elementary Photogrammetry: Frederick Ungar Publishing Co., New York, 348p.

Crowell, J.C., 1974, Origin of late Cenozoic basins in southern California: in Dickinson, W.R., (ed.), Tectonics and Sedimentation: Soc. Econ. Paleont. and Mineralogists Spec. Pub. 22, p. 190-204.

Dahlstrom, C.D.A., 1969, Balanced cross sections: Can. J. Earth Sci., v. 6, p. 743-757.

Dahlstrom, C.D.A., 1970, Structural geology in the eastern margin of the Canadian Rocky Mountains: Bull. Can. Petrol. Geologists, v. 18, p. 332-406.

Davis, D.M., and Engelder, T., 1985, The role of salt in fold-and-thrust belts: Tectonophys., v. 119, p. 67-88.

Davis, D.M., Suppe, J., and Dahlen, F.A., 1983, Mechanics of fold-and-thrust belts and accretionary wedges: J. Geophys. Res., v. 88, p. 1153-1172.

Davis, G.H., 1978, The monocline fold pattern of the Colorado Plateau: in Mathews, V. (ed.), Laramide Folding Associated with Basement Block Faulting in the Western United States: Geol. Soc. Am. Mem. 151, p. 215-233.

Davis, G.H., 1984, Structural Geology of Rocks and Regions: John Wiley & Sons, New York, 492 p.

Dennis, J.G., 1987, Structural Geology, An Introduction: Wm. C. Brown Publishers, Dubuque, 448 p.

Dennison, J.M., Analysis of Geologic Structures: W.W. Norton & Co., New York, 209 p.

Dennison, J.M., and Woodward, H.P., 1963, Palinspastic maps of central Appalachians: Am. Assoc. Petrol. Geologists Bull., v. 47, p. 666-680.

De Paor, D.G., 1986, Orthographic analysis of geological structures - II. Practical applications: J. Struc. Geol., v. 8, p. 87-100.

De Paor, D.G., 1988, R_f/ϕ_f strain analysis using an orientation net: J. Struc. Geol., in press.

De Sitter, L.U., 1958, Boudins and parasitic folds in relationship to cleavage and folding: Geol. en Mijnb., v. 20, p. 277-286.

Dewey, J.F., 1965, Nature and origin of kink bands: Tectonophys., v. 1, p. 459-494.

Dickinson, G.C., 1979, Maps and Air Photographs: John Wiley & Sons, New York, 348 p.

Donath, F.A., and Parker, R.B., 1964, Folds and folding: Geol. Soc. Am. Bull., v. 75, p. 45-62.

Donn, W.L., and Shimer, J.A., 1958, Graphic Methods in Structural Geology: Appleton-Century-Crofts, Inc., New York, 180 p.

Dunnet, D., and Siddans, A.W.B., 1971, Nonrandom sedimentary fabrics and their modification by strain: Tectonophys., v. 12, p. 307-325.

Durney, D.W., and Ramsay, J.G., 1973, Incremental strain measured by syntectonic crystal growths: in DeJong, K.A., and Scholten, R. (eds.), Gravity and Tectonics, John Wiley & Sons, New York, p. 67-96.

Dwerryhous, A.R., 1911, Geological and Topographical Maps: Edward Arnold, London, 133 p.

Dyson, J.L., 1967, Geology and mineral resources of the southern half of the New Bloomfield Quadrangle, Pennsylvania: Penn. Geol. Survey Atlas 137cd, 86 p.

Edmond, J.M., and Paterson, M.S., 1972, Volume changes during the deformation of rocks at high pressures: Int. J. of Rock Mech. and Mining Sci., v. 9, p. 161-182.

Elliot, D., 1970, Determination of finite strain and initial shape from deformed elliptical objects: Geol. Soc. Am. Bull., v. 81, p. 2221-2236.

Elliott, D., 1976, The energy balance and deformation mechanisms of thrust sheets: Phil. Trans. R. Soc. Lond, v. 283A, p. 289-312.

Elliott, D., 1983, The construction of balanced cross-sections: J. Struc. Geol., v. 5, p. 101.

Elliott, D., and Johnson, M.R.W., 1980, Structural evolution in the northern part of the Moine thrust belt, NW Scotland: Trans. Roy. Soc. Edinburgh, v. 71, p. 69-96.

Engelder, T., 1987, Joints and shear fractures in rock: in Atkinson, B.K., (ed.), Fracture Mechanics of Rock, Academic Press, London, p. 27-69.

Engelder, T., and Geiser, P., 1980, On the use of regional joint sets as trajectories of paleostress fields during the development of the Appalachian Plateau: J. Geophys. Res., v. 85, p. 6319-6341.

Engelder, T. and Marshak, S., 1985, Disjunctive cleavage formed at shallow depths in sedimentary rocks: J. Struc. Geol., v. 7, p. 327-343.

Etchecopar, A., Vasseur, G., and Daignieres, M., 1981, An inverse problem in microtectonics for the determination of stress tensors from fault striation analysis: J. Struc. Geol., v. 3, p. 51-65.

Faill, R.T., 1969, Kink band structures in the Valley and Ridge Province, central Pennsylvania: Geol. Soc. Am. Bull., v. 80, p. 2539-2550.

Faill, R.T., 1973, Kink band folding, Valley and Ridge Province, Pennsylvania: Geol. Soc. Am. Bull., v. 84, p. 1289-1314.

Faill, R.T., and Wells, R.B., 1974, Geology and mineral resources of the Millerstown quadrangle, Perry, Juniata, and Synder Counties, Pennsylvania: Penn. Geol. Survey Atlas, 136, 276 p.

Fleuty, M.J., 1964, The description of folds: Geol. Assoc. Lond. Proc., v. 75, p. 461-492.

Fleuty, M.J., 1975, Slickensides and slickenlines: Geological Mag., v. 112, p. 319-322.

Fleuty, M.J., 1987, Folds and folding: in Seyfert, C.K., (ed.), The Encyclopedia of Structural Geology and Plate Tectonics: Van Nostrand Reinhold Co., New York, p. 249-270.

Freund, R., 1970, Rotation of strike-slip faults in Sistan, southeast Iran: J. Geol., v. 78, p. 188-200.

Fry, N., 1979, Random point distributions and strain measurement in rocks: Tectonophys., v. 60, p. 89-105.

Geiser, P.A., 1988, The role of kinematics in the construction and analysis of geological cross sections in deformed terranes: Geol. Soc. Am. Mem., in press.

Geiser, P.A., and Engelder, T., 1983, The distribution of layer-parallel shortening fabrics in the Appalachian foreland of New York and Pennsylvania: in Hatcher, R.D., Jr., Williams, H., and Zietz, I., (eds.), Contribution to the Tectonics and Geophysics of Mountain Chains, Geol. Soc. Am. Mem. 158, p. 161-176.

Gibbs, A.D., 1983, Balanced cross-section construction from seismic sections in areas of extensional tectonics: J. Struc. Geol., v. 5, p. 153-160.

Gibbs, A.D., 1984, Structural evolution of extensional basin margins: J. Geol. Soc. Lond., v. 141, p. 609-620.

Gill, W.D., 1953, Construction of geological sections of folds with steep limb attenuation: Am. Assoc. Petrol. Geologists Bull., v. 37, p. 2389-2406.

Gilotti, J.A., 1987, The role of ductile deformation in the emplacement of the Sarv thrust sheet, Swedish Caledonides: Ph.D. dissertation, Johns Hopkins Univ., Baltimore, 223 p.

Gilotti, J.A., and Kumpulainen, R., 1986, Strain softening induced ductile flow in Sarv thrust sheet, Scandinavian Caledonides: J. Struc. Geol., v. 8, p. 441-455.

Goguel, J., 1962, Tectonics: W.H. Freeman and Co., San Francisco, 384 p.

Gray, D.R., 1977, Morphological classification of crenulation cleavage: J. Geol., v. 85, p. 229-235.

Gray, D.R., 1981, Cleavage-fold relationships and their implications for transected folds: an example from southwest Virginia, U.S.A.: J. Struc. Geol., v. 3, p. 265-277.

Gray, D.R., and Durney, D.W., 1979, Investigations on the mechanical significance of crenulation cleavage: Tectonophys., v. 58, p. 35-79.

Greenhood, D., 1964, Mapping: University of Chicago Press, Chicago, 289 p.

Gwinn, V.E., 1964, Thin-skinned tectonics in the Plateau and northwestern Valley and Ridge provinces of the central Appalachians: Geol. Soc. Am. Bull., v. 75, p. 863-900.

Gwinn, V.E., 1970, Kinematic patterns and estimates of lateral shortening, Valley and Ridge and Great Valley Provinces, central Appalachians, south-central Pennsylvania: in Fisher, G.W., et al., (eds.), Studies in Appalachian Geology: Central and Southern, Interscience, New York, p. 127-146.

Hamblin, W.K., 1965, Origin of 'reverse drag' on the down-thrown side of normal faults: Geol. Soc. Am. Bull., v. 76, p. 1145-1164.

Hamblin, W.K., and Howard, J.D., 1986, Exercises in Physical Geology, 6th Ed.: Burgess Publishing, Edina, 191 p.

Hancock, P., 1985, Brittle microtectonics: principles and practice: J. Struc. Geol., v. 7, p. 437-458.

Handin, J., 1966, Strength and Ductility: in Clark, S.P., Jr., (ed.), Handbook of Physical Constants, Geol. Soc. Am. Mem. 97, p. 223-289.

Handin, J., 1969, On the Coulomb-Mohr failure criterion: J. Geophys. Res., v. 74, p. 5343-5348.

Handin, J., and Hager, R.V., Jr., 1957, Experimental deformation of sedimentary rocks under confining pressure: Tests at room temperature on dry samples: Am. Assoc. Petrol. Geologists Bull., v. 41, p. 1-50.

Handin, J., Hager, R.V., Jr., Friedman, M., and Feather, J.N., 1963, Experimental deformation of sedimentary rocks under confining pressure: Pore pressure tests: Am. Assoc. Petrol. Geologists Bull., v. 47, p. 717-755.

Hansen, E., 1971, Strain Facies: Springer-Verlag, New York, 207 p.

Harris, L.D., 1979, Similarities between the thick-skinned Blue Ridge anticlinorium and the thin-skinned Powell Valley anticline: Geol. Soc. Am. Bull., v. 90, Part I, p. 525-539.

Harris, L.D., and Milici, R., 1977, Characteristics of thin-skinned style of deformation in the southern Appalachians and potential hydrocarbon traps: U. S. Geol. Survey Prof. Paper 1018, 40 p.

Harwood, D., 1983, Stratigraphy of upper Paleozoic volcanic rocks and regional unconformities in part of the northern Sierra terrane, California: Geol. Soc. Am. Bull., v. 94, p. 413-419.

Haughton, S., 1856, On slaty cleavage and the distortion of fossils: Philos. Mag., Series 4, v. 12, p. 409-421.

Hay, A.M., and Abdel-Rahman, M.A., 1974, Use of Chi-square for the identificaion of peaks in orientation data: Comment: Geol. Soc. Am. Bull., v. 85, p. 1963-1966.

Heard, H.C., 1960, Transition from brittle fracture to ductile flow in Solenhofen limestone as a function of temperature, confining pressure and interstitial fluid pressure: Geol. Soc. Am. Mem. 79, p. 193-226.

Heard, H.C., 1963, Effect of large changes in strain rate in the experimental deformation of Yule Marble: J. Geol., v. 71, p. 162-195.

Hewett, D.F., 1920, Measurements of folded beds: Econ. Geol., v. 15, p. 367-385.

Hobbs, B.E., Means, W.D., and Williams, P.F., 1976, An Outline of Structural Geology: John Wiley & Sons, New York, 571 p.

Hodgson, R., 1961, Regional study of jointing in the Comb Ridge-Navajo Mountain area, Arizona and Utah: Am. Assoc. Petrol. Geologists Bull., v. 45, p. 1-38.

Hoeppener, R., 1955, Tektonik im schiefergebirge eine einfuhrung: Geol Rund., v. 44, p. 26-58.

Hossack, J.R., 1979, The use of balanced cross-sections in the calculation of orogenic contraction: A review: J. Geol. Soc. Lond., v. 136, p. 705-711.

Hubbert, M.K, 1931, Graphic solution of strike and dip from two angular components: Am. Assoc. Petrol. Geologists Bull., v. 15, p. 283-286.

Hudleston, P.J., 1973, Fold morphology and some geometrical implications of theories of fold development: Tectonophys., v. 16, p. 1-46.

Hull, J., Koto, R., and Bizub, R., 1986, Deformation zones in the Highlands of New Jersey: Geol. Assoc. New Jersey Guidebook 3, p. 19-66.

Huntoon, P.W., 1974, The post-Paleozoic structural geology of the eastern Grand Canyon, Arizona: *in* Geology of the Grand Canyon, Museum of Northern Arizona, and Grand Canyon Natural History Assoc., p. 82-115.

Jaeger, J.C., and Cook, N.G.W., 1979, Fundamentals of Rock Mechanics, 3rd ed.: Chapman and Hall, London, 593.

Jamison, W.R., 1987, Geometric analysis of fold development in overthrust terranes: J. Struc. Geol., v. 9, p. 207-219.

Jones, P.B., 1971, Folded faults and sequence of thrusting in Alberta foothills: Am. Assoc. of Petrol. Geologists Bull., v. 55, p. 292-306.

Journal of Structural Geology, 1983, Balanced cross-sections and their geological significance, Elliott Memorial Volume, v. 5, 223 p.

Judson, S., Kauffman, M.E., and Leet., L.D., 1987, Physical Geology, 7th ed.: Prentice-Hall, Inc., Englewood Cliffs, New Jersey, 484 p.

Kalsbeek, F., 1963, A hexagonal net for the counting out and testing of fabric diagrams: Neues Jahr. fur Miner., Monatshefte, v. 7, p. 173-176.

Kamb, W.B., 1959, Petrofabric observations from Blue Glacier, Washington, in relation to theory and experiment: J. Geophys. Res., v. 64, p. 1908-1909.

Karcz, I., and Dickman, S.R., 1979, Determination of fracture intensity: Tectonophys., v. 56, p. T1-T7.

Kay, M., 1945, Paleogeographic and palinspastic maps: Am. Assoc. Petrol. Geologists Bull., v. 29, p. 426-450.

Kilby, W.E., and Charlesworth, H.A.K., 1980, Computerized downplunge projection and the analysis of low-angle thrust faults in the Rocky Mountains of Alberta, Canada: Tectonophys., v. 66, p. 287-299.

King, P.B., 1964, Geology of the central Great Smoky Mountains, Tennessee: U.S. Geol. Survey, Prof. Paper 349-C, 148 p.

King, P.B., 1977, The Evolution of North America: Princeton University Press, Princeton, 197 p.

Klein, J., 1981, Sequential development of fold orders in ptygmatic structures (Damara orogenic belt, Namibia): Geol. Rund., v. 70, p. 925-940.

Kligfield, R., Geiser, P., and Geiser, J., 1986, Construction of geologic cross sections using microcomputer systems: Geobyte, Spring issue, p. 60-66.

Kligfield, R., Owens, W.H., and Lowrie, W., 1981, Magnetic susceptibility anisotropy, strain and progressive deformation in Permian sediments from the Maritime Alps (France): Earth Planet. Sci. Lett., v. 55, p. 181-189.

Kohlbeck, R., and Scheidegger, A., 1977, On the theory of the evaluation of joint orientation measurements: Rock Mechanics, v. 9, p. 9-25.

Kuenen, P.H., 1953, Significant features of graded bedding: Am. Assoc. Petrol. Geologists Bull., v. 37, p. 1044-1066.

Kulander, B., Barton, C., and Dean, S., 1979, The application of fractography to core and outcrop fracture investigations: Technical report METC/SP-79/3 for the U.S. Dept. of Energy, Morgantown, WV, 174 p.

Lamerson, P.R., 1982, The Fossil Basin area and its relationship to the Absaroka thrust fault system: *in* Powers, R.B., (*ed.*), Geologic Studies of the Cordilleran Thrust Belt, Rocky Mountain Assoc. Geologists, p. 279-340.

Langenberg, W., 1985, The geometry of folded and thrusted rocks in the Rocky Mountain foothills near Grande Cache, Alberta: Can. J. Earth. Sci., v. 22, p. 1711-1719.

Langenberg, W., Charlesworth, H., and La Riviere, A., 1988, Computer-constructed cross-sections of the Morcles Nappe: Eclog. Geol. Helv., in press.

LaPointe, P., and Judson, J., 1985, Characterization and interpretation of rock mass joint patterns: Geol. Soc. Am. Spec. Pap. 199, 37 p.

Laubscher, H.P., 1962, Die Zwiephasenhypothese der Jurafaltung: Eclog. Geol. Helv., v. 55, p. 1-22.

Laubscher, H.P., 1977, Fold development in the Jura: Tectonophys. v. 37, p. 337-362.

Levens, A.S., 1959, Nomography: John Wiley & Sons, New York, 415 p.

Lisle, R.J., 1977, Estimation of the tectonic strain ratio from the mean shape of deformed elliptical markers: Geologie en Mijnbouw, v. 56, p. 140-144.

Lisle, R.J., 1980, A simplified work scheme for using block diagrams with the orthographic net: J. Geol. Education, v. 29, p. 81-83.

Lisle, R. J., 1985, Geological Strain Analysis: A Manual for the R_f /f technique: Pergamon Press, Oxford, 99 p.

Lister, G.S., and Price, G.P., 1978, Fabric development in a quartz-feldspar mylonite: Tectonophys., v. 49, p. 37-78.

Lister, G.S., and Snoke, A.W., 1984, S-C mylonites: J. Struc. Geol., v. 6, p. 617-638.

Lobeck, A.K., 1958, Block Diagrams: Emerson-Trussell Book Company, Amherst, MA, 212 p.

Lumino, K.M., 1987, Deformation within the Diana Complex along the Carthage-Colton mylonite zone: M.S. thesis, University of Rochester, Rochester, NY, 104 p.

Mackin, J.H., 1950, The down-structure method of viewing geologic maps: J. Geol., v. 58, p. 55-72.

Malavieille, J., 1987, Extensional shearing deformation and kilometer-scale "a"-type folds in a Cordilleran metamorphic core complex (Raft River Mountains, Northwestern Utah): Tectonics, v. 6, p. 423-448.

Marshak, S., 1986, Structure and tectonics of the Hudson Valley fold-thrust belt, eastern New York State: Geol. Soc. Am. Bull., v. 97, p. 354-368.

Marshak, S., et al., 1986, Structure of the Hudson Valley Fold-Thrust belt between Catskill and Kingston, New York: A Field Guide: prepared for Geol. Soc. Am., Northeast Section, 21st meeting, Kiamesha Lake, NY, 70 p.

Marshak, S., and Engelder, T., 1985, Development of cleavage in a fold-thrust belt in eastern New York: J. Struc. Geol., v. 7, p. 345-359.

Marshak, S., Geiser, P.A., Alvarez, W., and Engelder, T., 1982, Mesoscopic fault array of the northern Umbrian Apennine fold belt, Italy: Geometry of conjugate shear by pressure-solution slip: Geol. Soc. Am. Bull., v. 93, p. 1013-1022.

McEachran, D.B., and Marshak, S., 1985, Teaching strain theory in structural geology using graphics programs for the Apple Macintosh Computer: J. Geol. Ed., v. 34, p. 191-195.

McIntyre, D.B., and Weiss, L.E., 1956, Construction of block diagrams to scale in orthographic projection: Proc. Geologists Assoc. v. 67, p. 142-155.

Means, W.D., 1976, Stress and Strain - Basic Concepts of Continuum Mechanics for Geologists: Springer-Verlag, New York, 339 p.

Means, W.D., 1987, A newly recognized type of slickenside striation: J. Struc. Geol., v. 9, p. 585-590.

Mertie, J.B., Jr., 1922, Graphic and mechanical computation of thickness of stratum and distance to a stratum: U.S. Geol. Survey Prof. Paper 129-C, 52 p.

Mertie, J.B., Jr., 1940, Stratigraphic measurements in parallel folds: Geol. Soc. Am. Bull., v. 51, p. 1113-1122.

Mertie, J.B., Jr., 1947, Delineation of parallel folds and measurement of stratigraphic dimensions: Geol. Soc. Am. Bull., v. 58, p. 779-802.

Mitra, G., 1978, Ductile deformation zones and mylonites. The mechanical processes involved in the deformation of crystalline basement rocks: Am. J. Sci., v. 278, p. 1057-1084.

Mitra, G., 1984, Brittle to ductile transition due to large strains along White Rock thrust, Wind River Mountains, Wyoming: J. Struc. Geol., v. 6, p. 51-61.

Mitra, G., and Frost, B.R., 1981, Mechanisms of deformation within Laramide and Precambrian deformation zones in basement rocks of the Wind River Mountains: Contrib. to Geol. Univ. Wyo., v. 19, p. 161-173.

Mitra, G., and Yonkee, W.A., 1985, Relationship of spaced cleavage to folds and thrusts in the Idaho-Utah-Wyoming thrust belt: J. Struc. Geol., v. 7, p. 361-373.

Mitra, S., 1979, Deformation at various scales in the South Mountain anticlinorium of the central Appalachians: Geol. Soc. Am. Bull., v. 90, pt. I, p. 227-229.

Mitra, S., and Datta, J., 1978, Ptygmatic structures: an analysis and review: Geol. Rund., v. 67, p. 880-895.

Mosher, S., 1987, Stretched pebbles: in Seyfert, C.K., (ed.), The Encyclopedia of Structural Geology and Plate Tectonics: Van Nostrand Reinhold Co., p. 753-757.

Nelson, R.A., 1979, Natural fracture systems; description and classification: Am. Assoc. Petrol. Geologists Bull., v. 63, p. 2214-2221.

Nickelsen, R.P., 1956, Geology of the Blue Ridge near Harpers Ferry, West Virginia: Geol Soc. Am. Bull., v. 67, p. 239-269.

Nickelsen, R., 1979, Sequence of structural stages of the Allegheny orogeny at the Bear Valley strip mine, Shamokin, Pennsylvania: Am. J. Sci., v. 179, p. 225-271.

Nickelsen, R., and Hough, V., 1967, Jointing in the Appalachian Plateau of Pennsylvania: Geol. Soc. Am. Bull., v. 78, p. 609-630.

Nicolas, A., and Poirier, J.P., 1976, Crystalline Plasticity and Solid State Flow in Metamorphic Rocks: John Wiley & Sons, New York, 444 p.

Owens, W. H., 1984, The calculation of a best-fit ellipsoid from elliptical sections on arbitrarily oriented planes: J. Struc. Geol., v. 6, p. 571-578.

Owens, W.H., and Bamford, D., 1976, Magnetic seismic, and other anisotropic properties of rocks: Phil. Trans. R. Soc. Lond., v. 282A, p. 55-68.

Palmer, H.S., 1919, New graphic method for determining the depth and thickness of strata and the projection of dip: *in* Shorter Contributions to Geology - 1918, U.S. Geol. Survey Prof. Pap. 120, p. 122-128.

Passchier, C.W., and Simpson, C., 1986, Porphyroclast systems as kinematic indicators: J. Struc. Geol., v. 8, p. 831-843.

Perry, W.J., 1978, Sequential deformation in the central Appalachians: Am. J. Sci., v. 278, p. 518-542.

Phillips, F.C., 1954, The Use of Stereographic Projection in Structural Geology: E. Arnold, London, 86 p.

Phillips, F.C., 1971, An Introduction to Crystallography, 4th ed.: John Wiley & Sons, New York, 351 p.

Platt, J.P., and Vissers, R.L.M., 1980, Extensional structures in anisotropic rocks: J. Struc. Geol., v. 2, p. 397-410.

Poirier, J.-P., 1985, Creep of Crystals: Cambridge University Press, Cambridge, 260 p.

Powell, C.McA., and Vernon, R.H., 1979, Growth and rotation history of garnet porphyroblasts with inclusion spirals in a Karakoram schist: Tectonophys., v. 54, p. 25-43.

Powell, C.McA., Cole, J.P., and Cudahy, T.J., 1985, Megakinking in the Lachlan Fold Belt, Australia: J. Struc. Geol., v. 7, p. 281-300.

Price, N.J., 1966, Fault and Joint Development in Brittle and Semi-Brittle Rock: Pergamon Press, Oxford, 176 p.

Price, R.A., 1981, The Cordilleran foreland thrust and fold belt in the southern Canadian Rocky Mountains: *in* Coward, M.P., and McClay, D.R., (*eds.*), Thrust and Nappe Tectonics, Geol Soc. Lond., Spec. Pub. 9, p. 427-448.

Price, R.A., 1967, The tectonic significance of meso-scopic subfabrics in the southern Rocky Mountains of Alberta and British Columbia: Can. J. Earth Sci., v. 4, p. 39-70.

Profett, J.M., Jr., 1977, Cenozoic geology of the Yerrington district, Nevada, and implications for the nature and origin of Basin and Range faulting: Geol. Soc. Am. Bull., v. 88, p. 247-266.

Ragan, D.M., 1985, Structural Geology: An Introduction to Geometrical Techniques, 3rd ed.: John Wiley & Sons, New York, 393 p.

Raisz, E., 1962, Principals of Cartography: McGraw-Hill Book Company, New York, 315 p.

Ramsay, J.G., 1962, Interference patterns produced by the superposition of folds of similar type: J. Geol., v. 60, p. 466-481.

Ramsay, J.G., 1965, Structural investigations in the Barberton Mountain Land, Eastern Transvaal: Geol. Soc. S. Africa Trans. v. 66, p. 353-401.

Ramsay, J.G., 1967, Folding and Fracturing of Rocks: McGraw-Hill Book Co., New York, 568 p.

Ramsay, J.G., 1974, The Development of chevron folds: Geol. Soc. Am. Bull., v. 85, p. 1741-1754.

Ramsay, J.G., 1980, The crack-seal mechanism of rock deformation: Nature, v. 284, p. 135-139.

Ramsay, J.G., 1980, Shear zone geometry, a review: J. Struc. Geol., v. 2, p. 83-89.

Ramsay, J.G. and Huber, M.I., 1983, The Techniques of Modern Structural Geology, Volume 1, Strain Analysis: Academic Press, London, p. 1-308.

Ramsay, J.G., and Huber, M.I., 1987, The Techniques of Modern Structural Geology, Volume 2, Folds and Fractures: Academic Press, London, p. 309-700.

Reches, Z., 1978, Analysis of faulting in three-dimensional strain field: Tectonophys., v. 47, p. 109-129.

Reches, Z., 1983, Faulting of rocks in three-dimensional strain fields; II, theoretical analysis: Tectonophys., v. 95, p. 133-156.

Reches, Z., and Johnson, A.M., 1978, Development of monoclines, Part II: Theoretical analysis of monoclines: *in* Matthews, V., (*ed.*), Laramide Folding Associated with Basement Block Faulting in the Western United States: Geol. Soc. Am. Mem. 151, p. 273-311.

Reck, B., 1985, Deformation and metamorphism in the southwestern Narragansett Basin and their relationship to granite intrusion: M.A. thesis, University of Texas, Austin, 76 p.

Redmond, J.L., 1972, Null combination in fault interpretation: Am. Assoc. of Petrol. Geologists Bull., v. 56, p. 150-166.

Reks, I.J., and Gray, D.R., 1983, Strain patterns and shortening in a folded thrust sheet: An example from the southern Appalachians: Tectonophys., v. 93, p. 99-128.

Reks, I.J., and Gray, D.R., 1982, Pencil structure and strain in weakly deformed mudstone and siltstone: J. Struc. Geol., v. 4, p. 161-176.

Rettger, R.E., 1929, On specifying the type of structural contouring: Am. Assoc. Petrol. Geologists Bull., v. 13, p. 1559-1560.

Rich, J.L., 1934, Mechanics of low-angle overthrust faulting as illustrated by Cumberland thrust block, Virginia, Kentucky, and Tennessee: Am. Assoc. Petrol. Geologists Bull., v. 18, p. 1584-1596.

Roberts, J.L., 1974, The structure of the Dalradian rocks in the SW highlands of Scotland: J. Geol. Soc. Lond., v. 130, p. 93-124.

Roberts, J.L., 1982, Introduction to Geological Maps and Structures: Pergamon Press, Oxford, 332 p.

Robin, P-Y.F., 1977, Determination of geologic strain using randomly oriented strain markers of any shape: Tectonophys., v. 42, p. T7 - T16.

Robinson, A.H., and Sale, R.D., 1969, Elements of cartography: John Wiley & Sons, New York, 415 p.

Rodgers, J., 1949, Evolution of thought on structure of middle and southern Appalachians: Am. Assoc. Petrol. Geologists Bull., v. 33, p. 1643-1654.

Rodgers, J., 1963, Mechanics of Appalachian foreland folding in Pennsylvania and West Virginia: Am. Assoc. Petrol. Geologists Bull., v. 47, p. 1527-1536.

Roeder, D., Gilbert, O.E., Jr., and Witherspoon, W.D., 1978, Evolution and macroscopic structure of Valley and Ridge thrust belt, Tennessee and Virginia: University of Tennessee Dept. Geol. Sci. Studies in Geology 2, Knoxville, 25 p.

Rosenfeld, J.L., 1970, Rotated garnets in metamorphic rocks: Geol. Soc. Am. Spec. Pap. 129, 105 p.

Royse, F., Jr., Warner, M.A., and Reese, D.L., 1975, Thrust belt structural geometry and related stratigraphic problems, Wyoming-Idaho-Northern Utah: in Rocky Mountain Assoc. Geol., Symposium on Deep Drilling Frontiers in central Rocky Mountains, p. 4-54.

Sanderson, D.J., 1974, Patterns of boudinage and apparent stretching lineation developed in folded rocks: J. Geol., v. 82, p. 651-661.

Satin, L.R., 1960, Apparent-dip computer: Geol. Soc. Am. Bull., v. 71, p. 231-234.

Schmid, S.M., 1975, The Glarus overthrust: Field evidence and mechanical model: Eclog. Geol. Helv., v. 68, p. 247-280.

Schmid, S.M., 1983, Microfabric studies as indicators of deformation mechanisms and flow laws operative in mountain building: in Hsu, K.J. (ed.), Mountain Building Processes: Academic Press, London, p. 95-110.

Schmid, S.M., Boland, J.N., and Paterson, M.S., 1977, Superplastic flow in fine-grained limestone: Tectonophys., v. 43, p. 257-291.

Schryver, K., 1966, On the measurement of the orientation of axial planes of minor folds: J. Geol., v. 74, p. 83-84.

Secor, D.T., Jr., 1965, Role of fluid pressure in jointing: Am. J. Sci., v. 263, p. 633-646.

Shackleton, R.M., 1957, Downward facing structures of the Highland Border: J. Geol. Soc. Lond., v. 113, p. 361-392.

Shimamoto, T., and Logan, J.M., 1981, Effects of simulated fault gouge on the sliding behavior of Tennessee Sandstone: Nonclay gouges: J. Geophys. Res., v. 96, p. 2902-2914.

Sibson, R.H., 1975, Generation of pseudotachylyte by ancient seismic faulting: Geophys. J. R. Astron. Soc., v. 43, p. 775-794.

Sibson, R.H., 1977, Fault rocks and fault mechanisms: J. Geol. Soc., London, v. 133, p. 191-213.

Siddans, A.W.B., 1972, Slaty cleavage, a review of research since 1815: Earth Sci. Rev., v. 8, p. 205-232.

Simpson, C., 1986, Determination of movement sense in mylonites: J. Geol. Ed., v. 34, p. 246-261.

Simpson, C., and Schmid, S., 1983, An evaluation of criteria to deduce the sense of movement in sheared rocks: Geol. Soc. Am. Bull., v. 94, p. 1281-1288.

Smith, R.B., 1975, Unified theory of the onset of folding, boudinage, and mullion structure: Geol. Soc. Am. Bull., v. 86, p. 1601-1609.

Spry, A., 1969, Metamorphic Textures: Pergamon Press, Oxford, 350 p.

Stanley, R.S., and Ratcliffe, N.M., 1985, Tectonic synthesis of the Taconian orogeny in western New England: Geol. Soc. Am. Bull., v. 96, p. 1227-1250.

Stearns, D.W., 1968, Faulting and forced folding in the Rocky Mountain foreland: in Mathews, V. (ed.), Laramide Folding Associated with Basement Block Faulting in the Western United States: Geol. Soc. Am. Mem. 151, p. 1-37.

Stromgard, K.E., 1973, Stress distribution during formation of boudinage and pressure shadows: Tectonophys., v. 16, p. 215-248.

Suppe, J., 1980, A retrodeformable cross-section of northern Taiwan: Proc. of the Geol. Soc. of China, n.23, p. 46-55.

Suppe, J., 1983, Geometry and kinematics of fault-bend folding: Am. J. Sci., v. 283, p. 684-721.

Suppe, J., 1985, Principles of Structural Geology: Prentice-Hall, Englewood Cliffs, N. J., 537 p.

Suppe, J., and Medwedeff, D.A., 1984, Fault-propagation folding: Geol. Soc. Am. Abstr. w. Pgms. v. 16, p. 670.

Thiessen, R.L., and Means, W.D., 1980, Classification of fold interference patterns: A reexamination: J Struc. Geol. v. 2, p. 311-316.

Thompson, R.I., 1981, The nature and significance of large "blind" thrusts within the northern Rocky Mountains of Canada: in McClay, K.R., and Price, N.J., (eds.), Thrust and Nappe Tectonics, Geol Soc. Lond., Spec. Pub. 9, Blackwell Scientific Publications, p. 449-462.

Tillman, J.E., and Barnes, H., 1983, Deciphering fracturing and fluid migration histories in the northern Appalachian basin: Am. Assoc. Petrol. Geologists Bull., v. 67, p. 692-705.

Turner, F.J., and Weiss, L.E., 1963, Structural Analysis of Metamorphic Tectonites: McGraw-Hill Book Co., New York, 545 p.

Usdansky, S., and Groshong, R.H., Jr., 1984, Comparison of analytical models for dip-domain and fault-bend folding: Geol. Soc. Am. Abst. w. Pgms., v. 16, p. 680.

Vernon, R.H., 1976, Metamorphic Processes. Reactions and Microstructure Developoment: George Allen & Unwin, Boston, 247 p.

Vistelius, A.B., 1966, Structural Diagrams: Pergamon, Press, Oxford, 178 p.

Watkinson, A.J., and Cobbold, P.R., 1981, Axial directions of folds in rocks with linear/planar fabrics: J. Struc. Geol., v. 3, p. 211-217.

Weiss, L.E., 1972, The Minor Structures of Deformed Rocks: A Photographic Atlas: Springer-Verlag, New York, 431 p.

Weiss, L.E., 1980, Nucleation and growth of kink bands: Tectonophys., v. 65, p. 1-38.

Wellman, H. W., 1962, A graphical method for analysing fossil distortion caused by tectonic deformation: Geological Mag., v. 99, p. 348-352.

Wernicke, B., and Burchfiel, B.C., 1982, Modes of extensional tectonics: J. Struc. Geol., v. 4, p. 104-115.

Wheeler, R., and Dixon, J., 1980, Intensity of systematic joints: methods and application: Geology, v. 8, p. 230-233.

Wheeler, R., and Holland, 1981, Style elements of systematic joints: an analytic procedure with a field example: in O'Leary, D., Podwysocki, M., and Earle, J., (eds.) Proc. of the Third Int. Conf. on Basement Tectonics, p. 393-404.

Wheeler, R., and Stubbs, J., 1979, Style elements of systematic joints: statistical analysis of size, spacing and other charcteristics: in O'Leary, D., Podwysocki, M., and Earle, J., (eds.) Proc. of the Third Int. Conf. on Basement Tectonics, p. 491-499.

White, S.H., Burrows., S.E., Carreras, J., Shaw, N.D., and Humphreys, F.J., 1980, On mylonites in ductile shear zones: J. Struc. Geol., v. 2, p. 175-187.

White, W.S., and Jahns, R.H., 1950, The structure of central and east-central Vermont: J. Geol., v. 58, p. 179-220.

Williams, P.F., 1976, Relationships between axial plane foliation and strain: Tectonophys., v. 30, p. 181-196.

Williams, P.F., 1977, Foliation: A review and discussion: Tectonophys., v. 39, p. 305-328.

Williams, P.F., 1984, Multiply deformed terrains - problems of correlation: J. Struc. Geol., v. 7, p. 269-280.

Wilson, G., 1953, Mullion and rodding structures in the Moine Series of Scotland: Proc. Geol. Assoc. Lond., v. 64, p. 118-151.

Wilson, G., 1982, Introduction to Small-scale Geological Structures: George Allen and Unwin, Boston, 128 p.

Wintsch, R.P., and Knipe, R.J., 1983, Growth of a zoned plagioclase porphyroblast in a mylonite: Geology, v. 11, p. 360-363.

Wise, D., 1982, Linesmanship and the practice of linear geo-art: Geol. Soc. Am. Bull., v. 93, p. 886-888.

Wise, D., Funiciello, R., Parotto, M., and Salvini, F., 1985, Topographic lineament swarms: clues to their origin from domain analysis of Italy: Geol. Soc. Am., Bull., v. 96, p. 952-967.

Wise, D., and McCrory, T., 1982, A new method of fracture analysis: azimuth versus traverse distance plots: Geol. Soc. Am. Bull., v. 93, p. 889-897.

Wojtal, S., 1986, Deformation within foreland thrust sheets by populations of minor faults: J. Struc. Geol., v. 8, p. 341-360.

Wojtal, S., and Mitra, G., 1986, Strain hardening and strain softening in fault zones from foreland thrusts: Geol. Soc. Am Bull., v. 57, p. 674-687.

Wood, D.S., 1974, Current views of the development of slaty cleavage: An. Rev. Earth Planet. Sci., v. 2, p. 369-401.

Wood, D.S., Oertel, G., Singh, J., and Bennett, H.F., 1976, Strain and anisotropy in rocks: Phil. Trans. R. Soc. Lond., v. 283A, p. 27-42.

Woodward, N. B., (ed.), 1985, Valley and Ridge Thrust Belt: Balanced structural sections, Pennsylvania to Alabama: Univ. of Tennessee, Dept. of Geol. Sci. Studies in Geology 12, Knoxville, 64 p.

Woodward, N. Boyer, S.E., and Suppe, J., 1985, An Outline of Balanced Cross-Sections: Univ. of Tennessee Dept. of Geol. Sci. Studies in Geol. 11, 2nd ed., Knoxville, 170 p.

Zadins, Z.Z., 1983, Structure of the northern Appalachian thrust belt at Cementon, New York: M.S. thesis, Univ. of Rochester, Rochester, NY, 137 p.

Zar, J.H., 1984, Biostatistical Analysis, 2nd ed.: Prentice-Hall, Inc., Englewood Cliffs, NJ.

Zwart, H.J., 1962, On the determination of polymetamorphic mineral associations and its application to the Busost area (central Pyrenees): Geol. Rund., v. 52, p. 38-65.

INDEX